ANIMAL BIOTECHNOLOGY
RECENT CONCEPTS AND DEVELOPMENTS

MJP
PUBLISHERS

ANIMAL BIOTECHNOLOGY
RECENT CONCEPTS AND DEVELOPMENTS

P Ramadass

Professor and Head (Retd.)
Department of Animal Biotechnology
Madras Veterinary College
Tamil Nadu Veterinary and Animal Sciences University
Chennai

MJP
PUBLISHERS

Chennai   Trichy   Tirunelveli   NewDelhi

MJP Publishers

No. 44, Nallathambi Street,
Triplicane, Chennai 600 005

MJP 034   © Publishers, 2020

Publisher :   **C. Janarthanan**

Project Editor :   **C. Ambica**

PREFACE

Biotechnology is a highly multidisciplinary subject and has got its foundation in many fields including biology, microbiology, biochemistry, molecular biology, genetics, chemistry and chemical and processing engineering. Application of biotechnology in medicine and agriculture has been a recent phenomenon. Modern biotechnological processes now encompass a wide range of new products including antibiotics, recombinant and nucleic acid vaccines, monoclonal antibodies, recombinant therapeutic products like recombinant insulin, growth hormones, prolactin and gene therapy, production of transgenic animals and plants and use of embryo biotechnological methods and stem cells to augment animal production and human therapy, respectively.

Animal biotechnology is in its infancy and only during the past ten years, much work has been done in animal biotechnology in few isolated laboratories throughout the world. There is an increasing need to train manpower in animal biotechnology. Even though many colleges are offering courses in Biotechnology for the students, there is no single text book available covering all the aspects of animal biotechnology for the students. This book on Animal Biotechnology has been written to meet out the requirements of both undergraduate and postgraduate students on the subject of biotechnology.

There are seventeen chapters in this book covering different aspects of animal biotechnology including enzyme technology, gene therapy, biotechnology in medicine, Intellectual Property Rights and biosafety in biotechnology. Many up-to-date references on most of the topics have been included so that it would be a reference book for postgraduate students studying biotechnology and molecular biology. This would be a useful book for students who are writing competitive examinations for fellowship.

With my extensive experience in teaching and research in Animal Biotechnology I have compiled this book to provide students the basic principles of animal biotechnology, current information on different topics of biotechnology, as well as information on Intellectual Property Rights and biosafety guidelines to be adopted in the laboratories.

I do not claim that this book is my original contribution or research work. It is merely an outcome of the compilation of information from different books and journals, so that the students could get access to all the information in one book. List of references are given at the end of each chapter for students who would like to have more information on the topics.

I thank my colleagues Dr. A. Senthil Kumar, Dr. Dhinakar Raj, Dr. K. Kumanan, Dr. A. Palaniswamy and Dr. Wilson S. Aruni for their technical input.

If there has been any error in the publication which has occurred inadvertently, I beg pardon. Suggestions from the readers to improve the book in further editions would be highly appreciated.

P Ramadass

CONTENTS

1. Introduction — 1

What is Biotechnology? — 1
Biotechnology—Past, Present and Future — 3
Single Cell Protein (SCP) — 7
    SCP from Wastes — 9
Activities of Biotechnologists — 10
    Recombinant DNA and Genetic Engineering — 12
    Mammalian Cell Culture — 14
    Reproductive Techniques — 14
    Transgenic Animals — 15
    Somatic Cell Nuclear Transfer — 16
    Plant Cell Culture — 16
    Fuels — 18
    Biocatalysis — 19
    Waste Treatment and Utilization — 19
    Fermentation — 20
    Process Engineering — 20
    Biotechnology and Global Health — 21
    Review Questions — 21
    References — 21

2. Cell Culture and Fermentation Technology — 23

Introduction — 23
Culture Media — 23
    Natural Media — 24
    Defined Media — 24
Primary Culture — 26
    Enzymatic Disaggregation — 27

Mechanical Disaggregation	27
Separation of Viable and Non-viable Cells	27
Cell Lines	27
Maintenance of Cultures—Cell Lines	29
Large-scale Cell Cultures	30
Homogeneous Bioreactor Systems	33
Heterogeneous Bioreactor Systems	33
Tissue and Organ Cultures	34
Tissue Culture	35
Organ Culture	35
Cryopreservation	36
Cryopreservatives	36
Freezing Mixture	36
Freezing down Cells	36
Thawing Procedure	37
Usefulness of Cell Cultures in Veterinary Research	37
Disease Diagnosis	37
Virus Vaccines	38
Monoclonal Antibodies	38
Hormones	39
Immunoregulators	39
Tumour-specific Antigens	39
Fermentation Technology	39
Downstream Processing	41
Applications of Fermentation	45
Genetic Improvement of Product Formation	49
Solid-state Fermentation (SSF)	50
Applications of SSF	50
Fermented Foods	50
Review Questions	53
References	53
3. Biotechnological Approaches to Vaccine Production	**55**
Introduction	55
Recombinant Subunit Vaccines	59
Recombinant Vaccines against Viral Diseases	60

Recombinant Vaccines against Bacterial Diseases ... 64

Expression of Target Antigens in Prokaryotic Hosts ... 65

Expression of Target Antigens in Eukaryotic Hosts ... 66

Peptide Vaccines ... 69

Fusion Protein Vaccines ... 70

Synthetic Peptide Vaccines ... 71

Genetically Engineered Reassortants
and Deletion Mutant Vaccines ... 72

Anti-idiotype Antibody Vaccines ... 73

Nucleic Acid Vaccines ... 75

Bacterial Plasmid DNA-delivery Systems ... 79

Review Questions ... 80

References ... 80

4. Hybridoma Technology ... 87

Introduction ... 87

Early Research ... 87

Polyclonal Antiserum vs. Monoclonal Antibody ... 88

Advantages of Monoclonal Antibodies ... 88

Disadvantages of Monoclonal Antibodies ... 89

Monoclonal Antibody Production Strategy ... 89

Purity and Form of Immunogen ... 91

Choice of Animals ... 91

Immunization Procedures ... 91

Choice of Myeloma ... 92

Growth of Myeloma Cell Lines ... 92

Spleen Cells ... 92

Cell Fusion ... 92

Cloning of Hybrids ... 93

Cryopreservation of Cells ... 94

Large-scale Production of Monoclonals ... 94

Screening Assays for MAb Activity ... 94

Uses of Monoclonal Antibodies ... 95

Immunodiagnostic Reagents ... 96

Antigen Detection ... 96

Antibody Detection ... 102

Anti-idiotype Antibody 103
Experimental uses of Monoclonal Antibodies 103
Cells of Immune System 103
Pregnancy and Sex Determination 104
Molecular Structure of Antigens 104
Affinity Chromatography 105
Immunochemical Application of MAbs 106
Monoclonals as *in vivo* Reagents for Animals 107
Therapeutic Monoclonal Antibodies 108
Inhibition of Alloimmune Reactivity 108
Inhibition of Autoimmune Reactivity 110
MAbs against Cancers 111
Anti-platelet Therapy 111
MAbs in Infectious Diseases 112
MAbs in Parasitic Infections 112
MAbs in Bacterial Infections 112
MAbs in Viral Infections 114
Review Questions 118
References 118

5.   Nucleic Acid Probes and Hybridization 125

Introduction 125
Nucleic Acid Probes 126
Probe Development 128
Hybridization 137
Hybridization Strategies 139
Immobilization of Nucleic Acid on Filters 144
Prehybridization 144
Hybridization 145
Washing 145
Detection of Radioactive Hybrids 145
Detection of Non-radioactive Hybrids 146
Types of Hybridization 147
DNA Arrays and Chips 152
Hybridization in Diagnosis 156
Diagnosis of Bovine Herpesvirus Infection
    using Biotinylated cDNA Probes 157

Diagnosis of Enterotoxigenic *E. coli* Infections   158

Diagnosis of Sickle Cell Anaemia   159

DNA Fingerprinting for Genotype Analysis   159

Review Questions   160

References   160

6.   Gene Cloning   165

Introduction   165

Cloning Vectors for Recombinant DNA   165

Plasmids as Vectors   166

Yeast Plasmid Vectors   168

Bacteriophages as Vectors   170

Commercial Cloning and Expression Vectors   175

Restriction Endonucleases in Cloning   176

Cloning Strategies   181

Steps in Cloning   181

Preparation of DNA Fragments for Cloning   182

Restriction Enzyme Digestion   184

Ligation of DNA   185

Transformation   186

Identification, Selection, Screening and
Analysis of Recombinants   187

Genetic Methods   188

Expression of Cloned DNA in *E. coli* Plasmids   190

The *lac* Promoter   190

The *trp* Promoter   191

The *tac* Promoter   191

Expression of Cloned Genes in Animal Cells   191

SV-40 Vectors   192

Adenovirus Vectors   192

Bovine Papilloma Virus Vectors   193

Retroviral Vectors   193

Vaccinia Virus Vector   194

Applications of Genetic Engineering   194

Production of Recombinant Pharmaceuticals   195

Identification of Genes Responsible for
Human and Animal Diseases   197

Recombinant Vaccines   201
Gene Therapy   204
Review Questions   205
References   205

7.   Hybridoma Technology   209

Introduction   209
Embryo Transfer (ET) Technology   209
Selection of Donor   211
Induction of Superovulation   211
Embryo Collection   212
Evaluation of Embryos   212
Selection of Recipients   213
Transfer of Embryos   213
Cryopreservation of Embryos   214
Embryo Splitting   214
Embryo Sexing   215
Sex Chromosomal Analysis   216
Demonstration of H-Y antigen   217
Metabolic Activity of X-linked Enzymes   218
DNA-based Methods   218
In vitro Fertilization   221
Preparation and Collection of the Oocyst   223
Preparation of the Spermatozoa   225
Embryo Transfer (ET)   226
In vitro Fertilization in Farm Animals   226
Review Questions   229
References   229

8.   Transgenic Animals   233

Introduction   233
Transfer of Genes   234
Microinjection   238
Use of Embryonic Stem Cells   240
Retroviral Vectors   241
Transmission of Transgenes   243
Nuclear Transfer (NT)   244
Applications of Transgenic Mice   248

Disease Models 248

Oncogenes 250

Applications of Transgenic Livestock 252

Growth and Carcass Composition 252

Biochemical Pathways and Quality of Products 255

Production of Pharmaceuticals and Biomolecules 258

Gene Pharming 261

Disease Resistance 263

Environmentally Friendly Farm Animals 267

Xenotransplant 267

Transgenic Animals as a Model for Human Diseases 268

Transgenic Chicken 269

Review Questions 270

References 270

9. DNA Fingerprinting and RFLP in Domestic Animals 281

Introduction 281

Applications of DNA Fingerprints
in Domestic Animals 282

Minisatellite Probes used in Domestic Animals 283

Microsatellite Probes 284

Oligonucleotide Fingerprinting 285

DNA Fingerprints for Detection of Major Genes 286

Restriction Fragment Length Polymorphisms (RFLP) 286

RFLPs in Cattle 287

RFLPs in Sheep and Goats 291

RFLPs in Pigs 293

RFLPs in Horses 294

RFLPs in Poultry 294

RFLPs in Dogs and Cats 296

Major Histocompatibility Complex (MHC) in Cattle 297

Review Questions 298

References 298

10. Biotechnology in Animal Production 303

Manipulation of Growth 303

Insulin 303

Growth Hormone Releasing Hormone, Somatotropin,
  Growth Hormone and Somatomedins   303

Somatostatin (Somatotropin Release Inhibiting Factor, SRIF)   305

Thyroid Hormones   305

Reproductive Steroids   305

Probiotics as Growth Promoters   308

Manipulation of Lactation   315

Mammogenesis   315

Lactogenesis and Galactopoiesis   316

Manipulation of Wool Growth in Sheep   316

Manipulation of the Rumen Microbial Digestive System   317

Review Questions   318

References   318

11. Manipulation of Rumen Microorganisms   321

Introduction   321

The Target Organisms   322

Gene and Protein Structure, and
  Expression in Rumen Bacteria   323

Application of Recombinant DNA
  Technology to Rumen Organisms   324

Non-genetic Manipulation of Rumen Microbes   325

Addition to Antibiotics   326

Elimination of Rumen Ciliate Protozoa (Defaunation)   326

Addition of Fats   327

Addition of Protein Degradation Protectors   328

Addition of Buffer Substances   328

Addition of Bacteria or Rumen Content   328

Addition of Branched-chain Volatile Fatty Acids   329

Feed Distribution Method   329

Review Questions   330

References   330

12. Polymerase Chain Reaction   333

Introduction   333

Standard PCR   335

Template DNA   335

Primers	336
Deoxynucleoside Triphosphates (dNTPs)	337
PCR Buffers and MgCl$_2$	337
PCR Enzymes	338
Inhibitors and Enhancers of PCR	341
Steps in PCR	342
Detection and Analysis of PCR Products	344
Modifications of PCR	344
Inverse PCR (Chromosome Crawling)	344
Anchored PCR	345
Reverse Transcriptase PCR	346
Nested PCR	348
Asymmetric PCR	349
Multiplex PCR	349
Arbitrarily Primed PCR	352
Quantitative PCR	353
Single-strand Conformation Polymorphism PCR (SSCP-PCR)	353
Vectorate PCR (vPCR)	353
Restriction Site-specific PCR (RSS-PCR)	354
Degenerate PCR	354
Allele-specific PCR	355
Rep-PCR	355
In situ PCR	356
Real-time PCR	356
Major Advantages of PCR	359
Speed and Ease of Use	359
Sensitivity	359
Robustness	359
Applications of PCR	359
Detecting Pathogens	360
Study of DNA Polymorphism using PCR	373
PCR and RAPD Markers	374
Screening of Uncharacterized Mutations	375
PCR, VNTR and SSR Loci	375
Molecular Mapping using PCR	376
Gene Tagging using PCR	376

PCR for Confirming the Presence of Transferred Gene	376
Human Genetics using PCR	377
DNA Fingerprinting using PCR	377
Future Development in PCR	377
Review Questions	378
References	378

13. Nucleic Acid Sequencing — 389

Introduction	389
Chain Termination Sequencing	390
Preparation of Single-stranded DNA Template	390
Strand Synthesis Reaction for Chain Termination Sequencing	391
Running the Gel and Reading the Sequence	395
Direct Cycle Sequencing	396
Chemical Degradation Sequencing	397
Preparation of End-labelled DNA	398
Chemical Degradation Reactions	399
Running the Gel and Reading the Sequence	400
Automated Nucleic acid Sequencing	400
Applications of Gene Sequencing	401
Review Questions	409
References	410

14. Enzyme Technology — 413

The Nature of Enzymes	413
Novel Applications and Future Uses	415
Biomedical Applications of Enzyme Technology	416
Enzyme Sources	416
Clarification of the Soluble Enzymes	419
Enzyme Concentration	420
Precipitation	420
Ultrafiltration and Reverse Osmosis	420
Other Methods	421
Enzyme Purification	421
Gel Chromatography	421
Ion-exchange Chromatography	422

Affinity Purification 422
Applications of Enzymes in Biotechnology 422
Uses of Enzymes in Starch Hydrolysis 423
Uses of Enzymes in Detergents 426
Uses of Proteases in the Food Industry 428
Uses of Proteases in the Leather and Wool Industries 431
Uses of Lactases in the Dairy Industry 431
Uses of Enzymes in the Fruit Juice, Wine, 432
Brewing and Distilling Industries 432
Uses of Glucose Oxidase and Catalase in the Food Industry 434
Medical Applications of Enzymes 434
Advantages of using Enzymes
in Manufacture of Products 437
Genetic Engineering and
Protein Engineering of Enzymes 437
The Technology of Enzyme Production 438
Enzyme Immobilization 439
Methods of Enzyme Immobilization 440
Enzymatic Electrocatalysis 443
Basic Principle of Biosensors 443
Types of Biosensors 445
Calorimetric Biosensors 445
Potentiometric Biosensors 446
Amperometric Biosensors 446
Optical Biosensors 447
Immunosensors 448
Enzyme Engineering 450
Artificial Enzymes 451
Review Questions 452
References 452

15. Gene Therapy **457**

Introduction 457
Ex vivo Gene Transfer 458
In vivo Gene Transfer 458
Principles of Gene Transfer 459

Genes Integrated into Chromosomes 459
Non-integrated Genes 460
Clinical Studies 460
Techniques in Gene Therapy 461
Candidate Diseases for Gene Therapy 462
Gene Transfer Systems 463
Retrovirus 464
Adenovirus 465
Adeno-associated Virus 467
Naked DNA 468
Herpes Simplex Virus Vectors 468
Lentiviruses 468
Non-viral Vector Systems for Gene Therapy 469
Liposomes 469
Direct Injection/Particle Bombardment 469
Receptor-mediated Endocytosis 469
Gene Therapy for Inherited Disorders 470
Recessively Inherited Disorders 470
The First Gene Therapy Trial in 1990 470
Gene Therapy Trials for a Few Inherited Disorders 471
Familial Hypercholesterolaemia (FH) 472
Cystic Fibrosis 473
Duchenne Muscular Dystrophy 473
Recent Developments and Future Prospects 474
Arguments in Favour of Gene Therapy 474
Arguments against Gene Therapy 475
Review Questions 475
References 476

16. Biotechnology and Medicine 479

Introduction 479
Pharmaceuticals 480
Antibiotics 480
Vaccines and Monoclonal Antibodies 481
Biopharmaceuticals 482
Gene Therapy 482
Genetic Manipulation 483

Transgenic Animals   483
Knock-out Animals   484
Potential Animal Organ Donors   486
Somatic Gene Transfer   487
Applications of Animal Biotechnology in Medicine   488
Animal Biotechnology as a Scientific Tool   488
Biotechnology to Produce Recombinant Pharmaceuticals   489
Genetically Engineered Animals as Therapeutic Tools   491
Somatic Gene Transfer   491
Environmental Biotechnology   492
Waste-water and Sewage Treatment   493
Better Treatments for Solid Waste and Wastewater   493
Bioremediation—Cleaning up Contamination   494
Tracking the Health of the
          Environment through Biomonitoring   494
Biomass Energy   494
Clever Plants   495
Genetic Engineering for Environmental Solutions   495
Biotechnology in the Agriculture
          and Forestry Industries   496
Plant Biotechnology   496
Input Traits   497
Output Traits   497
Value-added Traits   497
Animal Biotechnology   498
Genetic Engineering for Transgenic Animals   498
Genetically Engineered Hormones and Vaccines   500
Diagnostics in Agriculture   501
Review Questions   501
References   502

**17. Intellectual Property Rights and
Biosafety in Biotechnology**   **505**

Intellectual Property Rights   **505**
Introduction   505
Patents   506
Patent Protection   507

Patentable Inventions	509
Biotechnology Patents in Australia, USA and Europe	511
Indian Perspective	511
Establishment of Patent Administration in India	513
What is Patentable in Biotechnology?	513
What is not Patentable in Biotechnology?	514
Obligations with Patent Applications	515
Implications of Patenting	516
Persons Entitled to Apply for a Patent in India	516
Where to Apply?	516
Examples of Patents	516
Biosafety in Biotechnology	**517**
Introduction	517
Laboratory Practice and Technique	518
Safety Equipment (Primary Barriers)	518
Facility Design and Construction (Secondary Barriers)	518
Biosafety Levels	519
Biosafety Level 1 (BSL-1)	520
Biosafety Level 2 (BSL-2)	522
Biosafety Level 3 (BSL-3)	525
Biosafety Level 4 (BSL-4)	528
Biosafety Cabinets (BSC)	538
Genetic Engineering—Safety, Social, Moral and Ethical Considerations	541
Release of Genetically Manipulated Organisms to the Environment	541
Review Questions	543
References	543
Glossary	545
Index	585

ABBREVIATIONS

16s rRNA	Subunit ribosomal RNA
AAV	Adeno-associated virus
ACAD	Acyl-coenzyme A dehydrogenase
AC-ELISA	Antigen-capture ELISA
ACNPV	*Autographa californica* nuclear polyhedrosis virus
ACTH	Adrenocorticotropic hormone
Ad	Adenovirus
ADA	Adenosine deaminase deficiency
AFE	Age at first egg
AGP	Agar gel precipitation test
AI	Artificial insemination
AIDS	Acquired immunodeficiency syndrome
AIV	Avian influenza virus
ALL	Acute lymphoblastic leukaemia
ALV	Avian leucosis virus
AMV	Avian myeloblastosis virus
AOX1	Alcohol oxidase 1 promoter
AP	Alkaline phosphatase
APA	Aminopenicillanic acid
APO	Australian Patent Office
ART	Advanced Reproductive Technologies
ARV	Avian reovirus
ATCC	American type culture collection
AVR	Acute vascular rejection
AWC	African wild cat
BCG	Bacillus Calmette–Guerin
BCIP/NTP	5-bromo 4-chloro 3-indolyl phosphate/nitroblue tetrazolium
BcoV	Bovine corona virus
BHV	Bovine herpesvirus
BIA	*Bacillus licheniformis* alpha-amylase
bLF	Bovine lactoferrin
BMOC-3	Branchat and Olivhant medium for oocyte collection
BoLA	Bovine leucocyte antigen

BPV	Bovine papilloma virus
BSC	Biological safety cabinet
BSE	Bovine spongiform encephalopathy
BSL	Biological safety Level
BST	Bovine somatotropin
BT	Bluetongue
BTV	Bluetongue virus
CaMV	Cauliflower mosaic virus
CC	Cumulus cells
CDGS	Combined defensin genotypes
cDNA	Complementary DNA
CDV	Canine distemper virus
CEA	Carcino-embryonic antigen
CGH	Chicken growth hormone
CHO	Chinese hamster ovary
CIA	Collagen-induced arthritis
CJD	Creutzfeldt–Jakob disease
CL	Corpus luteum
CMV	Cytomegalovirus
CNS	Central nervous system
COC	Cumulus–oocyte complexes
CPA	Cryoprotective agent
CPGODN	Cytosine-phosphate-guanine oligodeoxy nucleotide
CPV	Canine parvovirus
CS-REV	Chicken syncytial strain of reticuloendotheliosis virus
CTL	Cytotoxic T-lymphocyte
DAS ELISA	Double antibody sandwich ELISA
dATP	Deoxyadenosine triphosphate
dCTP	Deoxycytidine triphosphate
ddNTP	Dideoxy nucleoside triphosphate
DFM	Dark-field microscopy
DHBV	Duck hepatitis B virus
DM	Dexamethasone
DMD	Duchenne muscular dystrophy
DMEM	Dulbecco's modified Eagle's medium
DMSO	Dimethyl sulphoxide
dNTP	Deoxynucleoside triphosphate
DP	Degree of polymerization
DRT-PCR	Duplex reverse transcription-PCR

DWG	Daily weight gain
EBL	European Bat Lyssa virus
EC	Embryonic carcinoma cells
ECG	Equine chorionic gonadotrophin
ECL	Enhanced chemiluminescence
EDTA	Ethylenediamine tetraacetic acid
EGF	Epidermal growth factor
EHDV	Epizootic haemorrhagic disease virus
ELISA	Enzyme-linked immunosorbent assay
EP	Electroporation
EPO	Erythropoietin
ERIC	Enterobacterial repetitve intergenic consensus sequence
ES cells	Embryonic stem cells
EST	Expressed sequence tags
ET	Embryo transfer
ETEC	Enterotoxigenic *E. coli*
EWT	Egg weight
Fab	Fragment antibody
FAT	Fluorescence antibody test
FAV	Fowl adenovinus
FCS	Foetal calf serum
FDA	Food and Drug Administration
FH	Familial hypercholesterolaemia
FISH	Fluorescence *in situ* hybridization
FMD	Foot-and-mouth disease
FMDV	Foot-and-mouth disease virus
FPCS	Fusion protein cleavage site
FPV	Feline parvovirus
FPV	Fowlpox virus
FSH	Follicle stimulating hormone
FTAI	Fixed-time artificial insemination
GAT	Gene augmentation therapy
gc	Glycoprotein C
GH	Growth hormone
GHRH	Growth hormone releasing hormone
GLP-1	Glucagon like peptide
Glut-1	Glucose transporter-1
GM-CSF	Granulocyte–macrophage colony stimulating factor
GMO	Genetically modified organism

GOT	Glutamic-oxaloacetic transaminase
GPD	Glyceraldehyde 3-phosphate dehydrogenase
GPID(50)	Guinea pig 50% infection dosage
HA	Haemagglutination assay
HAR	Hyperacute rejection
HAT	Hypoxanthine aminopterin and thymidine medium
HBsAg	Hepatitis B surface antigen
HBV	Hepatitis B virus
HD	High dosage
HDL	High density lipoprotein
HEPA	High efficiency particulate filter
HEPES- N	(2- hydroxy ethyl) piperazine N´ (2- ethane sulphonic acid)
HF	Holstein–Friesian
hGH	Human growth hormone
HGPRT	Hypoxanthine-guanine phosphoribosyl transferase
HIV	Human immunodeficiency virus
hLF	Human lactoferrin
HMG	Human menopausal gonadotrophin
HN	Haemagglutinin–neuraminidase
HPI	Hour post-infection
HPRT	Hypoxanthine phosphoribosyl transferase
HRP	Horseradish peroxidase
HSV	Herpes simplex virus
HuMAbs	Human monoclonal antibodies
HVT	Herpes virus of Turkeys
IBD	Infectious bursal disease
IBDV	Infectious bursal disease virus
IBV	Infectious bronchitis virus
ICM	Inner cell mass
ICPI	Intracerebral pathogenicity indices
Id	Idiotype
IFA	Indirect fluorescent antibody
IF-tau	Interferon-tau
IGF-1	Insulin-like growth factor 1
Igf2r	Insulin-like growth factor 2 receptor
ILT	Infections laryngotracheitis
IPP	Intellectual Property Protection
IPR	Intellectual Property Rights
IPTG	Isopropyl thiogalactoside

ISH	*In situ* hybridization
IVC	*In vitro* culture
IVF	*In vitro* fertilization
IVM	*In vitro* maturation
KLH	Keyhole limpet haemocyanin
LAMP	Loop-mediated isothermal amplification
LD	Low dosage
LDL	Low density lipoprotein
LF	Lactoferrin
LOPU	Laproscopic ovum pick-up
LT	Labile toxin
LTR	Long terminal repeats
Lv	Lentiviruses
MAb	Monoclonal antibody
MAC	*Mycobacterium avium* complex
MAPPing	Message amplification phenotyping
MBP	Maltose-binding protein
MC	Microcarrier
MCK	Muscle-specific creatine kinase
MCS	Multicloning site
MD	Medium dosage
MDCK	Madin Darby canine kidney cells
MDV	Marek's disease virus
MEM	Minimum essential medium
ME-RSAT	Mercaptoethanol rapid slide agglutination test
MG	*Mycoplasma gallisepticum*
MHC	Major histocompatibility complex
MID	Minimal infecting dose
MIT	Mouse inoculation test
MLST	Multilocus sequence typing
MMLV	Moloney murine leukaemia virus
mPCR	Multiplex PCR
MS	*Mycoplasma synoviae*
MSG	Monosodium glutamate
MV	Measles virus
NBT	4-Nitroblue tetrazolium chloride
NC	Nitrocellulose
NCM	Nitrocellulose membrane
NDV	New Castle disease virus

NF	Nylon filters
NIL	Near isogenic lines
NMGB	*Neisseria meningitidis* serogroup B
nPCR	Nested PCR
NSP	Non-structural protein
NT	Nuclear transfer
ORF	Open reading frame
ORS	Oral rehydration solution
PA	Protective antigen
PAGE	Polyacrylamide gel electrophoresis
pBR 322	plasmid Bolivar Rodriguez 322
PBR	Plant breeder's rights
PCR	Polymerase chain reaction
pDNA	Plasmid DNA
PDV	Phocine distemper virus
PEG	Polyethylene glycol
PG	Prostaglandin
PG	Placebo group
PGF	Prostaglandin F2 alpha
PKU	Phenylketonuria
p.n.	post natum
PMSG	Pregnant mare serum gonadotrophin
PNK	Polynucleotide kinase'
PPR	Peste des petitis
PPRV	Peste des petitis virus
PRL	Prolactin
PRP	Prion-related peptide
PSS	Porcine stress syndrome
PVA	Polyvinyl alcohol
pYAC	Plasmid yeast artificial chromosome
QTL	Quantitative trait loci
RAPD	Random amplified polymorphic DNA
RCV	Replication-competent virus
rDNA	Recombinant DNA
REP	Repititive extragenic palindromic sequence
RFFIT	Rapid fluorescent focus inhibition test
RFLP	Restriction fragment length polymorphism
rhLF	Recombinant human lactoferrin
RIA	Radioimmunoassay

RIG	Rabies immunoglobulins
RPMI 1640	Rosewell Park Memorial Institute 1640 medium
RSP	Restriction site polymorphism
RSS-PCR	Restriction site-specific PCR
RSV	Respiratory syncytial virus
RT-PCR	Reverse transcription polymerase chain reaction
RT	Reverse transcriptase
Rv	Retrovirus
SBT	Sequence-based typing
SCC	Somatic cell count
SCID	Severe combined immune deficiency
SCNT	Somatic cell nuclear transfer
SCP	Single cell protein
SDM	Serologically detectable male antigen
SDS	Sodium dodecyl sulphate
SMGT	Sperm mediated gene transfer
SMT	Sulphamethazine
SNP	Single nucleotide polymorphism
SPF	Specific pathogen-free
sqr	Simple quadruplet repeats
SR	Superovulatory response
SRIF	Somatotropin release inhibiting factor
SSC	Sodium chloride–sodium citrate buffer
SSCP	Single-strand conformation polymorphism
SSF	Solid-state fermentation
SSPE	Sodium chloride –sodium phosphate–EDTA buffer
SSR	Simple sequence repeats
ST	Sequence type
ST	Stable toxin
STH	Somatotrophic hormone
str	Simple tandem repeats
STR	Short tandem repeats
STRP	Short tandem repeat polymorphism
SV-40	Simian virus-40
SVDV	Swine vesicular disease virus
T1D	Type I diabetes
T2D	Type II diabetes
TcoV	Turkey corona virus
TCR	T-cell receptors

TGEV	Transmissible gastroenteritis virus
TGF	Transforming growth factor
tK	Thymidine kinase
T_m	Melting temperature
TMV	Tobacco mosaic virus
tPA	Tissue plasminogen activator
TPH	Tissue print hybridization
Tr	Regulatory T cells
UTC	Uterine tubal cells
VAS	Visual analog scale
VEGF	Vascular endothelial growth factor
VI	Viral isolation
VNT	Virus neutralization test
VNTR	Variable numbers of tandem repeats
VPC	Vector producing cells
VPCR	Vectorate PCR
VRRV	Vampire-related rabies virus
VSV	Vesicular stomatitis virus
VSV-IN	Vesicular stomatitis virus - Indiana
VSV-NJ	Vesicular stomatitis virus-New Jersey
VT	Vero toxin
VTEC	Vero toxin *Escherischia coli*
vvIBDV	Very virulent infectious bursal disease virus
YCp	Yeast centromere plasmid
YEp	Yeast episomal plasmid
YIp	Yeast integrative plasmid
YRp	Yeast replicative plasmid

1
INTRODUCTION

WHAT IS BIOTECHNOLOGY?

Biotechnology is a buzzword in today's world, when everyone wants to use biotechnology for betterment in their sphere of activity. The governments of developed and developing countries are committing substantial funds for research in biotechnology. Even though man has exploited biotechnological methods for thousands of years in brewing, wine making, bread making and food preservation, its industrial application in large scale for development of better products are gaining momentum only recently. Application of genetic engineering techniques via recombinant DNA technology is responsible for the current "Biotechnology boom" occurring world over. Modern biotechnology has played a substantial role in the development of the health care and chemical industries. It has made possible the availability of several diagnostics, prophylactic and therapeutic products.

The term biotechnology was coined during late 1970s. Biotechnology has been defined in various ways, mostly unsatisfactorily. Biotechnology has been defined as, "the application of biological organisms, systems or processes to manufacturing and service industries" (Coombs, 1984). It is also defined as "the application of scientific and engineering principles to the processing of materials by biological agents to produce goods and services" (Coleman, 1986). In this, the word "agent" denotes a wide range of biological things such as enzymes, whole cells or multicellular organisms. Services and goods mean such processes as waste-water treatment. The scientific and engineering principles are chiefly microbiology, biochemistry, genetics, biochemical and chemical engineering. Biotechnology is the commercial exploitation of living organisms or their components (Primrose, 1987). It is also defined as "the industrial exploitation of biological systems or processes and it is largely based upon the expertise of biological systems in recognition and catalysis" (Higgins, 1985).

While biotechnology has been defined in many forms, in essence, it implies the use of microbial, animal or plant cells or enzymes to synthesize, breakdown or transform materials.

Biotechnology is the integrated use of biochemistry, microbiology and engineering sciences in order to achieve technological (industrial) application capabilities of microorganisms, cultured tissues, cells and parts thereof. The extraordinary multidisciplinary nature of biotechnology is shown by an anecdote, that the fruits of biotechnology are born on a tree whose roots are the biological sciences, in particular microbiology, genetics, molecular biology and biochemistry, whose trunk is chemical engineering in its widest sense (Figure 1.1).

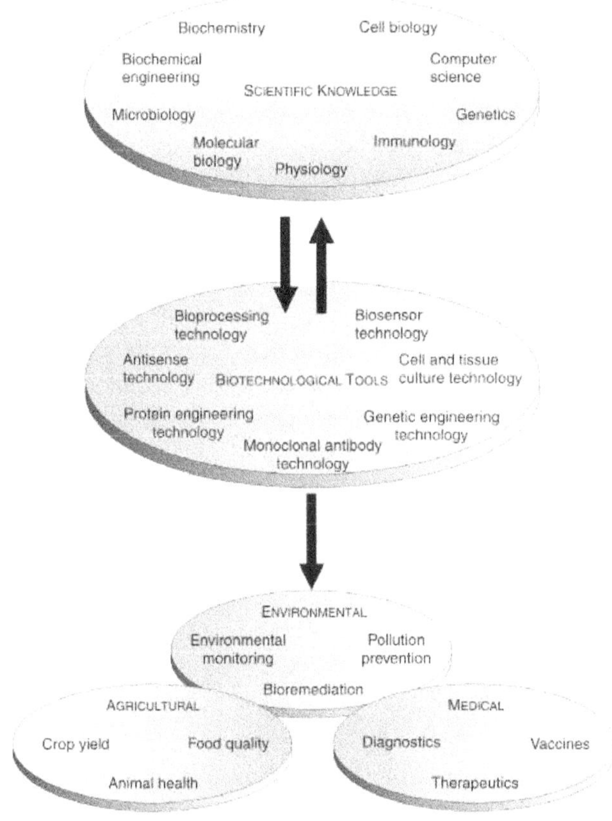

Figure 1.1 Origin of biotechnology and its different applications

A key factor in the distinction between biology and biotechnology is their scale of operation. The biologist usually works in the range of nanograms to milligrams. Biotechnologists working on the production of vaccines may be satisfied with milligram yields, but many other projects aim at kilogram or tonnes. Thus, one of the main aspects of biotechnology consists of scaling up of biological processes.

BIOTECHNOLOGY—PAST, PRESENT AND FUTURE

It should be recognized that biotechnology is not something new, but represents a developing and expanding series of technologies dating back thousands of years, when humans first began unwittingly to use microbes to produce foods and beverages such as bread and beer and to modify plants and animals through progressive selection for desired traits. Biotechnology encompasses many traditional processes such as brewing, baking, wine making, cheese production, the production of oriental foods such as soya sauce and sewage treatment, where the use of microorganisms has been developed somewhat empirically over countless years. However, it was the discovery of antibiotics in 1929 and their subsequent large-scale production in the 1940s that created the greatest advances in fermentation technology. Since then we have witnessed a phenomenal development in the technology, not only in the production of antibiotics but in many other useful, simple or complex biochemical products, e.g. organic acids, polysaccharides, enzymes, vaccines and hormones. Thus, biotechnology is not a sudden discovery but rather a coming of age of a technology that was initiated several decades ago. Microorganisms have been used to produce beer, vinegar, yoghurt and cheese for over eight millennia; the ancient Sumerians, produced beer and ale houses which were popular in Roman civilization. Wine was also popular with the Romans. References to wine and vinegar are scattered throughout the Bible, which is an indication that their production, dates back to early times.

Ethanol was the first chemical to be produced with the aid of biotechnology. The origin of distillation was not clear but by the fourteenth century it was widely used to increase the alcoholic content of wines and beers.

Louis Pasteur was the first to observe that yeast convert sugars into alcohol in the absence of air. Such an anaerobic process is known as **"fermentation"**. Souring and spoilage occur later and are due to the activities of a group of bacteria, the acetic acid bacteria, which converts alcohol into vinegar (acetic acid). Pasteur's intention was to heat the alcohol just enough to kill most of the microorganisms present, a process that does not mostly affect the flavour of the wine and beer. This process is known as **"Pasteurization"**.

Another major milestone in biotechnological generation of valuable products was the development of the antibiotic industry, arising initially from the discovery of the chemotherapeutic properties of penicillin by Fleming, Flory and Chain in 1940.

In the late 60s, considerable excitement was generated by the prospect of using microbial cells as a source of protein, the so called **"single cell proteins"** or SCP. The dried cells of certain algae, bacteria, yeasts, moulds, and mushrooms serve as food collectively called as single cell proteins. However the introduction of SCP has not been successful. The developed countries did not need SCP as they had a plentiful supply of proteins from conventional sources; whereas the under developed countries cannot afford to buy SCP or even to build and run SCP plants.

The treatment of waste products like manure from animals by anaerobic digestion by mixed microflora, eventually generating biogas (mainly methane and CO_2) has become an increasingly important process. It is highly efficient in terms of conserving and concentrating the energy available in the waste (over 80% of the free energy is recovered in the gas).

During 1980s, biotechnology became the major growth area. This change came about through a single development; the ability to explain together *in vitro* DNA molecules derived from different sources. This gene splicing ability is referred to as "gene manipulation". This is also known as **recombinant DNA technology**. Recombinant DNA technology and genetic engineering techniques found several useful applications in the area of vaccines, foods, antibiotics, alcohols, hormones and monoclonal antibodies. It has been possible to diagnose genetic defects by use of the restriction mapping technique. This is based on the fact that, the base sequence in defective gene differs from that in normal genes, leading to the production of different-sized DNA fragments when a gene is cut with a restriction endonuclease. Figure 1.2 shows the production of recombinant vaccines using *E. coli* system.

One of the important applications of genetic engineering is the production of insulin and somatostatin using bacteria. It is also a new process to insert foreign genes into cells and produce transgenic plants/animals with desired traits.

Animal health can be improved with new biotechnology methods of diagnosis, prevention and control of animal diseases. Diagnostic tests based on the use of antibodies and new vaccines against viral and bacterial diseases are also particularly relevant for developing countries and have a wide application for prevention of cattle epidemics and diseases. Genetically engineered vaccines against *E. coli* (pigs), Feline leukaemia virus, rinderpest virus, rabies virus, Aujeszky's disease virus and against many other microbes have been prepared.

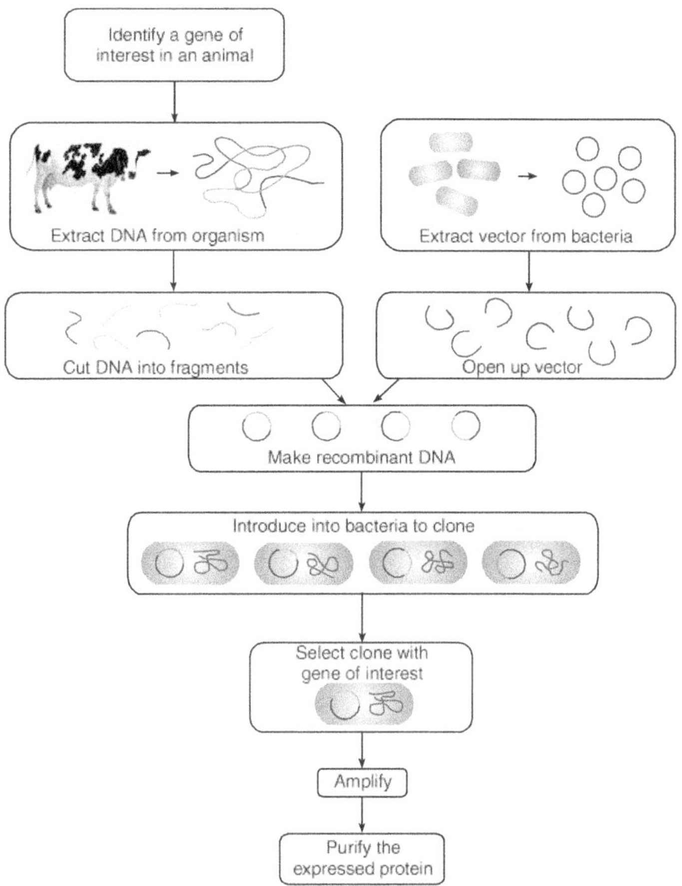

Figure 1.2   Recombinant DNA technology procedure showing the production of recombinant protein

The emergence of recombinant DNA technology has also given rise to diagnostic DNA/RNA probes. Along with this technology, hybridoma technology leads to the development of monoclonal antibodies against very many diseases, which has enhanced diagnostic sensitivities and specificities in therapeutics. The MAb production steps are shown in Figure 1.3.

In the field of reproduction, new biotechniques such as embryo transfers, *in vitro* fertilization, cloning and sex determination of embryos have been developed for different types of livestock. Superovulation techniques with gonadotrophic hormones, led to embryo transfer on an increasing scale in 1980s when embryo freezing techniques became available.

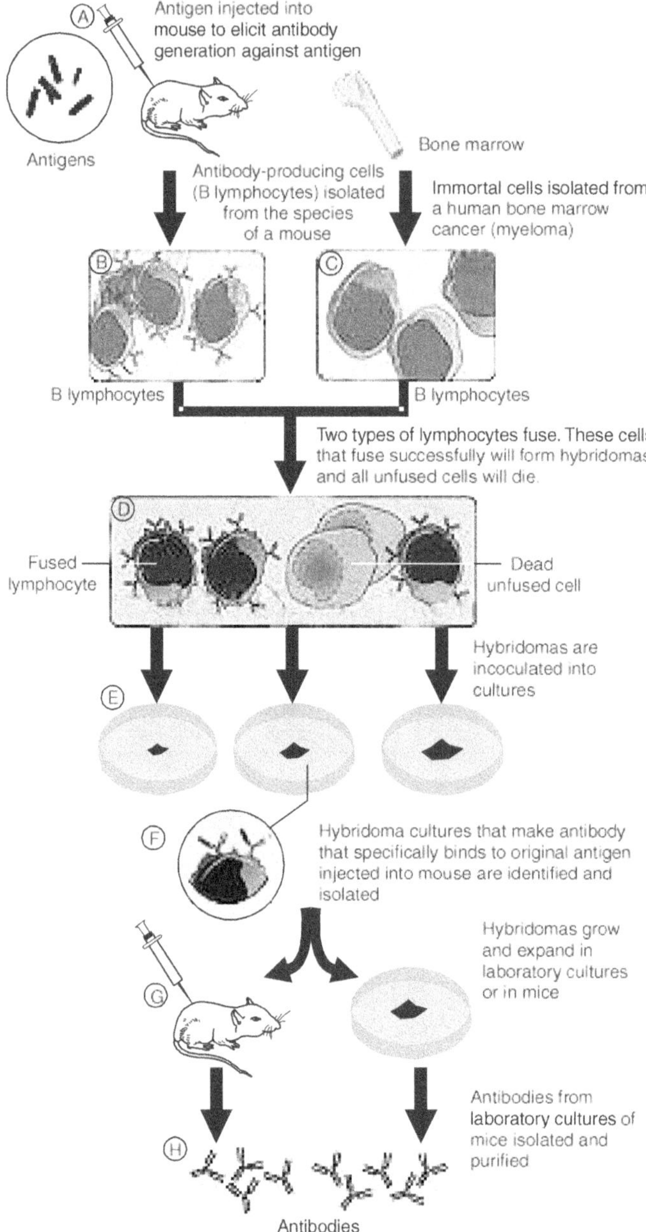

Figure 1.3 Hybridoma technology—steps involved in the production of monoclonal antibodies

A new DNA amplification system, known as polymerase chain reaction (PCR) has greatly facilitated DNA analysis and has found several applications.

This works with intact or broken DNA pieces (as small as 50 base pairs). It is essentially an *in vitro* method for copying simultaneously the two corresponding DNA strands which make up a gene sequence. The specific DNA segment to be amplified is selected by using primers (short DNA). By heating the DNA samples, the two strands separate, allowing the primers to bind to the flanking sequences, one on each strand. Thereafter, the primers initiate the synthesis of the daughter strands, complementary to the parental strands in the presence of DNA polymerase. This cycle of heating, annealing of primers and the synthesis of new strand can be repeated several times. Roughly twenty cycles can amplify the DNA by a factor of about a million.

Biotechnology will, in the future make an important contribution to the quality of life. In medicine, many organizations are involved in the large-scale production of human interferon. Genes for human insulin and growth hormone have been cloned and expressed in bacteria and the genes for many other human proteins of value for diagnosis or treatment are being cloned to facilitate their large-scale production. At present, microbially derived hormone insulin is already being marketed and being used for therapy in diabetics.

Biotechnological innovations will produce efficient vaccines and antibiotics for veterinary use. If vaccines from genetically engineered microorganisms live up to expectations, we will witness the eventual eradication of disseminating diseases such as foot-and-mouth disease and rinderpest. Growth hormones may be useful to increase the meat and milk yield.

Many of our energy needs could be solved by conversion of light energy from sun and waste materials. Electricity could be produced from these sources.

Biotechnology has profound impact on food and beverage industries. This may be achieved by use of industrial enzymes to manufacture high fructose corn syrups which are used as sweeteners in soft drinks. Recombinant DNA techniques will be employed to make rennin which is used to clot milk for making cheese. Till recently, most of these rennin have been extracted from the stomach of young calves.

Biotechnology can provide methods to the improvement of whole crops, both in terms of quality and yield. It can provide supplements or alternatives to expensive chemical fertilizers and pesticides.

SINGLE CELL PROTEIN (SCP)

The use of microbes as protein producers has gained wide experimental success. This field of study has become known as single cell protein or SCP production, referring to the fact that most of the microorganisms used as producers grow as single or filamentous individuals, rather than as complex multicellular organisms such as plants or animals.

During the last two decades there has been a growing interest in using microbes for food production, in particular for feeding domesticated food-producing animals such as poultry. It has been argued that the use of SCP derived from low value waste materials for animal feed would improve human nutrition by taking protein-rich vegetable foods out of the human verses animal competition and making them more freely available for human consumption in the producer countries, which are often developing countries.

The high protein levels, bland odour and taste of SCP, together with ease of storage confer considerable potential to SCP in food and food outlets. Its high protein makes its use attractive in aquaculture, e.g. farming of shrimps, prawns, trout, salmon, etc.

Microorganisms produce protein much more efficiently than any farm animal. The protein-producing capacities of a 250 kg cow and 250 g of microorganisms are often compared. Whereas the cow will put on 200 g of protein per day, the microbes, in theory, could produce 25 tonnes in the same time under ideal growing conditions. However, the cow also has the unique ability to convert grass into protein-rich milk. After decades of research no rival method for that conversion process has been developed. The cow has recently been described as "a live, self-reproducing and edible bioreactor". The advantages of using microbes for SCP production are as follows:

1. Microorganisms can grow at remarkably rapid rates under optimum conditions; some microbes can double their mass every 0.5 to 1 hour.

2. Microorganisms are more easily modified genetically than plants and animals; they are more amenable to large-scale screening programmes to select for higher growth rate, improved acid content, etc., and can be more easily subjected to gene transfer technology.

3. Microorganisms have relatively high protein content and the nutritional value of the protein is good.

4. Microorganisms can be grown in vast numbers in relatively small continuous fermentation processes, using relatively small land area, and growth is also independent of climate.

5. Microorganisms can grow on a wide range of raw materials, in particular low value wastes, and some can also use plant-derived cellulose.

The acceptability of SCP when presented as a human food does not depend only on its safety and nutritional value. In addition to the general reluctance of people to consume material derived from microbes, the eating of food has many subtle psychological, sociological and religious implications.

SCP from Wastes

The materials that make up wastes should normally be recycled back into the ecosystem, e.g. straw, bagasse, citric acid, olive and date wastes, whey, molasses, animal manures and sewage. The amount of these wastes can be locally very high and may contribute to a significant level of pollution in water courses. Thus, the utilization of such materials in SCP processes serve two functions— reduction in pollution and creation of edible protein. Of particular interest have been the extensive programmes to convert cattle and pig wastes to feed protein.

In Britain, Rank-Hovis-McDougall in conjunction with ICI (Marlow Foods) is now commercially marketing another fungal protein, mucoprotein or Quorn, derived from the growth of a *Fusarium* fungus on simple carbohydrates. Unlike almost all other forms of SCP, mucoprotein is produced for human consumption. The final fermentation product is pale-buff in colour, bland to taste, highly nutritious and fibrous in composition. It is now textured by food scientists into meat-like analogues and with correct culinary ingredients, which seem remarkably similar to meat and chicken dishes. A typical composition of mucoprotein compared to beef is shown in the Table 1.1.

Table 1.1 Typical composition of mucoprotein in comparison to beef

Component	Mucoprotein	Raw lean beef steak
Protein	47	68
Fat	14	30
Dietary fibre	25	Trace
Carbohydrate	10	0
Ash	3	2
RNA	1	Trace

There can be little doubt that cellulose from agriculture and forestry sources and from wastes must constitute the future major feedstock for many biotechnology processes including SCP. Cellulose, in its natural association with lignin, is by far the most prevalent organic material available for biotechnological conversion. Next to cellulose, lignin is the earth's second most abundant natural biopolymer found in plants. Approximately 30% of most woody plants is composed of lignin and its catabolism and utilization as a renewable resource are of great commercial interest.

Mushroom cultivation provides one of the few examples of successful, commercial biotechnology processes based on lignocellulose as a substrate. On the worldwide basis, mushroom growing is one of the fastest growing biotechnological industries. The cultivation of the common white mushroom *Agaricus bisporus* has expanded worldwide and accounts for over 70% of all mushrooms produced and consumed. The USA continues to be the world's largest producer. Mushroom production is in principle, a fermentation process. In the case of *Agaricus,* the substrate for growth is straw while for *Lentimula* it is wood. For *Agaricus* cultivation, the straw is composed of animal manures and other organic nitrogen compounds, over a period of one to two weeks and the final product is a unique substrate, suitable for the rapid growth of the *Agaricus* inoculum. *Lenticula edodes* is the second most cultivated mushroom in the world and has been farmed for over 2000 years. Currently, over 90% of its production occurs in Japan. The principle method of cultivation has been to inoculate wooden logs with spore inoculum or mycelial plugs, which allow the logs to stand for, up to nine months to achieve colonization by the fungus and then during subsequent early summer and autumn periods the mushrooms will grow out and be harvested.

The greatest advantage of these mushroom fermentation is the possibility of converting industrial, urban and wood wastes into a product that is directly edible for humans.

ACTIVITIES OF BIOTECHNOLOGISTS

A biotechnologist can utilize techniques derived from chemistry, microbiology, biochemistry, chemical engineering and computer science. The main objectives will be the innovation, development and optimal operation of processes in which biochemical catalysis has a fundamental and irreplaceable role. Biotechnologists must also aim to achieve a close working cooperation with the experts from other related fields such as medicine, nutrition, pharmaceutical and chemical industries, environmental protection and waste process technology. The main types of company categories involved with biotechnology are as follows:

Therapeutics   Pharmaceutical products for the cure or control of human diseases including antibiotics, vaccines, gene therapy

Diagnostics   Clinical testing and diagnosis, food, environment, agriculture

Agriculture/forestry/horticulture Novel crops or animal varieties, pesticides

Food Wide range of food products, fertilizers, beverages, ingredients

Environment Waste treatment, bioremediation, energy production

Chemical intermediates Reagents including enzymes, DNA/RNA, speciality chemicals

Equipment Hardware, bioreactors, software and consumables supporting biotechnology

The main areas of application of biotechnology are as follows:

1. **Bioprocess technology** Historically, the most important area of biotechnology are brewing, antibiotics and mammalian cell culture. Extensive development is in progress with new products envisaged, namely polysaccharides, medically important drugs, solvents, protein-enhanced foods, and novel fermenter designs to optimize productivity.

2. **Enzyme technology** It is used for the catalysis of extremely specific chemical reaction; immobilization of enzymes; to create specific molecular converters (bioreactors). Products formed include L-amino acids, high fructose syrup, semi-synthetic penicillin, starch and cellulose hydrolysis, etc. enzyme probes for bioassay.

3. **Waste technology** Long historical importance but more emphasis is now being made to couple these processes with the conservation and recycling of resources—food and fertilizers, biological fuels.

4. **Environmental technology** Great scope exists for the application of biotechnological concepts for solving many environmental problems— pollution control, removing toxic wastes; recovery of metals from mining wastes and low-grade ores.

5. **Renewable resources technology** The use of renewable energy sources, in particular, lignocellulose to generate new sources of chemical raw materials and energy, ethanol, methane and hydrogen. Total utilization of plant and animal materials.

6. **Plant and animal agriculture** Genetically engineered plants to improve nutrition, disease resistance, maintenance of quality, improved yields and stress tolerance will become increasingly and commercially available. Improved productivity, etc. for animal pharming, improved food quality, flavour, taste and microbial safety.

7. **Health care** New drugs and better treatment for delivering medicines to diseased parts, improved disease diagnosis and understanding of the human genome.

Recombinant DNA and Genetic Engineering

The DNA contained in genes determines inherited characteristics. Modifying DNA to remove, add, or alter genetic information is called genetic modification or genetic engineering. Recombinant DNA allow fragments of DNA from an animal/plant/microbe to be transferred to a host bacterium, which in turn incorporates the fragments into its own genome, thereby giving new capabilities for synthesis or biochemical reactions. Genetic engineering refers to "the isolation of individual or group of genes and transforming them to other organisms". In the host, the transformed gene replicates and is expressed as part of their new host genome.

The ability to extract a gene coding for a desired product and transfer it to another organism has opened the way either to the more effective production of useful proteins or to the introduction of novel characteristics in the host organisms. Thus, the large-scale production of hormones, vaccines, blood clotting factor or enzymes by friendly bacterium becomes possible. But why go to all these troubles, why not just extract the desired protein from useful source? There are four reasons.

The first reason is, it is often not possible or practical to grow certain types of cells on a large-scale. For example, mammalian cells, particularly of human origin may be difficult to obtain, grow slowly and are not amenable to the simple culturing techniques available for the growth of microorganisms. It has been speculated that the production of interferon from cultured human cell is likely in the long term, to be supplanted by production of genetically engineered microorganisms.

Secondly, availability of the natural source material may be strictly limited and may not be available all the time. Thirdly, the natural supply may be unavoidably contaminated. Haemophiliacs receiving factor VIII isolated from the occupational groups are exposed to the risk of receiving hepatitis or the AIDS virus. The fourth reason is its cost. Since the natural source material availability is limited, it is expensive.

In addition to these reasons, there is also an additional possibility for the production of truly novel proteins. Let us take the example of enzymes. The specificity, the catalytic activity and stability of enzymes are governed by the precise structure of the enzyme molecule. By selectively modifying the gene coding for the enzyme before it is introduced into the host organisms, the structure and hence properties of the enzymes, may be advantageously modified, hence a new brand of super enzymes. Recombinant insulin is now available for use in man, who are intolerant to pig insulin.

There is also a great potential for the modification of genome of economically important plants. Among the first commercial applications of genetically engineered food was a tomato in which the gene that produces the enzyme responsible for softening was turned off. The tomato could then be allowed to ripen on the vine without getting too soft to be packed and shipped. As of 2002, over forty food crops had been modified using recombinant DNA technology, including pesticide-resistant soyabeans, virus-resistant squash, frost-resistant strawberries, corn and potatoes containing a natural pesticide and rice containing beta-carotene. The introduction into crops of the ability to fix nitrogen from the atmosphere would not only save the cost of applying nitrogenous fertilizers, but would also eliminate the potential problem of water pollution from nitrates, washed out agricultural land. It is estimated that the nitrogen fixing Brussel sprouts could be produced at half the cost of the traditional variety. The levels of storage proteins in seeds could be increased to give high protein wheat. It is also possible that crops could be engineered for a greater resistance to herbicides or infections.

Diagnostic procedures based on nucleic acid probes One of the significant impacts of biotechnology on clinical medicine is the development of various diagnostic procedures based on nucleic acid probes. The test sample can be prepared either as a whole or sectioned and the DNA examined directly from organisms using chosen species or subspecies—specific or cross reacting DNA probes, i.e., *in situ* hybridization. The DNA can also be purified from the sample either as whole DNA fixed on to a nitrocellulose membrane or first digested with restriction enzymes, thus breaking it into variously sized small fragments which can be separated by electrophoresis on agarose or polyacrylamide gel, transferred to nitrocellulose membrane and then probed by labelled DNA to pick out possible restriction enzyme polymorphisms. The use of DNA probes for taxonomic studies such as the identification of subspecies and variations within species with geographical location is of epidemiological interest.

Hybridoma technology Hybridomas are hybrids between myeloma tumour cells and antigen stimulated lymphocytes which can be cloned, grown in large quantities and for indefinite periods of time and secrete high concentration of monoclonal antibodies and hence monospecific antibodies. A monoclonal antibody is an antibody directed against one antigenic determinant or epitope of an antigen. It is a single isotope. By testing a number of different clones it is possible to select those producing monoclonal antibody of an appropriate specificity and isotope suitable for a particular serological assay. The homogeneity, reproducibility and permanent availability of monoclonal antibodies are the attributes arousing the great interest in hybridoma technology.

Monoclonal antibodies have been used as immunodiagnostic reagents for detection of the causative agents or antibodies produced by causative agents; for experimental purposes for characterization of antigenic epitopes, and as vaccine in the case of monoclonal anti-idiotype antibody and for immunoprophylaxis or immunotherapeutics to infectious disease.

Recombinant vaccines With the development of recombinant DNA technology, novel vaccines like recombinant vaccines, subunit/synthetic vaccines, nucleic acid vaccines and peptide vaccines have been produced, which are more antigenic than conventional vaccines and can be produced in large amount and stored at room temperature.

The first recombinant vaccine produced for animal diseases is against foot-and-mouth disease virus. The FMD virus has got four viral proteins, viz., VP1, VP2, VP3 and VP4. Of these proteins, viral neutralizing property resides in VP1 fraction. Even in this portion, the amino acid residues 141–160 and 200–213 are the portions which are responsible for virus neutralizing ability. Once this specific amino acids have been identified, the synthetic peptide could be produced corresponding to the amino acid residues of 141–160 and used as vaccine against FMD.

Vaccinia-recombinant rinderpest vaccine has been produced using haemagglutinin and/or fusion protein genes. A recombinant vaccinia virus expressing the immunogenic rabies glycoprotein, has been produced and used successfully to protect foxes against rabies in Europe (Kieny *et al.*, 1984). Fowl pox recombinant Newcastle disease virus vaccine expressing haemagglutinin–neuraminidase (*HN*) gene has been designed and used to protect chickens (Morrison *et al.*, 1990).

Mammalian Cell Culture

Some mammalian proteins can be produced from cultured mammalian cells. The two most likely candidates in this category are monoclonal antibodies, for reasons of the complexity of the transcription and translation of their genetic material and interferon, for reasons of cost and effectiveness. Animal cell culture products include enzymes, lymphokines and insecticides. Human leucocyte cell lines are used for production of interferon. Dog kidney cell lines are used for FMD virus vaccine production.

Reproductive Techniques

The embryo manipulation techniques are purely biotechnological. This includes the splitting and freezing of embryos, *in vitro* fertilization and embryo transfer. The embryo splitting is done to produce identical twins and is now routine in cattle. Successful nuclear transplantation in sheep makes cloning (making large groups of genetically identical animals) more likely. Splitting of embryos are also done to increase the number of embryos.

In farm animals, *in vitro* fertilization combined with embryo transfer has become a common technique. These methods make it possible to use valuable animals repeatedly as donors, to supply many oocytes for improvement and manipulation in animal production.

The first successful embryo transfer in farm animals was done in 1934 in sheep and goats. The first offspring after transfer of embryo in cattle and sheep was reported in 1951.

The embryo transfer includes, (a) selection of donors (b) induction of superovulation (c) embryo collection (d) evaluation of embryos (e) selection of recipients and (f) transfer of embryos. Since embryo transfer techniques become available in livestock, embryos can be maintained *in vitro* and this makes them accessible to further manipulation prior to transplantation.

Transgenic Animals

Transgenic animals are produced by transfer of genetic material from one animal to another. The animals into which foreign DNA is integrated are called **"transgenics"**. The foreign genetic material will generally consists of a structural gene with regulatory sequences that are required for the transcription of the introduced genes (fusion genes). The first, direct evidence for transfer of traits from one generation to another through genes, was given by Gregor Mendel.

The classical example of genetic engineering was given by Palmiter and colleagues (1982), when they showed that mice injected with a fusion gene containing the metallothionein promoter and rat growth hormone gene. Transgenic livestock and transgenic aquatic species have been generated with increased growth rates, enhanced lean muscle mass, enhanced resistance to disease or improved use of dietary phosphorus to lessen the environmental impacts of animal manure. Transgenic poultry, swine, goats and cattle also have been produced that generate large quantities of human proteins in eggs, milk, blood or urine, with the goal of using these products as human pharmaceuticals. Examples of human pharmaceutical proteins include enzymes, clotting factors, albumin and antibodies. Transgenic calves were produced in Australia with genes to produce 25% more milk proteins that normal cows (Figure 1.4). Transgenic pig was produced by microinjection of embryos with fusion genes consisting of metallothionein promoter linked to human growth hormone expressed in vector pBR 322.

Figure 1.4   Transgenic calves produced in Australia with extra gene to produce milk by 25% more than normal animals

Somatic Cell Nuclear Transfer

Another application of animal biotechnology is the use of somatic cell nuclear transfer to produce multiple copies of animals that are nearly identical copies of other animals. This process has been referred to as cloning. Till date, somatic cell nuclear transfer has been used to clone cattle, sheep, pigs, goats, horses, mules, cats, rats and mice. The technique involves culturing somatic cells from an appropriate tissue (fibroblasts) from the animal to be cloned. Nuclei from the cultured somatic cells are then microinjected into an enucleated oocyte obtained from another individual of the same or a closely related species. The nucleus from the somatic cells is reprogrammed to a pattern of gene expression suitable for directing normal development of the embryo. After further culture and development *in vitro*, the embryos are transferred to a recipient female and ultimately will result in the birth of living offspring. Somatic cell nuclear transfer scheme is shown in Figure 1.5.

Plant Cell Culture

The plants, quite apart from their key role in producing foods, are also important source of other raw materials. Most of the bulk plant products are starch and sugar; 90% of the cars in Brazil now use a mixture of petrol and alcohol, the latter derived from fermentation of corn sugars. Sugar has also been used as feedstock for its conversion into ethylene oxide. Plants are also important sources of high value drugs, some 25% of the drugs are of plant origin.

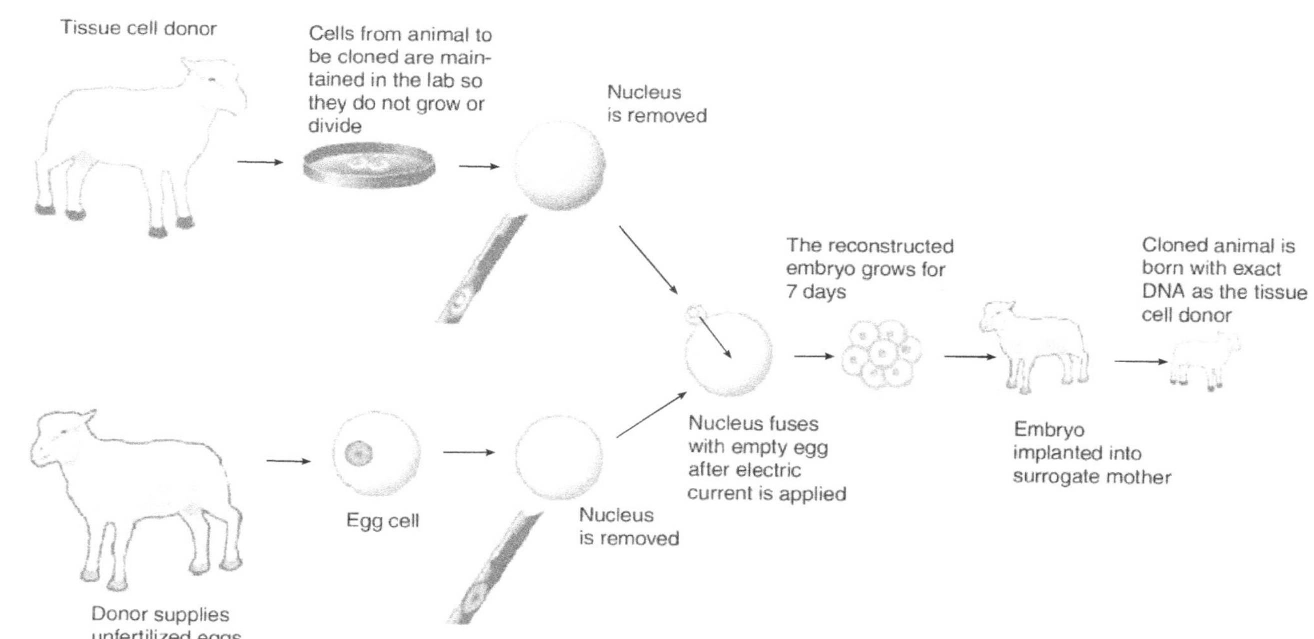

Figure 1.5 Somatic cell nuclear transfer—steps involved in the production of cloned sheep using somatic cell nuclear transfer method

Plant cell culture can regenerate entire plants. Cell culture provides a good way to extend studies in plant pathology and somatic cell hybridization. Some of the substances produced by plant cell suspension cultures include alkaloids, quinines, antibiotics, anti-viral agents, cardiac glycosides, amylases, carotenoids, etc. The ability to culture plant cells on a large scale, either for the production of biomass or in order to extract the desired product from cell cultures is becoming a highly desirable technology. Regenerated *Pelargonium* plantlet from protoplast culture is shown in Figure 1.6.

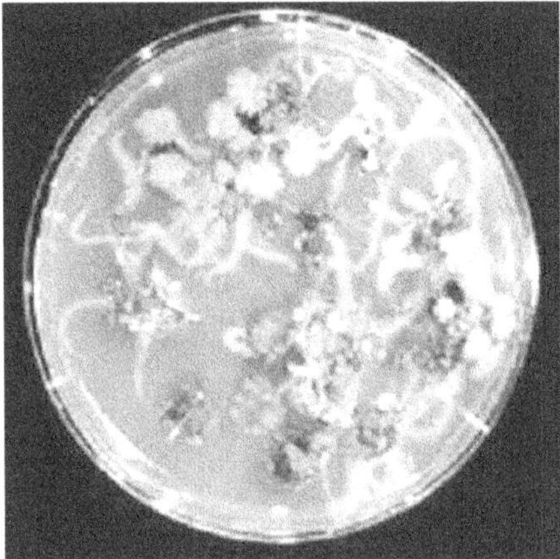

Figure 1.6   Plant cell culture—regenerated *Pelargonium* plantlet from protoplast culture

Fuels

The world is running short of combustible fuels, in particular mineral oil. Biotechnology might provide all the solutions, new fuels and alternative carbon feedstocks. In the economics of Brazilian fuel-alcohol process, sugar cane waste is used as fuel for distillation waste material such as pulp mill sludge for producing alcohol in many countries.

Another potential biotechnological fuel is methane. All that is needed is some pig slurry or animal waste material like dung and a hole in the ground with a lid on top; nature does the rest. It is one example of biotechnology that can be transferred to third world agricultural societies without much difficulty. When methane is produced by the fermentation of animal dung the gaseous products are usually referred to as biogas and the installations called biogas plants or bioreactors. Biogas is a flammable mixture of 50–80% methane, 15–45% CO_2, 5% water and some trace gases. Biogas is produced via biomethanation and is in fact, a self-regulating symbiotic microbial process operating under anaerobic conditions, and functions best at temperatures around 30°C. The organisms involved are all found naturally in ruminant manures. In such systems, the animal dung is mixed with water and allowed to ferment in near-anaerobic conditions. The production of biogas by such methods goes back to antiquity and is of particular importance in India, China and Pakistan.

The most aesthetically pleasing biotechnological fuel would be hydrogen derived from the biophotolysis of water. Biophotolytic production of hydrogen from water is based on the photosynthesis of plants with bacteria-derived hydrogenase enzymes and light. The greatest advantage with a fuel of hydrogen derived from water is that when burnt, it produces no pollution and regenerates its source material.

Biocatalysis

Enzymes are nature's supreme catalysts, exhibiting great specificity and enormous catalytic power. At present about 150 of the 2000 known enzymes find commercial applications and another 200 are available for use in genetic engineering. The latter include restriction endonucleases and ligases. The enzymes have been in use for centuries particularly in food processing (for e.g. cheese making, hair removal from hide). More recently they are used in chemical production, analytical and diagnostic systems and in the treatment of diseases. Animals, plants and microbes are the three important biological sources of enzymes. Over one half of the industrial enzyme market is accounted for, by proteolytic enzymes, for e.g. milk clotting enzymes for production of cheese, detergent protease, animal rennets and protease of *Aspergillus* used in baking.

Waste Treatment and Utilization

Sewage disposal is a problem long faced by mankind. Today's sewage plants are good examples of simple biotechnology—a fixed bed of microbes degrading the sewage products trickling over them.

In cheese production, milk is curdled and liquid whey is a waste material. This whey contains few proteins, minerals and 4% lactose. Two-thirds of the world's population cannot digest it. However, it does find some application in the manufacture of ice creams, packet soups and desserts, but its use can be widely extended if the lactose is split into its constituent sugars, glucose and galactose. Methods have been developed for splitting this with the enzyme β-D-galactosidase.

Cellulose is another waste material particularly in the form of straw, from cereal crops. This waste cellulose could be biologically degraded and used as a feedstock for the production of microbial proteins. It has been estimated that sufficient protein can be produced in this way from agricultural waste alone to feed the entire world population. The ability of some fungi to convert cellulose has been commercially exploited for producing edible mushrooms for human food.

Fermentation

Fermentation starts with biocatalysis, the distinction of being the oldest form of biotechnology. Traditionally, fermentation has meant the production of palatable alcohol for carbohydrates. However, fermentation, i.e., the application of microbial metabolism to transform simple raw materials into valuable products—can produce amazing range of useful substances, e.g. chemicals, such as citric acid, antibiotics, biopolymers and single cell proteins. What is needed is the knowledge of these microorganisms, the control of their metabolism and growth, and the ability to handle them on a large scale.

The current uses of fermentation process includes:

i. production of biomass, e.g. spirulina and yeast

ii. production of enzymes and nucleic acids

iii. production of metabolites, both primary (e.g. ethanol, lactic acid) and secondary (e.g. antibiotics, anti-metabolites)

iv. enzymatic conversion of specific substrates (e.g. glucose to fructose) and

v. enzymatic conversion of multiple substrates, e.g. biological wastes.

Idlies and dosas which are the staple diet of south Indians are also the products of fermentation. When the ground rice and urad dhal (black gram) are left overnight with salt, it becomes acidic and it is leavened by carbon dioxide, primarily produced by *Leuconostoc mesenteroides*. This activity is attributed to the presence of *Streptococcus faecalis*. The high salt content has excellent preservative action and is valuable as a condiment.

Process Engineering

It is one thing for the laboratory scientist to clone a novel gene, discover a new antibiotic or invent a new catalysed process, but it is quite another to transfer this technology to the scale of operation required to make useful product in a significant quantity. This last is the responsibility of an engineer, be he chemical, biochemical, or whatever. Harvesting, pretreatment, filtration of the raw material, reactor design and control, recovery/reuse of biocatalyst or organism, product extraction and analysis, effluent treatment, water recycling—all these are his concerns. Chemical engineers have shown this to be very adept at handling chemical processes on a large scale. However, biological processes differ from chemical processes in many ways, e.g. the requirement for sterility or axenic productions or extraction of liable materials diluted in large volume.

Biotechnology and Global Health

The World Health Organization estimates that more than 8 million lives could be saved by 2010 by combating infectious diseases and malnutrition through developments in biotechnology. The following are few of the biotechnological methods which could have the greatest potential to improve global health.

1. Hand-held devices to test for infectious diseases
2. Genetically engineered vaccines that are cheaper, safer and more effective in fighting common infectious diseases
3. Drug delivery alternatives to needle injections, such as inhalable or powered drugs
4. Genetically modified bacteria and plants to clean up contaminated air, water and soil
5. Genetically modified foods with greater nutritional value

REVIEW QUESTIONS

1. How can biotechnology improve the health of humans and animals?
2. What are the activities of biotechnologists?

REFERENCES

Coleman, R.F. (1986). *Biotechnology: A Plain Man's Guide*. Laboratory of the Government Chemist, Dept. of Trade and Industry, London.

Coombs, J. (1984). *The International Biotechnology Directory*. The Nature Press, New York.

Higgins, I.J. (1985). In: *Biotechnology, Principles and Applications*. (eds.). Higgins, I.J., Best, D.J. and Jones, J. Blackwell Scientific Publications, Oxford, London.

Kieny, M.P., Drillien, R., Spehner, D., Dkory, S., Schmitt, D., Wiktor, T., Koprowskin, H. and Lecocq, J.P. (1984). "Expression of rabies virus glycoprotein from a recombinant vaccinia virus." *Nature*. 312 : 163–166.

Morrison, T., Hinshaw, V.S., Sheerer, M., Cooley, A.J., Brown, D., CMcQuaia and McGinnes, L. (1990). *Microbial Pathogenesis*. 9: 387–396.

Palmiter, R.D., Brinster, R.L., Hammer, R.E., Trumbauer, M.E., Rosenfeld, M.G., Brinbger, N.C. and Erams, R.M. (1982). "Dramatic growth of mice that developed from eggs microinjected with metallothionein—growth hormone fusion genes." *Nature*. 300:611–615.

Primrose, S.B. (1991). *Molecular Biotechnology*. Blackwell Scientific Publications, London-Edinburgh-Melbourne.

2

CELL CULTURE AND FERMENTATION TECHNOLOGY

INTRODUCTION

One of the earliest workers in cell and tissue culture was Harrison who in 1907 developed a reproducible technique for tissue culture using frog. Later in 1912, Carrel used tissue embryo extracts as culture media. They successfully demonstrated that animal cells can be grown indefinitely *in vitro* just like other microorganisms. Initially cold-blooded animals like frog were used, so that no incubation was required, but later warm-blooded animals like chick and rodents were used and eventually mammals became a favourite material.

Mammalian cells are grown *in vitro* for a number of reasons including the production of viral vaccines. Under standard tissue culture techniques, cells are immersed in a pool of medium containing essential nutrients and metabolic products. The concentration of both constituents changes as the cell population grows, thus influencing cell survival and function.

CULTURE MEDIA

Culture medium is the single most important factor for culturing cells and tissues. It provides the optimum conditions of factors like pH, osmotic pressure, etc., and the chemical constituents, which the cells or tissues are incapable of synthesizing (unlike microorganisms, which can synthesize them from simple inorganic substances). Animal cells need either a completely natural medium or an artificial medium supplemented with some natural products. For some tissues, the natural medium is preferred, since it is the cheapest and most convenient.

Natural Media

The natural media used in cell culture systems fall in the following three categories:

1. coagula, such as plasma clots
2. biological fluids, such as serum and
3. tissue extracts of which embryo extract is the most common.

Plasma clots have been in use for a long time and are now available commercially either as liquid plasma in **siliconized ampoules** or as **lyophilized plasma**, to be reconstituted by the addition of distilled water saturated with carbon dioxide.

The most commonly used **biological fluid** is serum, which is commonly obtained from human adult blood, placental cord blood, horse blood or calf blood. Of these, human placental cord serum and foetal calf serum seem to be particularly satisfactory. The serum is tested for sterility and toxicity before use.

Embryo extract is the most commonly used **tissue extract**, although effective substitutes have now been developed to replace the embryo extract as a natural medium. The most important component of the substitutes is a mixture of amino acids. Embryo extract is often prepared from chick or bovine embryos of different ages (up to 10 days).

Defined Media

Media with serum   Although natural media are very useful and convenient for a wide range of uses, they suffer from the disadvantage of poor reproducibility due to lack of knowledge of the exact composition. For this reason, synthetic media have been designed for a variety of uses and purposes including the following:

- media for immediate survival,
- media for prolonged survival,
- media for indefinite growth and
- media for specialized functions.

For immediate survival, a "balanced salt solution" with defined osmotic pressure and pH is added. This requirement is met by the combination of certain inorganic ions, salts and glucose. For longer survival, serum may be used or the balanced salt solution may be supplemented with amino acids, oxygen, vitamins and serum proteins. One such medium was developed and modified by Eagle (1955) and is described as 'minimum essential medium' (MEM). MEM was developed for mammalian cells grown on monolayers with more fastidious requirements. This has been used for primary mammalian cells and established cell lines. This contains essential amino acids, vitamins and salts. Other more complex synthetic media include the following: (a) 199, (b) CMRL 1066, (c) RPMI 1640 and (d) F12. All these media are supplemented with 5–20% serum. **RPMI** 1640 (Roswell Park Memorial Institute 1640 medium) is commonly used for human, murine normal and neoplastic white blood cells; these grow as suspension cultures and have a reduced calcium requirement.

Role of serum in cell culture   Serum is an extremely complex mixture of many small and large biomolecules with different physiologically balanced growth-promoting and growth-inhibiting activities.

Some of the major functions of serum are to provide the following:

1.   basic nutrients, in solution and bound to proteins,
2.   hormones and growth factors stimulating cell growth and specialized functions.
3.   attachment and spreading factors.
4.   binding proteins (albumin, transferrin) carrying hormones, vitamins, minerals, lipids, etc.
5.   Non-specific protection factors against mechanical damage and viscosity (shear forces during agitation of cell suspension)
6.   Protease inhibitors
7.   pH buffer

Disadvantages of using serum in cell culture

1.   Serum may contain inadequate levels of cell-specific growth factors which have been supplemented, and an overabundance of others which may be cytotoxic.
2.   Risk of contamination with virus, fungi or mycoplasma.

Serum-free media   The use of serum in culture media has several disadvantages, which led to the development of many serum-free media. Some of the disadvantages as listed in the earlier section are

1.   Serum varies from batch to batch and deteriorates within one year
2.   Changing serum in a batch requires fresh testing, so that the replacement is as close to the previous batch as possible
3.   If more than one cell types are used, which may require different serum concentrations, so that a number of batches may need to be maintained and co-culturing may be different

4. The demand for serum usually exceeds the supply for a variety of reasons

5. When cell culture is used for downstream processing to recover cell products, the presence of serum is an obstacle to purification.

6. Serum increases the cost of the medium, since its cost is ten times the cost of its constituents if used as substitute and

7. For most cells, serum is not a physiological fluid which they contact in the original tissue except during wound healing and blood clotting processes, where serum promotes fibroblast growth, but suppresses epidermal keratinocyte growth.

8. Selective inhibitors, bacterial toxins and lipids found in FCS and human pregnant serum may react wth polyamines to form cytotoxic polyaminoaldehydes. Hence, preference is given for horse serum. This may explain reports on the suppressor activity of these sera, and the preference of some workers for horse serum.

The major advantage of serum-free media is the ability to make medium selective for a particular cell type, since each cell type appears to require a different recipe. Examples of serum-free media include MCDB 110, MCDB 402, MCDB 153, Iscov's and LHC media.

PRIMARY CULTURE

Primary cell culture can be obtained either by allowing cells to migrate out from the tissue, which is adhering to a substrate or by disaggregating the tissue mechanically or enzymatically to produce a suspension of cells. Most of the normal untransferred cells survive and proliferate to produce a primary culture and when attached to a substrate, these cells need to be obtained by disintegration. Primary cell cultures retain most of the characteristics of the cells from which they originated. Primary cell cultures exhibit a phenomenon called contact inhibition which results in the cells lining up in strongly oriented parallel strands. With regard to chromosomal number, primary cultures usually retain their diploid karyotype. The most commonly employed tissues for preparing primary cultures are those from kidney, lung, liver, thyroid and testis. Primary cells are often used in preference to established cell lines and are regarded as a better representation of cells *in vivo*. Primary cells are more likely to reflect the true activity and functions that they display in their natural environment.

Disaggregation is achieved by any one of the following 3 methods: (i) physical disruption, (ii) enzymatic digestion and (iii) treatment with chelating agents. Physical disruption may often be combined with other methods. Of the many enzymes that are used for disaggregation, trypsin and pronase are the most commonly used. Similarly, tissues like epithelium (which needs Ca^{++}, Mg^{++} ions for its integrity) can be treated with chelating agents, such as citrate and ethylene diamine tetra acetic acid (**EDTA**).

Enzymatic Disaggregation

Trypsin is the most common enzyme used for disaggregation for the following reasons.

i. It is tolerated by a variety of cells.

ii. It is effective for many tissues.

iii. Its residual activity is neutralized by serum of the medium or by a trypsin inhibitor in the case of a serum-free medium.

Disaggregation by trypsin can be damaging to epithelial cells or ineffective to fibrous tissues. Since intracellular matrix contains collagen, collagenase has proved effective for disaggregation of several normal and malignant tissues, which may be rather insensitive to trypsin. Crude collagenase is often used with a finely chopped tissue in complete medium.

Mechanical Disaggregation

Enzymatic disaggregation is labour-intensive and involves damage of cells. Therefore, mechanical disaggregation of cells is sometimes preferred. In this method, tissue is carefully sliced and the cells that spill out are collected. Alternatively, the cells are either pressed through the sieves of a gradually reducing mesh or forced through a syringe and needle, or even repeatedly pipetted. Although the method may cause mechanical damage, the cell suspension is more quickly obtained than in the enzymatic disaggregation.

Separation of Viable and Non-viable Cells

The dissociated cells grow well when seeded on culture in plates at high density. In the adherent culture, non-viable cells will be removed at the first change of medium. In suspension culture, on the other hand, non-viable cells are gradually diluted out, when cell proliferation starts. However, non-viable cells can also be removed from primary disaggregate by centrifuging the cells in a mixture of Ficoll and sodium metrizoate, when viable cells are collected from interface after centrifugation.

CELL LINES

Cell lines can be divided into two types: adherent (monolayer cells) and non-adherent (suspension cells). Adherent cells attach to the plastic surface of a flask or plate and therefore need to be detached from this surface before they can be used. Suspended cells do not normally attach to the surface of the culture vessel.

The major reason for the use of cell lines is that they are often easier to handle than primary cells, grow continuously and a large number of cells can be obtained. The cultured cells are of three types.

1. **Precursor** or **stem cells** which are capable of proliferation, but remain undifferentiated until the correct inducing conditions are applied so that some or all of the cells mature to differentiated cells.

2. **Undifferentiated cells** also known as committed precursor cells.

3. **Mature differentiated cells.**

This equilibrium may shift according to the environmental conditions. High serum, growth factors and low cell density will promote cell proliferation and a low serum, appropriate hormones and high cell density will promote differentiation.

The source of the culture will also determine which of the above three types will be present in the culture. Cell lines derived from the embryo may contain more stem cells (or precursor cells), which will be capable of greater self renewal than the cultures from adults; the cultures from tissues which are undergoing continuous renewal *in vivo* (e.g. epidermis, intestinal epithelium, haemopoietic cells) will still contain stem cells and these may survive indefinitely, and cultures from tissues which renew only under stress (e.g. fibroblasts, muscles, glia) may only contain committed precursor cells with a limited culture lifespan. From any of the 3 kinds of cells derived from primary explant, cell lines may be developed. After the first subculture, the primary culture becomes a cell line and may be propagated and subcultured several times. With each successive subculture, the competent population with the ability to proliferate most rapidly will gradually predominate and non-proliferating or slowly proliferating cells will be diluted out.

The established cell lines will have altered chromosome number, shorter doubling time and should be subcultured indefinitely. Established cell lines do not exhibit the property of contact inhibition. The cell lines show a great variation in karyotype. Vero, BHK21 and MDCK are some of the cell lines commonly used in the field of veterinary virology (Figure 2.1).

The cell lines may be propagated in an unaltered form for a limited number of cell generations beyond which they may either die out or give rise to **continuous cell lines**. The continuous cell lines are often aneuploid and have larger variation in chromosome number. The alteration in culture, giving rise to a continuous cell line is commonly called *in vitro* transformation.

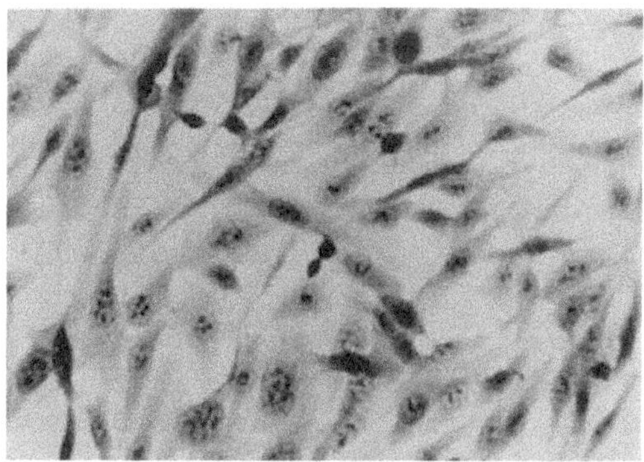

Figure 2.1   Normal MDBK monolayer cell line used for isolation and propagation of viruses like bovine herpesvirus 1, rotavirus and Newcastle disease virus

Maintenance of Cultures—Cell Lines

When a primary culture is grown in a medium, as they grow and increase in number, the available medium is used up and there arises a need to subculture it from a heterogeneous primary culture containing many types of cells derived from the original tissues. During subculturing, a very homogeneous **cell line** emerges. The culture now called a **cell line**, can be propagated, characterized and stored. The term 'cell line' implies the presence of several cell lineages either similar or distinct. A cell line may be finite or continuous, depending upon whether it has limited culture lifespan or is immortal in culture. Some commonly used cell lines are listed in Table 2.1.

Table 2.1   Cell lines in regular use

Name	Morphology	Origin
Finite cell lines		
MRC 5	Fibroblast	Human embryo lung
W 138	Fibroblast	Human embryo lung
IMR 90	Fibroblast	Human embryo lung
Continuous cell lines		
BHK 21	Fibroblast	Newborn Syrian hamster kidney cells
CHOK 1	Fibroblast	Adult Chinese hamster ovary
HeLa	Fibroblast	Adult human
Vero	Fibroblast	Adult monkey kidney

Most of the primary cultures or continuous cell lines grow as monolayers. These cultures will need a periodic change of medium, whether or not cells are proliferating. In cultures where cells are proliferating, the usual practice in subculturing adherent cell line involves the following steps:

1. removal of the medium and
2. dissociation of the cells in the monolayer with trypsin or other enzymes.

Intervals between change of medium and subculturing vary from one cell line to another depending upon the rate of growth or metabolism. Rapidly growing cell lines like HeLa are subcultured once in a week and the medium is changed 4 days later. More slowly growing cell lines are required to be subcultured every 2, 3 and 4 weeks and the medium changed weekly between subcultures. Cell culture flasks with medium used for culture of cells and tissue are shown in Figure 2.2.

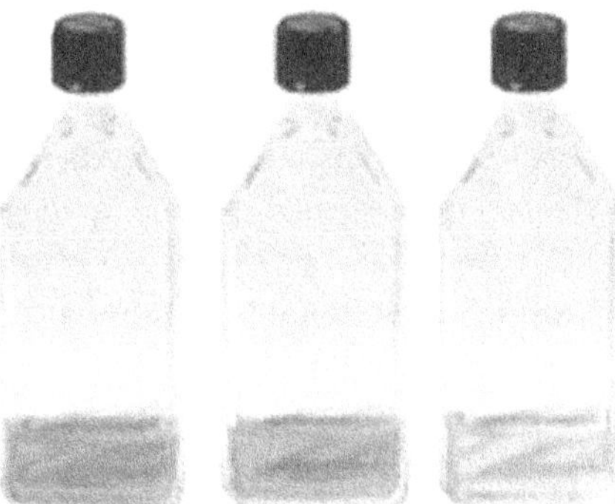

Figure 2.2 Cell culture flasks with medium used for culture of cells and tissues

Maintenance of aseptic conditions is one of the most difficult tasks in the tissue culture technique. Bacteria, yeasts, fungi, moulds, protozoa and mycoplasma appear as contaminants in tissue culture. To avoid contamination of cultures, antibiotics should be included in the medium at the recommended concentration. The commonly used antibiotics are given in the Table 2.2.

Large-scale Cell Cultures

Availability of genetically modified cells for the production of recombinant compounds has increased the ability of large-scale cell cultures in industry. Specific promoters like human metallothionein II A and baculovirus IE1 promoters have been utilized to provide high expression of foreign genes, so that the large-scale cell cultures may become commercially profitable. In a

large-scale production unit, for completion of reaction, cell biomass must be separated from supernatant, which usually contain the desired product. Cell separation is usually achieved by centrifugation or by cross-flow filtration.

Table 2.2   Details of some of the most commonly used antibiotics

Name of antibiotic	Storage temperature	Active against	Working concentration	Mode of action
Gentamicin	4°C	Gram-positive and gram-negative bacteria, *Mycoplasma*	50 mg/l	Inhibits bacterial protein synthesis
Penicillin	–20°C	Gram-positive and gram-negative bacteria	100,000 U/l	Interferes with bacterial cell wall synthesis
Strepto-mycin	–20°C	Gram-positive and gram-negative bacteria	50–100 mg/l	Interferes with protein synthesis
Tylosin	–20°C	Gram-positive and gram-negative bacteria, *Mycoplasma*	6–10 mg/l	Interferes with protein synthesis
Nystatin	–20°C	Yeasts, moulds	10–200 U/ml	Permeability of cell membrane affected

The cell culture products are enzymes, hormones, animal vaccines, monoclonal antibodies, interferons, etc. (Table 2.3). Some of the important enzymes obtained from animal cells are asparaginase, collagenase, hyaluronidase, pepsin, rennin, trypsin, hydroxylase and urokinase. Similarly four important hormones obtained from cultured cells include luteinizing hormone, follicle stimulating hormone, chorionic hormone and erythropoietin.

Table 2.3 Cell lines and the products obtained from them

Cell line	Product
Human leucocytes	Interferons
Mouse fibroblasts	Interferons
Human kidney	Urokinase
Dog kidney	Canine distemper vaccine
Chick embryo fluid	Foot-and-mouth disease vaccine
Chick embryo fluid	Vaccines for influenza, measles and mumps
Duck embryo fluid	Vaccines for rabies and rubella
Cell lines of mouse, rat or human origin	Monoclonal antibodies
Chinese hamster ovary cells (CHO)	Tissue-type plasminogen activator (tPA), beta-gamma interferons, factor VIII

Production of FMD vaccines is the most important example of the use of large-scale cell cultures. There are several other vaccines including polio vaccines, bovine leukaemia virus vaccines, rabies vaccines, etc., which are produced at a commercial scale, using cell cultures. Due to heavy demands for a variety of interferons for cancer therapy, recently, interferons α and β have been made in large quantities. For production of interferon α, 8000 litres of cell culture virus have been used. Similarly interferon β was produced by the cell grown on microcarriers at scales greater than 1000 litres.

Production of monoclonal antibodies with the help of cultured cells has also achieved significance in industry, due to the increasing utility of the antibodies in

i. *in vitro* diagnostics

ii. *in vivo* imaging

iii. therapy in humans and animals and

iv. industrial applications like immunoprecipitation.

The approaches followed for the production of monoclonal antibodies include *in vivo* production using ascites tumours in mice or rats and *in vitro* production through cell culture techniques.

The second method is economically viable in view of the increasing demand. It has the following additional advantages.

i. It can be scaled up, for instance, one kg of antibody (which can be produced through cultured cells in single fermenter) would require 20,000 mice if made by ascites tumours.

ii. The risk of contamination is greatly reduced.

Homogeneous Bioreactor Systems

Suspension cultures   The classical system is the suspension culture using a stirred tank with different impeller types and installations, equipped with or without a spin filter. In large-scale bioreactors, slight modifications of several internal parts of bioreactors used for bacterial fermentation are made in order to adopt them for culturing animal cells. The modifications are in the agitation system. Marine type impeller, vibromixer or rotating flexible sheets replace the turbine type impeller widely used in microbial fermentation. Perfusion systems were also developed for submerged cultivation of animal cells.

Microcarrier cultures   Among the several cell systems available, some of them are anchorage-dependent and require a surface for their attachment and subsequent growth. In microcarrier (MC) systems, large surface is provided for the cells to grow in unlimited volume. MC culture is the growth or maintenance of anchorage-dependent cells on the small beads suspended in a stirred tank. The advantage of MC cultures are based on the provision of a large surface area in relation to the volume of the vessel used. This may be understood by comparing roller bottles to MCs in terms of the surface area afforded. A standard 500 cm^2 roller bottle requires 100 ml of nutrient medium wherein 1 ml provides a minimum of 30 cm^2 of growth surface. One gram of beads of 0.2 mm diameter will have a surface area of more than 6000 cm^3. MC beads have been manufactured from different synthetic materials like dextran, polyacrylide and polystyrene.

Heterogeneous Bioreactor Systems

In this system, cells are immobilized and separated from the medium. This system has been developed to accommodate the growth of surface-adherent cells. The following are some immobilization techniques employed widely.

Absorption   This system consists of a ceramic cylinder with a uniform sequence channel along its length. It provides a surface area of 4.25 m^2. There are two types of materials, one with a smooth surface for anchorage-dependent cells and rough surface for suspension grown cells. To achieve immobilization, cells are injected into the ceramic cylinder and allowed to settle and attach before the circulation of the medium is started. This system is used for the production of a protein over a long period of time.

Entrapment   Entrapment of cells in various polymers is a gentle means of immobilization which affords protection from mechanical stress. The cells are mixed with an aqueous solution of sodium alginate which is then dropped into a solution of calcium chloride. Calcium alginate is insoluble and forms beads in which the cells are entrapped. The cells can also be mixed with an aqueous solution of agarose, the temperature is decreased below the gel-forming temperature and the block thus formed is dispersed in paraffin oil. The force with which the dispersion is made controls the size of the beads. As with alginate, agarose entrapment is more suitable for suspension-grown cells.

Collagen beads are also used as matrix. Cells penetrate the pores and colonize the beads. Collagen promotes adhesion of anchorage-dependent cells. Fibrin is another bead which is used for entrapment. Fibrinogen is converted to insoluble fibrin through the action of an enzyme thrombin. This reaction has been used to entrap the animal cells in beaded fibrin. The matrix is suitable for both suspension-grown and anchorage-dependent cells.

Encapsulation   This involves two methods, one based on the membrane formation by "polyacrylate anions–polyacrylate cations" and the other on the membrane formation by cellulose sulphate (poly-anion). Cells may be physically entrapped in different configuration of hollow fibres by simply injecting them in the extracapillary space.

Mammalian cell technology demands more and more sophisticated, effective and economical methods and devices for cell production and concentration. The trend in the large-scale production system will be towards cell retention systems, because of their large advantages over low-density cultivation systems.

TISSUE AND ORGAN CULTURES

For development of primary culture and cell lines, a variety of tissues and disaggregation methods are used to give a good yield of separated cells. For obtaining cell cultures, the tissue is cultured using **primary explantation technique.** This technique is used for cultivation of pieces of fresh tissues derived from the organisms and this was almost the exclusive technique used for animal tissue culture till about 1945. Different forms of primary explanation techniques are still widely used. These techniques differ only in the type of vessel (flasks, test tubes, etc.) used for growing the tissue, but are uniform in principle.

The primary explanation technique is also used for embryo and organ culture. The different explantation techniques are

1. slide cultures
2. carrel flask cultures
3. roller test tube cultures

Tissue Culture

Slide or coverslip culture   In this technique, slides or coverslips are prepared by placing a fragment of tissue (explantation) on a coverslip, which is subsequently inverted over the cavity of a depression slide. This is very useful for morphological studies through the use of time lapse cine micrographic investigations. This technique is simple and relatively inexpensive. Cells in living state are spread out in a manner suitable for microscopy and photography. Cells grow directly on coverslips and can be fixed and stained to make permanent slides. Few of the limitations of this technique are

i. supply of oxygen and nutrients is rapidly exhausted, so that the medium quickly becomes acidic and requires transfer into fresh medium for rapidly growing tissues
ii. sterility cannot be maintained for a long period
iii. only very small amounts of tissue can be cultured.

Flask cultures   The main use of flask cultures is in the establishment of a strain from fresh explants of tissue. A good carrel flask has excellent optical properties for microscopic examination, even though polystyrene culture flasks can also be used. This method has few advantages.

i. tissue can be maintained in the same flask for months or even years;
ii. large numbers of cultures can be easily prepared and more amounts of tissue can be grown with large amount of medium.

Test tube cultures   Test tubes are cheap and convenient vessels for tissue culture and can be used for preparing a large number of cultures, which can be placed in stationary racks or roller drums. However, this technique has disadvantages like poor optical property for microscopy, difficulty in quantitation due to the curvature and high risk of contamination.

Organ Culture

Organ culture usually implies culturing pieces of an organ *in vitro* and its objectives are to maintain the architecture of the tissue and direct it towards normal development such as occurs *in vivo*. The media used for growing organ culture are generally the same as those used for tissue culture. Organ cultures are grown in both solid medium and liquid medium.

CRYOPRESERVATION

Cryopreservation involves the storing of cells at a very low temperature ($-180°C$) and in liquid nitrogen in a state of suspended animation until they are needed. A large amount of time, effort and money would be wasted if cells could not be preserved when they are not needed.

Cryopreservatives

Healthy cells are suspended in a solution of either glycerol or dimethyl sulphoxide (DMSO)—the cryopreservatives—with high concentration of serum, cooled at a defined rate in liquid nitrogen vapour and then placed in liquid nitrogen. The function of the cryopreservatives is to reduce the water content of the cells. The DMSO enters cells extremely quickly by diffusion across the lipid layer of the plasma membrane. In the presence of DMSO, ice crystals, which would otherwise rupture cell membranes and cause them to lyse, do not form. The high serum concentration probably contributes to cell integrity by maintaining the intracellular protein concentration of cells rendered permeable by the DMSO. Glycerol has much the same effect as DMSO.

Freezing Mixture

A double concentrated freezing mixture consists of 40% (v/v) growth medium (containing 10% serum), 40% (v/v) FCS and 20% (v/v) DMSO (or glycerol). The mixture should be made in the order stated and mixed well by inversion. The mixture can be filtered through a 0.2-μm sterile disposable filter into a sterile universal tube and can be stored frozen at $-20°C$. When mixed with equal volume of cell suspension in complete medium, the effective freezing mixture concentration is obtained.

Freezing Down Cells

Cells should be cryopreserved only when they are healthy and growing in exponential phase. Cultures should be checked carefully for contamination before cryopreservation. The cells are centrifuged at 150–200 g at 4°C for 5 minutes. The supernatant should be carefully discarded, taking care not to disturb the pellet. The cells are then resuspended in the residual medium by gently tapping the side of the tube near the pellet, until no cells remain stuck to the bottom and adjusted to double final required concentration with fresh, ice-cold growth medium (antibiotic-free) containing 10% (v/v) FCS. Adherent cells should be resuspended to 2×10^6/ml and suspension cells to 1×10^7 ml. The cell suspension should be placed on ice and an equal volume of the freezing mixture is added and mixed thoroughly. Aliquots (1 ml) of cell suspension should be put into cold cryotubes. A freezing plug allows the cells to be cooled at a defined rate in the vapour that forms over the liquid nitrogen within the storage tank; the cooling rate should be 1°C/minute. If a freezing plug is not available, a polythene box is an acceptable alternative.

Thawing Procedure

A water bath should be maintained at 37°C and the complete medium is warmed up to the temperature. The vials should be collected from the liquid nitrogen storage with care. The vial may be thawed as quickly as possible in the 37°C water bath. The contents of the vial should be pipetted into a 10-ml sterile centrifuge tube containing 9 ml complete medium, the cap secured, and the contents mixed by inversion. The cells must be removed from the freezing mixture in which they were stored by centrifugation at 150–200 g for 5 minutes and then resuspended in 10 ml of complete medium. A sample should be taken for counting and to check the viability.

USEFULNESS OF CELL CULTURES IN VETERINARY RESEARCH

Cell culture system has become an important tool in the diagnosis and control of viral disease of animals and birds. The various biological products produced through cell culture systems include viral vaccines, monoclonal antibodies, hormones, immunoregulators and tumour-specific antigens.

Disease Diagnosis

Isolation and subsequent identification of the causative agent is the only way to prove the aetiology of any viral infection. Since viruses require a living system for their replication, the virologists are left with only three options, employing experimental animals, embryonated eggs or cell culture systems. Although experimental animals/birds are best suited for the isolation and identification, cell culture system is cheap and easy to handle, while embryonated eggs can carry adventitious agents and contaminants. Cell cultures can be made devoid of them. Primary cultures like calf kidney monolayer, pig kidney monolayer, sheep kidney monolayer and chicken embryo fibroblast monolayer, and cell lines like Vero, BHK21 and MDBK are routinely used for virus isolation and identification.

Madin Darby canine kidney (MDCK) cells were adapted to serum-free RPMI 1640 medium and used for cultivation of canine viruses like canine distemper virus, canine parvovirus, canine adenovirus and canine parainfluenza virus (Mochizuki, 2006). RPMI 1640 medium was supplemented with a soybean peptone, L-glutamine and antibiotics. Canine viruses grew in the MCK-SP cell culture as efficiently as the parental MDCK cells cultured in the conventional Eagle's MEM containing foetal bovine serum.

A velogenic Newcastle disease virus isolate was passaged 50 times in Vero cell culture and the increasing passage levels decreased the virulence of the virus (Mohan *et al.*, 2007).

Virus Vaccines

Cell culture system also plays a vital role in the development of viral vaccines. These viral vaccines seem to be the main animal cell products with high market value. Animal cells are used with the substrate for vaccine production. The best examples for cell culture vaccines are rinderpest vaccine, foot-and-mouth disease vaccine, sheep pox vaccine, rabies vaccine and Marek's disease vaccine. Continuous Vero cell lines are found to be more suitable for large-scale production of rabies vaccine (Prem Kumar *et al.*, 2005).

Monoclonal Antibodies

Somatic cell hybrids have become an important source of cellular products which cannot be obtained from short primary cultures. The best example for such system is the hybrid myeloma used in the production of MAbs against the antigen of choice. The production of MAbs involves the inoculation of the antigen of choice into BALB/c mice, the spleen cells could be removed from the mice and put into culture. These cells in the culture will divide and produce antibodies directed against the antigen, but will last for only a short period in culture. However, cell hybridization makes it possible to combine the spleen cells producing antibodies with mouse myeloma cells which can be cultured continuously. The mouse myeloma cells, viz., SP2/0 cells are fused with the spleen cells and the resulting hybrids are screened for the production of desired antibodies. Such hybrids will provide a continuous supply of MAbs raised against specific epitopes and will be useful in differentiating several strains of viruses such as Newcastle disease virus, Marek's disease virus, infectious bronchitis virus, etc. MAbs are useful in differentiating vaccine virus from field viruses.

Jeff *et al.* (2002) used monoclonal antibodies directed against recombinant *Mycobacterium bovis* antigen 85 (Ag85) for early diagnosis of Johne's disease in cattle. The authors concluded that monoclonal antibodies could be useful for

reliable diagnosis for early detection of *Mycobacterium paratuberculosis* infections in ruminants as well as for identifying contaminated dairy products.

Hormones

Glycoprotein hormones are produced in cultured cells and have therapeutic use. Several clones of pituitary tumour cells are available which synthesize and secrete hormones like prolactin, growth hormone and adrenocorticotrophic hormone. These include rat GH cells and mouse AtT20/D16 cells. The GH cells, previously known as MtTW5 were established from rat pituitary tumour. These cells apart from producing growth hormone also synthesize prolactin. The AtT20/D16 cells are clonal strains isolated from radiation-induced pituitary tumour in mouse and produce large amounts of ACTH during serial propagation in culture. Islets of Langerhans of rat, pig and human foetus are grown *in vitro* for the production of insulin employing hollow fibre cell culture and systems.

Immunoregulators

Animal cells are employed in the production of several non-antibody immunoregulators like interferons, interleukins, colony stimulating factors, B-cell growth factor, macrophage activating factor, T-cell replacing factor and migration inhibition factor.

Tumour-specific Antigens

The best example of this is the carcino-embryonic antigen (CEA). This is a tumour-associated antigen produced by adenocarcinoma cells which are found in sera or other body fluids indicating a state of malignancy. Production of CEA is achieved by its extraction from cultured adenocarcinoma cell lines. This tumour-specific antigen is more useful in screening for anti-tumour agents and detecting malignancy in patients using immunoassays.

FERMENTATION TECHNOLOGY

The very beginning of fermentation technology or, as it is now better recognized, bioprocess technology were derived in part from the use of microorganisms for the production of foods such as cheese, yoghurts, sauerkraut, fermented pickles and sausages, soy sauce and other oriental products and beverages such as beers, wines and derived spirits.

The term **fermentation** is derived from the Latin word *fervere*, meaning "to boil", which describes the action of yeast on extracts of fruit or malted grain during the production of alcoholic beverages. However, "fermentation" is interpreted differently by microbiologists and biochemists. To a microbiologist, fermentation means "any process for the production of a product by the mass culture of microorganisms". To a biochemist, it means "an energy-generating process in which organic compounds act as both electron donors and acceptors", that is, an anaerobic process where energy is produced without the participation of oxygen or other inorganic electron acceptors. In this chapter, 'fermentation' is used in its broader, microbiological context.

Bioprocessing in its many forms involves a multitude of complex enzyme-catalysed reactions within specific microorganisms and these reactions are critically dependent on the physical and chemical conditions that exist in their immediate environment.

Although the traditional forms of bioprocess technology related to foods and beverages still represent the major commercial by-products, new products are increasingly being derived from microbial fermentations for the following purposes.

1. To overproduce essential primary metabolites such as acetic and lactic acids, glycerol, acetone, butyl alcohol, organic acids, amino acids, vitamins and polysaccharides.

2. To produce secondary metabolites (metabolites that do not appear to have an obvious role in the metabolism of the producer organism) such as penicillin, streptomycin, cephalosporin, gibberellins, etc.

3. To produce many forms of industrially useful enzymes, e.g. exocellular enzymes such as amylases, pectinases and proteases, and intracellular enzymes such as invertase, asparaginase, restriction endonucleases, etc.

More recently, bioprocess technology is increasingly using cells derived from higher plants and animals to produce many important products. Plant cell culture is aimed largely at secondary product formation such as flavours, perfumes and drugs, whereas mammalian cell culture has been concerned with vaccine and antibody formation and the production of protein molecules such as interferons and interleukins.

The central component of the fermentation process is the fermenter, (Figure 2.3), in which the organism is grown under conditions optimum for product formation. Operation of upstream and downstream processes of the fermenter is also important. All the operations before starting the fermenter are collectively called "upstream" processes and those after the fermentation has occurred are called "downstream" processes. Examples of upstream operations include the preparation and sterilization of the culture medium, the sterilization of the fermenter itself, and the preparation and growth of a suitable inoculum culture of the microbial strain. Downstream operations involve the collection or harvesting of the product, its concentration, purification and further processing such as separation from contaminants or unwanted components and also a proper treatment of the spent culture medium or effluent. Before the fermentation is started, the medium must be formulated and sterilized, and the fermenter sterilized and inoculated with a variable metabolically active culture which is capable of producing the required product. Product recovery or downstream processing involves the extraction and purification of the biological products and further processing and treatment of effluents produced by fermentation. The most widely used industrial scale fermenters are stirred, baffled, aerated tanks provided with systems of temperature, pH and foam formation control.

Figure 2.3 Fermenter used for large-scale production of bacterial cultures

Downstream Processing

It is not enough to grow the required cells in a bioreactor. Extraction and purification of the desired end product (the so-called downstream processing) from the bioprocesses is equally important, calling on the skills of the chemists and chemical engineers as well as those of bioscientists and process engineers.

Downstream processing will be primarily concerned with initial separation of the bioreactor broth into a liquid phase and a solid phase and subsequent concentration and purification of the product (Table 2.4). Processing will normally involve more than one stage.

The fermenter is also known as bioreactor. It provides an optimum environment in which organisms or their enzymes can interact with a substrate, forming the desired product. Some fermenters allow continuous processing with substrates entering at one end and products leaving at the other end. Most fermenters are of closed type, i.e., processing is done in batches, as in the case of producing antibiotics, since these are synthesized only after mycelial growth of the fungi has slowed down.

Table 2.4 Downstream processing operation

Separation
Filtration
Centrifugation
Floatation
Disruption
Concentration
Solubilization
Extraction
Thermal processing
Membrane filtration
Precipitation
Purification
Crystallization
Chromatography
Modification
Drying

A bioreactor is a device in which substances are treated to stimulate biochemical transformation by living cells or by *in vitro* cellular components such as enzymes. Bioreactors are commonly employed in the food and fermentation industries, in waste treatment, and in some biomedical operations. In fact, the bioreactor constitutes the heart of any industrial fermentation process. Bioreactors are of two main types:

1. for fermentations, employing living cells, and
2. for enzyme transformations, using cell-free systems such as enzymes.

The type 1 bioreactor is also called a fermenter. In this, cell growth is promoted or regulated to allow formation of products—a metabolite (e.g. alcohol, antibiotic), biomass (e.g. single cell protein), transformed substrate (e.g. steroid) or purified solvent (e.g. in water reclamation). System involving cultures of animal or plant cells are called tissue cultures, while those based on dispersed, non-tissue forming cultures of microorganisms are often referred to as microbial fermenters or microbial reactors.

Several kinds of fermenters are being used for various purposes. A few are mentioned below.

1. Stirred tank fermenter, for making antibiotics.
2. Tubular tower fermenter, for making beer, wine, vinegar.
3. Recycle airlift reactors for producing yeast from oil or bacterial biomass from methanol.
4. The Nathan fermenter used in breweries.

Primary separation involves those separation activities which are carried out at the start of downstream processing operation, at the fermenter interface. The separation processes cover such techniques as whole broth extraction and broth conditioning, in addition to cell and cell debris recovery from cell and cell debris suspensions. Another technique to separate out cells and cell debris is flocculation, using synthetic polyelectrolytes.

The terms fermentation can be defined in two different ways: (a) It is any process for the production of a product by the mass culture of microbial or other cells. In this the emphasis is on the mass culture of microbe or plant cell or animal cell; (b) It can be any energy-generating process in which organic compounds act as an anaerobic process where energy is produced without the participation of oxygen or other inorganic electron acceptors.

Microbial fermentation may be classified into the following categories:

1. Those that produce microbial cells or biomass as the product.
2. Those that produce microbial enzymes.
3. Those that produce microbial metabolites.
4. Those that alter or transform some compound that is added to the fermenter.

The growth of the microorganisms may result in the production of a range of metabolites, but the predominant type of metabolite synthesized depends upon the nature of the organism, the cultural conditions employed and the growth rate of the culture. If a microorganism is introduced into a nutrient medium which supports its growth, the inoculated culture will pass through a number of stages and the system is termed as batch culture. Initially, a growth does not occur and this period is referred to as lag phase and may be considered as the period of adaptation. Following an interval during which the growth rate of cells gradually increases, the cells grow at a constant maximum rate and this period is referred to as the log or exponential phase. As nutrients are exhausted or as toxic metabolites accumulate, the growth rate of the cells deviate from the maximum and eventually growth ceases and the culture is said to enter the stationary phase. After a further period of time, the viable cell number begins to decline and the culture enters the death phase.

The term **trophophase** was used to describe the log or exponential phase of a culture during which the sole products of metabolism are those that are either essential to growth such as amino acids, nucleotides, proteins, nucleic acids, lipids, carbohydrates, etc., or the by-products of energy-yielding catabolism such as ethanol, acetone and butanol. The metabolites produced during the trophophase are described as **primary metabolites**. The term **idiophase** was used to describe the phase of a culture during which products other than primary metabolites are produced and these are termed as **secondary metabolites**.

Secondary metabolites have been defined as compounds which are synthesized by slow-growing or non-growing cells, which play no obvious role in cell growth and which are taxonomically limited in their distribution.

The microorganisms may be grown in batch, fed-batch or continuous culture. In **batch culture**, the microorganisms are grown on a limited amount of medium until either one essential nutrient component is exhausted or toxic by-products accumulate to growth-inhibiting levels. Another method of culture is **continuous culture** wherein there is a continuous addition of fresh medium to the culture vessel. This procedure is repeated several times until the vessel is full. However, if an overflow provision is installed in the side of the fermenter, the addition of fresh medium displaced an equal volume of culture, then continuous production of cell could be achieved. In **fed-batch culture,** batch culture are fed continuously or sequentially with fresh medium without the removal of cultural medium.

The earliest example of the commercial use of **fed-batch culture** is the production of baker's yeast. It was recognized as early as 1915 that an excess of malt in the production medium would result in a high rate of biomass production and an oxygen demand which could not be met by the fermenter. This resulted in the development of anaerobic conditions and the formation of ethanol at the expense of biomass.

The penicillin fermentation provides a very good example of the use of fed-batch culture for the production of a secondary metabolite. The penicillin process is a two-stage fermentation in which an initial growth phase is followed by the production phase or idiophase. During the production phase, glucose is fed to the fermenter at a rate which allows a relatively high growth rate (and therefore rapid accumulation of biomass) yet maintains the oxygen demand of the culture within the aeration capacity of the equipment. During the production phase, the biomass must be maintained at a relatively low growth rate and thus, the glucose is fed at slower dilution rate. Phenylacetic acid is a precursor of the penicillin molecule, but it is also toxic to the producer organism above a threshold concentration. Thus, the precursor is also fed into the fermenter continuously, thereby maintaining its concentration below the inhibitory level.

Continuous culture offers the most control over the growth of the cells. However the commercial adoption of continuous culture is confined to the production of biomass and to a limited extent the production of potable and industrial alcohol. Fed-batch culture method is commonly used in fermentation technology.

Applications of Fermentation

The technology utilized in fermentation processes are designed to obtain maximum growth of an organism under the optimum physical conditions in a specific medium for the production of desired end products of commercial value. This includes the cultivation of microbes for the production of cells themselves, e.g. biomass or the production of useful metabolites or proteins. Many of these processes are not new, but are currently undergoing rapid development as a result of recombinant DNA technology.

Microbial biomass Microbial biomass is produced commercially as a single cell protein (SCP) by both uni-and multicellular bacteria, yeast, filamentous fungi or algae which can be used as food or feed additives. SCP is of great nutritional value because of its high protein (45–50%), vitamins and lipid contents and the general presence of a complete array of all essential amino acids. The advantages of using microbial biomass are

1.   Microorganisms in general have a high rate of multiplication.
2.   They can utilize a large number of different carbon sources, some of which are waste products.
3.   The strains with high yield and good composition can be selected and produced relatively easily.
4.   Microbial biomass production is independent of seasonal and climatic variation.

ICI Pruteen process for the production of bacterial SCP for animal feed was a milestone in the development of the fermentation industry. The main disadvantage is that many microorganisms produce toxic substances and it has to be made sure that the biomass does not contain any such substances. However, the economics of the production of SCP as animal feed were marginal, which eventually led to the discontinuation of the Pruteen process.

Microbial metabolites   The metabolites produced during the trophophase of culture are referred to as **primary metabolites** and include amino acids, nucleotides, proteins, nucleic acids, lipids and carbohydrates, and the by-products of energy-yielding metabolism are ethanol, acetone and butanol. Some of the commercially important primary metabolites are listed in the Table 2.5

Table 2.5   Some commercially important microbial primary metabolites

Primary metabolite	Producing microbe	Commercial significance/application
Ethanol	*Saccharomyces cerevisiae*	Production of alcoholic beverages
Acetone, butanol	*Clostridium acetobutyricum*	Solvents
Lysine	*Corynebacterium glutamicum*	Feed supplement
Glutamic acid	*Corynebacterium glutamicum*	Flavour enhancement
Polysaccharides	*Xanthomonas* spp.	Food industry, enhanced oil recovery

The metabolites produced during the idiophase are referred to as the **secondary metabolites.** Secondary metabolites are synthesized from the intermediates and end products of primary metabolism. Few of the secondary metabolites include antibiotics (penicillin, cephalosporin, tetracyclines, streptomycin and griseofulvin), anti-tumour agents (actinomycin), anticancer agents (krestin and bestatin) and immunosuppresent (cyclosporin A).

The production of microbial metabolites may be achieved in continuous, as well as in batch systems. The secondary metabolites are produced by slow-growing organisms in continuous cultures.

Microbial enzymes   Microbial enzymes are most widely used in the food and beverage industries and to a lesser extent in clinical and analytical laboratories and as protease detergents in washing powders. The most economical and convenient method of producing these enzymes is by microbial fermentation. *Bacillus stearothermophilus* produces amylases as secondary metabolites, but most other microbes produce enzymes as primary metabolites, during exponential growth.

The major commercial utilization of microbial enzymes is in the food and beverage industries (Jeenes *et al.*, 1991), in pharmaceutical textile and leather industries, in analytical processes as well as in detergent and dairy industry. Most enzymes are synthesized in the logarithmic phase of the batch culture and may, therefore, be considered as primary metabolites. However, some, for example, amylases of *Bacillus stearothermophilus* are produced by idiophase-type cultures and may be considered as equivalent to secondary metabolites (Manning and Campbell, 1961). Enzymes may be produced from animals and plants as well as microbial sources, but the production by microbial fermentation is the most economical and convenient method. For example, microbial rennin is being used instead of calf rennin and proteases used in detergents are produced by microbes. It is now possible to engineer microbial cells to produce animal or plant enzymes.

Most of the enzymes in industrial use are extracellular proteins produced by *Aspergillus* sp. or *Bacillus* sp. and include α-amylase, β-glucanase, cellulase, dextranase, lactase, lipase, pectinase, proteases and others. The extent of purification required for most of these enzymes is minimal, and they can be produced in tonnes without serious problems. Some of the important microbial enzymes, their source and applications are given in Table 2.6.

Table 2.6 Some important microbial enzymes and their applications

Source (genus)	Enzymes	Reaction	Application
Bacillus	α-Amylase	Starch hydrolysis	Converts starch to glucose or dextran in food industry
	Proteases	Protein digestion	Help laundering
Escherichia	Penicillin acylase	Benzoyl cleavage	Production of 6-APA
	L-Asparaginase	Removal of L-asparagine involved in tumour growth	Leukaemia/cancer treatment

(Contd.)

Table 2.6 (Continued)

Source (genus)	Enzymes	Reaction	Application
Aspergillus	Amyloglucosidase	Dextrin hydrolysis	Glucose production
	β-Galactosidase	Lactose hydrolysis	Lactose hydrolysis in milk or whey
	Aminoacylase	Hydrolysis of acylated L-amino acids	Resolution of racemic mixtures
	Glucose oxidase	Oxidation of glucose	Glucose detection in blood
Streptomyces	Glucose isomerase	Conversion of glucose to fructose	Production of high-fructose syrups
Several marine bacteria	Luciferase	Bioluminescence	Assay for ATP
Several bacteria and cyanobacteria	Nucleases (restriction endonucleases)	Hydrolysis of phosphodiester bonds in nucleic acids	Genetic engineering

Transformation process The microbial cells can be used to catalyse the conversion of a compound into a structurally similar, but financially more valuable compound. Such fermentations are termed transformation processes, biotransformation, or bioconversions. The production of vinegar is the oldest and most well established transformation process (the conversion of ethanol into acetic acid). Other examples of transformation are sorbitol to sorbose and production of new antibiotics that are more effective than the existing ones. The anomaly of the transformation process is that a large biomass has to be produced to catalyse a single reaction.

Recombinant products The advent of recombinant DNA technology has extended the range of potential microbial fermentation products. It is possible to introduce genes from higher organisms into microbial cells such that the recipient cells are capable of synthesizing foreign (or heterologous) proteins.

Examples of the 'hosts' for such foreign genes include *E. coli*, *Saccharomyces cerevisiae* and other yeasts and fungi. Products produced by genetically manipulated organisms include interferon, insulin, human serum albumin, factors VIII and IX, epidermal growth factor and bovine somatostatin.

Genetic Improvement of Product Formation

Microorganism usually produce commercially important metabolites in very low concentration. Thus, to improve the potential productivity, the organism's genome must be modified and this may be achieved in two ways, by mutation or by recombination.

Mutation   Each time a microbial cell divides, there is a small probability of an inheritable change occurring. A strain exhibiting such changed characteristics is termed a mutant and the process giving rise to it is mutation. The probability of a mutation occurring may be increased by exposing the culture to a mutagenic agent such as UV light, ionizing radiation and various chemicals, for example nitrosoguanidine and nitrous acid. Such an exposure results in the death of the vast majority of the cells. The survivors of the mutagen exposure may then contain some mutants, a very small proportion of which may be improved producers.

The synthesis of a primary microbial metabolite (such as an amino acid) is controlled such that it is only produced at a level required by the organism. The control mechanisms involved are the inhibition of enzyme activity and the repression of enzyme synthesis by the end product, when it is present in the cell at a sufficient concentration. Thus, these mechanisms are referred to as feedback control. It is obvious that a good "commercial" mutant should lack the control system, so that 'overproduction' of the end product will result.

The isolation of mutants of *Corynebacterium glutamicum* capable of producing lysine will be used to illustrate this approach. The first enzyme in the lysine synthesis pathway is aspartokinase, which is inhibited only when both lysine and threonine are synthesized above a threshold level referred to as concerted feedback control. A mutant which could not catalyse the conversion of aspartic semi-aldehyde into homoserine would be capable of growth only in homoserine-supplemented medium. If such organisms are grown in the presence of very low concentration of homoserine, the endogenous level of threonine would not reach the inhibitory level for aspartokinase control, and thus aspartate would be converted into lysine which could accumulate in the medium.

Recombination   Recombination is a process which helps to generate new combination of genes that were originally present in different individuals. Techniques are now widely available which allow the use of recombination as a system improvement. *In vivo* recombination may be achieved in asexual fungi (for example *Penicillium chrysogenum*, used for the commercial production of penicillin) using the parasexual cycle. The technique of protoplast fusion has increased greatly the prospects of combining together characteristics found in different production strains. Protoplast fusion has been achieved with the filamentous fungi, yeast, streptomyces and bacteria and is an increasingly used technique. Tosaka *et al.* (1982) improved the rate of glucose consumption (and therefore lysine production) of a high lysine-producing strain of *B. flavum* by fusing it with another *B. flavum* strain which was a non-lysine producer, but consumed glucose at a high rate. Among the fusants, one strain exhibited high lysine production with rapid glucose utilization.

Threonine production by *E. coli* has been improved by incorporating the entire threonine operon of a threonine-analogue-resistant mutant into a plasmid which was then introduced back into the bacterium. The plasmid copy number in the cell was approximately 20 and the activity of the threonine operon enzymes was increased 40–50 times. The organism produced 30 gdm^{-3} threonine compared with the 2–3 gdm^{-3} of the non-manipulated strain. Backman *et al.* (1990) described the construction of an *E. coli* strain capable of synthesizing commercial level of phenylalanine (an important fermentation product as it is a precursor in the manufacture of the sweetener, aspartame). Several of the phenylalanine genes are subject to control by the repressor protein of the *tyrR* gene. *In vitro* techniques were used to generate *tyrR* mutations and introduce them into the production strain. The promoter of the *pheA* gene was replaced to remove repression and attenuation control.

SOLID-STATE FERMENTATION (SSF)

The growth and metabolism of microorganisms on moist, solid substrates lacking free water is called solid-state fermentation (SSF). SSF contrasts with submerged fermentation in which free water is present. Historically, SSF processes have been more popular in Oriental, Asian and African countries, whereas submerged fermentations were popular in Europe. The presence of some moisture (about 15%) is necessary for SSF to occur, but there should be no free water. Though many microbes can grow on solid substrates, only filamentous fungi can grow to a significant extent in the absence of free water.

APPLICATIONS OF SSF

Fermented Foods

Thousands of kinds of fermented foods are being produced industrially in Japan, Korea, China and other oriental countries. *Miso, shoyu, ontjam, khimchi, beet, tempeh* and fermented fish and meat are good examples. Fermentation often makes the food more nutritious, more digestible, safer, or having better flavour. It also tends to preserve the food, increasing its shelf life and lowering the need for refrigeration. Fermentation can be applied to all kinds of foods. Eight classes of fermented food may be recognized, viz., beverages, cereal products, diary products, fish products, fruit and vegetable products, legumes, meat products and starchy products. Of these, dairy products, cereal products and beverages are the most common. Beer is produced from cereal grains which have been malted, dried and ground into fine powder. The powder is washed in warm water. Fermentation of the washed powder is mediated either by a bottom-yeast (e.g. *Saccharomyces uvarum*) or by a top-yeast (*S. cerevisiae*). The final product (beer) has up to about 8% alcohol. Grapes can be directly fermented by yeasts to wine. Wine is made by distillation of the alcoholic broth formed by fermentation of grape juice.

Cereal products   Three popular cereals are wheat, rice and maize. Bread is the commonest type of fermented cereal product. Wheat dough is fermented by *S. cerevisiae* along with some lactic bacteria. Idli, dosa, vada, dhoka and papadam are some common Indian examples of fermented cereal foods. These use mixtures of wheat and legume flours, which are fermented by *Streptococcus*, *Pediococcus* or *Leuconostoc*.

Dairy products   Cheese constitutes one of the largest groups of fermented dairy products, besides yoghurt. Cheese is formed when the casein in milk is coagulated as the pH drops to 4.5. This happens when acid is produced by the lactic acid bacteria which convert the milk lactose to lactic acid. *Streptococcus lactis* subsp. *diacetylactis* is the principal microorganism involved. Yoghurt is made from milk by lactic acid bacteria, often a mixture of *Lactobacillus bulgaricus* and *Streptococcus thermophilus*.

Fruit and vegetable products   Fermentation is usually used, along with salt and acid, to preserve pickle, fruits and vegetables. In these products, salt-resistant lactic acid bacteria, initially of *Leuconostoc* spp. and *Lactobacillus brevis*, are soon replaced by *L. plantarum* and *Pediococcus* spp. *Klebsiella* spp. are also involved. With the release of acids and a drop in pH, yeasts become prominent.

Organic acids   Citric acid is being industrially produced by SSF. Itaconic acid and gallic acid can also be so produced. For citric acid, *Aspergillus niger* is grown on moistened wheat or rice bran at pH 4.0–5.0 at 28°C for about a week.

Mushrooms   The quality and flavour of mushrooms produced by SSF are better than those by submerged fermentations. Composting of the substrates, spawn preparation, and mushroom cultivation all involve SSF. Two mushrooms being widely produced by SSF are the button mushroom, *Agaricus bisporus* and shiitake, *Lentinus edodes*.

Cheese   Throughout the world, cheese are ripened by fungi through SSF to impart distinct flavours. Soft cheese is formed by growing *Penicillium camemberti* on the surface of curd cake. When *P. roqueforti* grown through the body of raws, processed curd, marbled cheeses such as green and blue-veined varieties are formed. A list of industrial chemicals produced by fermentation are given in Table 2.7.

Table 2.7   Industrial chemicals produced by fermentation

Chemicals	Microbial source	Some uses
Formic acid	*Aspergillus*	Textile dyeing, leather treatment, electroplating, rubber manufacture
Ethanol	*Saccharomyces*	Industrial solvent; intermediate for vinegar, esters and ethers; beverages
Oxalic acid	*Aspergillus*	Printing, dyeing, bleaching; cleaner; reducing agent
Glycerol	*Saccharomyces, Dunaliella*	Solvent; plasticizer, sweetener, manufactures of explosives; printing; cosmetics; soaps
Propylene glycol	*Bacillus*	Antifreeze; solvents; synthetic resin manufacture; mould inhibitor
Acetone	*Clostridium*	Industrial solvent; intermediate for several organic chemicals
Malonic acid	*Penicillium*	Manufacture of barbiturates
Lactic acid	*Lactobacillus, Streptococcus*	Food acidulant; dyeing; intermediate for lactates; leather treatment
Acrylic acid	*Bacillus*	Industrial intermediate for plastics
Butanol	*Clostridium*	Industrial solvents; intermediate for several organic chemicals
Methylethylketone	*Chlamydomonas*	Industrial solvent; intermediate for explosives and synthetic resins
Succinic acid	*Rhizopus*	Manufacture of lacquers
Tartaric acid	*Acetobacter*	Acidulant, tanning, printing, esters for lacquers
Itaconic acid	*Aspergillus*	Textile and paper manufacture

REVIEW QUESTIONS

1. What are the different types of media available for cell and tissue culture? Enumerate their advantages and disadvantages.

2. Explain the uses of cell culture in animal and human sciences.

3. What is fermentation technology? Explain different downstream processing operations.

4. What are the different applications of fermentation technology?

REFERENCES

Freshney, R.I. (ed.) (1992). *Animal Cell Culture: A Practical Approach*. IRL Press, Oxford, New York, Tokyo.

Backman, K., O' Conner, M.J., Maruya, A., Rudd, E., McKay, D., Balakrishnan, R., Radjai, M., Dipasquantonio, V., Shoda, D., Hatch, R. and Venkatasubramanian, K. (1990). *Ann. N.Y. Acad. Sci.* 584:16.

Trevan, M.D., Boffey, S., Foulding, K.H. and Stanbury, P. (1987). *Biotechnology: The Biological Principles*. Tata McGraw-Hill edn. New Delhi.

Bu'Lock, J., Hamilton, D., Hulme, M.A., Powell, A.J., Shepherd, D., Smelley, H.M. and Smith, G.N. (1965). "Metabolic development and secondary biosynthesis in *Penicillium urticae*." *Can. J. Microbiol.* 11: 765.

Paul. J. (1975). *Cell and Tissue Culture*, 5th edn. Livingstrone, Edinburgh.

Gupta, P.K. (1994). *Elements of Biotechnology*. Rastogi and Co., Meerat, India.

Jeff, G.L., Hunter, D. and Speer, C.A. (2002). "Early diagnosis of Johne's disease in the American Bison by monoclonal antibodies directed against Antigen 85." *Ann. New York Acad. Sci.* 969: 66–72.

Jeenes, D.J., MacKenzie, D.A., Roberts, F.N. and Archer, D.B. (1991). "Biotechnology and genetic engineering review." Tombe, M.P. (ed.). In: *Intercept Andover*. 9:327.

Manning, A.B. and Campbell, L.L. (1961). *J. Biol. Chem.* 236: 2951.

Mochizuki, M. (2006). "Growth characteristics of canine pathogenic viruses in MDCK cells cultured in RPMI 1640 medium without animal protein." *Vaccine*. 24: 1744–1748.

Mohan, C.M., Dey, S., Kumanan, K., Manohar, B.M. and Nainar, A.M. (2007). "Adaptation of a velogenic Newcastle disease virus to vero cells: assessing the molecular changes before and after adaptation." *Vet. Res. Commun.* 31: 371–383.

Prem Kumar, A.A., Mani, K.R., Palaniappan, C., Bhau, L.N. and Swaminathan, K. (2006). "Purification, potency and immunogenicity analysis of vero cell culture-derived rabies vaccine: a comparative study of single-step column chromatography and zonal centrifuge purification." *Microbes Infect.* 7: 1110–1116.

Stanbury, P.F. and Whitaker. (eds.) (1984). *Principles of Fermentation Technology.* Pergamon Press, Oxford, UK.

Tosaka, O., Karasawa, M., Ikeda, S. and Yoshi, H. (1982). Abstracts of 4th International Symposium on Genetics of Industrial Microorganisms. p. 61.

3

BIOTECHNOLOGICAL APPROACHES TO VACCINE PRODUCTION

INTRODUCTION

At present, the majority of veterinary vaccines are produced by conventional methods similar to those implemented by Jenner or Pasteur. These include live, attenuated vaccines and killed or inactivated vaccines. Both of these types of vaccines have proven to be effective particularly in reducing the clinical manifestation following exposure to virulent field strains of the pathogens. One of the important impediments in the case of live vaccines is to ensure that the organism is attenuated sufficiently not to cause the disease, but still replicate to a sufficient level to induce an appropriate immune response. However, only a limited number of viral diseases can be prevented by live attenuated viral vaccines. Live viral vaccines carry the remote risk that they may revert to the virulent state and most DNA-containing viruses have the potential to establish persistent (or latent) infection. New viral strains may arise by recombination of the vaccine virus with other viral strains in animal populations; pregnant animals or their offspring may be adversely affected by the vaccine strain.

Certain apparently avirulent viral strains can revert to virulence, either by host-induced proteolysis or surface proteins of the phenotype, as in the activation of the HN and F_0 proteins of Newcastle disease virus (Nagai and Klenk, 1977) and of the HA protein of influenza virus or by mutations in the genotype, as in the reversion of attenuated oral Sabin poliovirus vaccines (Kew *et al.*, 1981). Similarly, the so-called inactivated foot-and-mouth disease (FMD) viral vaccines are found to be infectious far too frequently. For example, at least 44% of the outbreaks of FMD in Europe from 1968 to 1981 were caused by incompletely inactivated vaccines or due to escape of virus from vaccine-manufacturing facilities (FAO, 1981). Similarly, outbreaks of FMD in three South American countries in 1979 and 1980 were caused by three different vaccines that contained infectious virus (Casus, 1981). In addition, conventional whole-agent vaccines, particularly crude preparations, have been implicated in post-vaccinal pyrogenic and allergic reactions, abortions, Guillain–Barre neurological syndromes and adverse sequelae.

Conventional killed or inactivated viral vaccines also have potential risks, for example, incompletely or improperly killed batches of virus could result in contamination of the vaccines with active wild-type virus or with contaminating virus in the vaccine. Therefore, elaborate safety testing is a crucial and costly part of the production process.

Improvements in conventional biochemistry, recombinant DNA technology, peptide synthesis, molecular genetics and protein purification had laid the foundation for the development of new vaccines which should be more efficacious, cost-effective and which have fewer side effects.

Animal vaccines must be cheap, easy to administer and effective. To produce a vaccine, genetic engineers move genes around in the infectious organisms to separate the pathogenic factors (components that cause the disease) from the antigens (components that stimulate an immune response).

Five strategies are currently being applied to the generation of new types of vaccines:

1. Recombinant DNA cloning of immunogenic surface protein
2. Chemical synthesis of polypeptide vaccines
3. Construction of recombinant vaccines having guest genes for foreign surface proteins
4. Genetic engineering of non-pathogenic mutant agents
5. Production of monoclonal epitopes of the surface proteins of infectious agents
6. Nucleic acid vaccines

One approach is to remove virulent genes from the infectious organisms. Scours is a disease caused by *E. coli* that affects newborn calves and piglets. The resulting diarrhoea and severe dehydration can cause heavy mortality. Disease-causing strains of the bacteria produce one protein which allows the bacteria to adhere to the gut of the young animals and another which governs the water loss. If the gene for the water loss protein is removed, scours can be prevented. The organism sticks to the gut without causing diarrhoea and acts as a vaccine, stimulating an immune response against the adhering protein. The vaccine is often given to pregnant animals which pass immunity to their offspring.

Recently, recombinant DNA technology has helped to develop new generation vaccines, which are cheaper, safer and more effective. Some vaccines are made not by disarming the pathogen, but by transforming the genes coding for the antigens of pathogen into those coding for harmless characteristics. This method has been used to produce rabies vaccine. The gene for the surface glycoprotein of the rabies virus is inserted into the DNA of another virus vaccinia. Vaccinia virus causes cowpox, but is relatively harmless to dogs. It acts as a vector, transporting the piece of rabies virus RNA into a vaccinated individual and subsequently producing rabies antibodies. This stimulates a protective immune response against rabies. The steps involved in the production of recombinant vaccines are given in Figures 3.1a and b.

Subunit vaccines are produced by genetic engineering. They are purified single proteins from the surface of a pathogen which can be produced cheaply in fermenters. The great advantage of subunit vaccines is that they contain no live, potentially infectious organisms. The synthetic vaccines are advantageous because the immune system of the animal is challenged with only one antigen, thereby omitting other components of the virion that might adversely affect the immune response. The major drawback with subunit/peptide vaccines is that the antigenic mass cannot be greater than the amount injected. There is no amplification of the antigen. The conventional vaccine for FMD (killed vaccine) was responsible for nearly half of the outbreaks of the disease in Europe between 1968 and 1981, because it contained a low percentage of live organisms. It may now be replaced by a subunit vaccine.

Many important antigens are proteins and can be made in harmless organisms through genetic engineering. Genetic engineering also provides a way to produce vaccines even if the infectious organisms cannot be grown in animals or in artificial cultures. Genetic engineers have taken a gene that codes for a surface protein (hepatitis B surface antigen) and inserted it into *E. coli* or into yeast. When the proteins are produced by gene expression in new hosts, it is easy to purify the whole protein to obtain a genetically engineered vaccine. Only small regions of the proteins called epitopes are bound by antibodies.

Immunity of flu lasts for a year or so, not because the vaccines are no good, but because the virus keeps changing its protein coat. Scientists have investigated several strains of the flu virus, looking for regions that do not vary and were exposed sufficiently to stimulate an immune reaction. They found a peptide of 18 amino acids which they knew, from the protein structure studies, was located in an exposed part of a viral protein called haemagglutinin. This region was short enough to be produced chemically in a peptide synthesizer. Joining this synthetic peptide to a carrier protein provides a vaccine that protected mice from several different strains of the influenza virus.

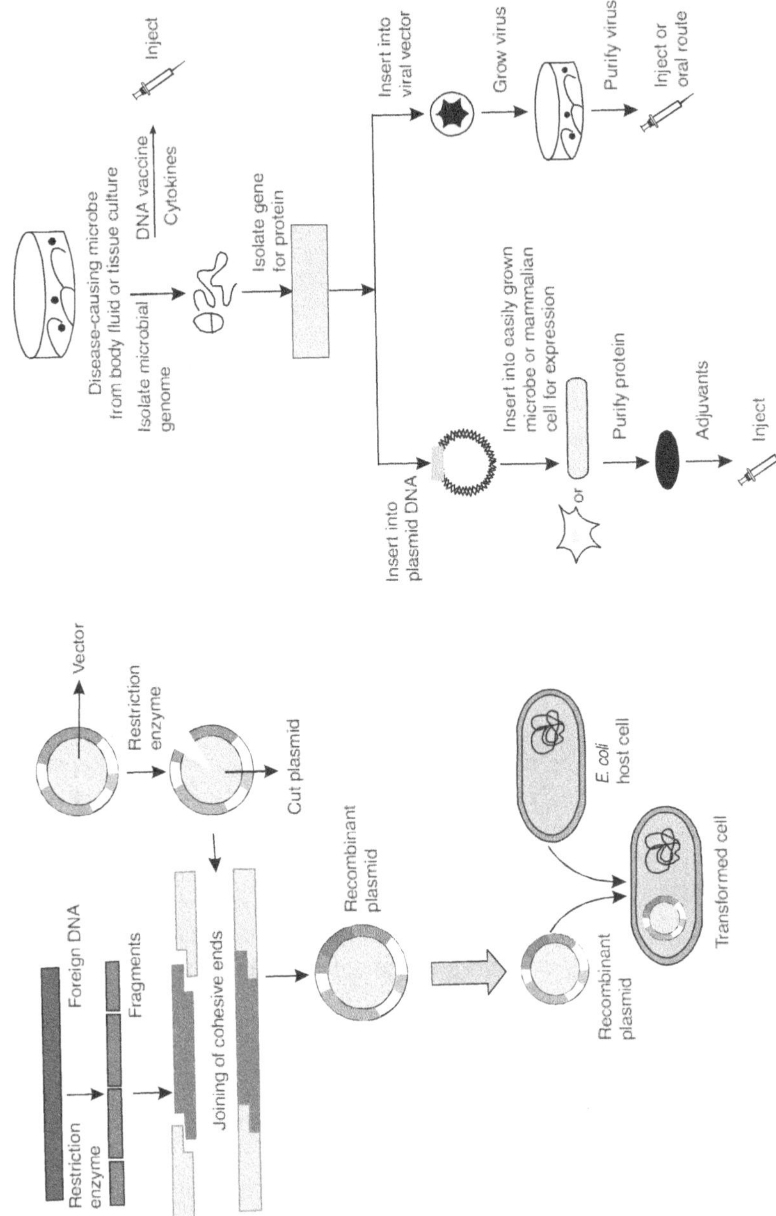

Figure 3.1   Steps involved in recombinant vaccine production

Protein and peptide preparations are "dead" vaccines. "Live" vaccines can also be produced by genetic engineering. If genes from other organisms are inserted into the DNA of vaccinia, the virus used as a vaccine will produce the corresponding proteins as it grows inside human cells. Experimental vaccines to protect animals against infection from rabies, herpes, hepatitis B and influenza have been produced in this way. Up to 25 genes can be inserted into vaccinia DNA and there are plans to produce multiple vaccines to confer immunity to several diseases simultaneously.

Anti-idiotype antibodies have been produced against very many antigens and are being used as an antigen. One of the biggest advantages of this is that the virulent/inactivated microbe need not be used, instead anti-idiotypic antibodies could be used for development of immune response. Anti-idiotypic antibody mimics the structure of the original antigen. Anti-idiotypic MAbs have also been evaluated as immunogen.

RECOMBINANT SUBUNIT VACCINES

Recombinant DNA (rDNA) technology, first described in 1970, deals with the splicing of the gene coding a known protein into a DNA vector, such as double-stranded DNA virus or bacterial plasmid, and subsequent transfer of the chimeric DNA into an alternate host cell for its replication and regulated expression of the guest gene. By means of recombinant DNA technology, genes (even those of RNA viruses after making complement DNA with reverse transcriptase) can be transplanted between plants, animals and microorganisms.

Modern advances in molecular biology and nucleic acid and protein chemistry have contributed to the development of rDNA technology. These advances include:

1. The elucidation of the structure and functions of bacterial plasmids and viruses;

2. The finding that retroviruses possess reverse transcriptase, which permits them to replicate by a recombinant DNA pathway;

3. The discovery and isolation of restriction endonucleases that cleave DNA generally where thymine, cytosine, adenine and guanine bases form palindromic sequences on opposite strands;

4. The isolation of DNA ligases, S1 nucleases and terminal deoxyribonucleotidyl transferases;

5. The development of rapid protein and DNA sequencing procedures; and

6. The finding that microorganisms such as *E. coli*, *Bacillus subtilis* and yeasts can serve as alternate hosts for chimeric DNA vectors containing guest genes encoding foreign proteins.

The first step in the production of recombinant subunit vaccine is the isolation of immunogenic genes, which are amplified by cloning. The specific genes of virions are purified from the preparation of DNA, or cDNA in the case of RNA viruses. The DNA is amplified by cloning and cleaved with restriction endonucleases to small fragments. The DNA fragments which code for immunogenic proteins are identified and used for the preparation of recombinant vaccines.

Vaccinia virus, a member of the pox virus family, is a large double-stranded DNA virus that replicates in the cytoplasm of infected cells. Vaccinia virus has a wide host range, including human, cattle, horses, swine, sheep, goats, mice and monkeys. The difficulty of using, vaccinia virus as a vector include its large genome, the non-infectious nature of the isolated DNA and the ability of homologous sequences in plasmid and virus DNA to recombine within infected cells.

Initial experiments indicate that vaccinia virus vectors retain infectivity, and accommodate at least 25,000 bp of exogenous DNA. Heterologous genes including hepatitis B virus surface antigens, influenza virus haemagglutinin, herpesvirus glycoprotein D, malaria sporozoite surface antigen and rabies glycoprotein have been expressed in this vector system.

Recombinant Vaccines Against Viral Diseases

Foot-and-mouth disease was the first disease against which effective vaccine was produced through gene splicing. The FMD virus genome is a single, positive-sense strand of infectious RNA, made of four viral proteins, VP1, VP2, VP3 and VP4. In this virus, the neutralizing immune response appears to be directed primarily against VP1. *VP1* gene carries type and strain specificities, receptor-binding sites, T-cell epitopes and immunogenic sites. Studies have shown that the peptide corresponding to residues 141–160 and 200–213 of VP1 are capable of neutralizing FMD virus. The neutralization elicited by residues 141–160 is much greater than that of antiserum to residues 200–213. These data demonstrate that VP1 of FMDV is an ideal target antigen for the development of a subunit vaccine against FMD.

A recombinant live vector vaccine was produced by insertion of cDNA encoding the structural proteins (P1) of foot-and-mouth disease virus (FMDV) into a replication-competent human adenovirus type 5 vaccine strain. (Sanz-Parra *et al.*, 1999). Groups of cattle were immunized twice, by the subcutaneous and/or intranasal routes, with either the Ad5 wt vaccine or with the recombinant FMDV Ad5-P1 vaccine. All animals were challenged by intranasal instillation of FMDV 4 weeks after the second immunization. Significant protection against viral challenge was seen in all of the animals immunized twice by the subcutaneous route with the recombinant vaccine. A tobacco mosaic virus-based vector was used to express in plants the complete open reading frame coding for VP1 protein of FMDV (Wigdorovitz *et al.*, 1999). Foliar extracts prepared from infected leaves were injected intraperitoneally into mice and all of the immunized animals developed a specific antibody response to both the complete virus particle and the major immunogenic region. All immunized mice developed a protective immune response against experimental challenge with virulent FMDV.

Wu *et al.* (2003) expressed two immunogenic dominant epitopes of FMD virus serotype O in tobacco plant using a vector based on a recombinant tobacco mosaic virus (TMV). The recombinant viruses TMVF11 and TMVF14 contained peptides of 11 and 14 amino acid residues respectively, from FMDV VP1 fused to the open reading frame of TMV coat protein gene. Most guinea pigs were protected against FMDV challenge after parenteral injection with TMVF11 and TMVF14. The protective effect of TMVF11 was also demonstrated in swine, where preliminary tests showed that nine pigs immunized with TMVF11 in three experiments were protected against FMDV challenge.

Rhabdoviridae possess an immunogenic surface glycoprotein G (haemagglutinin). Rabies virus subunits or the isolated glycoprotein G alone elicited neutralizing antibody and induced immunity in animals. The glycoprotein G genes of rabies and vesicular stomatitis viruses have been cloned in *E. coli*. A recombinant vaccinia virus expressing the immunizing rabies G glycoprotein has been developed (Kieny *et al.*, 1984). This vaccine has been successfully used to protect foxes against rabies. The administration of recombinant rabies vaccine to adult foxes, raccoons and skunks elicited high levels of rabies-virus-neutralizing antibodies and long-term protection against rabies.

A recombinant canine adenovirus type 2 expressing the rabies virus glycoprotein (SRV9 strain) was used to immunize dogs (Hu *et al.*, 2006). All vaccinated dogs produced effective neutralizing antibodies after one inoculation and a stronger anamnestic immune response was produced after booster injection. The immunized dogs could survive the challenge with virulent rabies virus.

Recombinant vaccines have been produced against rinderpest virus. Dr. T.D. Yilma and his colleagues in USA were the first in the field to construct the Wyeth strain of vaccinia virus recombinants expressing haemagglutinin (*H*) and fusion (*F*) genes cloned from the original highly virulent Kabate 'O' strain of rinderpest virus. This vaccine was administered to cattle intradermally in two doses, four weeks apart. All the vaccinated cattle developed neutralizing antibody, but the antibody response in the cattle receiving the recombinant expressing the rinderpest *F* gene were significantly lower than the response in cattle given the recombinant expressing the rinderpest *H* gene. Solid immunity with recombinant vaccine requires two doses.

Dr. T. Barret and his colleagues, working at UK inserted the F-protein gene from Plowright's cell-cultured rinderpest vaccine virus into the vaccinia virus. When this vaccine was injected intradermally into cattle and pigs, the antibody response was higher in pigs than in cattle. Two doses were administered.

In Japan, Dr. Yamanouchi's group inserted the H-protein gene into the tissue culture strain of vaccinia virus. The vaccine was given intradermally into rabbits and they produced neutralizing antibodies.

Sinnathamby *et al.* (2001) developed a recombinant baculovirus expressing H protein of rinderpest virus and studied its protective properties in cattle. Single administration of low doses of purified recombinant extracellular virus with or without adjuvant induced virus-neutralizing antibody responses and bovine leucocyte antigen class II restricted helper T-cell responses in cattle.

Newcastle disease recombinant vaccine had been produced using different virus vectors. The haemagglutinin–neuraminidase (*HN*) gene and the phosphoprotein (*P*) gene of Newcastle disease virus (NDV) were inserted into an avian leucosis virus vector. These recombinants were used to immunize four-week-old chicken. Birds receiving the virus containing the *HN* gene developed low levels of serum haemagglutination inhibition titres. Upon challenge, all birds vaccinated with *HN*-gene-containing virus were protected from the disease. Birds immunized with the *P* gene containing retrovirus developed more severe clinical signs (Morrison *et al.*, 1990). In another study, a cDNA copy of the RNA encoding the fusion (F) protein of NDV was inserted into a fowlpox virus vector. Ocular or oral administration of recombinant fowlpox virus gave partial protection, whereas both intramuscular or wing web routes of inoculation gave complete protection after a single injection (Taylor *et al.*, 1990).

A recombinant Newcastle disease virus (NDV) expressing VP2 protein of IBDV was developed (Huang *et al.*, 2004). The gene encoding the VP2 protein of the IBDV was inserted into the most 3´-proximal locus of a full-length NDV cDNA for high-level expression. Vaccination with the rLaSota/VP2 generated antibody response against both NDV and IBDV and provided 90% protection against NDV and IBDV. Booster immunization induced higher levels of antibody responses against both NDV and IBDV and conferred complete protection against both viruses.

Three glycoproteins (gp 115/110, gp 63 and gp 51) from herpesvirus of turkeys and Marek's disease virus would protect chicks against Marek's disease (Ikuta *et al.*, 1984). The surface glycoprotein of infectious bovine rhinotracheitis (Lupton and Reed, 1980) and pseudorabies virions have been reported to induce neutralizing antibodies and protective immunity in cattle and swine, respectively. A β-galactoside pseudorabies glycoprotein fusion protein from *E. coli* is reported to protect mice from challenge with pseudorabies virus.

VP2 protein from highly virulent IBDV expressed in *E. coli* provided protection against a hvIBDV challenge in specific pathogen-free (SPF) chickens (Omar *et al.*, 2006). Six out of seven chickens that were injected with crude VP2 protein, three times, developed significant antibody titre against IBDV.

The genome of bluetongue virus is divided into 10 DNA segments. VP2 protein of type 10 bluetongue virus induce both neutralizing antibodies and protective immunity in sheep. RNA segment 2 which represents the VP2 neutralizing antigen, has been inserted into an insect baculovirus (*Autographa californica*, AcNPV). Infection of *Spodoptera frugiperda* cell lines (which supports the baculovirus to grow in high titre) by the recombinant virus produced VP2 protein that was precipitated with BTV-10 and to a lesser extent BTV-serotype 11 and 17 viruses (Roy and Inumam, 1989).

Canine distemper virus (CDV) and measles virus (MV) are members of the morbillivirus genus of the family paramyxoviridae and contain a non-segmented single-stranded RNA genome. These viruses have membrane glycoprotein haemagglutinin (HA) which is responsible for haemagglutination and attachment of the virus to the host cell and the fusion glycoprotein (F) which causes membrane fusion between the virus and the infected adjacent cells. Vaccinia virus-MV, HA and F recombinants were produced and inoculated into dogs. Of these two recombinants, vaccinia virus-MVHA gave better neutralizing antibody production than VV-MVF protein (Taylor and Paloettic, 1991).

VP2 of canine parvovirus which is a major component of the capsid protein was expressed in baculovirus. This recombinant virus was able to elicit good protective response (Lopez *et al.*, 1992). Dogs immunized with the inactivated recombinant plant virus, displaying a peptide derived from the CP2 capsid protein of canine parvovirus (CPMV-PARVO1) elicited high titres of peptide-specific antibody (Langeveld *et al.*, 2001). Since plant-virus-derived vaccines have the potential for cost-effective manufacture and are not known to replicate in mammalian cells, they represent a viable alternative to current replicating vaccine vectors for development of both human and veterinary vaccines.

Recombinant Vaccines Against Bacterial Diseases

The first work of rDNA bacterial protein vaccines involved the cloning of somatic pili of enterotoxigenic *E. coli* (ETEC) for use in preventing diarrhoea in livestock. The ETEC have somatic pili composed of 14 to 22-KDa protein molecule that adhere to the mucosal surface of the small intestines, initiating colonization by the bacteria. Pili isolated from ETEC have been used as vaccines to prevent diarrhoeal diseases in both humans and animals.

An alternate method for protecting domestic animals against ETEC has been achieved through genetic engineering. Monoclonal antibody to K99 pilus antigen produced in hybridomas and amplified in mice significantly reduced the severity of scours and the mortality of calves so treated (Sherman *et al.*, 1983). This product was successfully tested in Canada and had since been approved by the USDA for use in the US.

Lee *et al.* (2001) developed an edible vaccine against bovine pneumonic pasteurellosis using transgenic white clover expressing a *Mannheimia haemolytica* A1 leukotoxin 50 fusion protein Lkt50-GFP. An extract of white clover containing Lkt50-GFP was able to induce an immune response in rabbits.

Inactivated rabies virus envelope protein was used as carrier molecule for *Bacillus anthracis* protective antigen (PA) (Smith *et al.*, 2006). Mice immunized with the inactivated recombinant RV virions containing anthrax PA exhibited seroconversion against both RV glycoprotein and anthrax PA, and a second inoculation greatly increased these responses.

Synthetic peptides derived from predicted loop 2 and loop 5 of outer membrane protein of *Pasteurella multocida* strain X-73 (serotype 1) was tested on chicken. Results showed that synthetic peptide mimicking the predicted loop 2 induced 70% protection in chicken against strain X-73 challenge (Luo *et al.*, 1999).

Production of rDNA protein vaccine against parasitic diseases are in progress, including those for avian coccidiosis and malaria. Two recombinant *Taenia solium* oncosphere antigens, designated TSOL18 and TSOL45-1A, were investigated as vaccines to prevent transmission of the zoonotic disease cysticercosis through pigs (Flisser *et al.*, 2004). Both antigens were effective in inducing very high levels of protection (up to 100%) in three independent vaccine trials in pigs against experimental challenge infection with *T. solium* eggs. These two oncosphere antigens provided the basis for development of a highly effective practical vaccine that could assist in the control and, potentially, the eradication of human neurocysticercosis.

Protective immunity was developed against *Eimeria acervulina* following *in ovo* immunization with a recombinant subunit vaccine and cytokine genes (Ding *et al.*, 2004). A purified recombinant protein from *Eimeria acervulina* (3-1E) was used to vaccinate chickens *in ovo* against coccidiosis both alone and in combination with expression plasmids encoding the interleukin 1 (IL-1), IL-2, IL-6, IL-8, IL-15, IL-16, IL-17, IL-18, or gamma interferon (IFN-gamma) gene. When used alone, vaccination with 100 or 500 µg of 3-1E resulted in significantly decreased oocyst shedding compared with that in non-vaccinated chicken. Simultaneous vaccination of the 3-1E protein with the IL-1, -15, -16, or -17 gene induced higher serum antibody responses than 3-1E alone. These results provide the first evidence that *in ovo* vaccination with the recombinant 3-1E *Eimeria* protein induces protective intestinal immunity against coccidiosis, and this effect was enhanced by co-administration of genes encoding immunity-related cytokines.

Expression of Target Antigens in Prokaryotic Hosts

Escherichia coli   *E. coli* is the most commonly used prokaryotic host system for expression of eukaryotic genes. The first recombinant expression systems were developed using *E. coli* bacteria. Some of the promoters used to express foreign eukaryotic genes included the bacterial promoters like *trp-lac* UV5, hybrid *trp-lac* and phage lambda PL and other phage T7 and T5 promoters.

One of the earlier works on recombinant vaccine was with VP1 protein of FMD virus. A DNA sequence coding for VP1 from FMDV A12, prepared from virion RNA, was ligated to a plasmid designed to express a chimeric protein from the *E. coli* tryptophan (*trp*) promoter–operator system. When *E. coli* transformed with this plasmid was grown in trp-depleted media, approximately 17% of the total cellular protein was found to be insoluble and stable chimeric protein.

E. coli expression has produced the first recombinant retroviral subunit vaccine, which protects cats against challenge with feline leukaemia virus. The recombinant protein includes the entire envelope protein gp 70 plus the first 34 amino acids from a transmembrane protein p15E and FeLV serotype A. A full-length DNA copy of the glycoprotein gene of ERA strain of rabies virus was obtained by oligonucleotide-primed synthesis from the genomic RNA template. The rabies glycoprotein gene was inserted into an expression plasmid containing the *lac* UV5 promoter.

The genes for the protective antigens of the tripartite protein toxin of *Bacillus anthracis* has been cloned and expanded in an *E. coli*/pBR 322 system for the purpose of developing a vaccine.

E. coli offers a useful expression host for the production of unmodified prokaryotic and certain unmodified eukaryotic proteins. However, the major disadvantage of using *E. coli* as a host cell for gene expression for vaccine development is its inability to carry out proper post-translational modifications such as glycosylation, proteolytic cleavage and assembly of multimeric proteins. As a result, the non-glycosylated proteins expressed in *E. coli* often fold improperly and generally lose their antigenicity as well as immunogenicity.

Expression of Target Antigens in Eukaryotic Hosts

Yeasts   Yeasts offer certain advantages over prokaryotic hosts for the production of foreign proteins of eukaryotic origin. These primitive eukaryotic cells have the ability to glycosylate proteins, an important post-translational modification that may be important for the integrity of protein structure. Yeasts can provide a suitable environment for the correct folding of recombinant products. Baker's yeast, *Saccharomyces cerevisiae*, was first used for this purpose. A general-purpose yeast expression vector has been designed which contains a promoter for the highly expressed yeast glyceraldehyde 3-phosphate dehydrogenase (GPD) gene. The plasmid also contains selective markers and origins of replication for both *E. coli* and yeast. Other yeast promoters used frequently are alcohol dehydrogenase I promoter, 3-phosphoglycerate kinase promoter and the acid phosphatase promoter.

The surface antigen of hepatitis B virus has been synthesized in the yeast *S. cerevisiae* by using an expression vector that employs the 5´-flanking region of yeast alcohol dehydrogenase I as a promoter to transcribe the surface antigen-coding sequences.

Despite these promising results, several problems have been encountered when *S. cerevisiae* was used for large-scale production. First, most of the plasmid vectors which are used to express the heterologous genes are present in low copy number and are unstable. Secondly, some promoters are constitutive so that strains which express the product can be at a selective disadvantage relative to non-expressing strains.

In the light of these problems, a second generation yeast expression system based on the methylotrophic yeast *Pichia pastoris* has been developed. Methanol-regulated promoters from *P. pastoris*, such as that controlling the alcohol oxidase 1 (AOX1) gene have been shown to be highly expressed and tightly regulated. *HbsAg* gene has been expressed using AOX1 promoter and methanol in *P. pastoris*.

Mammalian cells   Continuous lines of mammalian cells have been used to express cloned foreign genes under the control of various promoters from mammalian viruses. The most frequently used are SV-40 early and late promoters, bovine papilloma virus promoter and retrovirus promoters. The murine metallothionein promoter is also a frequently used cellular promoter for expression of cloned genes. An advantage of mammalian cells is that cloned eukaryotic genes are often expressed as fully functional and processed proteins. However, the level of expression is relatively low.

The first to be developed involved SV-40 DNA virus vectors to deliver genetic information into the cells. Subsequent retrovirus and adenovirus vectors offered more flexibility in terms of the amount of genetic information that could be transferred.

HbsAg has been successfully prepared in an animal cell system. The antigen has been synthesized, glycosylated and assembled as 22-nm particles. The particles were secreted into the cell supernatant without lysis of the cells. It was therefore possible to prepare large quantities of the HbsAg particles. For expression of the *HbsAg* gene, an SV-40 vector was constructed in which the HbsAg gene replaced the early region, T-antigen for the viral genome. This chimeric genome was then used to transfect monkey kidney cells that constitutively express the SV-40 T-antigen functions allowing the replication of the defective virus. Stock virus from the transfection was used for subsequent infection of monkey kidney cells and production of HbsAg. The HbsAg was found in the extracellular culture medium as 22-nm particles.

Insect cells   Recent advances in the genetics of invertebrate viruses and cells have allowed development of viral/cellular systems which achieve both a high level of synthesis and complex processing of recombinant products. The baculovirus, *Autographa californica* nuclear polyhedrosis virus (AcNPV), is an extremely useful helper-independent eukaryotic expression vector which is easily engineered. The system is based on a cell line established in the late 1970s from pupal ovarian cells of the moth *Spodoptera frugiperda*. When infected with baculovirus carrying a foreign gene, these cells can secrete recombinant products complete with post-translational modification if the gene product is a secretory protein. The system generates large amounts of extracellular product. The high productivity of the baculovirus system derives from the high efficiency of a viral promoter of a gene encoding a protein called polyhedron, the sole component of a crystalline matrix that acts as a protective shield for viral particles outside their insect host. Late in lytic infection, polyhedron can account for as much as 60% of total cellular protein in S. *frugiperda* cells. Since the polyhedron protein is not required for virus replication, expression vectors have been developed using the strong AcNPV polyhedron promoter.

Large amounts of several eukaryotic proteins have been successfully produced in this system, including human β-interferon, human interleukin 2, lymphocytic choriomeningitis virus S coded antigens, influenza virus haemagglutinin, human rotavirus major capsid antigen and human hepatitis B virus surface antigen. Kuroda *et al.*, (1986) expressed the fowl plague virus *HA* gene using the baculovirus expression vector. A DNA sequence of the influenza (fowl plague) virus *HA* gene was inserted into the *Bam* HI site of the pAc37 polyhedrin vector.

A recombinant baculovirus containing a cDNA which encodes haemagglutinin–neuraminidase (*HN*) of NDV was constructed. *Spodoptera frugiperda* cells infected with this recombinant virus produced a large amount of HN glycoprotein. The recombinant HN glycoprotein was localized on the surface of the infected cells and conserved its haemadsorption and neuraminidase activities. Chicken inoculated with the cells infected with the recombinant virus developed haemagglutination-inhibition and virus neutralization antibodies and were completely protected from the NDV challenge (Nikura *et al.*, 1991).

The *VP2* coding gene of canine parvovirus was engineered to be expressed by a recombinant baculovirus under the control of the polyhedron promoter. A transfer vector that contains the *lacZ* gene under the control of the p10 promoter was used in order to facilitate the selection of recombinants. The expressed VP2 was found to be structurally and immunologically indistinguishable from authentic VP2. These virus particles have been used to immunize dogs with different doses and combinations of adjuvants. A dose of ca. 10 μg of VP2 was able to elicit a good protective response, higher than that obtained with a commercially available, inactivated vaccine. The results indicate that these virus like particles can be used to protect dogs from CPV infection (Lopez de Turiso *et al.*, 1992).

Herpesvirus of turkeys   Recombinant strains of herpesvirus of turkeys (HVT) were constructed that contained either the fusion protein gene or the haemagglutinin–neuraminidase gene of Newcastle disease virus (NDV) inserted into a non-essential gene of HVT. Expression of the NDV antigens was regulated from a strong promoter element derived from the Rous sarcoma virus long terminal repeat. Recombinant HVT strains were stable and fully infectious in cell culture and in chicken. Chicken receiving a single intra-abdominal inoculation at 1 day of age with recombinant HVT expressing the NDV fusion protein had an immunological response that protected the birds against challenge (Morgan *et al.*, 1992).

Avian fowlpox virus   The avipox virus, fowlpox virus (FPV) has recently been developed as a live viral vector for the delivery of specific poultry disease antigens. In an earlier study, the haemagglutinin protein of virulent avian influenza virus strains were expressed in an FPV recombinant. After inoculation of the recombinant into chicken and turkeys, an immune response was induced which protected the animals against either a homologous or a heterologous virulent influenza virus challenge (Taylor *et al.*, 1988). In this study, a cDNA copy of the RNA encoding the fusion (F) protein of NDV virus strain Texas, a velogenic strain of NDV was obtained and inserted into a fowlpox virus vector. A glycosylated protein was expressed and presented on the surface of infected chicken embryo fibroblast cells. The F protein expressed by the recombinant fowlpox virus was cleaved into two polypeptides. When inoculated into susceptible birds by a variety of routes, an immunological response was induced. Ocular or oral administration of the recombinant fowlpox virus gave partial protection, whereas both intramuscular and wing-web routes of inoculation gave complete protection after a single inoculation (Taylor *et al.*, 1990).

Avian leucosis virus　The haemagglutinin–neuraminidase *(HN)* gene and the phosphoprotein (P) gene of NDV were inserted into a replication-competent avian leucosis virus vector. The *HN* gene and *P* genes were expressed in chick embryo cells. The retroviruses were used to immunize 4-week-old chicken. Birds receiving the virus containing the *HN* gene developed low levels of serum HI titres and NDV neutralization titres. Upon challenge, all birds vaccinated with the *HN*-gene-containing virus were protected from disease. In contrast, birds immunized with the *P* gene containing retrovirus developed more severe clinical signs of disease earlier than birds receiving no immunization or retrovirus alone (Morrison *et al.*, 1990).

PEPTIDE VACCINES

The recent interest in synthetic peptide vaccines is due primarily to the remarkable advances in protein and nucleic acid chemistry and in immunochemistry. It is now possible to define many biologically important proteins in terms of their amino acid sequence. This is particularly true for the disease-inducing viruses, since determination of the genome structure by direct cDNA sequencing or after recombinant DNA cloning is relatively easy and the nucleotide sequence so obtained can be used to predict the structure of immunogenic virion proteins.

Synthetic peptide vaccines offer the advantage of safety, purity and low cost as compared to live or inactivated vaccines. It is now possible to induce virus-neutralizing antibody response using peptides of specific amino acid sequences. The peptides can be chemically synthesized in pure form or made by bacterial expression using rDNA technology. In the latter case, the peptides may be fused into other expressed proteins (fusion peptides).

Fusion Protein Vaccines

FMD virus particles contain a positive-strand genome of molecular weight 2.6×10^6 in association with 60 copies of each of the 4 proteins namely VP1, VP2, VP3 (molecular weight of around 24 kDa) and VP4 (molecular weight of 7 kDa). The capsid proteins can be isolated from disrupted virus by PAGE. There is considerable evidence that neutralizing activity is elicited primarily by VP1, based on protection of swine and cattle against virus challenge.

Results from studies have demonstrated that VP1 was located on the surface of the FMD virion. The antigenic determinants which elicit virus-neutralizing antibodies as well as the virus attachment site for susceptible cells were located on VP1. The finding that animals injected with VP1 or a fragment of this protein induced protection in animals against infection when exposed to live virus represented an important observation. As a result, several groups decided to make peptides based on the structure of these fragments in large amounts by rDNA technology. Identification of the DNA fragment encoding VP1 was established by sequence analysis of restriction endonuclease fragments and prior knowledge of the amino acid sequences of portions of VP1. VP1 has been expressed in the K12 strain of *E. coli*. Such cells grown in culture express 1–2 million copies per cell of VP1 fusion protein. The extracted and purified fusion protein has been shown to elicit neutralizing antibodies and protective immunity against type A12 FMD in both swine and cattle (Kleid *et al.*, 1981). Two vaccinations with fusion protein vaccine were required before the immunity of the animals was challenged a month later. Cattle vaccinated on day one and again at 3.5 or 7.5 months with a similarly cloned and expressed 26-kDa fusion protein vaccine containing residues 1 to 211 of VP1 were protected against virus challenge at 7 and 10.5 months (Mc Kercher *et al.*, 1985).

In most instances, two doses of the fusion protein vaccine have been required to stimulate protection of cattle against contact exposure. In studies of a 29-kDa protein, consisting of a 190 amino acid fusion protein linked to amino acids 137–168 of VP1 of type A12 virus, protection was demonstrated in 5 of 6 cattle vaccinated with a single 250-μg dose. The cattle showed high levels of neutralizing antibodies 25 days after vaccination, and on day 35, their immunity was challenged by contact exposure to infected cattle. Five of the six vaccinated animals were protected against infection (Kleid *et al.*, 1985).

Synthetic Peptide Vaccines

Some of the advantages of synthetic vaccines in terms of cost, ease of storage and transport include:

1. The product would be stable indefinitely at ambient temperature, thus obviating the need for cold storage and saving the cost of transporting large volumes of vaccine under refrigerated conditions.

2. There would be no need for downstream processing.

3. The production facility would be very simple compared with that used for biological products.

4. The product would be defined in precise chemical terms and could be altered readily to elicit the appropriate responses in different situations.

The first indication that a synthetic peptide can elicit an antitoxin activity was reported by Audibert *et al.* (1981), who demonstrated that a tetradecapeptide of diphtheria toxin induced protective antibodies when administered to guinea pigs.

Strohmaier *et al.* (1982) found out that two regions of VP1 protein of FMD virus were responsible for eliciting immunogenic response. One of these included amino acids 146–154 and the other at the carboxy terminus of the molecule included amino acids 201–213. Synthetic peptides were produced linked to a carrier protein, most commonly keyhole limpet haemocyanin (KLH) and the conjugate was tested as an immunogen. Usually, equal weights of peptides and carrier proteins have been coupled, which is equivalent to about 30 copies of peptide per protein molecule, when KLH is used as carrier. Bittle *et al.* (1982) in rabbits tested the immunogenicity of 7 peptides linked to KLH representing amino acids 9 to 24, 17 to 32, 25 to 41, 1 to 41, 141 to 160, 151 to 160 and 200 to 213 of VP1 from O_1 strain of FMD virus. Antibodies were generated against all 7 peptides and those made against peptides 141 to 160 and 200 to 213 were able to neutralize the homologous virus. When the peptide 141 to 160 was inoculated into guinea pigs, the animals were protected by a single dose of 200 µg. Subsequently, it was shown that subtype serological specificity could largely be minimized by sera made against the A_{10} or A_{12} virus 141 to 160 peptide.

Cattle were protected against FMD type O_1 by a single injection of an organically synthesized peptide. A 4-kDa peptide residue was prepared from the two previously identified immunogenic regions (amino acids 141 to 160 and 200 to 213) connected by a diproline spacer and terminating on each end with cysteine residues and was administered in Freund's complete adjuvant (Di Marchi *et al.*, 1986). The serum-neutralizing titre rose rapidly and by day 14 reached a plateau that was maintained up to day 32. Two of the three cattle given a single dose (5 mg) of the peptide were protected when exposed to live virus and three others given two doses of 1 mg each were similarly protected. Thus far, this 40-residue synthetic peptide is the smallest agent in molecular size found capable of inducing protection in cattle.

Two peptides derived from highly conserved regions of bovine rotavirus outer capsid proteins (VP7 and VP4) were tried as rotavirus vaccine (Ejaz *et al.*, 1991). The VP4-VP6 peptide conjugate provided better protection, than VP7-VP6 peptide conjugate. A mixture of VP4 peptide, VP6 and VP7 peptide conjugates provided better heterologous protection than immunization with bovine rotavirus.

Research is in progress for chemical synthesis of peptide vaccines for bacterial diseases, for example, diphtheria and cholera. A synthesized hexadecapeptide corresponding to amino acid residues 186–201 of the A chain of diphtheria toxin covalently bound to both synthetic muramyl dipeptide adjuvant and synthetic multichain poly-DL-alanine induced protective anti-toxic immunity in guinea pigs. For cholera, several peptides ranging between 11 and 18 amino acid residues in length corresponding to regions of the 103 amino acid chain of the B subunit of *Vibrio cholerae* have been synthesized and linked either to tetanus toxoid or to multichain poly-DL-alanine.

GENETICALLY ENGINEERED REASSORTANTS AND DELETION MUTANT VACCINES

The ability of genetic engineers to manipulate genes allows for the removal of pathogenic genes from infectious agents without destroying their ability to propagate and express immunodominant epitopes.

Genes reassorting and deletion techniques are being applied to both viruses and bacteria. Highly attenuated temperature-sensitive mutants of influenza virus or avian strains (that do not replicate efficiently in mammals) are being used with virulent wild-type virus to co-infect cell cultures. Some reassortants can be isolated that contain the six internal genes from an avirulent parental virus and the two genes for the surface proteins, HA and NA, of the wild-type virus. Humans immunized with such reassortants from temperature-sensitive mutants exhibit lower rates of infection and shed less virus than those who received inactivated influenza vaccine (Murphy *et al.*, 1982). In human trials, an avian-human influenza A reassortant virus had the immunogenicity of wild virus and was not transmissible between individuals.

The 11 genes of rotaviruses, which cause diarrhoeal disease in young animals and humans, can also undergo reassortment in mixed infections. A reassortant with its surface antigen *VP1* genes from a humans rotavirus and its other genes from bovine or rhesus rotavirus have diminished growth characteristics in humans and appear to have potential use as immunizing agent. Because of this high degree of cross protection, an attenuated bovine rotavirus vaccine is being considered as a vaccine for humans.

Deletion mutants of several bacteria have been engineered and their potential usefulness as vaccines is being explored. Included in these studies are deletion mutants of *Salmonella*, *Shigella* and *Vibrio cholerae*. The gal-E strain of *Salmonella typhi* Ty21a (lacking UDP galactose 4-epimerase activity and highly deficient in two additional enzyme activities as well), was able to proliferate long enough following oral administration to confer immunity against wild type strains of *Salmonella typhi*. Similarly, mutant strains of *Salmonella* defective in aromatic biosynthesis (they require amino acid dihydroxybenzoate, which are not available in host tissues) given in two doses either intramuscularly or orally to calves resulted in induction of resistance to oral challenge with more than 10^{11} organisms of a virulent UDC strain of *S. typhi* (Smith *et al.*, 1984).

A hybrid strain of *Salmonella typhi/Shigella sonei* appears to have a potential as a vaccine against both typhoid and dysentery. A plasmid encoding a gene for *Shigella sonei* transformed by conjugation of the Ty21a attenuated strain of *Salmonella typhi* produces a hybrid bacterium which elicits antibodies in animals to antigens of both organisms (Tramont *et al.*, 1984).

Using overlapping cosmid clone-based technology, an MDV vIL-8 deletion mutant virus, rMd5/delta vIL-8 was developed which had attenuated virulence and conferred protection against challenge (Cui *et al.*, 2005). These results confirm the potential of partially attenuated viruses as vaccines against very virulent plus strains and the usefulness of recombinant DNA technology to generate the next generation of MDV vaccines.

ANTI-IDIOTYPE ANTIBODY VACCINES

Anti-idiotype antibody has the potential to act as vaccine for animals and humans against infectious diseases. Animals modulate their response to a foreign antigen by the production of a cascade of idiotype, anti-idiotype, anti-anti-idiotype and higher anti-idiotype antibodies. However, they have shown generally only the ability to prime the animals for anamnestic-like response to infectious agents. The anti-idiotype antibody of particular interest is anti-paratope antibody. The paratope is the binding site in the Fab portion of the antibody and the antibody to it will be the mirror image of the epitope of the original antigen. The advent of monoclonal antibodies and recombinant DNA techniques have made the development of anti-idiotype antibody vaccines a realistic goal. Thus, anti-idiotype antibodies can now be produced in large quantities in hybridomas or by cloning and expression of DNA transcribed from anti-idiotype antibody-specific mRNA.

Anti-idiotypic antibodies that have been examined as potential immunogens include those mimicking epitopes of HbsAg, poliovirus, reovirus, rabies virus, bacteria, trypanosomes and cancer antigens. Anti-idiotype antibodies have been prepared against five monoclonal antibodies to rabies glycoprotein G. Two of the five anti-idiotype antibodies injected into mice elicited specific virus-neutralizing antibody response (Reagen *et al.*, 1983). Protective immunity against African trypanosomiasis has been produced in mice using anti-idiotype antibodies (Sacks and Sher, 1983). Anti-idiotype antibodies have been produced against parvovirus (Rimmelzwann *et al.*, 1991), Newcastle disease virus (Tanaka *et al.*, 1989) and Listeria (Kaufmann *et al.*, 1985).

Using HmenB3, a protective and non-autoreactive monoclonal antibody (MAb) to *Neisseria meningitidis* serogroup B (NMGB) capsular PS, an anti-idiotypic MAb, Naid60 was produced, which mimicked the capsular PS of NMGB (Park *et al.*, 2005). They produced an anti-anti-idiotypic MAb, MoB34, by using the immunogenic site on Naid60 responsible for inducing the anti-NMGB PS antibody response. MoB34 elicited the complement-mediated killing of representative strains of serogroup B meningococci. Naid60 has an immunogenic site that elicits antibodies with bactericidal activity against NMGB.

A monoclonal antibody TGCM5 (isotype G, subclass 1, kappa light chain) against SAG2, a major surface antigen of *Toxoplasma gondii* tachyzoites, was produced and its Fab-containing idiotype (Id) was injected into rabbits for production of rabbit anti-Id sera (Yang *et al.*, 2003). Large amounts of gamma interferon were detected in supernatants of splenocyte cultures prepared from mice immunized with aId-IgG. Of the mice immunized with aId-IgG, 75–87.5% survived at least 28 days after a lethal challenge of 1×10^3 live tachyzoites, which killed all non-immunized control mice within 10 days. The results showed that the SAG2 internal image presented by aId-IgG indeed elicited a protection against the infection of *T. gondii* in mice.

Mice were immunized with anti-hepatitis B virus (HBV) mouse monoclonal and anti-HBV goat polyclonal antibody to produce anti-idiotypic antibodies (Yildirim *et al.*, 2004). Two mouse monoclonal antibodies (6C9, 6H9) were obtained from the fusions, and the immunogenic properties and specificity of antibodies were analysed. BALB/c mice were immunized with varying concentrations of anti-idiotypic antibodies (25, 50, 75, and 100 micrograms of anti-Id), and it was shown that anti-idiotypic antibodies generated hepatitis B surface antigen (HBsAg), as well as a BSA-specific antibody response.

NUCLEIC ACID VACCINES

A completely new field of vaccination was opened by the pioneering works of Lie and colleagues, who in 1993 reported that direct injections of a gene from influenza A virus induced protective immune response in immunized mice. Since then, the technology of DNA vaccination has become well established and widely spread in the research community as a method for infectious disease prophylaxis. In contrast to vaccines that employ recombinant bacteria or viruses, genetic vaccines consist only of DNA (as plasmids) or RNA (as mRNA), which is taken up by cells and translated into protein. In the case of gene-gun delivery, plasmid DNA is precipitated on to an inert particle (generally gold beads) and forced into the cells with a helium blast. Transfected cells then express the antigen encoded on the plasmid resulting in an immune response. Like live or attenuated viruses, DNA vaccines effectively engage both MHC-I and MHC-II pathways allowing for the induction of $CD8^+$ and $CD4^+$ T cells, whereas antigen present in soluble form, such as recombinant protein, generally induces only antibody response. In addition to the use of plasmid DNA for prophylactic vaccination against infectious diseases, DNA vaccines may be useful in the treatment of individuals chronically infected with viruses, e.g. HIV and hepatitis virus strains B and C.

Genetic vaccination through the delivery of RNA has also been investigated, but to a lesser extent than DNA vaccination. RNA expression is short-lived, and is thus less effective in inducing an immune response. The preparation and administration of RNA is troublesome because of the low stability of the RNA. One advantage of the RNA strategy is that there is no risk of integration of the delivered gene into the host genome.

DNA vaccines consist of plasmid DNA expression vectors of *E. coli* origin, which encode the antigen or antigens of interest under the control of strong viral promoters recognized by the mammalian host. When the plasmid DNA is administered into an animal, the antigen is expressed *in situ*, leading to an antigen-specific immunity. This genetic vaccination method offers a number of attractive qualities: the simplicity of producing large quantities of pure DNA, its wide applicability to various pathogens, the ability to induce cellular immune responses through MHC class I presentation, and the potential to manipulate the immune response through the co-delivery of genes encoding immunologically relevant molecules.

Because genetic vaccines are relatively inexpensive and easy to manufacture and use, their immunogenicity and efficacy have been analysed in a large number of systems. Indeed, DNA vaccines can circumvent many of the problems associated with recombinant protein-based vaccines, such as high costs of production, difficulties in purification, incorrect folding of antigen and poor induction of CD8$^+$ T cells. DNA also has clear advantages over recombinant viruses, which are plagued with the problems of pre-existing immunity, risk of insertion-mutagenesis, loss of attenuation or spread of inadvertent infection.

Genetic vaccines can be delivered into the host by several routes and methods. The main methods of plasmid-DNA delivery to the skin, by needle injection or by gene-gun, differ in several aspects. While needle injection requires relatively large amounts of plasmid (similar to the 50–100 µg dose used in intramuscular immunization), the amount of plasmid required for gene-gun immunization has been titrated down to a few nanograms. When delivered by gene-gun, the plasmid solubilizes when the plasmid-coated gold bullet penetrates the cells in the skin. Thus, plasmid is directly deposited into cells transfecting up to 20% of the cells in the target area. The gold particles directly penetrate the skin due to the force of delivery, thereby increasing the rate of transfection without having to rely on the uptake of DNA by the host cell itself. When compared to plasma DNA injection by syringe and needle, at least 100-fold less plasmid DNA is required for induction of protective immune response when DNA adsorbed to gold particles is delivered using a gene gun. As little as 0.4 µg of DNA can induce protective immunity in mice when compared with hundreds of micrograms of DNA in saline administered without a gene gun. The steps involved in the production of DNA vaccine is shown in Figure 3.2.

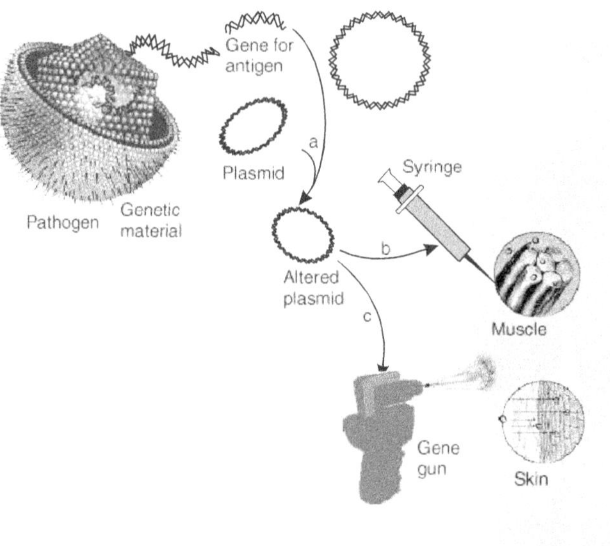

Figure 3.2   DNA vaccine production

Robinson *et al.* (1993) immunized chicks with DNA containing the *HA* gene of influenza virus strain H7N7. Chicks were given primary immunization with 100 µg of the *HA* DNA vaccine, followed by a booster dose of another 100 µg of the same DNA vaccine one month later and challenged with 100 lethal doses of the homologous virus 1–2 weeks post boost. A significantly high protection was observed in immunized chicks compared to non-immunized chicks.

Vaccination with plasmid DNA containing IBDV gene encodes complete polyprotein (VP2/4/3) induced protective immune responses. Oral DNA vaccination with the polyprotein gene of IBDV delivered by the attenuated *Salmonella enterica* sv. *Typhimurium* elicited protective immune response in chickens (Li *et al.*, 2006). The efficacy of recombinant plasmid DNA vaccine encoding *VP2* gene of very virulent strain of infectious bursal disease virus (vvIBDV) isolated from Pakistan was investigated with or without co-administration of cytosine-phosphate-guanine oligodeoxynucleotide (CpG ODN) to protect the chicken against the disease (Mahmood *et al.*, 2006). The results revealed that co-administration of recombinant plasmids with *CpG ODN* could protect chicken efficiently from IBDV challenge.

Vaccination of young ducks with DNA vaccines expressing the duck hepatitis B virus (DHBV) pre-surface (pre-S/S) and surface (S) proteins induced rapid immune responses that reduced the extent of initial DHBV infection in the liver and prevented the development of chronic infection in a virus dose-dependent manner (Miller *et al.*, 2006).

Pigs were immunized with plasmid vector containing genes for immunogenic envelope glycoproteins C or D (gC or gD) of pseudorabies virus under the control of the major immediate early promoter of human cytomegalovirus (Gerdts *et al.*, 1997). Vaccinated animals developed serum antibodies which recognized the respective antigen in immunoblot and exhibited neutralizing activity.

Jiang *et al.* (1998) used a plasmid DNA vaccine containing *VP1* gene of canine parvovirus to immunize dogs. The vaccinated dogs were completely protected against a virulent homologous CPV challenge.

Cedilla-Barron *et al.* (2001) induced protective response in swine vaccinated with DNA encoding FMD virus empty capsid proteins and the 3D RNA polymerase. A plasmid (pcDNA 3.1/P1-2A3C3D) was constructed containing FMDV DNA sequences encoding the viral structural protein precursor P1-2A and the non-structural proteins 3C and 3D. The 3C protein was included to ensure cleavage of the P1-2A precursor to VP0, VP1 and VP3, the components of self-assembling empty capsids. The non-structural protein 3D was also included in the construct in order to provide additional stimulation of CD4$^+$ T cells. When swine were immunized with this plasmid, antibodies to FMDV and the 3D polymerase were synthesized. Neutralizing antibodies were detected and after three sequential vaccinations with DNA, some of the animals were protected against challenge with live virus.

Beagles were immunized by intramuscular injection with a plasmid encoding the rabies virus (PV strain) glycoprotein. This was inserted into the pClneo vector (Promega, pGPV) under the control of the cytomegalovirus, an immediate early promoter and enhancer. Neutralizing antibodies against both wild type rabies virus and European Bat Lyssavirus (EBL1 & EBL2) were detected after a single injection and a boost, but levels of neutralizing antibodies against EBL1 were low. Moreover, all vaccinated dogs were protected against a lethal challenge with a wild type rabies strain (Perrin *et al.*, 2000). This is one of the first studies to demonstrate that dogs can be protected by DNA vaccines.

Dogs immunized with plasmid containing the nucleocapsid, the fusion and the attachment protein genes of a virulent canine distemper virus strain, developed virus-neutralizing antibodies. Vaccinated dogs were protected against challenge with virulent CD (Cherpillod *et al.*, 2000).

Suarez and Schultz-Cherry (2000) tested four different promoters in five different plasmids to express the haemagglutinin proteins of an H5 avian influenza virus, including two different immediate early cytomegaloviruses (CMVs), Rous sarcoma virus, chicken actin and simian virus 40 promoters. All five constructs expressed detectable haemagglutinin proteins in cell culture, but the pClneo HA plasmid with the CMV promoter provided the best response in chicken when vaccinated intramuscularly at 1 day of age on the basis of antibody titre and survivability after challenge with a highly pathogenic avian influenza virus at 6 weeks post-vaccination.

A DNA vaccine expressing glycoprotein C (gC) of bovine herpesvirus-1 (BHV-1) was evaluated for inducing immunity in bovines (Gupta *et al.*, 2001). The plasmid encoding gC of BHV-1 was injected six times intramuscularly or intradermally into calves at monthly intervals. After immunization by both routes, neutralizing antibody and lympho-proliferative responses developed. However, the immunity developed was not sufficient to protect calves completely from BHV-1 challenge.

A DNA vaccine coding for the *Brucella* outer membrane protein 31 conferred protection against B. *melitensis* and B. *ovis* infections by eliciting a specific cytotoxic responses (Cassataro *et al.*, 2005).

A combined DNA vaccine-prime, BCG-booster strategy resulted in better protection against *Mycobacterium bovis* challenge (Cai *et al.*, 2006). Calves were vaccinated with a combined DNA vaccine encoding Ag85B, MPT-64 and MPT-83 antigens from the *Mycobacterium tuberculosis* for the priming and subsequently boosting with BCG prior to experimental challenge with virulent *Mycobacterium bovis*, resulted in improved immune responses. Vaccinations with the combined DNA/BCG induced higher levels of antigen-specific gamma interferon in whole blood cultures and gave protection against an intra-tracheal challenge with virulent M. *bovis*. The combined DNA/BCG increased the protective efficacy by more than 10–100-fold.

Sechi *et al.* (2006) studied immune response to DNA vaccines encoding different mycobacterial antigens in lambs against *Mycobacterium avium* subspecies *paratuberculosis*. Plasmid vaccines containing p-85A-Mav, p-85A-BCG and p-Hsp65 were used. Immune responses were evaluated by measuring the expression of INF-gamma (Th1 type response) and IL-10 (Th2 type response) by real-time PCR. Plasmids p-85A-BCG and p-Hsp65 elicited a stronger protective immune response against M. *paratuberculosis* by evaluating the expression level of INF-gamma and evaluating the presence of M. *paratuberculosis* and animal cell organ damage post-mortem.

Bacterial Plasmid DNA-delivery Systems

The use of bacteria as delivery system for antigen-encoding plasmid DNA to mammalian cells has recently been investigated as an alternative to intramuscular or intradermal immunization of DNA. One advantage of using bacteria is the possibility of using non-parenteral immunization routes, and thereby stimulating mucosal immunity. Attenuated *Salmonella flexneri* was used as a DNA delivery vehicle, which invaded mammalian cells and delivered mammalian expression plasmids to the cytoplasm of the infected cells.

DNA vaccines afford a number of advantages and possible disadvantages when compared with alternative vaccination strategies. The advantages of DNA vaccines include the fact that they can encode multiple immunogenic epitopes and evoke both humoral and cell-mediated immune responses. Some concerns and potential disadvantages of DNA vaccines also exist. These include the potential for integration of the DNA into the host chromosome. Plasmid DNA-based vectors often contain nucleic acid sequences from oncogenic viruses, and the possibility for chromosomal integration exists. A second concern of DNA vaccination is the possibility of generating antibodies to DNA. A third concern is the effect that long-term expression of injected DNA into muscle cells may have on immune responses to subsequent vaccination with different DNA. The fourth disadvantage is that DNA vaccination strategies are unsuccessful when evaluating non-protein-based antigens, such as bacterial polysaccharides and lipids.

Chambers *et al.* (2000) used a plasmid DNA encoding the *Mycobacterium bovis* antigen MPB83 to vaccinate mice and cattle. The mice responded to vaccination with a mixed IgGl/gG2a response to the antigen and were protected from intravenous challenge with virulent M. *bovis* to a similar extent as those vaccinated with Bacillius Calmette–Guerin. Cattle vaccinated with the DNA vaccine gave strong whole-blood proliferative responses after stimulation with recombinant MPB83 protein.

REVIEW QUESTIONS

1. What are the different types of vaccines available from recombinant DNA technology?

2. How can a recombinant vaccine for a viral pathogen be prepared?

3. What are nucleic acid vaccines, how are they produced and what are their advantages and disadvantages over other types of vaccines?

REFERENCES

Audibert, Jolivet, F.M., Chedid, M.I., Alouf, J.E., Boquet, P., Rivaille, P. and Siffert, S.O. (1981). "Active anti-toxic immunization by a diphtheria toxin synthetic oligopeptide." *Nature.* 289: 593–594.

Bittle, J.L., Houghten, R.A., Alexander, H., Shinnick, T.M., Sutdiffe, J.G., Lerner, R.A., Rowlands, D.R. and Brown, F. (1982). "Protection against FMD by immunization with a chemically synthesized peptide produced from the viral nucleotide sequence." *Nature.* 298: 30–33.

Cai, H., Yu, D.H., Hu, X.D., Li, S.X. and Zhu, Y.X. (2006). "A combined DNA vaccine-prime, BCG-boost strategy results in better protection against *Mycobacterium bovis* challenge." *DNA Cell Biol.* 25: 438–447.

Cassataro, J., Velikovsky, C.A., de la Barrera, S., Estein, S.M., Bruno, L., Bowden, R., Pasquevich, K.A., Fossati, C.A. and Giambartolomei, G.H. (2005). "A DNA vaccine coding for the *Brucella* outer membrane protein 31 confers protection against *B. melitensis* and *B. ovis* infection by eliciting a specific cytotoxic response." *Infect. Immun.* 73: 6537–6546.

Casus, R. (1981). Reports of Pan American Health Organization (PAHO) Foot and Mouth disease Centre, Director Informations. (1980). 1st Half 1981 and Situations in Uruguay, 1979. Rio de Jeneiro.

Cedilla-Barron, L., Foster-Fuevas, M., Belsham, G.J., Lefevre, F. and Park House, R.M.E. (2001). "Induction of protective response in swine vaccinated with DNA encoding foot and mouth disease virus empty capsid proteins and the 3D RNA polymerase." *J. Gen. Virol.* 82: 1713–1724.

Chambers, M.A., Vordermeier, H.M., Whelan, A., Commander, N., Tascon, R., Lowrie, D. and Hewinson, R.G. (2000). "Vaccination of mice and cattle with plasmid DNA encoding the *Mycobacterium bovis* antigen MPB83." *Clin. Infect. Dis.* 30(Suppl. 3): S283–287.

Cherpillod, P., Tipoid, A., Griot-Wenk, M., Cardozo, C., Schmid, I., Fatzer, R., Vandevelde, M., Wittek, R. and Zurbriggen, A. (2000). "DNA vaccine encoding nucleocapsid on surface proteins of wild type canine distemper virus protects its natural host against distemper." *Vaccine.* 18: 2927–2936.

Cui, X., Lee, L.F., Hunt, H.D., Reed, W.M., Lupiani, B., Reddy, S.M. (2005). "A Marek's disease virus vIL-8 deletion mutant has attenuated virulence and confers protection against challenge with a very virulent plus strain." *Avian Dis.* 49:199–206.

Di Marchi, R., Brooker, G., Gale, C., Cracknell, V., Doel, T. and MOwat, N. (1986). "Protection of cattle against FMD by a synthetic peptide." *Science.* 232: 639–641.

Ding, X., Lillehoj, H.S., Quiroz, M.A., Bevensee, E. and Lillehoj, E.P. (2004). "Protective immunity against *Eimeria acervulina* following *in ovo* immunization with a recombinant subunit vaccine and cytokine genes." *Infect. Immun.* 72: 6939–6944.

Ejaz, M.K., Attah-Potu, S.K., Redmond, M.J., Parker, M.D., Sabara, M.I. and Babiuk, L.A. (1991). "Heterotypic passive protection induced by synthetic peptides corresponding to VP7 and VP4 of bovine rotavirus." *J. Virol.* 65: 3106–3113.

FAO (1981). European Commission Report for the Control of Foot and Mouth Disease, 24th Session FAT and Outbreak, Isle of Wight.

Flisser, A., Gauci, C.G., Zoli, Z., Martinez–Ocana, J., Garza-Rodriguez, A., Dominguez-Alpizar, J.L., Maravilla, P., Rodriguez-Canul, R., Avila, G., Aguilar-Vega, L., Kyngdon, C., Geerts, S. and Lightowlers, M.W. (2004). "Induction of protection against porcine cysticercosis by vaccination with recombinant oncosphere antigens." *Infect. Immunol.* 72: 5292–5297.

Garmendia, A.E., Morgan, D.O. and Baxt, B. (1989). "Foot and mouth disease virus-neutralizing antibodies induced in mice by anti-idiotype antibodies immunology." 68: 265–271.

Gerdts, V., Jons, A., Makoschey, B., Visser, N. and Mettenleiter, T.C. (1997). "Protection of pigs against Aujeszky's disease by DNA vaccination." *J. Gen. Virol.* 78: 2139–2146.

Gupta, P.K., Saini, M., Gupta, L.K., Rao, V.D., Bandyopadhyay, S.K., Butchaiah, G., Garg, G.K. and Garg, S.K. (2001). "Induction of immune responses in cattle with a DNA vaccine encoding glycoprotein C of bovine herpesvirus-1." *Vet. Microbiol.* 78: 293–305.

Hu, R., Zhang, S., Fooks, A.R., Yuan, H., Liu, Y., Li, H., Tu, C., Xia, X. and Xiao, Y. (2006). "Prevention of rabies virus infection in dogs by a recombinant canine adenovirus type-2 encoding the rabies glycoprotein." *Microbes Infect.* 8: 1090–1097.

Huang, Z., Elankumaran, S., Yunus, A.S. and Samal, S.K. (2004). "A recombinant Newcastle disease virus (NDV) expressing VP2 protein of IBDV protects against NDV and IBDV." *J. Virol.* 78: 10054–10063.

Ikuta, K., Ueda, S., Kato, S. and Hirai, K. (1984). "Identification of MAbs of glycoproteins of Marek's disease virus and Herpes virus of turkeys related to virus neutralization." *J. Virol.* 49: 1014–1017.

Jiang, W., Baker, H.J., Swango, L.J., Schorr, J., Self, M.J., Smith, B.F. (1998). "Nucleic acid immunization protects dogs against challenge with virulent canine parvovirus." *Vaccine.* 16: 601–607.

Kaufmann, S.H.F., Eichmann, K., Muller, I. and Wrazel., L.J. (1985). "Vaccination against the intracellular bacterium *Listeria monocytogenes* with clonotypic antiserum." *J. Immunol* 134: 4123–4127.

Kew, O.M., Nottay, B.K., Hatch., M.H., Nakano, J.H. and Obijeski, J.F. (1981). "Multiple genetic changes can occur in the oral poliovaccines upon replication in humans." *J. Gen. Virol.* 56: 337–347.

Kieny, H.P., Lathe, R., Drillen, R., Spehner, D., Skory, S., Schmitt, D., Wiktor, T., Koprowski, H. and Lecocoq. J.P. (1984). "Expression of rabies virus glycoprotein from a recombinant vaccine virus." *Nature.* 312: 163–169.

Kleid, D.G., Yanasura, D., Small, B., Dowbenko., D., Moore, D.M., Grubman, M.J., Mc Kercher, P.D., Morgan, D.O., Robertson, B.H. and Bachrach, H.L. (1981). "Cloned viral protein vaccine for FMD responses in cattle and swine." *Science.* 214: 1125–1129.

Kleid, D.G., Dowbenko, D.J., Bock, L.A., Hoatlin, M.E., Jackson, M.L., Petzer, E.J., Shire, S.J., Weddell, G.N., Yansura, D.G., Morgan, D.O., Mc Kercher, P.D. and Moore, D.M. (1985). "Production of recombinant vaccines from microorganisms." In: *Microbiology.* (ed.). Leive, L. Amer. Soc. for Microbiology, Washington DC. pp. 405–408.

Kurado, K., Hauser, C., Rott, R., Klenk, H.D. and Doerfler, W. (1986). "Expression of the influenza virus haemagglutinin in insect cells by a baculovirus vector." *EMBO. J.* 5: 1359–1365.

Li, L., Fang, W., Li, J., Huang, Y. and Yu, L. (2006). "Oral DNA vaccination with the polyprotein gene of IBDV delivered by the attenuated *Salmonella* elicits protective immune responses in chickens." *Vaccine.* 24: 5919–5921.

Lopez de Turiso, J.A., Cortes, E., Martinez, C., Ruiz de Ybaner, R., Dimarro, C., Vela, C. and Casal, I. (1992). "Recombinant vaccine for canine parvovirus in dogs." *J. Virol.* 66: 2748–2753.

Lupton, H.N. and Reed, D.E. (1980). "Evaluation of experimental subunit vaccines for infectious bovine rhinotracheitis." *Am. J. Vet. Res.* 41: 383.

Mahmood, M.S., Siddique, M., Hussain, I., Khan, A. and Mansoor. M.K. (2006). "Protection capability of recombinant plasmid DNA vaccine containing VP2 gene of very virulent infectious bursal disease virus in chickens adjuvanated with CpG oligodeoxynucleotide." *Vaccine.* 24: 4838–4846.

Mc Kercher, P.D., Moore, D.M., Morgan, D.O., Roberson, B.H., Callis, J.J., Kleid, D.G., Shire, S.J., Yansura, O.G., Dowbenko, D., and Small, B. (1985). "Dose response evaluation of a genetically engineered FMD virus polypeptide immunogen in cattle." *Am. J. Vet. Res.* 66: 587–590.

Miller, D.S., Kotlarski, I. and Jilbert, A.R. (2006). "DNA vaccines expressing the duck hepatitis B virus surface proteins lead to reduced numbers of infected hepatocytes and protect ducks against the development of chronic infection in a virus dose-dependent manner." *Virology.* 351: 159–169.

Morgan, R.W., Gelb, Jr., J., Schreur, C.S., Lutticken, D., Rosenberger, J.K. and Sondermeijer, P.J.A. (1992). "Protection of chickens from Newcastle and Marek's disease with a recombinant herpesvirus of turkeys vaccine expressing the Newcastle disease virus from fusion protein." *Avian Dis.* 36: 858–870.

Morrison, T., Hinshaw, V.S., Sheerar, M., Cooley, A.J., Brown, D., McQuain, C. and Mc Ginnes, L. (1990). "Retroviral expressed haemagglutinin-neuraminidase protein protects chickens from Newcastle disease virus induced disease." *Microbial Pathogenesis.* 9: 387–396.

Murphy, B.R., Sly, D.L., Tierney, E.L., Hosier, N.T., Massicot, J.D., London, W.T., Chanococ, R.M., Webster, R.G. and Hinshaw, V.S. (1982). "Reassortant virus derived from avian and human influenza A virus is attenuated and immunogenic in monkeys." *Science.* 218: 1330–1332.

Nagai, Y. and Klenk, H.D. (1977). "Activation of precursors of both glycoproteins of Newcastle disease virus by proteolytic cleavage." *Virology.* 77: 125–134.

Niikura, M., Matsuura, Y., Hattori, M., Onuma, M. and Mikami, T. (1991). "Characterization of haemagglutinin-neuraminidase glycoprotein of Newcastle disease virus expressed by a recombinant baculovirus." *Virus Res.* 20: 31–43.

Omar, A.R., Kim, C.L., Bejo, M.H. and Ideris, A. (2006). "Efficacy of VP2 protein expressed in *E. coli* for protection against highly virulent infectious bursal disease virus." *J. Vet. Sci.* 7: 241–247.

Park, I.H., Youn, J.H., Choi, I.H., Nahm, M.H., Kim, S.J. and Shin, J.S. (2005). "Anti-idiotypic antibody as a potential candidate vaccine for *Neisseria meningitidis* serogroup B." *Infect. Immun.* 73:6399–6406.

Perrin, P., Jacob, Y., Aguilar-Setien, A., Loza-Rubio, E., Jallet, C., Desmezieres, E., Aubert, M., Cliquet, F. and Tordo, N. (2000). "Immunization of dogs with a DNA vaccine induces protection against rabies virus." *Vaccine.* 18: 479–486.

Reagan, K.J., Wunner, W.H., Wiktor, T.J. and Koprowski, H. (1983). "Anti-idiotypic antibodies induce neutralizing antibodies to rabies virus glycoprotein." *J. Virol.* 48: 660–666.

Rigden, R.C., Jandhyala, D.M., Dupont, C. Crosbie-Caird, D., Lopez-Villalobos, N., Maeda, N., Gicquel, B. and Murray, A. (2006). "Humoral and cellular immune responses in sheep immunized with a 22 kilodalton exported protein of *Mycobacterium avium* subspecies *paratuberculosis*." *J. Med. Microbiol.* 55: 1735–1740.

Rimmelzwann, G.G., Van Eys, J.H., Drost, G., Vgtdehaag, F.M.C.M. and Osterhans, A.D.M.E. (1991). "Induction and characterization of monoclonal anti-idiotype antibodies reactive with idiotype of CPV neutralizing MAbs." *Vet. Immunol. Immunopathol.* 26: 139–150.

Robinson, H.L., Hunt, L.A. and Webster, R.G. (1993). "Protection against a lethal influenza virus challenge by immunization with a haemagglutinin-expressing plasmid DNA." *Vaccine.* 11: 957–960.

Roy, P. and Inumaru, S. (1989). In: *Biotechnology for Livestock Production.* FAO, Plenum Press, NY. p.347.

Sacks, D.L. and Sher, A. (1983). "Evidence that anti-idiotype induced immunity to experimental african trypanosomiasis in genetically restricted and sequence recognition of combining site related idiotypes." *Immunol. J.* 131: 1511–1515.

Sechi, L.A., Mara, L., Cappai, P., Frothingam, R., Ortu, S., Leoni, A., Ahmed, N. and Zanetti, S. (2006). "Immunization with DNA vaccines encoding different mycobacterial antigens elicits a Th1 type immune response in lambs and protects against *Mycobacterium avium* subspecies *paratuberculosis* infection." *Vaccine.* 24: 229–235.

Sherman, O.M., Aeres, Sadouski, P.L., Springer, J.A., Bray, B., Raybould, T.J.G. and Muscoplat, C.C. (1983). "Protection of calves against fatal enteric colibacillosis by orally administered *Escherichia coli* K99-specific monoclonal antibody." *Infect. Immun.* 42: 653–658.

Smith, G.L., Fodson, G.N., Nussenzweig, V., Nussengweig, R.S., Barnwell, J. and Moss, B. (1984). "*Plasmodium knowlesi* sporozoite antigen: expression by infectious recombinant vaccinia virus." *Science.* 224: 397–399.

Smith, M.E., Koser, M., Xiao, S., Siler, C., McGettigan, J.P., Calkins, C., Pomerantz, R.J., Bietzschold, B. and Schnell, M.J. (2006). "Rabies virus glycoprotein as a carrier for anthrax protective antigen." *Virology.* 353: 344–356.

Strohmaier, K.R., Franze and Adam, K.W., (1982). "Location and characterization of the antigenic portion of the FMD virus immunizing protein." *J. Gen. Virol.* 59: 295–306.

Suarez, D.L. and Schultz-Cherry, S. (2000). "The effect of eukaryotic expression vectors and adjuvants on DNA vaccines in chickens using an avian influenza model." *Avian Dis.* 44: 861–868.

Tanaka, M., Sasaki, N. and Seto, A. (1986). "Induction of antibodies against Newcastle disease virus with syngeneic anti-idiotype antibodies in mice." *Microbiol. Immunol.* 30: 323–331.

Taylor, J., Edbauer, C., Rey Senelonge, A., Bonquet, J.F., Norton, E., Goebel, S., Desmettre, P. and Paoletti, E. (1990). "Newcastle disease virus fusion protein expressed in a fowlpox-virus recombinant confers protection in chickens." *J. Virol.* 64: 1441–1450.

Taylor, J., Weinberg, R., Kawaoka, Y., Webster, R.G. and Paoletti, E. (1998). "Protective immunity against avian influenza induced by a fowlpox virus recombinant." *Vaccine.* 6: 504–508.

Taylor, J. and Paoletti. S. (1991). "Vaccinia virus recombinants expressing either the measles virus fusion or haemagglutinin glycoprotein protect dogs against canine distemper virus challenge." *J. Virol.* 65: 4263–4274.

Tramont, E.C., Chung, R., Berman, S., Keren, D., Kapfer, C. and Formal, S.B. (1984). "Safety and antigenicity of typhoid-*Shigella sonei* vaccine (Strain 5076-1c)." *J. Infect. Dis.* 149: 133–136.

Wu, L., Jiang, L., Zhou, Z., Fan, J., Zhang, Q., Zhu, H., Han, Q. and Xu, Z. (2003). "Expression of FMDV epitopes in tobacco by a tobacco mosaic virus-based vector." *Vaccine.* 21: 4390–4398.

Yang, C.D., Chang, G.N. and Chao, D. (2003). "Protective immunity against *Toxoplasma gondii* in mice induced by the SAG2 internal image of anti-idiotype antibody." *Parasitol. Res.* 91:452–457.

Yildirim, T., Basalp, A., Yücel, F., Manav, A.I. and Sezen, I.Y. (2004). "Generation of anti-idiotypic antibodies that mimic HBsAg and vaccination against hepatitis B virus." *Hybrid Hybridomics.* 23:192–197.

4

HYBRIDOMA TECHNOLOGY

INTRODUCTION

Increased research and accumulation of knowledge in the fields of somatic cell genetics, tissue culture technology and immunology resulted in the development of a technique for *in vitro* production of antibodies. The construction of antibody-secreting cell lines—hybridomas are produced by fusing antibody-secreting lymphocytes with appropriate tumour cell lines—myelomas. Monoclonal antibodies (MAbs) are the first products to move rapidly from the research laboratories through industry to the market place. MAbs will find use chiefly as research reagents, diagnostics *in vitro* and *in vivo*, as therapeutic agents and in the purification process and vaccine production.

Hybridoma technology is a form of genetic engineering resulting in the production of specific antibodies by specialized tissue culture lines. MAb is an antibody directed against one antigenic determinant or epitope of an antigen. It is single isotype.

EARLY RESEARCH

Hybridoma technology, like recombinant DNA technology, is rooted in basic biology and is the result of years of basic research in cell fusion. In 1973, Jerold Schwaber and Ed Cohen working at the University of Chicago's Liribida Institute were the first to produce antibody-secreting hybridomas by using normal antibody-producing human cells (B cells) and myeloma cells.

But it was George Kohler and Cesar Milstein (1975) working at the Medical Research Council Laboratories in Cambridge, UK, who devised and demonstrated a deliberate and rational strategy for the construction of continuous cell lines, which secrete monoclonal antibodies of a desired specificity. That success, which has been widely reported, revolutionized immunology and created an industry. These workers were awarded Nobel Prize jointly with Niels Jerne of Denmark in Physiology and Medicine in 1984 for this work.

The mechanism of antibody production was understood in 1975, when Bernet proposed the clonal selection hypothesis. In the 1960s, it became well established that antibody diversity was a reflection of the diversification of lymphocyte clones which was better than an individual lymphocyte which produces only a single type of antibody. Even homologous antigens induce a heterologous antibody response because of the many different lymphocyte clones with appropriate specificity for antigen proliferation and secretion of antibody.

POLYCLONAL ANTISERUM VS. MONOCLONAL ANTIBODY

Animals respond to an antigenic stimulus by producing a large variety of antibody structures directed against the immunogen. Even a single antigenic determinant is likely to be recognized by a variety of different antibody structures. Thus, heterologous mixtures of antibody structures are produced in a polyclonal response, and the populations of antibodies continuously change within the same animal, since each antibody is secreted by all such individual cells into a common blood pool.

The composition of polyclonal antibodies will change from day to day and from animal to animal. Thus, polyclonal antisera, even when prepared employing well standardized procedures tend to differ from batch to batch. But MAbs produced by a single clone of cells which react with a single antigenic determinant are the most specific biological probes available making possible finer and more sophisticated differences.

In 1975, Kohler and Milstein gave the concept of "one lymphocyte-one antibody" in the form of **hybridomas**. Hybridomas are hybrids between myeloma tumour cells and antigen-stimulated lymphocytes which can be cloned and grown in large quantities and for indefinite periods of time, and secrete high concentration of monoclonal and hence monospecific antibodies. Each lymphocyte produces only one type of antibody, although the isotype may change. The homogeneity, reproducibility and permanent availability of MAbs are the attributes arousing the greatest interest in this quickly developing field.

ADVANTAGES OF MONOCLONAL ANTIBODIES

1. Single MAbs are chemically defined, single specificity molecules, and can be used easily for standardization of specific assays.

2. MAb provide a perpetual source of well defined homogeneous reagents.

3. All the antibodies produced (100%) are active antibodies. Therefore high specific activity of labelling is possible in radioimmunoassay, enzyme linked immunosorbent assay and fluorescent immunoassay.

4. MAbs could be advantageously used as immunosorbents for antigen purification using immunosorbent column chromatography.

5. Large amounts of MAbs (1–20 mg/ml of ascitic fluid) could be obtained with modest investments.

6. MAbs specific for a particular target antigen can be obtained even without prior purification of antigens.

7. MAbs are easily manipulatable. For example, bispecific antibodies can be prepared from two MAbs and used in novel assays, antigenic targeting in electron microscopy and cytological studies.

8. Distinct antigenic cross reactivities can be easily defined and hence most useful in diagnosis.

DISADVANTAGES OF MONOCLONAL ANTIBODIES

1. MAbs are too specific. Limited MAbs might miss important cross-reactive determinants. This could relate to the animal's immune response to the antigenic stimulus. Polyclonal sera may offer advantage of broad reactivity.

2. Biological effects of MAbs on binding do not represent situation in animals.

3. Single MAb is a single chemical, and physical or chemical treatment that affects one molecule will affect all the MAbs in that population. Treatments like freeze thawing, proteolytic enzyme contamination and ammonium sulphate precipitation may destroy all MAb activity.

4. It is time-consuming to produce MAbs. The entire process of producing MAbs takes 3–4 months for each fusion experiment.

MONOCLONAL ANTIBODY PRODUCTION STRATEGY

The "one cell one antibody theory" provided the conceptual basis for the practical advantage offered by the MAb technique. Each clone of lymphocytes produces a single antibody specificity. Nevertheless, the immune response to most antigens are polyclonal; it involves hundreds of clones of lymphocytes, each secreting a different antibody. These antibodies are markedly heterogeneous. They usually consist of antibodies directed at different antigenic determinants. MAb production involves the creation of hybrid cells derived from 2 parent cell types: normal antibody-producing cells and neoplastic myeloma cells. The hybrids are formed by fusing the normal and neoplastic cells. Hybrids have characters of both parents, viz., the infinite life of the myeloma cells and antibody-secreting functions of primed cells. Hybridization of antibody-forming cells with malignant myeloma cells results in hybridomas in which the advantages of both specific antibody secretion and continuous growth are combined. Further, selection and cloning of the hybrid cells allow the derivation of monoclonal antibodies of predefined specificity and desired activity.

The method of establishing permanent cell lines capable of producing antibodies directed to predefined immunogen is based on the fusion of immune lymphocytes (usually spleen cells from mice previously immunized with antigen) with myeloma cells adapted for growth in tissue culture conditions (Figure 4.1). The most common fusion agent is polyethylene glycol (PEG). The selection for growth of hybrid cells is based on the fact that (a) spleen cells die in the tissue culture medium and (b) the myeloma lines that serve for fusions are mutants lacking either the enzyme hypoxanthine-guanine phosphoribosyl transferase (HGPRT) (azaguanine-resistant) or thymidine kinase (bromodeoxyuridine-resistant). Such mutants die in the presence of aminopterin which blocks the main pathway of DNA synthesis because they cannot use the salvage pathway. Hypoxanthine, aminopterin and thymidine (HAT) allows the selective growth of hybrids. The parental myeloma cells that have not fused will not grow in the presence of HAT medium. Normal lymphocytes die after a short period in culture. Therefore, only the hybrid cells between the spleen cells and the myeloma cells will survive in the presence of aminopterin, but for their growth, they require hypoxanthine and thymidine that are utilized by the salvage pathway.

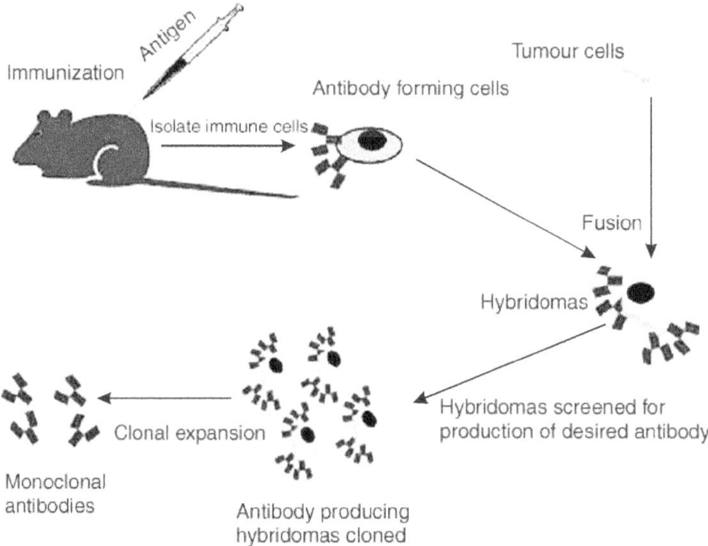

Figure 4.1 Production of monoclonal antibodies

The hybrid cells growing in HAT are then screened for their ability to produce and secrete the desired antibody. The positive hybridomas obtained after cloning are the source for monoclonal antibodies. Such cells can be cryopreserved and for the production of large amounts of MAbs, they can be grown in mass cultures, where they secrete 10–50 µg antibody/ml or propagated as ascitic tumour, where they usually produce 1–20 mg/ml of MAbs into the ascitic fluid.

Though the methodology involved in the production of MAbs appears simple, there are some crucial steps, especially in the screening and selection of positive hybridomas, that require pre-planned, sensitive, simple and fast assays. In addition, several precautions have to be taken and staunch methods have to be practised with regard to the tissue culture conditions, especially during the first phase of the hybridoma establishment.

Purity and Form of Immunogen

The purity of the antigen to be used as immunogen is crucial for generation of monoclonal antibodies. Molecules of low molecular weight (1000 daltons) are poor immunogens and have to be coupled to larger immunogenic molecules. Aggregated and particulate antigens elicit stronger responses. Adjuvants help to achieve stronger responses. The most common carriers are albumins, keyhole limpet haemocyanin, fowl gamma globulin and synthetic polypeptides.

Choice of Animals

The choice of the species and strain of the animal used as a source of spleen (donor) for fusion is largely dependent on the myeloma cells available and the origin of the immunogen. Mice are the most common species used for immunization primarily because there are more murine myeloma cell lines available. The BALB/c strain of mice is preferred since the myeloma cell lines available for fusion are derived from this strain.

Immunization Procedures

For most proteins, a dose of 1–50 µg per injection per mouse repeated two or three times is sufficient to evoke a strong antibody response. For soluble antigens, the first injection is given in the presence of the adjuvant, and the final booster is given in aqueous solutions. Freund's adjuvants (Difco Lab, USA) are commonly used for preparing antigen emulsion.

For most antigens, groups of 5–10 animals are sufficient. About 50 µl of the antigen, Freund's adjuvant emulsion (50 µg), is injected intraperitoneally (i/p) into each mouse. Two to 3 weeks later, the same amount of the emulsion is injected i/p. Ten to 14 days afterwards, individual animals are bled and the titre of the antibodies are determined. Animals that show the highest antibody titre are selected and at least 3–4 weeks after the last booster, they are injected i/p with 20–100 µg of soluble antigen on the first day and on the following day, they are injected with the same dose intravenously (i/v). Three days after the i/v booster, spleens are removed from two animals and their cells are used for fusion.

Choice of Myeloma

Most of the mouse myeloma cell lines are available from the American Type Culture Collection (ATCC), 12301 Parklwan Drive, Roskville, Maryland 20852, USA. The most common mouse myeloma cell line serving as parental cells for generation of B cell hybridomas is sp 2/0 Ag-14. This is a non-secreting cell line and is preferred since all of the antibodies produced and secreted will be dictated only by the immune cell. Other cell lines used include NS1/1, Ag 4.1 NSO, P3/X63, Ag8, FOX-NY from mouse and 43Ag 1.2.3 and YB 2/0 from rats.

Growth of Myeloma Cell Lines

The myeloma cells are generally grown in Dulbecco's Modified Eagle's Medium (DMEM) with high glucose (4.58 g/l) or in RPMI, both supplemented with glutamine (2 mM final concentration), antibiotics (usually 100 U/ml penicillin; 10 µg/ml streptomycin) and 10% foetal calf serum. Cells are maintained at 37°C in a humid atmosphere of 7–10% CO_2. The optimum conditions of a culture to be used for fusion are high viability and logarithmic growth for at least one week before fusion (not exceeding $3–8 \times 10^5$ cells/ml). A good practice is to grow the cells in the presence of 8-azaguanine (30 mg/ml) and to check whether they all die in HAT. These HGPRT negative cells in the log phase of growth are used for fusion.

Spleen Cells

The spleen is removed from the immunized mice under sterile conditions and transferred to a petri dish containing DMEM. Single cell suspension of splenocytes is prepared and B lymphocytes are separated by layering the suspension over Ficoll Hypaque gradient.

Cell Fusion

The spleen lymphocytes and myeloma cells at a ratio of 2 : 1 are mixed and placed in a centrifuge tube. After mild centrifugation, the supernatant is removed leaving the cell button. One ml of 50% pre-warmed PEG (4000 mol. wt.) is added dropwise with a wide-mouthed pipette for 90 seconds. After resuspending the pellet gently for 1 minure, pre-warmed DMEM without serum is added dropwise to dilute PEG with gentle mixing. The supernatant is removed after centrifugation and 5 ml of DMEM with HAT is added with gentle mixing and the final volume is made up to 50 ml, keeping the concentration of cells at 1.5×10^5 viable cells and distributed 0.1 ml aliquots in the wells of 96-well flat-bottomed microculture plates (5 plates can be prepared from 50 ml) and incubated at 37°C in 5–10% CO_2 incubator. After 3–4 days, hybrid clones are checked. 0.1 ml of DMEM with HAT is added and is repeated on the 7th day. When vigorous growth and change of colour to yellow are observed, usually 10–14 days following fusion, supernatants can be removed aseptically for screening for antibody activity. Wells are refilled with HAT medium. Positive wells are transferred into 24-well plate and then to small 25-cm^3 culture flasks. Thus, the positive cultures are expanded by repeated determination of antibody activity. Hybrid cells secreting the desired antibodies should be frozen in liquid nitrogen (LN_2) at each stage. The positive hybrid clones are again cloned for MAb production. A typical hybrid clone which secretes MAb is shown in Figure 4.2.

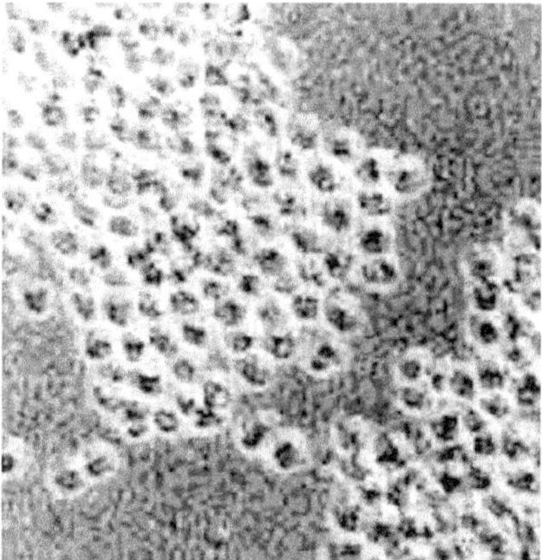

Figure 4.2 Hybridoma clone

Cloning of Hybrids

Cloning and recloning of hybrid cells is essential to ensure the monoclonality of the antibodies and to get rid of non-producer cells. Cloning under limiting dilution conditions is prepared when the fraction of specific antibody-secreting hybridoma is lower than one in 10^3 cells. Limiting dilution is a dilution in which hybrid cells are distributed at a concentration of 1 cell/well. Within 2 weeks, the "one clone" wells reach a cell density that allows screening MAb. To ensure monoclonality, the cloning procedure has to be repeated for a few cycles until all the subclones detected appear to secrete the specific antibody.

Cryopreservation of Cells

Hybridoma cells can be kept frozen in liquid nitrogen for many years in freezing medium containing 25–50% FCS and 10% DMSO. Ampoules containing hybrids are frozen in a programmable freezer at the rate of 1°C drop/min. from 4°C to –80°C and then the tubes are transferred into liquid nitrogen, where they can be kept for a long time.

Large-scale Production of Monoclonals

For large-scale production, hybridomas can be grown either in tissue culture— where they secrete up to 100 µg/ml (usually 10–50 µg/ml) or *in vivo*, as tumours in the peritoneal cavity of BALB/c mice, where they produce up to 40 mg of MAb/ml (usually 2–20 mg/ml). For large-scale growth, spinner bottles or roller bottles are useful. For maximal yield of MAbs, the supernatant is to be harvested at least one day after the cells have reached the stationary phase of growth. Recent development in biotechnology has led the industry to use hollow fibre technology for *in vitro* large-scale production of MAbs.

Ascitic fluid preparation and purification   Balb/c mice which are 6 weeks old are injected i/p with mineral oil, pristane (2, 6, 10, 14-tetramethyl pentadecane) for priming. Two weeks later, hybridoma cells from selected clones are injected i/p into pristane-primed mice. Seven to ten days after injection, the peritoneal cavities will appear swollen and an 18-gauge needle is inserted into the cavity to collect ascitic fluid in heparin. Ascitic fluid is collected 3 times in a 6-day period.

The ascitic fluid is centrifuged (1000 g for 15 min.) to remove cells and fibrin cells. The supernatant is centrifuged again at 52.000 g for 10 minutes to remove lipids and small fibrin clots. Antibodies are then precipitated from the supernatants by the addition of an equal volume of saturated ammonium sulphate to the clarified ascitic fluid. The resulting precipitate is dissolved in normal saline (0.85% sodium chloride) and dialysed extensively against normal saline for 12 to 16 hours in cold. The dialysate contains about 80% of antibodies.

Screening Assays for MAb Activity

The assay system used for the screening of antibody activity in the hybridoma culture supernatants is a key factor that determines success in obtaining the desired MAbs. The main factors to be considered are sensitivity, simplicity, ability to test many samples, speed and reliability. Solid phase radioimmunoassay (RIA) and enzyme linked immunosorbent assay (ELISA) are the frequently used techniques for screening antibody activity in MAbs. ELISA is used commonly for both soluble and cellular antigens. The main principle of the ELISA is that in the final step of the assay, the anti-immunoglobulin is enzyme-linked, instead of radio-iodinated. The degree of antibody-binding is evaluated by colour development that follows the addition of substrate to the system. The amount of colour developed is proportional to the level of enzyme-bound antibody present. A typical ELISA plate for MAb screening is shown in Figure 4.3.

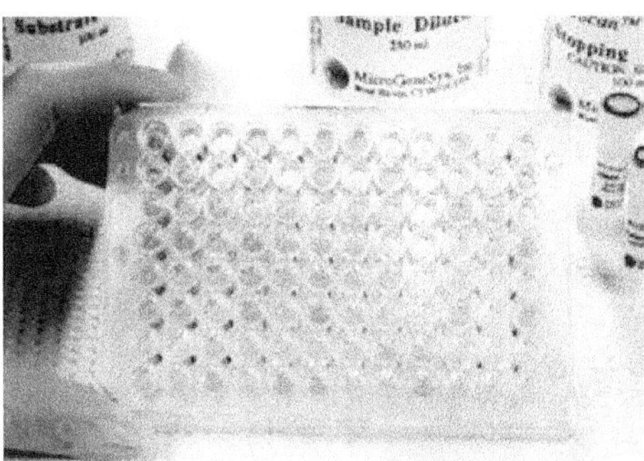

Figure 4.3   ELISA for screening the clone supernatants for finding out the isotypes

Although ELISA is somewhat less sensitive than RIA, the handling of samples, colour development and its measurements are faster, especially with automatic scanner that enables the reading, typing and computing of the results of the 96 wells within 1–2 minutes. Another practical advantage of ELISA, especially for the screening procedure, is that, it allows faster evaluation of results and selection of positive wells based on visual observation.

USES OF MONOCLONAL ANTIBODIES

There are three main areas in which MAbs are used in human and veterinary medicine. These areas are: (1) immunodiagnostic reagents, either for detection of the causative agents directly in tissues or body fluids or as a reagent used in indirect diagnosis such as serological detection of antibodies to the causative agent; (2) for experimental purposes ranging from molecular dissection of antigenic epitopes to monoclonal anti-idiotype antibody utilized as a vaccine to induce protective immunity and (3) immunoprophylaxis or immunotherapeutics applied to infectious diseases or as a vehicle for delivering toxic substances to, for example, tumours or as a tool to identify and locate target tumours.

IMMUNODIAGNOSTIC REAGENTS

Antigen Detection

With the development of MAbs of various specificities, advances have been made in the detection of infectious agents (or their antigens) directly in tissue or body fluid specimens. More work has been done on viruses because of the difficulties of viral culture.

MAbs have been developed against a large number of viruses including rabies virus (Koprowski and Wiktor, 1980), foot-and-mouth disease virus (Russel and Alexander, 1983), feline leukaemia virus (Jochen *et al.*, 1983), bovine leukaemia virus (Portetelle *et al.*, 1984), bovine enteric coronavirus (Crouch *et al.*, 1982) and rotavirus (Sona *et al.*, 1983).

MAbs have been produced against various parasites, including *Trichinella spiralis* (Gamble and Graham, 1983), *Babesia bovis* (Wright *et al.*, 1983), *Dirofilaria immitis* (Scott, 1983), and *Trypanosoma cruzi* (Araccio *et al.*, 1982). MAbs have also been raised against bacterial species, including *Mycobacterium* spp. (Kolk *et al.*, 1964, Morris and Ivanyi, 1985 and Young *et al.*, 1985), *Brucella* spp. (Holman *et al.*, 1983) and *Vibrio cholerae* (Gustafsoon *et al.*, 1982).

For detection of antigen, the following are the four commonly used methods using MAbs.

1. MAbs are attached to particles like latex or erythrocytes, which then detect the presence of an antigen in fluids by visible agglutination, usually in minutes. Disadvantage of this technique is the inability of small antigens to cause agglutination.

2. Direct demonstration of antigen in tissue sections, cultured cells or smears by the use of a MAb conjugated to a fluorochrome, enzyme (used in conjunction with a substrate that has an insoluble cleavage product) or radioisotope as a detecting agent. Figure 4.4 shows the use of fluorescent antibody method using MAb for detection of *Plasmodium* species.

3. A competitive-binding type of assay in which MAb is attached to a solid matrix such as the well of a polystyrene microtitre plate. Antigen conjugated with a fluorochrome, enzyme or radioisotope is added only in sufficient quantity to bind all the available antigen-binding sites of the antibody, then premixed with the known sample to be tested for the presence of an antigen. Decrease in binding of the known antigen indicates the presence and quantity of antigenic material in the unknown sample.

4. Antibody immobilized in polystyrene wells can 'trap' antigen in fluid preparations. The trapped antigen may then be detected using another antibody to the antigen. This second antibody may be conjugated with a fluorescent, radioactive or enzymic agent for easy detection. Alternatively, layers of anti-antibody can be added to increase sensitivity or an amplified system may be used. For this type of assay, MAbs may be used for both the trapping and detection, but they must have different specificities for the test to function properly unless the antigen is a single repeating unit such as the O-chain of some bacteria.

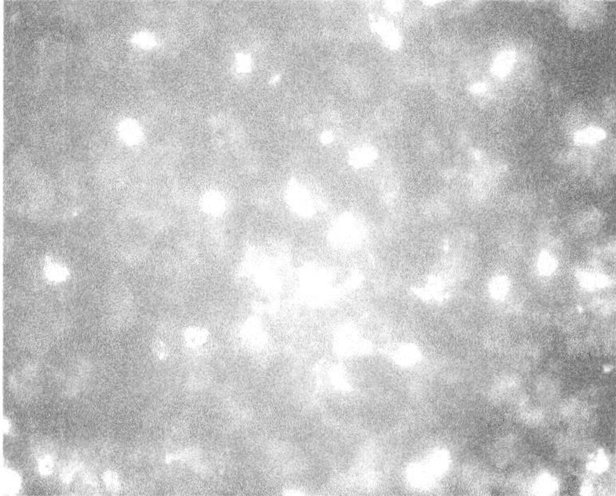

Figure 4.4 Fluorescent antibody technique with MAb for detection of *Plasmodium* species

Although MAbs are very useful in the detection of antigens or their components, caution should be exercised in that some MAbs may be too specific for general diagnostic use. Van Zaane (1983) reported that two MAbs were capable of distinguishing some strains of hog cholera virus from bovine viral diarrhoea virus; however, both antibody preparations failed to react with some strains of hog cholera virus by an immunofluorescence test. It is therefore essential to ascertain the reactivity of the MAb with the antigenic determinant present on all antigens to be tested for, before diagnostic use.

On the other hand, the exquisite specificity of MAbs allows the differentiation of antigens previously thought to be indistinguishable. For example, using a panel of MAbs, it is possible to differentiate various street strains (i.e., non-laboratory strains isolated from natural sources) and two vaccine strains of rabies virus using immunofluorescence technique (Webster *et al.*, 1986). Similarly, MAb to Marek's disease tumour associated antigen can be used to identify Marek's disease in chicken. As this antibody does not cross-react with antigen expressed by lymphoid leukosis virus, a differentiated diagnosis is possible

(Lee *et al.*, 1983). Because specific MAbs to rare and minor antigens could be produced, they have the potential to distinguish closely related strains of viruses, which cannot be differentiated by conventional serological methods. For example, MAbs that distinguish canine parvovirus from feline panleukopenia virus have recently been produced (Parish and Carmichael, 1983).

MAbs were produced against 42-kDa major structural proteins of rotavirus, which could be used for preliminary serotyping of strains from different animal species (Greensberg *et al.*, 1983).

Anderson (1984) described an enzyme immunoassay in which a blocking MAb was used to detect group-specific antibodies against bluetongue virus. This assay was capable of detecting antibodies against all 22 serotypes of bluetongue virus. The use of MAbs against bovine diarrhoea virus in immunofluorescent antibody techniques has aided preliminary serological characterization of strains (Peters *et al.*, 1986) and both indirect and competitive ELISA techniques utilizing MAbs have been used to detect antibodies against the virus in bovine sera.

He *et al.* (2007) developed a panel of monoclonal antibodies (MAbs) against the H5N1 avian influenza virus (AIV) and implemented an antigen-capture ELISA (AC-ELISA) to detect H5 viral antigen. Mice immunized with denatured haemagglutinin (HA) from A/goose/Guangdong/97 (H5N1), expressed in bacteria or with concentrated H5N2 virus, yielded a panel of hybridomas secreting MAbs specific for influenza HA. The reactivity of each MAb with several subtypes of influenza virus revealed that hybridomas 3D4 and 8B6 specifically recognized H5 HA. Therefore, purified antibodies from hybridomas 3D4 and 8B6, secreting IgG and IgM respectively, were used as the capture antibody and pooled hyperimmune guinea pig serum IgG served as detector antibody. Reconstituted clinical samples consisting of H5 AIVs mixed with normal chicken pharyngeal-tracheal mucus also yielded positive signals in the AC-ELISA and were confirmed by RT-PCR. The tracheal swab samples from H9N2-infected chickens did not give positive signals. Taken together, the newly developed MAb-based AC-ELISA offers an attractive alternative to other diagnostic approaches for specific detection of H5 avian influenza virus.

Accurate and rapid diagnosis of infectious agents is of considerable importance in veterinary medicine, particularly when dealing with highly infectious agents such as foot-and-mouth disease virus. MAbs have an important role in the diagnostic laboratory; however, if the antibody is not carefully selected, erroneous diagnosis may result. The following criteria should be followed for selection of MAbs for diagnostic purpose.

1. Antibody must have sufficient affinity for the antigen to permit efficient combination with low antigen concentrations and to displace efficiently host antibody already attached to the antigen.
2. The antibody should ideally be directed against an epitope of the antigen not recognized by host antibodies and therefore not masked by them.
3. The antibody should be directed against a repeating epitope which is readily accessible, such as lipopolysaccharides of gram-negative bacteria or virus coat proteins.
4. Multivalency of the antibody may increase test sensitivity. For example, IgM may be more effective than IgG.

In addition to these criteria, the antibody should be stable, should accept conjugation procedures without loss of antigen-binding capacity and should be of the appropriate specificity for its purpose. Care should be taken that MAb is not directed against a host cell or contaminant present in virus prepared from tissue culture.

MAb in cancer diagnosis   Although polyclonal antiserum (produced in animals) have been used in diagnosis of cancers, they are not very reliable, since they cross react with antigenic determinants of normal tissues. MAbs have helped to overcome these problems. They are used in three different ways in cancer diagnosis.

1. to detect tumour-related antigens or tumour markers in body fluids or on cells,
2. to demonstrate tumour antigens in histological section and
3. to serve as imaging agents.

Tumour markers   Blood levels of certain tumour products are used as tumour markers in the diagnosis and prognosis of cancers. When the tumour load is reduced, the blood level of the tumour marker is also reduced considerably. Hence, this can be used to monitor the tumour burden and judge the efficiency of therapy. Carcino-embryonic antigen is a tumour marker elevated in 91–100% of persons with colorectal cancer. CA-125, another marker, is elevated in 82% of women with ovarian cancer. Prostrate-specific antigen is a marker for late-stage prostate cancer.

MAbs raised against a variety of tumours have helped in the definition and characterization of tumour-specific antigens, thus helping in correct typing and diagnosis. Accurate diagnosis of the precise tumour type is essential since subsequent therapy depends on the type of tumour.

Diagnostic tumour pathology   The study of pathological change has been performed by microscopical examination of tissue section stained with MAbs using immunohistochemical techniques. A two step immunoperoxidase procedure, which gives a colour reaction at the end, is commonly used to demonstrate location of epitopes in a cell. MAbs are widely used in the diagnosis of haematopoietic malignancies. MAbs to lymphoid cell-surface markers are extensively used to assess the level of immunodeficiency in primary and acquired immunodeficiency syndromes (AIDS).

Immunoscintigraphy   MAbs with specificities to many types of human cancers are now available. These can be tagged with radioisotopes such as ^{131}I or ^{99}Technetium and injected intravenously into the patients; the antibody localizes in the tumour, which can then be detected by imaging the radioactivity. This technique of imaging is known as immunoscintigraphy although imaging techniques like CT scanning, ultrasound scanning and magnetic resonance imaging can detect small lesions, they are unable to differentiate between cancerous and non-cancerous growth. Being tumour-specific, antibody imaging has certain advantages. Small metastatic lesions which remain undetected by CT scan can be detected successfully in many cases with radio-labelled MAbs. Use of immunoscintigraphy in the diagnosis of bladder cancer is shown in Figure 4.5.

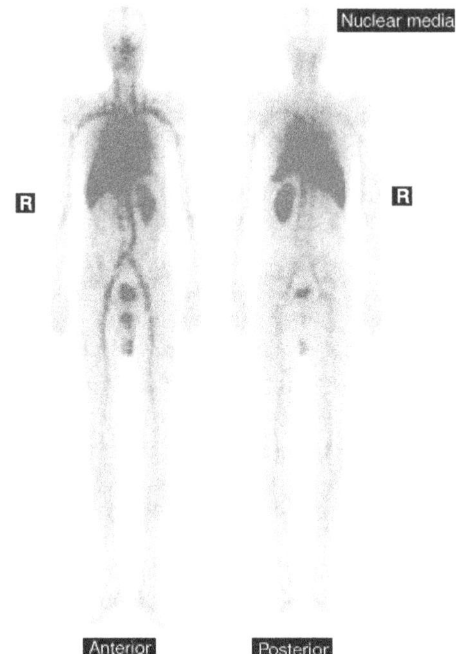

Figure 4.5   Immunoscintigraphy in diagnosis. Anterior (left) and posterior whole body images of a patient with primary bladder cancer 6 hours following intravenous injection of 800 MBq Tc-99m-C595.

MAbs as therapeutic agents   Mouse MAbs directed against the particular organism have been found to protect chimpanzees against challenge with hepatitis B virus of rodents and against infections with bacteria *Haemophilus influenzae, E. coli, Proteus* spp. and *Streptococcus* spp. MAbs against antigens present on the surface of cancer cells can exert site-specific cytotoxic effect brought about by mechanisms such as complement-mediated cytotoxic antibody-dependent cell-mediated cytotoxicity, and phagocytosis of antibody-coated cells by cells of reticuloendothelial systems.

In attempts to overcome rejection by cell-mediated immune response, patients undergo immunosuppressive therapy, following organ transplantation. This therapy is in the form of immunodepressive drugs such as cyclosporine plus prednisone, or patients are generally treated with glucocorticoid hormones in higher doses for short periods. MAbs specific for human T-cell-surface antigens (and for specific subpopulation of T cells) are now available. These MAbs can be used selectively to eliminate T-cell populations responsible for allograft (transplant) rejection; for example, a MAb reacting with mature intravascular T cells has been found to be useful in subjects who have undergone renal transplantation.

Rabies virus neutralizing human MAbs was evaluated *in vitro* and in a Syrian hamster model as a potential future alternative to rabies immunoglobulins (RIG) which is used for post-exposure prophylaxis (Hanlon *et al.*, 2001). Seven MAbs neutralized a representative rabies virus. In hamsters, one MAb resulted in protection that was comparable to human RIG.

Human monoclonal antibodies (HuMAbs) that neutralized rabies virus were developed (Sloan *et al.*, 2007). Transgenic mice carrying human immunoglobulin gene were used to isolate human MAbs that neutralized rabies virus. Several HuMAbs were identified that neutralized rabies virus variants from a broad panel of isolates of public health significance. HuMAb 17C7 was the most promising antibody identified because it neutralized all rabies virus isolates tested. HuMAb 17C7 protected hamsters from a lethal dose of rabies virus in a well-established *in vivo* model of post-exposure prophylaxis.

One of the best examples of use of MAb in therapy is MAb OKT3, which is directed against CD3 and is an antigen on the surface of T lymphocytes. CD3-activated T cells can play a crucial role in allograft (organ transplant) rejection. OKT3 can be used to prevent acute renal allograft rejection in humans.

For effective immune therapy against HIV infection, genetic engineering has been used to link the Fc portion of mouse MAb to human CD4 complex. CD4 is a cell-surface receptor on T helper cells and one which has the receptor site for HIV. HIV attaches to T cells prior to infecting them by specifically attaching to CD4 receptors using the gp-120 receptors on its own surface. After infection of a T cell, gp-120 antigens appear on the surface of that cell. If Fc-CD4 molecules are given to the sufferers, these bind to HIV-infected cells (displaying g-120) and the exposed mouse Fc fragment induces cell-mediated destruction of the infected cell. This strategy may prove useful in preventing the spread of HIV throughout the T-cell population.

Drug targeting   MAbs can be used as vehicles for the delivery of drugs *in vivo*. Conjugates of antibody with drugs are able to bind to an appropriate target cell and the antibody-linked drug is then introduced directly into that cell. This allows higher concentration of the drug to reach the desired site and minimize systemic toxicity. Drugs such as daunamycin, vindesin, methotrexate and adriamycin can be coupled to MAbs. Better survival rates can be observed in experimental mice with lymphomas given daunamycin attached to antilymphoma MAbs compared to those receiving the drug alone.

A recombinant chimeric MAb against fibrin coupled to tissue plasminogen activator (tPA) has been used in animal models. Because of its affinity for fibrin, the recombinant protein becomes concentrated in a clot, thereby locally raising the concentration of tPA. This helps in promoting an enhanced lysis of the clot. Such thrombolytic agents would have a crucial role to play in cardiovascular diseases.

Antibody Detection

The second use of MAbs in disease diagnosis is in the detection of antibody most commonly used as an antiglobulin reagent conjugated with a detection system (fluorochrome, enzyme or isotope) in a primary binding assay. The most commonly used method is an indirect assay in which the antigen is passively absorbed to a plastic (polystyrene) surface by hydrophobic interaction. The sample serum or other body fluid suitably diluted is applied to the immobilized antigen. After an incubation period, the monoclonal anti-species antibody conjugate is added. Each step is followed by a thorough washing cycle. These are essential steps in which measures are usually taken to eliminate non-specifically bound proteins to reduce "background" activity in negative samples. In the final step, if an enzyme conjugate is used, then a chromogenic substrate is added followed by spectroscopy, while radioisotope or fluorochrome conjugate binding may be assessed directly.

Primary binding assays are currently employed only to a small extent in veterinary medicine. However, with suitable apparatus becoming more common, primary binding assays and, in particular, enzyme immunoassay will replace current serological tests (complement fixation test, serum neutralization test, precipitation test and agglutination test). Standardized reagents are essential for conduct of these assays for obtaining consistent results. In this respect, MAb has an advantage that antibody of identical specificity can be prepared at will and the enzyme conjugation procedure results in no discernible differences from batch to batch.

Anti-idiotype Antibody

Anti-idiotype antibody is an antibody that recognizes the antigen-combining sites of other antibodies. Anti-idiotype antibody (Ab2) has antibody activity to the antigen-combining amino acid epitope sequence of the folded molecule of the hypervariable region of the antibody (Ab1) molecule. Anti-idiotype antibody can be prepared by immunization of a heterologous species or a homologous species with affinity purified Ab1. The resulting Ab2 should be able to compete with the original antigen for the antigen-binding sites of Ab1. If a polyclonal Ab2 is made, even in a homologous species, it generally has to be absorbed extensively to remove anti-species or anti-allotype activity. If Ab2 is monoclonal, it would have to be tested exhaustively with immunoglobulins with or without Ab1 activity and with or without the Fc portion to eliminate non-specific reactions that often occur between immunoglobulins. The monoclonal Ab2 can be used in a primary binding assay as a capture agent for Ab1. If Ab2 was raised in a homologous species, Fab and Ab2 should be used for capture, while an anti-Fc reagent conjugate should be used for detection. An alternative approach is to attach Ab2 chemically to particles such as erythrocytes or latex beads and passive agglutination should take place in the presence of Ab1. By either method, detection of antibody may be accomplished in the absence of an antigen, a technique which may prove useful to serology that otherwise involves antigens that may be very expensive, toxic or difficult to obtain. Anti-idiotype antibodies have been produced against *Staphylococcus* spp. and *Streptococcus* spp. (Arulanandam *et al.*, 1985) and *Brucella abortus* (Nielsen *et al.*, unpublished work.)

EXPERIMENTAL USES OF MONOCLONAL ANTIBODIES

Cells of Immune System

MAbs have been used extensively to characterize various cell populations involved in immune responses of mice and man. The Lyt system of antigens have been developed for functional differentiation of T lymphocytes. Lyb antigens have been used for differentiation of B lymphocytes. Similar studies in domestic animals are in infancy. Krawiec and Muscoplet (1984) developed MAbs specific for canine T lymphocytes and others capable of distinguishing medullary and cortical thymocytes. Equine lymphocyte surface antigens have been studied by Newman *et al.* (1984). MAbs against bovine lymphocytes have been produced (Pinder *et al.*, 1980). Three MAbs that distinguish T and B lymphocytes of cattle in peripheral blood were produced and used in a study of a bovine leukaemia virus-infected herd to show that no consistent differences in B cell number resulted from the virus infection. Davis *et al.* (1984) catalogued about 100 MAbs for their reactivity patterns with bovine, ovine, caprine, porcine, equine and human lymphocytes. Reactivity of the MAbs based on immunofluorescence analysed by a fluorescent antibody cell sorter resulted in patterns suggesting specificity with various T-lymphocyte subsets, including suppressor/cytotoxic cells, as well as T-lymphocyte reagents and both T and B cells, class I and class II MHC antigens, monocytes, granulocytes and erythrocytes, and IgM (B cells) and light chains. Interestingly, interspecies cross-reaction was extensive with reagents and limited with others, with monospecificity being relatively rare to some antigens, such as MHC-class I antigen. Lunney (1982) used monoclonal antibody to swine lymphocyte antigen complex to identify and purify Ia antigens and to study the immune response to various antigens including a synthetic peptide.

Pregnancy and Sex Determination

MAbs have been used to detect pregnancy by the presence of progesterone in milk. Using a direct double antibody solid phase enzyme immunoassay it was found that 1–2 pg of progesterone per ml of milk could be detected which was an excellent assay without extraction or centrifugation of the milk.

MAbs to a variety of other hormones have been prepared. Such hormones include somatotropin (Krivi and Rowold, 1984), human chorionic gonadotropin and bovine and porcine insulin (Schonherr *et al.*, 1984).

An interesting and at present, controversial application is in the area of embryo sexing. Antibody to H-Y antigen, the male-specific transplantation antigen, has been sought over the years to establish definitely the sex of an embryo. In general, polyclonal antisera were unsatisfactory because of their inconsistencies in detecting H-Y antigen and their low titres after absorption.

Molecular Structure of Antigens

Antigens are defined as substances capable of reacting with antibodies; however, the reactivity with antibody is confined to the antigenic determinants or epitopes present on the antigen molecule. The epitope complements the antigen-binding site of the variable region of the antibody molecule in its configuration. Epitopes of antigens fall into two broad categories—**sequential** and **conformational** determinants. Sequential determinants are sequences of amino acids of an unfolded or randomly folded peptide, whereas in conformational epitopes the amino acids are brought into close proximity by the conformation of the antigen; as a consequence, antibody to a true conformational determinant rarely reacts with the unfolded molecule.

A considerable number of protein antigens and viruses have been studied in detail with MAbs to determine their epitope structures. The molecular dissection approach has been used with MAbs to select antigenic variants of several viruses of veterinary importance. A single amino acid change—arginine at position 333—of a glycoprotein in the virulent form of the rabies virus was found to be the only change in non-pathogenic variants (Dietzschold *et al.*, 1983). Neutralizing antibodies to antigens such as VP2 antigens in bluetongue virus (Appletone and Letchworth, 1983) and foot-and-mouth disease virus VP1 antigen and its peptides have been described (Meloen *et al.*, 1983).

MAbs have been found to be useful in dissecting, mapping and identifying the antigenic epitopes of virus proteins that elicit the production of neutralizing antibodies. Two distinct serotypes of vesicular stomatitis virus (VSV) have been identified and characterized—VSV-New Jersey (VSV-NJ) and VSV-Indiana (VSV-IN). VSV-NJ possesses 4 major non-overlapping epitopes, whereas VSV-IN appears to have 5 epitopes, which exhibit varying degrees of antigenic similarities as detected by MAb. One MAb identified in these studies detected an epitope common to both VSV-IN and VSV-NJ, but neutralized only VSV-IN. This particular antibody has proved useful for distinguishing VSV from other vesicular diseases causing viruses such as foot-and-mouth disease, vesicular exanthema and swine vesicular disease viruses.

MAbs have been extensively used in the study of tumour-causing viruses and the survival and function of tumour virus proteins in host cells. MAbs have also been developed against the transformation-specific glycoproteins coded for by feline leukaemia oncogene *v-fms*, avian retrovirus reverse transcriptase, bovine leukaemia virus-transformed sheep cells and Marek's disease virus tumour-associated antigens.

Affinity Chromatography

Because of their exquisite specificity, MAbs are suitable for use in purification of antigens. This usually entails the immobilization of the MAb on an insoluble matrix, such as agarose beads, by a covalent linking agent, for example cyanogen bromide. The beads with MAb attached can be used either in a batch procedure or in a column chromatography for antigen purification. Antigen purification by affinity has been applied to substances ranging from Ia antigens of the swine lymphocyte antigen complex, glycoproteins of herpesviruses, and to antigens of *Trichinella spiralis*. The technique has great potential, especially for purification of antigens from complex mixtures containing related or physico-chemically similar substances, thus, isotypes of immunoglobulin for standardization and immunochemical studies would be the prime candidates.

MAbs are useful in enzyme purification, which cannot be purified by biochemical methods. By coupling MAb to cyanogen bromide activated Sepharose column, highly purified enzyme preparations with good yield can be obtained.

Immunochemical Application of MAbs

The studies on molecular structure of antigens have been greatly facilitated by application of MAbs to immunoblotting analysis. Digested, degraded or native antigen mixtures are separated by standard procedures such as one- or two-dimensional polyacrylamide gel electrophoresis. Usually more than one replicate is run, one gel is stained directly, while the other may be subjected to electroblotting, thereby transferring the separated materials onto the nitrocellulose. The nitrocellulose membrane is then blocked with a non-reactive protein and incubated with the MAb of choice. Mouse MAb binding to antigen may be detected by the antibody being conjugated with its own detection system, for example, an isotope or enzyme conjugated with a detection system as the second layer may be employed. Several methods are available for either purpose; however the immunoperoxidase technique has advantages in that it provides very low background activity, is highly sensitive and eliminates the use of isotopes. Western blotting technique has been widely applied to the study of protein antigens.

MAbs to animal immunoglobulins have been prepared. Antibodies to bovine IgG, IgG$_1$, IgG$_2$ and IgA have been prepared (Van Zaane and Ijzerman, 1984). MAb to porcine IgM (Paul, van Deusen and Mengeling, 1985) and to canine IgG subtypes have also been produced (Fuller and Hurrell, 1985).

There are three ways by which MAb specificity can be assayed; firstly, with the isotypes passively adsorbed onto polystyrene; secondly, with a panel of 500 normal sera (also passively adsorbed to polystyrene) and thirdly, with an antigen–antibody complex in which the antigen is immobilized and the antibody presumably in its native state to allow the MAb to react with the Fc end of the molecule exposed. The later assay invariably yielded the highest 'titres' with MAbs; however, it is unfortunately most expensive in terms of reagents.

Bifunctional antibodies have been produced by Milstein and Cuello (1984) by fusing two hybridoma cell lines, one producing antibody to peroxidase and one to somatostatin, resulting in a single cell that produces antibodies to peroxidase and somatostatin, resulting in a single cell of dual specificity. They found that some of the resulting hybrid hybridoma products were very useful for immunohistochemical staining when applied to sections containing somatostatin in the presence of peroxidase.

MAbs have been useful in several other immunochemical areas. The study of the complement cascade has been enhanced by the use of MAb of various isotypes in a haemolytic system. It has been shown that a mixture of IgC_{2a}, IgG_b and IgG_3 could effectively cause lysis, whereas IgD could not. IgG preparations resulted in weak haemolysis and IgM was highly haemolytic.

MONOCLONALS AS *IN VIVO* REAGENTS FOR ANIMALS

The current uses of monoclonals in domestic animals can be categorized as follows:

1. Passive antibody administered prophylactically or therapeutically for infectious diseases;
2. Passive antibody to target on particular cell markers; either alone or coupled to cytotoxic agents;
3. Passive antibody to enhance the clearance of toxic compounds;
4. Passive antibody to modulate the cellular or messenger components of the *in vivo* immune responses;
5. Anti-idiotype antibody administered as an immunogen.

The oral application of a MAb directed to K99 pilus antigen of enterotoxigenic *Escherichia coli* (ETEC) prevented severe fatal enteric disease in colostrum-fed and colostrum-deprived calves experimentally challenged with ETEC. The intravenous administration in sheep of a MAb to bluetongue virus neutralized the challenge virus, and prevented viraemia, development of precipitating antibodies and clinical disease (Letchworth and Appleton, 1983). The authors speculated that vaccines containing a single antigenic determinant might generate a similar neutralizing antibody and may both prevent bluetongue disease and perhaps interrupt virus transmission.

In rabies, antigenic variants, previously undetected with conventional polyclonal antisera, have been detected with MAb with specificity for either the nucleocapsid or glycoproteins of several fixed and street strains of the virus. These findings have been quite revealing in the sense, they have proved an explanation as to why current vaccines are not universally protective. They have also shown which additional variants must be included in the development of an effective vaccine.

Anti-idiotype (Ab2) has been used as an antigen to prepare anti-anti-idiotype (Ab3) which has same specificity of Ab1. In other words, Ab2 may be used as a vaccine to induce an antibody response to the original antigen. Thus, Ab2 has been used to induce a response to transplantation antigens in mice and to produce protective immunity in mice against trypanosomes. Ab2 has been made to rabies glycoprotein G and some of the Ab2s, when used to immunize mice, resulted in a specific neutralizing antibody response (Reagen *et al.*, 1983). These data clearly indicate that Ab2 can mimic an antigen epitope and cause the production of an immune response to it.

MAbs can be used as guided missiles in tumour therapy. The antibodies, when coupled to a toxic molecule and administered, can selectively kill the targeted tumour cell or virus-infected cell. This makes such therapy economically feasible in the animal world. Doctors hope to be able to use MAbs for purposes other than diagnostic tests, in particular, to attach to them toxic drugs that will bind cancer cells. Injected into a cancer patient, the combined molecule would circulate around the body in the blood until it finds the tumour and deliver the drug to its target. This treatment could greatly reduce the extremely traumatic side effects of the present cancer therapy. MAbs directed against various inducer and helper T cells might be desirable in controlling autoimmune disease.

THERAPEUTIC MONOCLONAL ANTIBODIES

The potential therapeutic application of MAb created a tremendous interest in the medical and pharmaceutical community. A recent survey suggested that over a quarter of all biotech drugs in development are MAbs. Within this group, more than 30 chimeric, humanized, or fully human antibodies account for more than 30 products being routinely used or investigated in the clinic for various indications. The most promising and advanced therapeutic strategies include inhibition of alloimmune and autoimmune reactivity, antitumour therapy, anti-platelet therapy and antiviral therapy. The products that obtained marketing approval by the registration authorities in the USA or Europe are discussed below.

Inhibition of Alloimmune Reactivity

After successful clinical trials by multicentre transplant study group in 1985, the first license for a MAb was granted by the US Food and Drug Administration for routine use of OKT3 (Muromonab CD3), a murine MAb for the prevention of graft rejection in renal transplant patients (OMTSG, 1985). As first-line or in steroid-resistant rejection therapy, OKT3 has proven efficacious and some studies have shown improved graft survival. In the search for more specific immunosuppression with MAb, the interleukin-2 receptor, which is expressed on activated T cells was chosen as a target. Chimeric and humanized MAb boasiliximab (Simulec) and daclizumab (Zenapax) were developed to bind to the interleukin-2 receptor (CD25). These MAbs provide immunosuppression by competitive antagonism of interleukin-2 receptor (CD25) or by elimination of activated T cells. In two large trials, the percentage of patients with biopsy-confirmed acute rejection episode for renal transplantation was significantly lower with basiliximab 20 mg (administered 2 hours before and then 4 days after transplantation surgery, 30 vs. 33%, respectively) than placebo (44 vs 46%) at 6 months after surgery. Daclizumab gave similar results: after 6 months use of two conventional immunosuppressants, acute rejection was 28 and 22% with daclizumab and 47 and 35% with placebo. Both therapies were well-tolerated and not associated with a lymphokine-release syndrome or with an increased frequency of infection. Some of the MAbs registered for therapeutic use are listed is Table 4.1.

Table 4.1 MAbs registered for therapeutic clinical use

Generic name	Trade name	Company	Indication	Dose	Route
Muromonab	Orthoclone OKT3	Janssen-Cilag	Renal graft rejection	5 mg per day	Intravenous
Basiliximab	Simulect	Novartis	Renal graft rejection	20 mg direct before and 4 days after transplantation	Intravenous
Daclizumab	Zenapax	Hoffman-la Roche	Renal graft rejection	1 mg/kg before and 2, 4, 6 and 8 weeks after transplantation	Intravenous
Infliximab	Remicade	Centocer	Rheumatoid arthritis, Crohn's disease	3–10 mg/kg every 4–8 weeks	Intravenous
Rituzimab		Genentech-Roche	Lymphoma	375 mg/m^2 in four weekly doses	Intravenous
Trastuzuma	Herceptin	Genentech	Metastatic breast cancer	4 mg/kg initially, followed by 2 mg/kg weekly	Intravenous
Abciximab	Reopro	Lilly	Anti-platelet	0.25 mg/kg initially followed by 0.125 mg/kg per min.	Intravenous
Palivizumab	Synagis	Abbot lab	Antiviral	15 mg/kg	Intravenous

Respiratory syncytial virus (RSV) is the leading cause of viral bronchiolitis and pneumonia in infants and children. Currently, palivizumab is the only approved monoclonal antibody (MAb) for prophylaxis of RSV. However, a small percentage of patients are not protected by palivizumab; in addition, palivizumab does not inhibit RSV replication effectively in the upper respiratory tract. Wu *et al.* (2007) reported the development and characterization of motavizumab, an ultra-potent, affinity-matured, humanized MAb derived from palivizumab. Motavizumab, binds to RSV F protein 70-fold better than palivizumab, and exhibits about a 20-fold improvement in neutralization of RSV *in vitro*. In cotton rats, at equivalent concentrations, motavizumab reduced pulmonary RSV titres to up to 100-fold lower levels than did palivizumab and, unlike palivizumab, motavizumab very potently inhibited viral replication in the upper respiratory tract.

Inhibition of Autoimmune Reactivity

MAbs have been investigated as a therapeutic approach for the treatment of autoimmune diseases to target elements of the immune response such as T or B lymphocytes, irrespective of their antigen specificity. This approach seeks to suppress excessive immunopathological responses by removing activated cells, blocking their function, or normalizing elevated levels of proinflammatory cytokines. The most encouraging clinical results emerged with TNF-blocking therapies in rheumatoid arthritis and Crohn's disease.

TNF is a cytokine produced primarily by activated monocytes and macrophages, with a broad spectrum of biological activities. This enzyme induces vasodilatation, increases vascular permeability, activates platelets, and regulates the production of acute phase proteins, other pro-inflammatory cytokines and mediators of inflammation. TNF is actively produced in various infectious diseases: sepsis, malaria, adult respiratory distress syndrome and AIDS. TNF is also important in autoimmune inflammatory diseases. The cytokine is actively produced at the synovial and mucosal sites of inflammation in rheumatoid arthritis and Crohn's disease. The chimeric (human-mouse) MAb, infliximab (CA2, Remicade), binds to TNF specifically and neutralizes its activity. Administration of infliximab was of substantial clinical benefit in both rheumatoid arthritis and Crohn's disease.

The first randomized trial of infliximab on rheumatoid arthritis was reported in 1994 (Elliot *et al.*, 1994). A single infusion of 1 or 10 mg/kg infliximab was compared with placebo in 73 patients. After 4 weeks, 8% of the placebo recipients fulfilled the response criteria compared with 44% and 79% of the low-dose and high-dose-treated patients. The efficacy of infliximab in Crohn's disease is based on a placebo-controlled dose-response study. In this trial, 108 patients with active disease received 0, 5, 10 or 20 mg/kg infliximab intravenously (Targan *et al.*, 1999). The combined response among all infliximab treatment groups was 65% which was significantly more than the 17% in the placebo group.

MAbs Against Cancers

Many therapeutic strategies have been explored that use MAb or their derivatives in the treatment of cancer, and some promising therapeutic possibilities have already emerged. Rituximab (Mabthera) is the first MAb approved for the treatment of cancer. It is a chimeric IgG_1 MAb directed against CD20, which is a transmembrane protein of pre-B and mature B lymphocytes. Multicentre studies have demonstrated its efficacy against relapsed low-grade and follicular non-Hodgkin lymphomas. *In vitro* experiments with rituximab showed its ability to medicate complement-dependent cellular cytotoxicity after binding to human Clq and to activate antibody-dependent cellular cytokines with humoral effector cells and to possess direct anti-proliferative and apoptotic activity in some CD20-positive cell lines. CD50 MAb (Campath-1H) has also been extensively evaluated for their capacity to lyse malignant lymphopoietic cells (Dyer, 1999). CD52 MAb provides effective therapy for chronic leukaemias of T cells or B cell origin that may be resistant to conventional chemotherapy.

Another strategy for anti-tumour therapy with MAb target growth factor receptors. Antibodies directed against epidermal growth factor receptors directly inhibit the growth of tumours bearing such receptors *in vitro* and *in vivo*. Trastuzumab (Herceptin), a humanized IgG_1 MAb targeting the extracellular domain of the HER2 receptor has been investigated in the treatment of breast cancer. *In vitro* and *in vivo* preclinical studies show that administration of trastuzumab alone or in combination with cytostatic drugs inhibits the growth of breast-tumour-derived cells.

Anti-platelet Therapy

Acute coronary syndromes and percutaneous coronary interventions share a common pathophysiological mechanism of intimal disruption and platelet aggregation. Glycoprotein IIb/IIIa receptor antagonists, which interrupt the final common pathway of platelet activation and aggregation, have clear benefit as acute therapy. Abciximab (Reo Pro) was the first antagonist to be evaluated in clinical studies. The EPIC trial examined whether an abciximab or placebo would reduce ischaemic complications in high-risk patients undergoing angioplasty (Topol *et al.*, 1994). At 30 days there was a 35% reduction of composite death, myocardial infarction and urgent revascularization compared with placebo (8.3 vs 12.8%).

MAbs IN INFECTIOUS DISEASES

A humanized MAb directed against respiratory syncytial virus (RSC) has been approved in the treatment of premature infants and infants with bronchopulmonary dysplasia. RSV replication was inhibited by palivizumab both *in vitro* and *ex vivo* in tracheal aspirates from infants receiving palivizumab 15 mg/kg (Malley *et al.* 1998). In a large multicentre trial in 1502 infants at high risk of RSV infection, palivizumab 15 mg/kg intravenously more than halved the frequency of RSV-attributable admissions to 4.8% compared with 10.6% in placebo recipients.

MAbs in Parasitic Infections

In the field of protozoan and helminthic parasites, MAbs have been used extensively to unravel the relationship between parasite and host in four main ways.

1. to probe for antigenic determinants;
2. to examine antigenic variability in parasite populations;
3. to detect expression of clonal DNA in various vectors and
4. to type parasites.

MAbs were used to characterize antigens of *Babesia bovis* and selected MAbs were employed to immunoaffinity-purify *B. bovis* antigen (Wright *et al.*, 1983). They further showed that active immunization with one of these purified antigens protected susceptible calves against challenge with virulent *B. bovis* infection. Protozoans studied include those of *Plasmodium, Trypanosomes, Leishmania, Theileria* and *Toxoplasma*. Helminths studied include nematodes (*Ascaris* sp., *Onchocerca* sp.), trematodes (*Schistosoma* sp., *Fasciola* sp.) and cestodes (*Taenia* sp. and *Echinococcus* sp.) Monoclonal antibodies were raised against the vaccine strain of *Anaplasma centrale*, which differentiated antigenic differences between cattle vaccinated with A. *centrale* and cattle naturally infected with A. *marginale* (Molloy *et al.*, 2001).

MAbs in Bacterial Infections

MAbs against bacteria have been used very successfully in classification for diagnostic purposes. The specificity of the MAbs has greatly reduced the cross-reactivity seen using polyclonal sera. Taxonomic studies involving the characterization of evolutionary relationship have been possible. MAbs have been used as structural probes and the recognition of specific antigens and parts of antigen that can elicit a protective response has been exploited in the preparation of vaccines. Rapid serodiagnostic methods for detection of antigens such as *Neisseria* or *Haemophilus* sp. have been developed.

Serotype-specific MAbs have been produced in leptospirosis. Surujballi and Elmgren (2000) produced MAbs suitable for use in a competitive ELISA for detection of specific antibodies to *Leptospira interrogans* serovar *pomona*. The epitopes recognized by the MAbs are of carbohydrate composition. For detection of *Brucella abortus* antibodies, MAbs have been raised against *B abortus* cell-surface antigen, which is employed with HRP for use in competitive ELISA. By using this technique, infected or immunized animals could be differentiated.

Monoclonal antibody was prepared against *E. coli* 0157 and ELISA specific for this pathogen was developed (Zhao and Liu, 2005). Spleen cells from BALB/c mice immunized with the somatic antigen of *E. coli* O157:H7 were fused with murine Sp2/0 myeloma cells. The hybridoma cell line specific for *E. coli* O157 was established after having been subcloned. Antisera specific for *E. coli* O157 was prepared by intravenous injection into New Zealand rabbits with a strain of *E. coli* O157:H7. The sandwich ELISA was developed with the polyclonal antibody as the capture antibody and the MAb 3A5 as the detection antibody. No cross-reactivity of the MAb was observed with strains of *Salmonella* spp., *Yersinia enterocolitica*, *Shigella dysenteriae*, etc. MAb 3A5 specific for *E. coli* O157 and O113:H21 can be produced by immunizing BALB/c mice with a strain of *E. coli* O157:H7. Then a sandwich ELISA can be developed with the polyclonal antibody as the capture antibody and the MAb 3A5 as the detection antibody. The method is proved to be a sensitive and specific technique to detect low number of *E. coli* O157 in food.

Monoclonal antibodies to purified K99 antigen of *Escherichia coli (E. coli)* were produced by the hybridoma technique, and a specific clone, NEK99-5.6.12, was selected for propagation in tissue culture (Varshney *et al.*, 2007). The antibodies, thus obtained, were affinity-purified, characterized and coated onto Giemsa-stained Cowan-I strain of *Staphylococcus aureus (S. aureus)*. The antibody-coated *S. aureus* were used in a co-agglutination test to detect K99+ *E. coli* isolated from faeces of diarrhoeic calves. The developed antibodies specifically detected purified K99 antigen in immunoblots, as well as K99+ *E. coli* in ELISA and co-agglutination tests. The co-agglutination test was specific and convenient for large-scale screening of K99+ *E. coli* isolates.

Swiecki *et al.* (2006) developed a variety of isotypes and specificities of MAbs that were able to distinguish *B. anthracis* spores from other *Bacillus* spores. The majority of MAbs were directed toward BclA, a major component of the exosporium, although other components were also distinguished. These MAbs did not react with vegetative forms. Some MAbs distinguished *B. anthracis* spores from spores of distantly related species in a highly specific manner, whereas others discriminated among strains that are the closest relatives of *B. anthracis*. These MAbs provide a rapid and reliable means of identifying *B. anthracis* spores, for probing the structure and function of the exosporium, and in the analysis of the life cycle of *B. anthracis*.

Suwinmonteerabutr *et al.* (2005) developed monoclonal antibodies and used in dot-blot ELISA for detection of *Leptospira* spp. in bovine urine samples. The results were compared with those of dark-field microscopy (DFM), microbial culture and PCR assay. All urine samples with positive results for DFM, microbial culture, PCR assay or >1 of these tests also had positive results when tested by use of the MAb-based dot-blot ELISA, except for 1 sample that had positive results only for the PCR assay. It was concluded that MAb-based dot-blot ELISA was suitable for detecting leptospires in urine samples of cattle.

LipL32, the major outer membrane protein of pathogenic *Leptospira*, in its recombinant form (rLipL32) was used to immunize BALB/c mice to develop murine MAbs (Fernandes *et al.*, 2007). Three MAbs against rLipL32 were produced, isotyped and evaluated for further use in diagnostic tests of leptospirosis using different approaches. Two MAbs were able to immunoprecipitate the native protein from live and motile leptospiral cells and adsorbed onto magnetic beads, captured intact bacteria from artificially contaminated human sera for detection by PCR amplification. Results from this study suggested that the MAbs produced could be useful for the development of diagnostic tests based on detection of LipL32 leptospiral antigen in biological fluids.

MAbs in Viral Infections

In virology, MAbs have been exploited in research and applied areas. Chemical studies (amino acid sequence data, X-ray crystallography) can be related to the binding of antibodies and the subsequent effect of that binding are monitored *in vitro* (tissue culture or *in vivo* (animal protection). This is leading to a better understanding in areas of vaccine formulations, particularly where novel peptide vaccines are being considered. The obvious use of MAbs is in the comparison of viruses from the field MAb panels and be prepared against distinct antigenic sites (epitomes) on viruses and used to examine the possession of these sites in immunoassays. This has an important role in understanding the epidemiology of disease (antigenic variation) and also in monitoring viruses throughout vaccine manufacture (quality control of antigenicity in rabies, polio and influenza vaccines). Because specific MAbs to rare and minor antigens can be produced they have the potential to distinguish closely related strains of viruses, which cannot be differentiated by conventional serological methods. For example,

MAb that distinguish CPV and feline panleukopenia virus have been recently produced.

MAbs have been extensively used in studying rinderpest and FMD viruses. FMD is caused by one of the seven serotypes. Within each serotype are many antigenic variations (subtypes) and the antigenic analysis of these subtypes is important since they may be sufficiently different to cause problems in animals vaccinated with antigenically dissimilar viruses. Presently, typing is best made with polyclonal sera and the use of MAbs is a disadvantage due to their specificity. However, in subtyping they offer tremendous advantages over polyclonal sera. The current methods used to subtype FMD viruses involve biochemical procedures such as PAGE, RNA T1 ribonuclease fingerprinting, isoelectric focusing of structural proteins and serological methods involving virus neutralizing test and complement fixation test. Such methods are time-consuming, that reliable results are not rapidly obtained.

Recently, MAbs against several FMDV serotypes have been prepared in European and American laboratories. These have been used as research tools to identify important neutralizing antibodies inducing epitopes and to evaluate the mechanisms by which virus is neutralized *in vitro* and *in vivo*. The panels of MAbs have also been used to antigenically "fingerprint" virus isolates.

Two foot-and-mouth disease virus (FMDV) monoclonal antibodies (MAbs) were produced from mice immunized with either FMDV serotype A, subunit (12S) or FMDV serotype O, whole virus (140S) (Yang *et al.*, 2007). Both MAbs (F1412SA and F21140SO) recognized all seven serotypes of FMDV in a double antibody sandwich (DAS) ELISA, suggesting that the binding epitopes of the two MAbs are conserved between serotypes. These MAbs are IgG1 isotype and contain kappa light chains. Both ELISA and Western blot results suggested that the polypeptide VP2 contributed to the immunodominant site. This MAb showed reactivity to VP2 peptide (DKKTEETTILEDRIL). The MAb, F21140SO, recognized an epitope which is trypsin-resistant and discontinuous. Because the use of MAbs increases the specificity, accuracy and efficiency of diagnostic tests compared to polyclonal antisera, these two MAbs with different specificities are suitable for type-independent diagnosis of FMDV, such as DAS ELISA, or could be adapted to immuno-chromatographic or flow-through rapid test.

MAbs produced in ascitic fluid can generate titres of over one million. With specific antibody available in such quantities, it is possible to envisage administering MAbs to parainfluenza-3 virus to beef cattle before transport to prevent outbreaks of shipping fever. On diagnosis of infectious bovine rhinotracheitis in 1 or 2 cows in a dairy herd, the rest of the animals in a dairy herd could be protected by an injection of the appropriate MAb until the danger of natural transmission of the virus has passed.

MAbs produced against 42-kDa major structural proteins of rotavirus could be used for preliminary serotyping of strains from different animal species (Greenberg *et al.*, 1983). MAbs against rabies virus have been employed for molecular characterization of street vaccine and laboratory strains (Smith *et al.*, 1984). The use of MAbs against bovine diarrhoea virus in immunofluroscent antibody technique has aided preliminary serological characterization of strains (Peter *et. al.*, 1986) and both indirect and competitive ELISA techniques utilizing MAbs to detect antibodies against the virus in bovine sera.

MAbs against bluetongue virus have been used for analysis of serotype and group-specific antibodies in cattle and sheep by a MAb-blocking (competitive) ELISA (Anderson, 1984) and for demonstration of viral antigens in infected tissues. Bovine leukaemia virus infection of cattle is another area where MAb-based ELISA system has enabled detection of virus-specific serum antibodies.

Singh *et al.* (2004) developed a monoclonal antibody-based competitive ELISA for detection and titration of antibodies to PPR virus. They used MAb to a neutralizing epitope of haemagglutinin protein of the virus. A total of 1668 sera samples from goats and sheep and 32 sera samples from cattle were screened by c-ELISA and virus neutralization test (VNT). Efficacy of cELISA compared very well with VNT having high relative specificity (98.4%) and sensitivity (92.4%). Results from this study indicated that cELISA developed could easily replace VNT for sero-surveillance, sero-monitoring, diagnosis from paired sera samples and end-point titration of PPR virus antibodies.

A competitive ELISA using two MAbs was developed and compared with the standard virus neutralization test (VNT) for detecting antibodies against canine distemper virus (CDV) and phocine distemper virus (PDV) in sera from dogs and various species of marine mammals (Saiki and Lehenbauer, 2000). The authors suggested that as cELISA proved to be nearly as sensitive and specific as the VNT while being simpler and more rapid, it would be an adequate screening test for suspect CDV or PDV cases. Moving *et al.* (1998) produced MAb reacted with the spike protein of seven field and laboratory strains of infectious bronchitis virus and used in blocking ELISA. They found this method to be more sensitive and specific than the commercially available indirect ELISA and haemagglutination inhibition test.

Monoclonal antibody-based capture ELISA was developed for the diagnosis of rabies-suspect specimens (Xu *et al.*, 2007). A combination of four MAbs directed against the rabies virus nucleocapsid was selected and used for the detection. The threshold of detection of Lyssavirus nucleocapsids is low (0.8 ng/ml). With a panel of 1030 specimens received for rabies diagnostic testing, this test was found to be highly specific (0.999) and sensitive (0.970) when compared to other recommended rabies diagnostic methods.

Canine parvovirus (CPV) is classified as a member of the feline parvovirus (FPV) subgroup. CPV isolates are divided into three antigenic types: CPV type 2 (CPV-2), CPV-2a, and CPV-2b. Recently, new antigenic types of CPV were isolated from Vietnamese leopard cats and designated CPV-2c(a) or CPV-2c(b). CPV-2c viruses were distinguished from the other antigenic types of the FPV subgroup by the absence of reactivity with several monoclonal antibodies (MAbs). To characterize the antigenicity of CPV-2c, a panel of MAbs against CPV-2c was generated and epitopes recognized by these MAbs were examined by selection of escape mutants (Nakamura *et al.*, 2003). Four MAbs were established and classified into three groups on the basis of their reactivities: MAbs which recognize CPV-2a, CPV-2b, and CPV-2c (MAbs 2G5 and 2G4); a MAb which reacts with only CPV-2b and CPV-2c(b) (MAb 21C3); and a MAb which recognizes all types of the FPV subgroup viruses (MAb 19D7). The reactivity of MAb 20G4 with CPV-2c was higher than its reactivities with CPV-2a and CPV-2b. These MAbs are expected to be useful for the detection and classification of FPV subgroup isolates.

The sigma C protein of avian reovirus (ARV) was involved in induction of apoptosis and neutralization antibody. This sigma C-His protein was expressed in Sf9 insect cells and purified by immobilized metal affinity chromatography (Hsu *et al.*, 2006). Eight monoclonal antibodies (MAbs) against sigma C-His and three MAbs against His were screened from hybridoma cells produced by fusion of splenocytes from immunized mice with NS1 myeloma cells. Among the eight MAbs against sigma C protein, all belonged to the IgG isotype except three for IgM. Compared with the commercial anti-ARV S1133 polyclonal antibody, MAb (D15) had universal reactivity to all serotypes or genotypes of ARVs tested. This monoclonal antibody may therefore be useful for the development of an antigen-capture enzyme-linked immunosorbent assay for rapid detection of field isolates.

Six clones of monoclonal antibodies (MAbs) to fowl adenovirus (FAV) serotype 1 were produced (Taharaguchi *et al.*, 2006). All MAbs reacted positively by enzyme-linked immunosorbent assay. Three MAbs recognized the putative 100-kDa hexon protein and reacted to serotype 1 specifically by western blot analysis but did not react to other FAV serotypes (2, 3, 4, 5, 6, 7, and 8a). These MAbs would be useful for immunodiagnosis of FAV serotype 1 infection in chicken with gizzard erosion and in further research studies involving the genomes and proteins of FAV serotype 1.

Monoclonal antibodies (MAbs) against the recombinant (poly100)S1 proteins of infectious bronchitis virus (IBV) strains were produced, characterized and used to analyse epitopes on the S1 subunit of IBV (Hu *et al.*, 2007). Immunocytochemistry indicated that all the MAbs to the (poly100) S1 proteins could react with the homologous S1 glycoprotein expressed in Vero cells. Moreover neutralization test demonstrated that only MAbs 6E2, 4F9 and 6G4 had neutralization activity for the homologous IBV. These MAbs to (poly100)S1 protein were potential candidates for detecting and distinguishing IBV strains, and also used to examine antigenic variation of the S1 protein.

A rapid diagnostic strip test for detection of chicken infectious bursal disease (IBD) was developed based on membrane chromatography using high affinity MAbs directed to chicken IBDV (Zhang *et al.*, 2005). The diagnostic strip has high specificity for detection of chicken IBDV antigen and recognizes a variety of the virus isolates, including virulent and attenuated strains, with no cross-reactivity to other viruses, such as Newcastle disease virus, Marek's disease virus, infectious bronchitis virus, infectious laryngotracheitis virus and egg drop syndrome virus. The results showed that its specificity was highly consistent with the agar gel precipitation test (AGP). In experimental infection, IBDV was detected in the bursa as early as 36 hour post-infection with the diagnostic strip before the clinical signs and gross lesions appeared. It takes only 1–2 minutes to do a strip test to detect chicken IBDV antigen.

REVIEW QUESTIONS

1. What are monoclonal antibodies and how do they differ from polyclonal antibodies? Enumerate their advangates and disadvantages.

2. How will you produce and characterize monoclonal antibodies against a pathogen?

3. What are the different uses of MAbs?

4. Write short notes on therapeutic monoclonal antibodies and anti-idiotypic monoclonal antibodies.

REFERENCES

Anderson, J. (1984). "Use of monoclonal antibody in a blocking ELISA to detect group specific antibodies to bluetongue virus." *J. Immunol. Meth.* 74:139–149.

Appleton, J.A. and Letchworth, G.J. (1983). "Monoclonal antibody analysis of serotype-restricted and unrestricted bluetongue viral antigenic determinants." *Virology.* 124: 286–299.

Arauio, F.G., Sharma, S.D., Tsai, V., Cox, P. and Remington, J.S. (1982). "Monoclonal antibodies to stages of *Trypanasoma cruzi*: Characterization and use for antigen detection." *Infect. Immun.* 37: 344–347.

Arulanandam, A.M. Sevoian and Holdsby, R.A. (1985). "A library of antibovine isotype monoclonal antibodies." In 66th Annual Meeting of the Conference of Research Workers in Animal Diseases, Chicago, Illinois.

Crouch, C.F., Raybould, T.J.G. and Aeres, S.D. (1984). "MAb capture ELISA for detection of bovine enteric coronavirus." *J. Clin. Microbiol.* 17: 388–393.

Davis, W.C., Perrman, L.E. and McGuire, T.C. (1984). "The identification and analysis of major functional population of differentiated cells." In: *Hybridoma Technology in Agriculture and Veterinary Research.* (eds.). Stern, N.J. and Gamble, H.R. Rowman & Allenheld, New Jersey. pp 121–150.

Dietzchold, B., Wunner, W.H., Wikot, T.J., Lopes, A.D., Lafon, M., Smith, C.L. and Koprowski, H. (1983). "Characterization of an antigenic determinant of the glycoprotein that correlates with pathogenicity of rabies virus." *Proc. Natl. Acad. Sci. USA.* 80: 70–74.

Dyer, M.J.S. (1999). "The role of CAMPATH-1 antibodies in the treatment of lymphoid malignancies." *Semin. Oncol.* 26: 52–47.

Elliot, M.J., Maini, R.N., Feldmann, M., Kalden, J.R., Antoni, C., Smolen, J.S., Leeb, B., Breedveld, F.C., Macfarlane, J.D. and Biji, H. (1994). "Randomised double-blind comparison of chimeric monoclonal antibody to tumor necrosis factor α (cA2) versus placebo rheumatoid arthritis." *Lancet.* 344: 1105–1140.

Fernandes, C.P., Seixas, F.K., Coutinho, M.L., Vasconcellos, F.A., Seyffert, N., Croda, J., McBride, A.J., Ko, A.I., Dellagostatin, O.A. and Aleixo, J.A. (2007). "Monoclonal antibodies against LipL32, the major outer membrane protein of pathogenic *Leptospira* production, characterization and testing in diagnostic applications." *Hybridoma.* 26: 35–41.

Fuller, S.A. and Hurrell, J.G.R. (1985). "Canine IgG subtypes defined by MAbs." Abst. 5369. 69th Annual Fed. Amer. Soc. Expl. Biol. Med. Meeting, Anaheim, California, FASEB, Baltimore, Maryland.

Gamble, H.R. and Graham, C.E. (1983). "A MAb purifed antigen for the immunodiagnosis of trichinosis." *Am. J. Vet. Res.* 45: 1143–1146.

Greenberg, H., McAvlittle, V., Valdeuso, J., Wyatt, R., Floves, J., Kalica, A, Hoshino, Y. and Singh, N. (1983). "Serological analysis of the subgroup protein of rotavirus using monoclonal antibodies." *Infect. Immun.* 39: 91–99.

Gustafsson, B., Rosen, A. and Holme, T. (1982). "MAbs against *Vibrio cholerae* lipopolysaccharide." *Infect. Immun.* 38: 449–454.

Hanlon,C.A., Demattos, C.A., DeMattos, C.C., Niezgoda, M., Hooper, D.C., Koprowski, H., Notkins, A. and Rypprecht, C.E. (2001). "Experimental utility of rabies virus neutralizing human monoclonal antibodies and post exposure prophylaxis." *Vaccine.* 19: 3834–3842.

He,Q., Velumani, S., Du,Q., Lim, C.W., Ng, F.K., Donis, R. and Kwang, J. (2007). "Detection of H5 avian influenza viruses by antigen capture ELISA using H5-specific monoclonal antibody." *Clin. Vaccine Immunol.* 14:617–623.

Holman, P.J., Adams, L.G., Hunter, D.M., Heck, F.C. and Nielsen, K.H.(1983). "Derivatives of MAbs against *Brucella abortus* antigens." *Vet. Immunol. Immunopath.* 4: 602–614.

Hsu, C.J., Wang, C.Y., Lee, L.H., Shih, W.L., Chang, C.I., Cheng, H.L., Chulu, J.L., Ji, W.T. Liu, H.J. (2006). "Development and characterization of monoclonal antibodies against avian reovirus sigma C protein and their application in detection of avian reovirus isolates." *Avian Pathol.* 35: 320–326.

Hu, J.Q., Li, Y.F., Guo, J.Q., Shen, H.G., Zhang, D.Y. and Zhou, J.Y. (2007). "Production and characterization of monoclonal antibodies to (poly100)S1 protein of avian infectious bronchitis virus." *Zoonoses Public Health.* 54: 69–77.

Jochen, R.F., Briddell, J.D. and Mia, A.S. (1983). "A simplified ELISA technique to detect FeLV using MAbs." *Mod. Vet. Pract.* 64: 827–829.

Kohler, G. and Milstein, C. (1975). "Continuous cultures of fused cells secreting antibody of predefined specificity." *Nature.* 256: 495–497.

Kolk, A.H.J., Ho, M.L., Klaster, P.R., Eggelib, T.A., Kujjper, S.,de Jonge and van Leeuwen, J. (1984). "Production and characterization of MAbs to *Mycobacterium tuberculosis, M. bovis* (BCG) and *M. leprae*." *Clin. Expl. Immunol.* 53: 511–521.

Koprowsik, H. and T. Wiktor, T. (1980). "Monoclonal antibodies against rabies virus." In: *Monoclonal Antibodies.* (eds.). Kemet, R.H., Mckearn, I.J. and Bechtol, K.B. Plenum Press N.Y. pp. 335–351.

Krawiec, D.R. and Muscoplat, C.C. (1984). "Development and characterization of a hybridoma derived antibody with specificity to canine thymocytes and peripheral T lymphocytes." *Am. J. Vet. Res.* 45: 491–498.

Krivi, G.G. and Rowlad, E. Jr. (1984). "MAbs to bovine somatotropin: preparation and characterization." In: *Hybridoma Technology in Agriculture and Veterinary Research.* (eds). Stern, N.J. and Gamble, H.R. Rowman and Allenheld. N.J. pp. 121–150.

Lee, L.F., Lie, X. and R.L. Wither (1983). "MAbs with specificity for three different serotypes of Marek's disease virus in chicken." *J. Immunol.* 130: 1003–1006.

Letchworth, G.H. and Appleton, J.A. (1983). "Heterogenicity of neutralization related epitopes within a bluetongue virus serotype." *Virology.* 124: 300–307.

Lunney, J.K. (1984). "Monoclonal antibodies in the analysis of immune response in swine." In: *Hybridoma Technology in Agriculture & Veterinary Research* (eds.). Stern, N.J. and Gamble, H.R., Rowman and Allenheld. Totowa, N.J. pp. 298–301.

Malley, R., De Vincenzo, J., Ramilo, O., Dennehy, P., Meissner, H.C., Gruber, W.C., Sanchez, P.J., Jefri, H., Balsey, J., Carlin, D., Buckingham, S., Vemacchio, L. and Ambrosino, D.M. (1998). "Reduction of respiratory syncytial virus (RSV) in tracheal aspirates in intubated infants by use of humanized monoclonal antibody of RSVF protein." *J. Infect. Dis.* 178: 1555–1561.

McCulough, V.C. and Butcher, R. (1982). "Monoclonal antibodies against foot and mouth disease virus 146S and 12S particles." *Arch. Virol.* 74: 1–9.

Meloen, R.H., Briaire, J., Woortmeyer, R.J. and Van Zaane, D. (1983). "The main antigenic determinant detected by neutralizing MAbs on the intact FMD virus particle is absent from isolated VP1." *J. Gen. Virol.* 64: 1193–1198.

Milstein, C. and Cuello, A.C. (1984). "Hybrid hybridoma and the production of bi-specific MAbs." *Immunol. Today.* 5: 299–304.

Morris, J.A and Ivanyi, J. (1985). "Immunoassay of field isolates and other mycobacteria by use of MAbs." *J. Med. I. Microbiol.* 19: 367–373.

Molloy, J.B., Bock, R.E., Templeton, J.M., Bruyeres, A.G., Bowles, P.M., Blight, G.W. and Jorgensen, W.K. (2001). "Identification of antigenic differences that discriminate between cattle vaccinated with *Anaplasma centrale* and cattle naturally infected with *Anaplasma marginale*." *Int. J. Parasitol.* 31: 179–186.

Moving, V., Czifra, G. and Renstrom, L. (1998). "A monoclonal antibody blocking ELISA for the detection of IBV antibodies in fowl." *Adv. Exp. Med. Biol.* 440: 495–499.

Nakamura, M., Nakamura, K., Miyazawa, T., Tohya, Y., Mochizuki, M. and Akashi, H.(2003). "Monoclonal antibodies that distinguish antigenic variants of canine parvovirus." *Clin. Diagn. Lab. Immunol.* 10: 1085–1089.

Nielsen, K.H., Henning, M.D. and Dunkan, J.R. Unpublished work.

Newman, M.J., Beegle, K.H. and Antezak, D.F. (1984). "Xenogeneic MAbs to cell surface antigens of equine lymphocytes." *Am. J. Vet. Res.* 45: 626–634.

Ortho Multicentre Transplant Study Group. (1985). "A randomized clinical trial of OKT3 monoclonal antibody for acute rejection of cadaveric renal transplants." *N. Engl. J. Med.* 313: 337–342.

Parish, C.R. and Carmichael, L.E. (1983). "Antigenic structure and variation of canine parvovirus type 2, feline panleucopenia virus and mink enteritis virus." *Virology.* 129: 401.

Paul, P.S., Van Deusen and Mengeling, W.L. (1985). "Monoclonal precipitating antibodies to porcine IgM." *Vet. Immunol. Immunopath.* 8: 311–328.

Peters, W., Greiser-Wike, I., Moenning, V. and Liess, B. (1986). "Preliminary serological characterization of bovine viral diarrhoea virus strains using monoclonal antibodies." *Vet. Microbiol.* 12: 195–200.

Pinder, M., Pearson, T.W. and Roelants, G.E. (1980). "The bovine lymphoid system II: derivation and partial characterization of MAbs against bovine peripheral blood lymphocytes." *Vet. Immunol. Immunopath.* 1: 303–306.

Portetelle, D.C. Bruck, Mammerick, M. and Burny, A. (1983). "Use of monoclonal antibody in an ELISA test for detection to bovine leukemia virus." *J. Virol. Meth.* 6: 19–29.

Reagon, K.J., Winter, W.H., Wiktor, J.J. and Koprowski, H. (1983). "Anti-idiotypic antibodies induce neutralizing antibodies to rabies virus glycoprotein." *J. Virol.* 48: 660–666.

Russell, P.H. and Alexander, D.J. (1983). "Antigenic variation of Newcastle disease virus strains by monoclonal antibodies." *Arch. Virol.* 75: 243–253.

Saliki, J.T. and Lehenbauer, T.W. (2001). "Monoclonal antibody-based competitive ELISA for detection of morbillivirus antibody in marine mammal sera." *J. Clin. Microbiol.* 39: 1877–1881.

Schonherr, O.T., Roelofs, H.W.M. and Hoziwink, E.H. (1984). "Development, screening and quality specifications of hybridomas synthesizing anti-polypeptide hormone antibodies to be used for diagnostic test and in process control." *Dev. Biol. Stand.* 55: 163–171.

Scott, A.L. (1983). "MAbs against somatic antigens of *Dirofilaria immitis*." *Fed. Proc.* 42: 1089.

Singh, R.P., Sreenivasa, B.P., Dhar, P., Shah, L.C. and Bandyaopadhyay, S.K. (2004). "Development of a monoclonal antibody based competitive-ELISA for detection and titration of antibodies to PPR virus." *Vet. Microbiol.* 98: 3–15.

Sloan, S.E., Hanlon, C., Weldon, W., Niezgoda, M., Blanton, J., Self, J., Rowley, K.J., Mandell, R.B., Babcock, G.J., Thomas, W.D.Jr., Rupprecht, C.E. and Ambrosino, D.M. (2007). "Identification and characterization of a human monoclonal antibody that potently neutralizes a broad panel of rabies virus isolates." *Vaccine.* 25: 2800–2810.

Sonza, S., Breschkin, A.M. and Holmes, I.H. (1983). "Derivation of neutralizing MAbs against rotavirus." *J. Virol.* 45: 1143–1146.

Smith, J.S., Summer, J.W. and Roumillat, L.F. (1984). "Enzyme immunoassay for rabies antibody in hybridoma culture fluids and its application in differentiation of street and laboratory strains of rabies virus." *J. Clin. Microbiol.* 19: 267–272.

Surujballi, O. and Elmgren, C. (2000). "Monoclonal antibodies suitable for incorporation into a competitive ELISA for detection of specific antibodies to *Leptospira interrogans* serovar *pomona*." *Vet. Microbiol.* 71: 149–159.

Suwimonteerabutr, J., Chaicumpa, W., Saengjaruk, P., Tapchaisri, P., Chongsanguan, M., Kalambaheti, T., Ramasoota, P., Sakolvaree, Y. and Virakul, P. (2005). "Evaluation of a monoclonal antibody-based dot-blot ELISA for detection of *Leptospira* spp. in bovine urine samples." *Am. J. Vet. Res.* 66: 762–766.

Swiecki, M.K., Lisanby, M.W., Shu, F., Turnbough, C.L. Jr. and Kearney, J.F. (2006). "Monoclonal antibodies for *Bacillus anthracis* spore detection and functional analyses of spore germination and outgrowth." *J. Immunol.* 176: 6076–6084.

Taharaguchi, S., Ito, H., Ohta, H. and Takase, K. (2006). "Characterization of monoclonal antibodies against fowl adenovirus serotype 1 (FAV1) isolated from gizzard erosion." *Avian Dis.* 50: 331–335.

Targen, S.R., Hanauer, S.B., Deventer, S.J.H., Mayer, L., Present, D.H., Braakman, T., DeWoody, K.L., Scaible, T.S. and Rutgeers, P.J. (1997). "A short term study of chimeric monoclonal antibody cA2 to tumour factor α for Crohn's disease." *N. Engl. J. Med.* 337: 1029–1035.

Topol, E.J., Califf, R.M., Weisman, H.F., Ellis, S.G., Tcheng, J.E., Worley, S., Ivanhoe, R., George, B.S., Fintel, D. and Weston, M. (1994). "Randomised trial of coronary intervention with antibody against platelet IIb/IIIa integrin for reduction of clinical responses: results at 6 months". *Lancet.* 343: 881–886.

Van Zaane, D. (1983). "Use of MAbs in virus diagnosis." In: *Recent Advances in Virus Diagnosis.* (eds.). Mc Nulty, M.S. and McFerran, J.B. Martinus Nijhoff Publishers, Boston. pp. 145–156.

Van Zaane, D. and Ijzerman, J. (1984). "MAbs against bovine immunoglobulins and their use in isotype-specific ELISAs for rotavirus antibody." *J. Immunol. Meth.* 72: 427–441.

Varshney, B.C., Ponnanna, N.M., Sarkar, P.A., Rehman, P. and Shah, J.H. (2007). "Development of a monoclonal antibody-based co-agglutination test to detect enterotoxigenic *Escherichia coli* isolated from diarrheic neonatal calves." *J. Vet. Sci.* 8: 57–64.

Webster, W.A., Casey, G.A. and Charlton, K.M. (1986). "Major genetic groups of rabies virus in Canada determined by anti-nucleocapsid monoclonal antibodies." *Comp. Immunol. Microbiol. Infect. Dis.* 9: 59–69.

Wright, I.G., White, M., Tracey-Patte, P.D., Donaldson, R.A., Goodger, B.V., Waltisbuhl, D.J. and Mahoney, D.F. (1983). "*Babesia bovis*: isolation of protective antigen by using monoclonal antibodies." *Infect. Immunol.* 41: 244–250.

Wu, H., Pfarr, D.S., Johnson, S., Brewah, Y.A., Woods, R.M., Patel, N.K., White, W.I., Young, J.F., Kiener, P.A. (2007). "Development of motavizumab, an ultra-potent antibody for the prevention of respiratory syncytial virus infection in the upper and lower respiratory tract." *J. Mol. Biol.* 368: 652–665.

Xu, G., Weber, P. Hu, Q., Xue, H., Audry, L., Li, C., Wd, J. and H. Bourhy, H. (2007). "A simple sandwich ELISA for the detection of Lyssavirus nucleocapsid in rabies suspected specimens using mouse monoclonal antibodies." *Biologicals.*

Yang, M., Clavijo, A., Suarez-Banmann, R., Avalo, R. (2007). "Production and characterization of two serotype independent monoclonal antibodies against foot-and-mouth disease virus." *Vet. Immunol. Immunopathol.* 115: 126–134.

Young, D.B., Fohn, M.J., Kharolkar, S.R. and Buchannan, T.M. (1985). "MAb to a 28,000 mol. wt. protein antigen of *Mycobacterium leprae*." *Clin. Exptl. Immunol.* 60: 46–552.

Zhao, Z.J. and Liu, X.M. (2005). "Preparation of monoclonal antibody and development of enzyme-linked immunosorbent assay specific for *Escherichia coli* O:157 in foods." *Biomed. Environ. Sci.* 18: 254–259.

Zhang, G.P., Li, Q.M., Yang, Y.Y., Guo, J.Q., Li, X.W., Deng, R.G., Xiao, Z.J., Zing, G.X., Yang, J.F., Zhao, D., Cai, S.J. and Zang, W.M. (2005). "Development of a one-step strip test to test the diagnosis of chicken infectious bursal disease." *Avian Dis.* 49: 177–181.

5

NUCLEIC ACID—PROBES AND HYBRIDIZATION

INTRODUCTION

Advances in recombinant DNA technology have directly led to the development of widespread application of molecular hybridization techniques in diagnostic and molecular biology laboratories. A wide variety of microbial infections as well as genetic disorders are now diagnosed using hybridization techniques (Haase *et al.*, 1984; Mulchay, 1986, Mc Dougall *et al.*, 1986; Caskay, 1987). In molecular hybridization, a specific nucleic acid probe is used to hybridize to complementary nucleic acid of a specific pathogen species directly in specimens or to analytes immobilized on solid phases. For genotype analysis, cloned genomic sequences are used to detect restriction fragment length polymorphisms in Southern blot hybridization protocols. Probes are labelled with either a radionucleotide or biotin in order to permit detection of specific hybrids.

Hybridization is particularly useful in the diagnosis of microbial infections in which the pathogen cannot be identified using conventional or classical diagnostic procedures. Some pathogens cannot be propagated in laboratory bioassays. Some viral pathogens may be present in latent states or integrated into host genomes. In such circumstances, transcriptional and translational processes are typically minimal and infectious agents may not be recoverable.

Two techniques are now widely used to diagnose infections by directly identifying the pathogen in clinical specimens: antigen detection using immunological techniques and nucleic acid detection using hybridization. Although both methods have proven to be sensitive and reliable for providing diagnostic results, molecular hybridization does permit diagnosis of certain infections and conditions where antigen detection cannot. For example, integrated or extrachromosomal nucleic acid of pathogens can be detected by hybridization, permitting diagnosis in the absence of gene expression.

NUCLEIC ACID PROBES

The nucleic acid probe technology is increasingly being used in basic research in veterinary microbiology and in the diagnosis of infectious diseases of veterinary importance. Advances in molecular cloning techniques that made possible the preparation of gene-specific probes led to the use of hybridization techniques for the detection of bacterial and viral pathogens in clinical samples. Nucleic acid hybridization can provide a confirmation of diagnosis before the results of culture and biochemical tests are available. One key advantage of hybridization techniques is detection sensitivity of fraction of a picogram (one millionth of a microgram) and possibly femtogram of viral or bacterial nucleic acids without the loss of specificity. No other viral or bacterial detection method has this degree of sensitivity. Probe technology has helped in the detection of virtually any organisms in food, clinical samples or environment. It is now possible to detect a wide variety of viral, bacterial and protozoan pathogens in urine, stools, tissues and sputum using radiolabelled or non-radiolabelled assay systems.

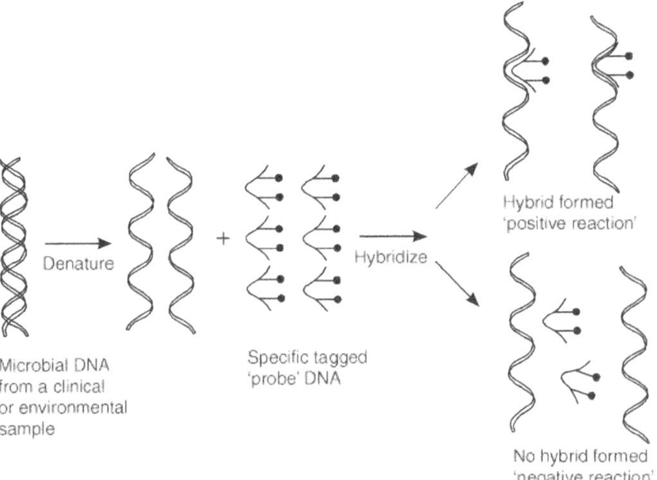

Figure 5.1 Steps involved in hybridization methods

The key to the development of nucleic acid probe is to identify nucleotide sequences which are unique to a particular organism of interest. The nucleic acid which contains such sequences is isolated, cleaned and labelled with a reporter molecule such as radioactive ^{32}P or ^{35}S, or non-radioactive biotin or digoxigenin molecules. The labelled DNA or RNA in single-stranded form is then hybridized to single-stranded nucleic acid (DNA and RNA) in tissues (*in situ* hybridization), in paper (Southern blot) or in solution. If the nucleotide sequences in the nucleic acid probes are complementary to those in the sample, hybridization occurs and results in the formation of double-stranded nucleic acid. Non-hybridized single-stranded probe is removed. Hybridization is monitored by autoradiography in the case of probes labelled with radioactive materials and colorimetrically or visually for probes with non-radioactive materials (Figure 5.1).

The most commonly used tag for the probe is phosphorus 32 (^{32}P) incorporated into the probe by enzymatic polymerization of nucleoside phosphates. This is an excellent tag for most research applications where short half-life and the hazards of radioactive materials are not the major concerns. ^{32}P has the highest specific activity and emits β-particles of high energy. This gives maximum sensitivity in the shortest time in autoradiography. Limitations are

♠ Short half-life of ^{32}P-labelled probe

♠ Cost and inconvenience of radioactivity disposals

♠ The hazards of working with the isotope. At present, there is no labelling system which has proven more sensitive than the use of ^{32}P. Different probes which are used in hybridization studies are given in Table 5.1.

Table 5.1 Commonly used radioisotopes with their half-life

Isotopes	Type of emission	Half-life	Characters and Uses
^{14}C	β particles	5568 years	Low sensitivity
^{3}H	β	12.6 years	Low sensitivity High resolution *In situ* hybridization
^{35}S	β	87.4 days	High sensitivity Good resolution filter *In situ* hybridization
^{125}I	γ and β	60 days	High sensitivity High resolution *In situ* hybridization
^{131}I	γ and β	8.6 days	High radiation Rarely used
^{32}P	β	14.2 days	High sensitivity Filter hybridization
^{33}P	β	24.5 days	Filter hybridization

Recently a new radioisotope, ^{33}P has been developed. This has a half-life of 24.5 days and emits β radiation with E_{max} of 0.249. The ^{33}P with its radiation characteristics in between those of ^{32}P and ^{35}S, strikes a reasonable balance between the sensitivity of ^{32}P and the resolution of ^{35}S. The maximum energy of ^{33}P is 1.5 times that of ^{35}S, but is 5 times weaker than ^{32}P. The specific activity of ^{33}P is in the same range as that of ^{32}P, but at the same time the resolution is on par with ^{35}S. ^{33}P is less sticky compared to ^{35}S and hence reduces the non-specific background in *in situ* hybridization. The relatively longer half-life of ^{33}P as compared to ^{32}P leads to increased shelf life of the labelled probe and hence they can be reused over longer periods of time.

Recently, non-isotopic reporter systems have been used for detection of microbes using biotin or digoxigenin probes. These techniques retain the exquisite sensitivity and specificity of nucleic acid hybridization without the drawbacks associated with radioisotopes. After hybridization, the reaction is detected immunologically by enzyme immunoassay or by immunofluorescence. Advantages of the non-isotopic detection system include the following.

- Biotinylated probes are stable for a year or more at 4°C.

- Picogram levels of nucleic acid can be detected.

- Hybridization can be detected rapidly within a matter of hours.

- Radioisotopes are not used, precluding problems with licensing, hazard and disposal problems.

- Hybridization tests are readily conducted in clinical or field situations.

Probe Development

For microbial diagnosis, three major types of probe constructs are commonly used. First, genomic sequences or complementary DNA (cDNA) of RNA species of the pathogen may be cloned into a vector. Second, the pathogen itself can be propagated, its nucleic acid extracted and labelled for use as the probe. Third, if the sequence of the analyte is known, a complementary probe can be synthesized. Several steps are involved in the production of a cloned probe:

1. Isolation of the desired nucleic acid sequences
2. Cloning the sequences into an appropriate vector system
3. Amplification of the sequences in a host vector system
4. Recovery and labelling of the sequences
5. Demonstration of the efficacy of the people in hybridization procedures.

The cloning of nucleic acid sequences for use as probes in hybridization has been made possible through the use of recombinant DNA technology. Restriction enzymes provide the mechanisms to cleave the DNA of pathogens to produce specific fragments that can be cloned into restriction enzyme sites of the chosen vector system. The DNA fragments are usually joined to cloning vector DNA molecules by the enzyme, DNA ligase, which is capable of joining DNA molecules with compatible ends. These DNA constructs are then introduced into bacteria where they replicate and can be easily amplified and purified.

Choice of probe   In order to interpret a hybridization experiment, hybrids must be distinguishable from nucleic acids that remain single-stranded. This is accomplished by labelling one of the reacting nucleic acids, the probe, with a reporter molecule. After hybridization, the probe that has not reacted is washed away and the hybrids are recognized by virtue of the reporter molecule. In principle, any nucleic acid—double-stranded DNA, single-stranded DNA, oligonucleotides, mRNA and RNA—can act as a probe. The choice is determined by the purpose of the experiment and by the materials available.

Characteristics of probes

DNA probes   Double-stranded DNA probes are very commonly used and consist of double-stranded DNAs that have been denatured and added quickly to the hybridization mix before they have time to reassociate. Single-stranded DNA probes are usually derived by reverse transcription of RNA. If the template is mRNA, the cDNA will consist of many different sequences and the complexity of the DNA will be high.

RNA probes   Single-stranded probes may be prepared by labelling isolated RNA, but it is more common to transcribe cloned sequences from recombinant plasmids in which the cloned sequences are flanked by promoters for bacteriophage RNA polymerases. The advantages of RNA probes are that they are relatively easy to prepare and RNA : DNA and RNA : RNA hybrids are more stable than the corresponding DNA : DNA hybrids. However their main disadvantage is that they are more difficult to handle than DNA probes because of the widespread presence of ribonucleases.

Oligonucleotide probes   Oligonucleotide probes are short (typically 17–50 nt in length) and are usually chemically synthesized. In theory, they may be oligodeoxyribonucleotides (more commonly used), oligoribonucleotides or molecules with a modified backbone such as peptide nucleic acids. One of the most useful properties of oligonucleotide probes is that they can discriminate between target sequences that differ by as little as a single nucleotide. The main disadvantage in their use is that hybrids containing oligonucleotides are much less stable than hybrids containing longer nucleic acids.

Labelling of probes   Double-stranded DNA probes are most typically labelled by nick translation, during which DNase I is used to nick the DNA strand.

Nicks are subsequently labelled by enzymatic incorporation of labelled deoxyribonucleotide triphosphates using DNA polymerase I. Biotinylated probes are also labelled by nick translation. DNA probes can also be labelled with either a radioisotope or a biotin by primer extension. To label a probe DNA cloned into M13, a commercially obtained primer adjacent to the cloning site is annealed to the M13 clone. After annealing, the appropriate cold dNTPs, α-^{32}P-dCTP or bio-11-dUTP and DNA polymerase I are added, and labelled ssDNA-probes are synthesized.

Radioactive probes   Traditionally, probes for filter hybridization have been labelled with radioactive isotopes and this procedure is still widely used. The most commonly used isotope for filter hybridization is ^{32}P as it is sensitive and gives reasonably good resolution. This means that it can be used to detect very small amounts of nucleic acid and it can resolve hybrids that are close to each other on the filter. Detection of hybrids is both easy and sensitive; single copy sequences can be detected in 1 μg DNA. Radioactive probes can be used with all types of filter and can be removed fairly efficiently to allow filters to be reprobed.

There are several disadvantages in using radioactive probes. Precautions have to be taken to prevent radiation being a health hazard. ^{32}P has a short half-life (14.3 days) which means that probes must be prepared frequently. They are expensive to prepare when taking into account the costs of the following isotope itself

- photographic film for detection
- licenses to use radioactivity
- disposal of waste material
- hand-held mini monitors
- badges to monitor emission of radioactivity and
- safety equipment such as Perspex/Lucite shields.

Labelling of DNA by random priming   The most common method for uniform labelling of DNA is random priming. DNA is denatured and incubated with short primers or random sequence which anneal to the DNA. The primers are extended by the Klenow fragment of DNA polymerase I in the presence of labelled dNTPs to create labelled products.

Labelling of DNA by nick translation   DNA can be uniformly labelled by nick translation. This procedure depends on the concerted action of two enzymes, DNase I and *E. coli* DNA polymerase I. DNase I introduces single-stranded nicks into a double-stranded DNA template at a frequency that depends on the amount of enzyme present. Starting at a nick, the 5´ to 3´ exonuclease activity of *E. coli* DNA polymerase I excises nucleotides from the newly created 5´ end. At the same time the 5´ to 3´ polymerase activity adds template-specific nucleotides at the 3´ end. The net result is that the nick is moved (translated) in the 5´ to 3´ direction as new nucleotides are added in the same direction. If one of the added nucleoside triphosphates is labelled, a uniformly labelled product is generated. The proportion of bases replaced by DNA polymerase I and hence the specific activity of the product depends on the number of nicks introduced into the DNA.

The advantage of random priming is that it tends to give higher specific activity of probes than nick translation, but nick translation gives better control of the length of the probe.

End labelling at the 5´ end   DNA, RNA or oligonucleotides can be radiolabelled at the 5´ ends by T4 polynucleotide kinase which catalyses the transfer of the γ-phosphate from [γ-^{32}P] ATP to a free 5´ hydroxyl group. Chemically synthesized oligonucleotides have a free 5´ hydroxyl group and radioactive phosphate can be attached directly with no pre-treatment of the oligonucleotide. The reaction is very efficient and high specific activities of $\sim 10^9$ cpm μg^{-1} oligonucleotide are readily achieved. DNA fragments derived by restriction enzyme digestion usually carry 5´ phosphates which must be removed with phosphatase to generate free 5´ hydroxyl groups. The phosphatase must then be completely inactivated or removed otherwise the [^{32}P] phosphate attached by the kinase will be removed by residual phosphatase activity. Calf intestinal alkaline phosphatase is used in preference to calf thymus phosphatase, because it is easier to inactivate.

End labelling at the 3´ end   Labelling DNA or oligonucleotides at the 3´ end depends on the ability of the enzyme terminal transferase to add nucleotides to the free 3´ hydroxyl in a template-independent manner. Short, labelled, homopolymer tails can be added by incubating the enzyme and DNA with a labelled reporter-dNTP (usally dATP). This is called DNA tailing. A useful feature of tailing is that it provides several reporter molecules per DNA or oligonucleotide molecule. This makes detection more sensitive than when a single reporter molecule is present.

Non-radioactive probes   Probes for filter hybridization are most commonly labelled with biotin, digoxygenin, fluorescein or enhanced chemiluminescence method. The enzymes alkaline phosphatase and horseradish peroxidase are fairly common reporter molecules.

Removal of unincorporated nucleotides   After the probe has been labelled, it is usual to remove unincorporated radioactive label because during hybridization free labelled nucleotides may bind non-specifically to the filter causing high backgrounds. However, removal of radioactive label may not be necessary if more than about 70% has been incorporated into the probe. Non-radioactive labelled nucleotides tend to have very low affinity for the commonly used types or filter and so unincorporated label is seldom removed.

There are several standard methods for removing unincorporated nucleotides. Size exclusion chromatography on Sephadex G-50 column will remove unincorporated dNTPs and very short oligonucleotides. Columns are easily set up in siliconized 15-cm Pasteur pipettes plugged with siliconized glass wool. Sephadex spin columns are simple, quick and reliable to use. They are available commercially, but it is easy to set one up using a 1-ml disposable syringe plugged with siliconized glass wool. Ethanol precipitation is another method which depends on the precipitation of nucleic acids greater than about 20 bases in length by ethanol in the presence of high concentration of salt.

Usefulness of nucleic acid probes   Some of the major uses of nucleic acid probes include the following.

- Detection of pathogens in clinical samples, especially those organisms which are fastidious and difficult to cultivate,

- Differentiation of virulent from avirulent organisms and vaccine strains from wild type isolates,

- Typing of microorganisms,

- Mapping genes, and

- Detection of latently infected or carrier animals in epidemiological studies and food safety.

Detection of pathogens in clinical samples   Nucleic acid probes provide a sensitive method for the detection of infectious agents. Nucleic acid probes have been used to detect bacteria, viruses and parasites in clinical samples. The DNA fragments which encode for heat-stable toxin (ST) and heat-labile toxin (LT) were radiolabelled and used as probes for detection of enterotoxic *E. coli* (Moseley *et al.*, 1980). ^{32}P-labelled DNA probes had been used for the detection of leptospires in blood, urine and liver specimens (Terpstra *et al.*, 1986). Specific ^{125}I-labelled single-stranded DNA probes complementary to rRNAs of the target organisms were used for rapid identification of *Mycobacterium avium*, *M. intracellulare* and *M. tuberculosis* (Kiehin and Edwards, 1987). This method was also used to differentiate *M. paratuberculosis* from *M. avium*, which are indistinguishable by cultural and biochemical test. St. Amand *et al.* (2005) described a set of rRNA-based oligonucleotide probes that specifically detected

either M. *intracellulare*, the two M. *avium* subspecies associated with human disease, or all members of the *Mycobacterium avium* complex (MAC).

The most important application of nucleic acid probes is the detection of fastidious organisms. Many organisms such as *Mycoplasma* and *Mycobacterium* spp. cannot be cultured easily, or require 3 weeks to 3 months to culture. Nucleic acid probes have been developed and are being successfully used for their detection either directly in clinical specimens or following short culture periods. A dot blot hybridization method was used for the detection of *Campylobacter* spp. in stools using ^{32}P-labelled probes (Gebbart *et al.*, 1989). *Mycoplasma hyorhinis* which causes chronic arthritis in swine was differentiated from other mycoplasma organisms using DNA probes in Southern blot hybridization. *Mycoplasma haemosuis* DNA was detected in experimentally infected splenectomized pigs by *in situ* hybridization (ISH) with a non-radioactive digoxigenin-labelled DNA probe (Ha *et al.*, 2005). An 839-base-pair DNA probe targeting a 16S rRNA gene was generated by the polymerase chain reaction. The ISH method developed in this study was useful for the detection of M. *haemosuis* DNA in formalin-fixed, paraffin wax-embedded tissues and may be valuable for studying the pathogenesis of M. *haemosuis* infection.

Detection of *Haemophilus parasuis* in naturally infected pigs was studied by *in situ* hybridization with a non-radioactive digoxigenin-labelled DNA probe (Jung and Chae, 2004). Twenty pigs were selected on the basis of bacterial isolation and histopathological lesions. An 821-base-pair DNA probe from the 16S small subunit ribosomal RNA (rRNA) was generated by the polymerase chain reaction. Hybridization signals were detected in formalin-fixed, paraffin-wax-embedded tissues (lung, heart, spleen and liver). It was concluded that *in situ* hybridization was a valuable tool for the diagnosis of H. *parasuis* infection. Specific DNA probes are also available for the detection of other bacteria like *Yersinia*, *Salmonella*, *Listeria* and *Clostridium* spp.

Among viral infections, ^{32}P-labelled probes prepared with polymerase sequences of FMD virus, detected the virus in oesophageal and pharyngeal fluids using dot blot hybridization (Rossie *et al.*, 1983). A 3-kb DNA fragment of porcine parvovirus was cloned and used as probe for detection of parvovirus (Krell *et al.*, 1988). Direct filter hybridization was used for the detection of pseudorabies virus in nasal and tonsillar cells (Linne, 1987). Bluetongue virus serogroup-specific and serotype-specific single-stranded DNA probes labelled with ^{35}S-CTP were used to identify bluetongue virus in infected cell lines (Dangler *et al.*, 1988). DNA probes have been produced for RNA viruses such as rotavirus and influenza A virus by synthesis of complementary DNA (cDNA) from the viral RNA using the enzyme reverse transcriptase. After synthesizing the homologous strand, cDNA can be inserted into a bacterial plasmid or phage and molecularly cloned to produce recombinant DNA probes. Specific DNA probes have been used for detection of other viruses like herpesvirus, canine parvovirus (Figure 5.2), infectious laryngotracheitis virus and adenovirus from clinical samples.

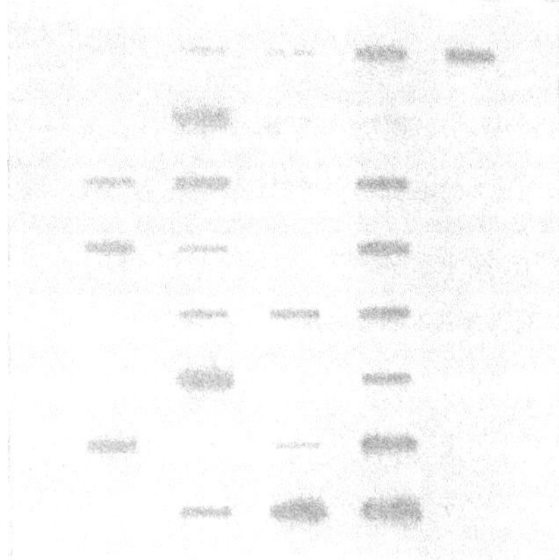

Figure 5.2   Autoradiograph of slot blot hybridization for detection of canine parvovirus in faecal sample using a ^{32}P-labelled PCR product as probe

Two non-radioactive probes using digoxigenin or biotin were developed for detecting canine herpesvirus (CHV) and compared for their sensitivities by *in situ* hybridization (ISH) in formalin-fixed, paraffin-embedded sections, which has been used routinely in veterinary fields (Kim, 2004). The best result was obtained by the optimized ISH protocol using digoxigenin-labelled probe for detection of CHV DNA. The optimized ISH assay, which developed in this study, may be a valid tool for the study of pathogenesis and diagnosis of CHV infection.

A digoxigenin-labelled probe was produced from the Pasteur virus strain for the detection of the rabies virus *N* gene (Carnieli Junior *et al.*, 2006). The probe hybridization was performed from amplified *N* gene obtained by reverse transcription polymerase chain reaction and the results from RT-PCR and hybridization showed 100% agreement. The hybridization, when carried out in products amplified by RT-PCR, increased the sensitivity of this technique even more and conferred specificity to the diagnosis. The technique described in this work would be useful in rabies diagnosis laboratories.

An improved method for the diagnosis of canine parvovirus using *in situ* hybridization in standard formalin-fixed, paraffin-embedded tissue sections was developed by Nho *et al.* (1997). A digoxigenin-labelled probe complementary to DNA sequences that coded for the entire sequence of the capsid protein VP-1 and the middle part of the sequence of the capsid protein VP-2 was designed. Specific histologic localization of canine parvovirus-infected cells was demonstrated in the small intestine, tonsil, lymph node, thymus, spleen, heart, liver, and kidney from dogs diagnosed at necropsy with canine parvovirus infection. The *in situ* hybridization accurately pinpointed the specific sites of viral infection. The detection of canine parvovirus in liver, kidney, and heart tissues together in the same pups could represent an enhanced virulence of this strain of canine parvovirus and suggests a broadened tissue tropism not seen before in Korean strains of canine parvovirus.

Mondal and Cardona (2003) used slot blot hybridization of the hypervariable regions of the S1 subunit of spike peplomer gene to identify and characterize infectious bronchitis virus (IBV) strains. Template DNA was created from six reference strain IBVs of different serotypes and immobilized on a nitrocellulose membrane. Digoxigenin-labelled probes were synthesized from reference and unknown field viruses and hybridized to template DNA. All reference strains could be distinguished and isolates identified by serotype if they were at least 95% identical to a reference strain. This slot blot hybridization procedure was specific and reproducible, and strain typing was consistent with the S1 sequencing of the IBV genome. This study thus provides a simple and rapid method for typing of IBV.

Among parasitic infections, DNA probes had been developed for diagnosis of anaplasmosis, babesiosis (Mc Laughlin *et al.*, 1986), cowdriosis and trypanosomiasis (Ambrosio *et al.*, 1988). The Dig-labelled probe specific to *Babesia bigemina* generated from monomorphic RAPD fragment of approximately 873 bp size amplified by a 10 mer CGGTGGCGAA, detected up to 100 ng of template DNA (Ravindran *et al.*, 2006). This non-radioactive probe also detected *B. bigemina* in preparations of larval tick DNA from two of the five samples on dot blot hybridization.

A reverse line blot hybridization (RLB) one-stage nested PCR (nPCR) for *Anaplasma centrale* and a nested PCR for *Anaplasma marginale* were used to detect infected cattle grazing within an endemic region in Israel (Molad *et al.*, 2006). A novel set of PCR primers and oligonucleotide probes based on a 16S ribosomal RNA gene was designed for RLB detection of both *Anaplasma* species, and the performance of the molecular assays compared. The RLB and the nested PCR procedures showed bacteraemia with sensitivity of 50 infected erythrocytes per millilitre.

Differentiation of virulent from avirulent organisms   The differentiation of virulent from avirulent organisms and vaccine strains from wild type isolated are extremely important in diagnostic medicine and epidemiological studies. This can be accomplished immunologically by monoclonal antibodies and genetically by restriction endonuclease analysis, sequence analysis and via probes. *Treponema hyodysenteriae*, causative agent of swine dysentery, cannot be easily distinguished from a non-pathogenic treponema, *Treponema innocens*, which is commonly present in normal swine. A nucleic acid probe prepared from a plasmid associated with *T. hyodysenteriae* was successfully used to differentiate pathogenic from non-pathogenic treponemas (Joens and Marquer, 1988).

Typing of microorganisms   Typing of microorganisms has usually been based on biological and antigenic characteristics. These classical procedures have been found to be unsatisfactory in many cases. Genetic tools like hybridization methods with specific probes have been used for typing microbes. Nucleic acid probes have been used to type pilus and enterotoxigenic gene types to differentiate *Campylobacter* spp. *E. coli*, *Leptospira* spp., *Mycobacterium* spp. and rotavirus group, subgroup and serotypes. Ramadass *et al.* (1990; 1992) used DNA homology studies with serovar-specific nucleic acid probes, and classified 40 *L. biflexa* serovars into six homology groups and 65 *L. interrogans* serovars into six homology groups.

Mapping genes   Nucleic acid probes provide a powerful tool for evaluating homology between related DNAs. Homology between polyoma and papovavirus was analysed by Southern blot hybridization (Howley *et al.*, 1979). Similarly, homology between porcine and canine parvovirus was evaluated using Southern blot hybridization. A complete map of bovine adenovirus was generated by Southern blot hybridization, at low stringency with probes prepared from similar adenovirus 3 DNA.

Epidemiological studies   The development of specific probes for organisms, groups of organisms or specific serotypes provide a valuable tool for epidemiological studies. Such tests can be developed in a few reference laboratories and distributed to diagnostic laboratories to determine prevalence and incidence of specific organisms. Probes will be especially useful for determining the epidemiology of microorganisms which are difficult to culture or type.

Food safety   There is increasingly more emphasis on the rapid detection and identification of microorganisms in food. In the past, this has been generally accomplished by the isolation and identification of bacteria using conventional methods that are time-consuming and require the presence of viable microorganisms. Nucleic acid probes provide a more rapid method for the identification of microorganisms in food. Nucleic acid probes have been developed and are commercially available for the detection of microorganisms of public health importance like toxigenic *E. coli*, *Listeria* spp., *Salmonella* spp. and *Yersinia enterocolitica* (Hill, 1989).

Detection of contaminants in the culture   Mycoplasma, parvovirus and bovine viral diarrhoea virus are frequently present as contaminants in cell culture. Detecting these viruses may be difficult task for inexperienced workers. Nucleic acid probes offer standardized reagents for their detection and such probes have been developed for these microorganisms (Mc Garrity and Kotani, 1986).

Study of mechanisms of pathogenesis   Nucleic acid probes are excellent tools for the determination of molecular mechanisms of pathogenesis. Mechanism of latency induction by herpesvirus have been studied by *in situ* hybridization. Rock *et al.* (1987) found that with bovine herpesvirus 1 (BHV-1), the causative agent of infectious bovine rhinotracheitis, the nucleic acid is present in ganglionic neurons of rabbits latently infected with BHV-1. Only selected regions of the viral genome were transcribed in neurons of latently infected rabbits, whereas all regions of the genome were transcribed in lytically infected cells. These studies indicate that transcription of BHV-1 is restricted to the latent phase of infection and suggest that transcription of specific genes may be important in the establishment and maintenance of latency.

Genetic analysis   Most of the genetic analyses are based on changes in DNA sequences that introduce or destroy restriction enzyme cleavage sites or cause restriction fragment length polymorphisms (RFLPs). These changes in the size of restriction fragments have been related to major genetic diseases. In a family, if one pattern is produced from the DNA of diseased individuals, but not from normal individuals, then this pattern or polymorphism serves as a marker for the disease gene. A 7.5-kb deletion within the human growth hormone gene results in dwarfism. Homozygous α-thalassemia (***hydrops fetalis***) is characterized by the absence of an *Eco* RI fragment which is approximately 14 kb in size and is present in the normal individuals' globin gene.

HYBRIDIZATION

The ability to diagnose infections by nucleic acid hybridization is based upon the ability of complementary strands of nucleic acid to anneal to form a double strand by hydrogen bonding between complementary base pairs. The duplex is stable under normal physiological conditions, but it will dissociate (denature) into the component single strands when heated or exposed to pH extremes. The melting temperature (T_m) is dependent upon the base pair composition of the molecules; an increased GC content confers high thermal stability. The principle involved in hybridization is shown Figure 5.3.

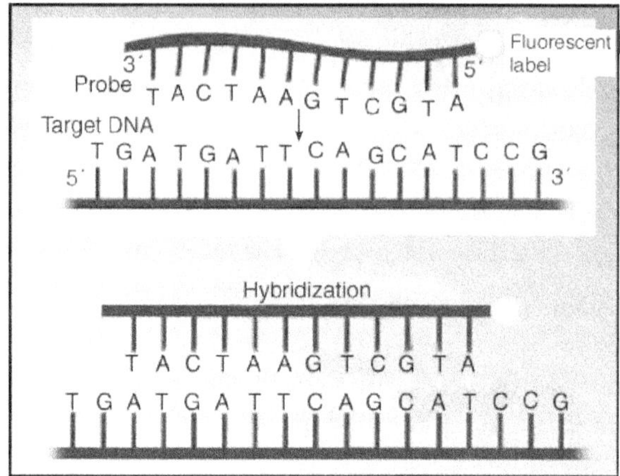

Figure 5.3 Nucleic acid hybridization. Probe labelled with a fluorescent dye

A single-stranded nucleic acid molecule will hybridize or anneal to a complementary nucleotide sequence with high avidity and with high degree of specificity. The parameters affecting this reaction in liquid solutions include reagent concentration, salt and formamide concentrations, temperature and G+C content of the nucleotide sequence in question. The concentration of organic solvents or denaturants in the reaction mixture also affects the T_m; for example, each 1% of formamide added reduces the T_m by 0.72°C. Ionic strength of the reaction mixture also directly affects T_m; decreasing the NaCl concentration from 0.1 to 0.01 M decreases the T_m by approximately 16°C.

Under conditions of low stringency (reaction temperature significantly lower than T_m), stable hybrids will be formed that may be composed of slightly dissimilar nucleotide sequence, whereas at high stringency (reaction temperature closer to T_m), stable hybrids will result only between nucleic acid molecules of very similar or identical sequences. Maximum rate of hybridization occurs at approximately 20–25°C below the melting temperature of the duplex. One key advantage of hybridization technique is the detection sensitivity of a DNA. Another advantage is the short time within which tests can be completed. Tests with radiolabelled probes take 36 hours, but tests with biotinylated probes take as little as 9–11 hours.

The hybridization methods have been used for prenatal and postnatal detection of gene defects and for the detection of a wide variety of human and plant pathogens. The virus causing avian influenza has got two glycoproteins called haemagglutinin and neuraminidase on its surface. For the avian influenza virus to be highly pathogenic, the haemagglutinin is to be cleaved into two compounds. Sequence studies of the haemagglutinin gene showed that there was a difference of only 4 amino acids between a non-pathogenic and a pathogenic form, and it was a result of a single point mutation in the haemagglutinin gene.

Hybridization Strategies

Hybridization occurs between two complementary molecules of DNA–DNA or RNA–RNA and the reaction can be carried out in solution, on a solid support or in cells (*in situ* hybridization).

Conditions of hybridization and washing have a profound influence in determining which hybrids form and which survive. The variables include:

i. length, concentration and base composition of the probe

ii. salt concentration

iii. temperature of hybridization

iv. length of time of hybridization

v. degree of mismatching of bases in the hybrid

Conditions under which only well-matched hybrids survive are said to be of high stringency, while those which allow mismatched hybrids to exist are said to be of low stringency.

The strategy chosen for filter hybridization depends on the purpose of the experiment and on many factors ranging from the availability of the target nucleic acid and the probe to the level of sensitivity of detection that is required and the possible need to detect mismatched sequences. The most likely parameters that will have to be modified to optimize signal-to-noise ratio are temperature, salt concentration, and the time period of hybridization.

Factors affecting the rate of hybridization   During filter hybridization, the single-stranded probe collides randomly with filter-bound sequences. When a correct match between complementary strands occurs (nucleation), the nucleic acid "zipper up" forming a duplex.

Concentration of nucleic acids   When filter-bound sequences are hybridized with a denatured double-stranded probe, there will be two competing events, the reassociation of the probe in solution and the hybridization of the probe to filter-bound sequences. The kinetics of hybridization differs depending on whether the sequences in solution or those bound to the filter are in excess.

1. For genomic Southern and northern blots and genomic DNA and RNA dot (or slot) blots, the probe sequences are likely to be in excess.

2. For screening recombinant plasmid or phage libraries or for plasmid, phage or PCR-generated genomic DNA or dot blots, the filter-bound sequences are likely to be in excess.

In practice, the amount of nucleic acid that should be bound to filters will depend on the purpose of the experiment and the availability of the nucleic acid. By using 1–10 µg mammalian DNA per lane of a Southern blot, single copy genes can easily be visualized with 20 ng/ml (typically 2×10^7 cpm/ml) of ^{32}P-labelled probe. For a restriction fragment of recombinant DNA, a band containing 5 ng is easily detected. For northern blots, 10–20 mg total RNA per lane of 1–2 mg poly(A) +RNA per lane is commonly used.

Length of the probe   The rate of hybridization in solution is proportional to the square root of the length of the smallest fragment. There are two different effects of the length of the probe on the rate of filter hybridization:

1. When C_f (conc. of filter-bound sequence) is low compared with C_s (conc. of probe in solution), the rate of hybridization is independent of the molecular weight (length) of the probe.

2. By contrast, when C_f is high, the higher the molecular weight of the probe, the lower will be the rate of hybridization.

Normally, DNA probes are about 400–600 bp in length. For nick-translated probes, the final size of the labelled probe is usually determined by the ratio of DNase I to DNA polymerase I used in the labelling reaction, while for probes generated by random priming, the size is determined by the ratio of probe primer. The use of long probes may lead to high, patchy backgrounds.

Salt concentration   At low ionic strength, the rate of hybridization changes significantly with changes in ionic strength. However, provided that the concentration of monovalent cations is above 0.4 M, there is no marked difference in DNA : DNA hybridization rates up to 1.5 M Na$^+$. For DNA : RNA hybridization, the rate of hybridization at 0.18 M NaCl is equivalent to that of DNA : DNA hybridization under the same conditions. However, increasing the salt concentration to 1 M NaCl increases the rate only two-fold. In practice, solutions containing 0.9 M Na$^+$ (e.g. 5 × SSC (Sodium chloride–sodium citrate buffer) or 5 × SSPE (Sodium chloride–sodium phosphate–EDTA buffer) are commonly used in DNA : DNA hybridizations.

Effect of washing   The stringency of washing is a function of the salt concentration and temperature.

    ♠ *Well-matched hybrids*   These are washed at high stringency (e.g. low salt concentration and high temperature). For well-matched

DNA : DNA hybrids, the salt concentration can be reduced to 0.1 × SSC. For DNA : RNA and RNA : RNA hybrids the salt concentration used is generally from 0.1 to 1 × SSPE.

♦ *Mismatched hybrids*   These are more stable at high salt concentrations than at low concentrations. This is very useful because varying the salt concentration can be used to stabilize or dissociate mismatched hybrids according to the requirements of the experiment. For preserving mismatched sequences, the salt concentration of the washing solution is generally 2.6 × SSC depending on the degree of homology required.

Base composition   The rate of hybridization increases as the %G+C content increases, but the effect is small and can be ignored in practice. A more important effect of base composition is its influence on the T_m of hybrids such that the higher the G+C content of the hybrid, the higher the T_m.

Temperature   At 0°C the rate of filter hybridization is very slow. As the temperature is increased, the rate rises markedly, reaching a maximum at about 20–25°C below the T_m of the hybrid, and then declines with further increase in temperature. In practice, the temperature used for hybridization is that giving the maximum rate of hybridization, i.e., about 20–25°C below the T_m for DNA : DNA hybridization and 10–15°C below the T_m for RNA : DNA hybridization. In practice, with probes of over 600 bp in length and for perfectly matched hybrids, an incubation temperature of between 60°C and 70°C is generally used for DNA : DNA hybridizations in aqueous solution and 37–45°C in solutions containing 50% formamide. DNA : RNA and RNA : RNA hybridization are usually carried out in 50% formamide and at an incubation temperature from about 42°C to 65°C.

Formamide   Formamide decreases the T_m of hybrids in solution by 0.75% per 1% formamide for poly(dA : dT) and by 0.5°C per 1% formamide for poly(dG : dC). An average value of 0.63°C per 1% formamide for T_m depression is usually adopted since filter hybridizations seldom use homopolymers. The depression of the T_m by formamide is the most useful property since, by incorporating 30–50% formamide in hybridization solutions, the incubation temperature can be reduced to between 30°C and 42°C. Practical advantages of using lower temperatures are that there is better retention of nucleic acid on the filter, there is less thermal degradation of the probe, and nitrocellulose filters are less fragile on repeated probings. Concentrations of formamide between 30% and 50% do not affect the rate of filter hybridization, and 20% formamide depresses the rate by about one-third.

Time period of hybridization   DNA : DNA hybridization is normally carried out for 6–18 hours, RNA : DNA hybridization for 8–18 hours and RNA : RNA hybridization for 10–12 hours. Filter hybridizations do not go to completion. This is, in part, because some of the filter-bound sequences are inaccessible to the probe. The rate of hybridization on filter is about 10 times slower than in solution. Even under optimized conditions, only about 80% of added probe appears in the hybrid.

Reaction volume   The lower the hybridization volume, the better, since this effectively concentrates the probe. This is especially important when using a scarce probe, but care should be taken to ensure that the filter is completely covered with hybridization solution or high background may occur.

Choice of filter matrix   Several types of membranes are available for immobilizing nucleic acids; the matrix of choice depends on the purpose of the experiment. The commonly used ones are nitrocellulose, nylon and charged nylon.

Nitrocellulose (NC) filters   NC binds both DNA and RNA efficiently (about 80 mg/cm^2). For large fragments of DNA, filters with a pore size of 0.45 mm are generally used, while for molecules less than about 500 nt long, a pore size of 0.2 μm gives better retention. For RNA, filters with 0.1–0.2 mm pore size give most efficient binding. For Southern and northern blots, NC must first be pre-wetted in a low-ionic-strength solution (e.g. deionized water or 2 × SSC) followed by wetting with the transfer solution. In Southern blots, solutions of 20 × SSC or 20 × SSPE are commonly used for the transfer of DNA to NC. It is particularly important to use high-ionic-strength solutions because the smaller the size of DNA, the higher the ionic strength needed for efficient binding. In Southern blots, the gel must be completely neutralized after alkali treatment; otherwise, the filter will turn yellow on baking, will be very brittle and will probably disintegrate. Immobilization of nucleic acid is traditionally done by baking for two hours at 80°C in a vacuum oven. The properties of different membranes are given in Table 5.2.

Uncharged nylon filters (NF)   Nylon filters have high tensile strength and are much more robust than NC. They are hydrophilic, do not require pre-wetting and have high nucleic acid binding capacity (about 500 mg/cm^2) for both DNA and RNA. Nylon filters can be reprobed repeatedly with no loss of sensitivity. Filters with pore size of 0.45 μm are commonly used, but for short nucleic acids a pore size of 0.22 μm has superior binding capacity. The transfer of DNA from Southern blots to nylon filters can be carried out in alkali, thus avoiding neutralization of the gel, and thus saving time. Fixing of nucleic acid on NF can be achieved by baking at 80°C for 2 hours, but the nucleic acid is not covalently bound. Covalent binding can be achieved by UV irradiation of the filter. This is much quicker than baking and is said to be more efficient.

The objective of UV irradiation is to link a small proportion of the thymidine residues of the DNA to the filter. If the filter is over-irradiated, too many thymidine residues will be linked and the DNA will become unavailable for hybridization, so the sensitivity will be reduced. If the filter is under-irradiated, the nucleic acid will dissociate from the filter and will be lost. It is usually recommended that UV irradiation be carried out at 0.15 J/cm^2 for damp membranes and 1.5 J/cm^2 for dry membranes.

Table 5.2 Properties of filters

Property	Nitrocellulose	Nylon	Charged nylon
Types of nucleic acid bound	DNA and RNA	DNA and RNA	DNA and RNA
Binding capacity	80–120 µg cm^{-2}	450–600 µg cm^{-2}	450–600 µg cm^{-2}
Pore size	0.45 µm for DNA 0.22 µm for <500 nt DNA 0.1–0.22 µm for RNAs	0.45 µm 0.22 µm for short nucleic acid	0.45 µm
Ionic conditions used for transfer	High salt concentration	Wide range salt concentration	Wide range salt concentration
Binding procedure	Baking at 80°C, non-covalent binding	Baking at 80°C, non-covalent binding, UV treatment, covalent binding, alkali treatment	Baking at 80°C non-covalent binding, UV treatment, covalent binding, alkali treatment
Advantages	Low backgrounds	High binding capacity	High binding capacity
Disadvantages	Fragility of unsupported nitrocellulose	–	Higher backgrounds

Charged nylon filters   Positively-charged NF have high capacity for binding both DNA and RNA (about 480 µg/cm^2). Nucleic acid can be transferred by vacuum blotting, electrophoresis or by capillary methods. Alkaline transfer solutions can be used for Southern or northern blotting to charged NF. Charged NF are particularly recommended by manufacturers for applications involving detection of hybrids by enhanced chemiluminescence.

Immobilization of Nucleic Acid on Filters

Nucleic acids transferred to nitrocellulose or nylon filters by any of the above methods are very loosely bound and would be readily lost from the filters if they were not attached more firmly. Immobilization of nucleic acids to filters can be achieved by baking, UV fixation or alkali treatment.

Baking Immobilization is traditionally achieved by baking in an oven for two hours at 80°C. A vacuum oven is recommended for baking nitrocellulose to reduce the possibility of explosion and fire, but is not necessary for nylon filters. It is important to note that nucleic acid baked onto any type of filter is firmly, but not covalently bound.

UV fixation Covalent binding of nucleic acid to nylon filters can be achieved by UV irradiation of the filter. This is quick, taking only a few minutes and is probably much more efficient than baking, particularly if used in conjunction with transfer in buffers of high ionic strength. UV irradiation links a small proportion of the thymine residues of the DNA to the filter. If the filter is over-irradiated, too many thymine residues will be linked and the DNA will have lost its ability to hybridize, so the sensitivity will be reduced. If the filter is under-irradiated, the nucleic acid will dissociate from the filter and will be lost. So, to achieve maximum sensitivity, the irradiation step must be optimized. Calibrated UV cabinets are commercially available, which emit a constant amount of energy (e.g. 120 mJ cm^{-2}) automatically.

Alkali fixation Capillary transfer of DNA or RNA in alkali immobilizes the nucleic acid covalently to positively charged nylon filters and some makes of ordinary nylon filters. A disadvantage with binding in alkali is that the background is often high on hybridization. If nucleic acid has been immobilized by one method, it is unnecessary and even detrimental to use a second method.

Prehybridization

Prehybridization serves two purposes: it equilibrates the filter with solution and it blocks sites at which the probes could bind non-specifically. Failure to prehybridize effectively can lead to patchy and unacceptably high background. Prehybridization solutions typically contain a detergent, a blocking agent and a heterologous nucleic acid. The detergent is usually 1% SDS. It facilitates even covering of the filter with solution and has weak RNase-inhibiting property. The most commonly used blocking agent is Denhardt's solution (i.e., 0.02% Ficoll (mol. wt. 400,000), 0.02% polyvinyl pyrrolidone (mol. wt. 40,000) and 0.02% bovine serum albumin). Denhardt's reagent tends to give lower backgrounds than other reagents, so it is widely used. The nucleic acid in prehybridization solutions is usually calf thymus or salmon sperm DNA that has been sheared or sonicated to about 600 nt in length. The prehybridization solution is preheated to the desired temperature before adding to the filter.

Hybridization

During hybridization, the labelled probe is incubated with the target sequences under conditions which allow the desired hybrids to form. The probe must be single-stranded. Double-stranded DNA probes are denatured by heat or alkali treatment. The denatured probe may be added directly to the prehybridization solution covering the filter. It is also quite common, and safer, to use a fresh solution for hybridization.

Washing

Filters are washed extensively after hybridization. This serves to remove both unreacted probe in solution and probe which binds loosely and non-specifically to the filter. Washing also dissociate unwanted, mismatched hybrids. Washing solutions contain salt and a detergent, usually SDS. Manipulation of salt concentration and temperature of washing are important means of controlling stringency, which determines which hybrids are allowed to persist. It is important that the filter is not allowed to dry out at any stage. It is also important that all solutions are pre-warmed to the appropriate temperatures before being added to the filter. Failure to preheat the solutions is a common source of variability in results.

Detection of Radioactive Hybrids

Autoradiography is the most commonly used technique for detecting ^{32}P-labelled hybrids. In autoradiography, a filter containing radioactive hybrids is placed in close contact with an X-ray film in a light-proof cassette. Decay of isotope results in the emission of β-particles which react with silver halide grains on the film to create a stable latent image. The latent image gives rise to a visible image when the film is developed. The position of the image coincides with the position of hybrid on the filter and the intensity of the image is a measure of the amount of hybrid present. To improve detection of ^{32}P, the X-ray film is sandwiched between an intensifying screen and the filter in the cassette. Intensifying screens are flexible plastic sheets coated with a scintillator such as calcium tungstate which emits light when activated. The energy emitted by ^{32}P passes through the film and strikes the intensifying screen where it is converted into light which is reflected back onto the film creating a latent image. This procedure increases detection of ^{32}P about 10-fold. If a second screen is present on the other side of the filter, a further two-fold enhancement is obtained. The latent image created by reflection of light is stabilized at low temperature, so it is customary to place the cassette in a freezer at –70°C when using intensifying screens.

Detection of Non-radioactive Hybrids

Methods for detecting hybrids containing non-radioactive hybrids depend on enzymatic activity. The enzymes most commonly used are alkaline phosphatase (AP) and horseradish peroxidase (HRP), but β-galactosidase can also be used.

Direct detection   If the enzymes themselves are the primary labels for the probe, detection simply involves reacting the enzymes with chromogenic or chemiluminescent substrates so that a coloured precipitate or flash of light is generated. The precipitate or flash of light is localized to the site at which the enzyme and consequently the hybrid is present. The coloured precipitate is visible by eye, whereas light can be detected on X-ray film that is sensitive to blue light. The disadvantage of direct detection is that the conditions of hybridization and washing must be very gentle in order to protect enzyme activity. This limits the stringency with which hybridization can be carried out. So direct detection is most often carried out with oligonucleotide probes which are normally hybridized at low temperature or with large probes, but in the presence of formamide or urea.

Indirect detection   Detection of hybrids containing digoxygenin, fluorescein and biotin requires an extra step in order to direct the enzyme to the site of hybrids. The reporter molecules used in filter hybridization are all highly immunogenic, so it is common to use label-specific antibodies conjugated to either AP or HRP to link the hybrid and enzyme. Detection involves sequential steps of adding antibodies conjugated to one of the enzymes and then supplying appropriate substrates.

Chromogenic substrates   The most sensitive chromogenic substrate system for alkaline phosphatase is 5-bromo 4-chloro 3-indolyl phosphate/nitroblue tetrazolium (BCIP/NTP). Colour formation is initiated when alkaline phosphatase catalyses the removal of phosphate from BCIP. A colour-producing cascade occurs that finally gives a precipitate of diformazan which appears brown on nylon filters and blue on nitrocellulose filters. An alternative and cheaper detection system is based on the use of different 2-hydroxy 3-naphthoic acid anilide (naphthol AS) phosphate substrates in the presence of different diazonium salts. In the presence of HRP and hydrogen peroxide, the chromogenic substrate 4-chloro 1-napthol gives rise to a bright purple precipitate.

Chemiluminescent substrates   The development of chemiluminescent substrates for AP and HRP has significantly improved the sensitivity with which hybrids can be detected. Chemiluminescence occurs when the energy released in a reaction is emitted as light. Stable, substituted dioxetane substrates have been developed that cause light to be emitted when they are hydrolysed by alkaline phosphatase. The most commonly used substrates for hybrid detection are sold under the trade names of AMPPD and CSPD and CDP-Star. The compounds are very stable until the phosphate groups are cleaved. This leads to formation of a metastable intermediate which decomposes and emits light at 477 nm. The light signal can be captured on X-ray film or by phosphor imaging using specialized photo-excitable storage phosphor screens. The intensity of light emitted is a measure of the amount of enzyme activity which in turn is a measure of the amount of hybrid. Chemiluminescence detection can be performed with both nylon and nitrocellulose filters, but longer exposure time is required for nitrocellulose and it may be necessary to use blocking agents for similar signal intensity.

Enhanced chemiluminescence (ECL) Horseradish peroxidase catalyses the oxidation of luminal to 3-aminophthalate via several intermediates. The reaction is accompanied by emission of low intensity light at 428 nm. However, in the presence of certain chemicals, the light emitted is enhanced up to 1000-fold making the light easier to detect and increasing the sensitivity of the reaction. The enhancement of light emission is called enhanced chemiluminescence (ECL). Several enhancers can be used, but the most effective are modified phenols, especially *p*-iodophenol. The intensity of light is a measure of the number of enzyme molecules reacting and thus of the amount of hybrid.

ECL is simple to set up and is sensitive, detecting about 0.5 pg nucleic acid in Southern blots and in northern blots. Detection by chemiluminescent substrates has several advantages over chromogenic substrates. The sensitivity is 10–100-fold greater and quantitation of light emission is possible over a wide dynamic range whereas that for coloured precipitates is much more limited, possibly only over one order of magnitude. Stripping filters are much easier when chemiluminescent substrates are used.

Types of Hybridization

The main types of hybridization used today are liquid (or solution) hybridization, filter hybridization, *in situ* hybridization and hybridization to 'chips'.

Solution hybridization In solution hybridization, the reacting species (denatured, single strands of nucleic acid) are free in solution and are incubated under conditions that favour hybrid formation. To detect and monitor hybrid formation, advantage is taken of the difference in physical properties between single and double-stranded nucleic acids. Duplex formation is usually measured by hypochromicity (reduction in the absorbance at 260 nm as double-stranded regions are formed) or selective binding of single and double strands to hydroxyapatite columns.

Solution hybridization is the standard method for studies which determine the number of copies of sequences and their relatedness. However, the procedures are tedious and not well-suited for carrying out many different analyses at once.

There are three main methods for following the progress of reassociation in solution hybridization.

Optical methods   Nucleic acid denaturation is accompanied by an increase in UV absorption (hyperchromicity). As reassociation takes place, there is a corresponding decrease in the UV absorption (hypochromicity). So DNA reassociation can be followed by measuring the change in absorption at 260 nm with time. Measurements are carried out in a temperature-controlled cuvette in a spectrophotometer which is capable of providing continuous read-out of UV absorption. The temperature is first raised to denature the DNA and to provide a maximum optical density reading against which all the other readings are compared. The temperature is then reduced rapidly to the incubation temperature and maintained there for the remainder of the experiment. The change in UV absorption with time is recorded.

Hydroxyapatite chromatography   Double-stranded nucleic acid binds firmly to hydroxyapatite (a form of calcium phosphate) at 50°C in 0.14 M sodium phosphate buffer or 60°C in 0.12 M sodium phosphate buffer, whereas single strands do not. So, to follow reassociation, aliquots of the reassociation mixture are removed at intervals and applied to a hydroxyapatite column which is maintained at 50°C or 60°C. The single-stranded nucleic acid does not bind and is collected in the flow through liquid. Double-stranded DNA is then eluted from the column by raising the salt concentration to 0.4 M sodium phosphate.

Nuclease S1 digestion   Under carefully controlled conditions, nuclease S1 digests single-stranded regions or single-stranded tails of nucleic acids and leaves duplexes intact. The nuclease-resistant nucleic acid remaining is a measure of duplex formation. Nuclease S1 treatment can therefore be used to monitor the progress of a hybridization reaction. At different times of the reaction, samples of the hybridization mixture are removed and digested with nuclease S1 so that only perfect double strands remain. The input nucleic acids are usually radioactively labelled so that after digestion the product can be quantified by liquid scintillation counting.

Filter hybridization   In filter hybridization, single-stranded DNA or RNA is bound to an inert surface and is hybridized to a nucleic acid probe added in solution. The probe carries a reporter molecule or label which is the basis for the subsequent detection of hybrids. After the filter-bound sequences have been incubated with the probe, non-reacting probe is removed by washing and the hybrids which remain on the filter are detected by means of the reporter molecule. The nucleic acid bound to the filter can come from various sources, for example, plates containing recombinant bacteriophage or plasmids, DNA or RNA size fractionated on gels, solutions of crude or purified nucleic acids applied in small dots/slots.

Filter hybridization is widely used to isolate phage and plasmid-containing sequences of interest from recombinant libraries. Through analysis of size-fractioned DNA and RNA, the size and number of hybridizing species in samples can be determined. Filter hybridization is capable of great discrimination and can detect single base changes in nucleic acid. This application is widely used in medical research to detect mutations that cause disease. Filter hybridization is very versatile though dot blot hybridization is ideally suited for analysing many samples at once. It can be used quantitatively in conjunction with a calibration curve for estimating copy number of sequences, but is more commonly used semi-quantitatively to compare the relative amounts of sequences in different samples. The main disadvantage of filter hybridization is that it is much slower than solution hybridization and the time required for hybrid formation to go to completion is often impossible to attain. Thus, it is a less useful technique than solution hybridization for analysing rare sequences.

There are four main variations of filter hybridization, viz., (a) probing recombinant libraries, (b) Southern blots, (c) northern blots and (d) dot/slot blots.

Probing recombinant libraries　Filter hybridization is used to search libraries of recombinant bacteriophages, phages or plasmids for sequences of interest. The recombinant phages are plated out on dishes and incubated under standard conditions so that phage plaques are produced in a background of unlysed bacteria. A sterile filter is placed on the surface of the plate. Some of the phages in each plaque stick to the filter so that when the filter is removed, it carries a replica of the pattern of plaques on the plate. The phages on the filter are lysed, and the DNA is made single-stranded by placing the filter on absorbent paper saturated with appropriate solutions. The filter is then neutralized and treated to bind the single strands firmly to the filter. Hybridization is carried out by incubating the filter with a single-stranded probe. The frequently used probes are labelled with ^{32}P nucleotides.

After hybridization, the probe is washed away and the filter is exposed to X-ray film. Where hybrids have formed, the decay of ^{32}P label in the hybridized probe expose the film giving rise to a black dot. By aligning the autoradiograph with the filter and the original plate, the plaque hybridizing to the probe can be identified.

Southern blots   The technique of Southern blotting is named after its inventor, Ed Southern. DNA that has been digested with restriction enzymes is applied to a gel and separated according to size on electrophoresis. The gel is treated with alkali to denature the DNA and a filter is placed over the gel. The DNA is transferred to the filter by capillary diffusion Figure 5.4. The pattern of fragments of DNA on the filter is a replica of that in the gel. The DNA is firmly fixed to the filter and hybridized. The position of the band (autoradiographic signal) is a measure of its size whereas the intensity of the band is a measure of the number of copies of sequences that hybridize.

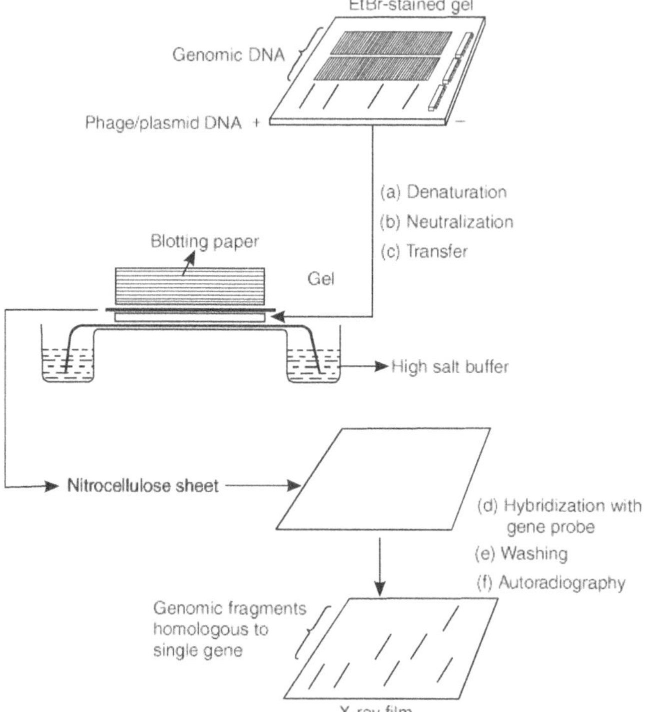

Figure 5.4   Steps involved in Southern blot hybridization

Northern blots   Northern blots are similar to Southern blots except that the nucleic acid being analysed is RNA instead of DNA. The RNA is denatured before electrophoresis, so the gel need not be pretreated before transferring the RNA to the filter. The probes are often cloned genes, so information regarding the expression of genes and the size and relative amounts of transcripts is obtained.

Dot/slot blots   In dot blots, DNA or RNA in solution is spotted onto filters and allowed to dry. For slot blots, the procedure is the same except that the nucleic acid is applied through a slot-shaped template. The nucleic acid is denatured and bound firmly to the filters. Hybridization is carried out with labelled probes. The autoradiograph shows which dots contain sequences complementary to the probe. Dot blots are faster to set up than Southern or northern blots because gels do not have to run, and the transfer of nucleic acid to the filter is far quicker. However, dot blots give no information on the size and number of different sequences contributing to the hybridization signal. Dot blots are well-suited for analysis of many samples at once and have the added advantage that it is easy to prepare replicate filters. A typical dot blot hybridized with radioisotopes in shown in Figure 5.5.

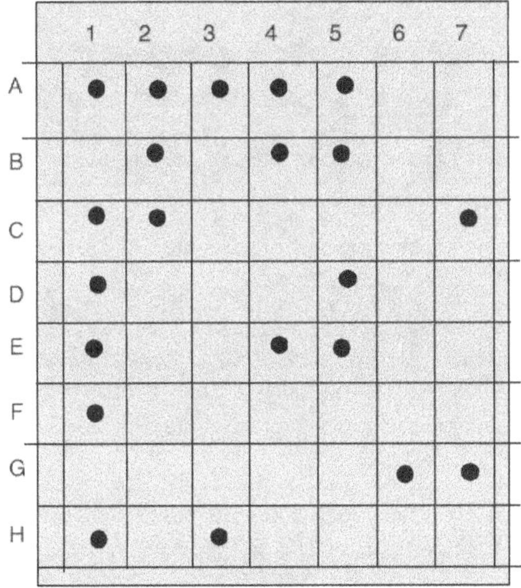

Figure 5.5   Dot blot hybridization

In situ hybridization   *In situ* hybridization is a powerful method used to locate nucleic acid sequences in histological and cytological preparations of tissues, organelles, cells and chromosomes. Before hybridization, samples are pretreated in such a way that, ideally

- The morphological features of the tissue/cell/chromosome are retained;

- The nucleic acid is neither extracted nor modified;

* The localization of the nucleic acid is unchanged;

* The probes and detection agents gain access to the nucleic acid.

Applications include identifying the location of genes on normal and aberrant chromosomes; identifying the sites of gene expression; determining levels of transcription and how these change with development and detecting viral pathogens. *In situ* hybridization showing *HER 2* gene amplification inside the cells is shown in Figure 5.6. *In situ* hybridization can be used to detect two or more target nucleic acids simultaneously and can detect both a particular transcript and the protein encoded by it. The method is particularly suited to detecting a sequence that is present in only a few cells within a large population, for example chromosomal translocations in residual disease. For many applications such as for detecting genes on chromosomes, sophisticated microscope is required and may be beyond the scope of many laboratories.

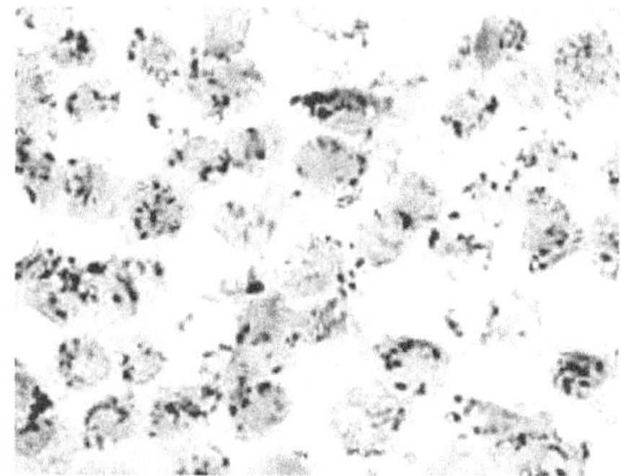

Figure 5.6 *In situ* hybridization showing HER 2 amplification

DNA Arrays and Chips

DNA microarrays are a new technology that allows the analysis of large numbers of genes at a high resolution by the hybridization of labelled DNA, which may be reverse-transcribed from mRNA, to a substrate containing thousands of spotted cDNA or oligonucleotides. The amount of hybridized target is analysed, giving information on gene expression, polymorphism of mutations present and allowing gene profiling of different serotypes of disease. They involve preparation of high-density arrays of sequences that are probed simultaneously. DNAs are attached to a solid surface in an ordered array so that the position or address of each DNA is known. The array is hybridized with a labelled probe and the hybrids are scanned and quantitated automatically. The term 'DNA chip' was originally used to describe oligonucleotide arrays only, but is now also used for cDNA arrays.

DNA chips consist of a large numbers of oligonucleotides or cloned DNA sequences attached to a small surface in such a way that each sequence has a known position. Many thousands of sequences can be immobilized on a surface smaller than a microscope slide. Hybridization usually takes place using probes labelled with fluorescent dyes. After hybridization, the position of hybrids is detected automatically by a computer-controlled reader. Applications include sequencing and detecting mutations. The procedure is well-suited to automation, but the equipment required is specialized and is expensive. So this technique may currently be beyond the resources of many laboratories.

Preparation of DNA arrays   The source of DNA is typically purified cDNAs that have been amplified by PCR in individual wells of a 96-well microtitre plate. Samples of about 0.05 µl are transferred by a computer-controlled print head to predetermined positions on a series of microscope slides. These have been coated with positively charged poly-L-lysine to enhance adhesion. The density of the array is about 1000 cDNAs cm^{-2}, so a microscope slide with an area of 10 cm^{-2} contains 10,000 hybridization targets.

Preparation of oligonucleotide arrays   Oligonucleotides of any sequence can be used. Oligonucleotides of specific sequence can either be synthesized *in situ* or prepared separately and attached post-synthetically. The maximum length attainable for oligonucleotides in an array is about 25 mers. A density of oligonucleotides of at least 400,000/1.6cm^2 can be achieved at present.

Label   Fluorescent labels are popular because they can be detected and quantitated automatically. Different probes can be labelled with different fluorophors, then mixed together and used to probe an array in a single hybridization reaction. Biotin can be attached to probes and subsequently detected in hybrids by binding to streptavidin conjugated to a dye such as phycoerythrin and laser light illumination. Particles like silicon can be attached to probes and detected by light scattering.

Applications   The principal applications of arrays are in genomics, diagnostics, gene expression analysis and DNA sequencing.

Gene expression analysis   Any given cell expresses only a limited number of its encoded genes. The number of genes expressed and the level of expression varies according to factors such as cell type, stage of development and response to external stimuli. Understanding gene expression and how it is regulated depends on analysing many transcripts simultaneously. Hybridization with labelled RNAs to targets on chips allows quantitative and qualitative measurement of transcripts to be made with a sensitivity of one to five transcripts per cell. Transcripts from two different cell types can be labelled with different fluorescent dyes. The probes are mixed and hybridized simultaneously to chip-bound targets. The two-colour fluorescence of hybrids allows differential gene expression to be monitored. A typical microarray signal showing different expression of *Bacillus subtilis* genes is shown in Figure 5.7.

Figure 5.7   Microarray showing expression of *Bacillus subtilis*

Detection of mutations and polymorphisms   Arrays can be prepared to include oligonucleotides that cover known mutations within a gene. An improved method is to include in the array, oligonucleotide sequences that have the other three bases at the sites of known mutations. The alternative bases are the centre of the oligonucleotides where mismatching has the most disruptive effect on the stability of hybrids. On hybridization, the mutant gene should hybridize to one of these mutant sequences.

Test arrays of 100,000 oligonucleotides of 20 mers have been used to screen for mutations in exon 11 of the human *BRCA1* gene. This exon codes for about 60% of the protein. Sequences covered the normal sequence as well as oligonucleotides covering insertions, deletions and substitutions. All known mutations were detected with no false positives. Diagnostic arrays are likely to be of increasing importance and already arrays of HIV oligonucleotides are available. In several conditions, such as the thalassemias and haemoglobinopathies, mutations in several different positions within the gene can give rise to disease. Overlapping oligonucleotide arrays are potentially well-suited to screening rapidly for the absence of mutations. Arrays of overlapping oligonucleotides of the gene to be screened are hybridized to DNA probes from normal and patient samples. By labelling the probes from each source with a different fluor, hybridization can be carried out with both probes at once.

DNA sequencing by hybridization   An array can contain all possible oligonucleotides of a chosen length, *n*. Hybridizing a probe to such an array provides information on all the constituent sequences of length *n* in the probe. Sufficient information can be generated to allow assembly into a reconstructed sequence of the probe. This approach is known as sequencing by hybridization. When the probe is hybridized to the array, some oligonucleotides will hybridize and some will not. The sequence of those oligonucleotides that have hybridized can be aligned by overlapping their sequences to reconstruct the complete sequence. Each non-terminal *n*-mer will overlap the next by *n*–1 nucleotides. This method of sequencing is useful in terms of speed, cost, ease of automation and quality of data. However, there are two main problems with the technique. First, it is a major operation to align overlapping sequences correctly. Second, the technique runs into difficulties if the probe contains repeated sequences because the overlaps will not be unique. The reconstruction stops at a repeated sequence. Confirmation of the sequence of the entire human mitochondrial genome has been obtained using oligonucleotide arrays (Chee *et al.*, 1996).

Microarrays for diagnosis   Baxi *et al.* (2006) developed a microarray-based test that used a FMD DNA chip containing 155 oligonucleotide probes, 35–45 base pairs (bp) long, and serotype-specific, designed from the VP3-VP1-2A region of the genome. A set of two forward primers and one reverse primer were also designed to allow amplification of approximately 1100 bp of target sequences from this region. The amplified target was labelled with Alexa-Fluor 546 dye and applied to the FMD DNA chip. A total of 23 different FMDV strains representing all seven serotypes were detected and typed by the FMD DNA chip. Microarray technology offers a unique capability to identify multiple pathogens in a single chip.

Sachse *et al.* (2006) developed a microarray assay for detection and differentiation of all currently defined chlamydial species belonging to the genera *Chlamydia* and *Chlamydophila* using the ArrayTube system, which they found to be particularly user-friendly and economical. The test included PCR amplification of a 23S rDNA target region with concurrent biotinylation and subsequent hybridization in the ArrayTube, a micro-reaction tube carrying the microarray chip on the bottom. In addition to high specificity, the assay was shown to allow detection and genetic characterization of single PCR-amplifiable target DNA copies.

A DNA microarray targeting O-serotype-specific genes was developed to detect 15 serotypes of *Shigella* and *E. coli*, including *Shigella sonnei*; *Shigella flexneri* type 2a; *Shigella boydii* types 7, 9, 13, 16, and 18; *Shigella dysenteriae* types 4, 8, and 10; and *E. coli* O55, O111, O114, O128, and O157 (Li *et al.*, 2006). The microarray was tested against 186 representative strains of all *Shigella* and

E. coli O serotypes, 38 clinical isolates, and 9 strains of other bacterial species that are commonly present in stool samples and was shown to be specific and reproducible. The detection sensitivity was 50 ng genomic DNA or 10(4) cfu per ml in mock stool specimens. This is the first report of a microarray for serotyping *Shigella* and pathogenic *E. coli*. The method has a number of advantages over traditional bacterial culture and antiserum agglutination methods and is promising for applications in basic microbiological research, clinical diagnosis, food safety, and epidemiological surveillance.

Wang *et al.* (2007) described a rapid (<4 h) high-throughput detection and identification system that used universal polymerase chain reaction (PCR) primers to amplify a variable region of bacterial 16S rRNA gene, followed by reverse hybridization of the products to species-specific oligonucleotide probes on a chip. This procedure was successful in discriminating 204 strains of bacteria from pure culture belonging to 13 genera of bacteria. When this method was applied directly to 115 strains of bacteria isolated from foods, 112/115 (97.4%) were correctly identified; two strains were indistinguishable due to weak signal, while one failed to produce a PCR product. The array was used to detect and successfully identify two strains of bacteria from food poisoning outbreak samples, giving results through hybridization that were identical to those obtained by traditional methods. The sensitivity of the microarray assay was 10(2) cfu of bacteria. Thus, the oligonucleotide microarray is a powerful tool for the detection and identification of pathogens from food.

HYBRIDIZATION IN DIAGNOSIS

Nucleic acid probes and hybridization provide an alternate technique for the diagnosis of microbial infections and genetic disorders. Selected applications of this technique have been described to illustrate the potential and unique capabilities and the nature of complexicities of these techniques.

Takahashi *et al.* (1999) developed a sensitive probe using a repetitive sequence of *Leptospira interrogans* serovar *icterohaemorrhagiae* strain Ictero No. 1 for demonstration of *Leptospira interrogans* strains. Hybridization experiments showed positive reactions with the strains of *L. interrogans* and *L. kirschneri*, but not with any strains of *L. borgpetersenii*, *L. weilli* and *L. meyeri* or *L. biflexa*.

O'Brien *et al.* (2000) developed a cloned probe from the region of the *Mycobacterium bovis* genome, pUCD, which generated a clear, highly polymorphic banding pattern when used as an RFLP probe on *Alu* I restriction-digested *M. bovis* genomic DNA. When used to type 60 Irish *M. bovis* isolates, pUCD exhibited greater discriminatory power than the commonly used mycobacterial RFLP probes IS6110, PGRS (highly repetitive polymorphic GC rich repeat sequence) and DR (Direct repeat) and detected an equivalent number of strain types to a combination of these three probes.

Zhang and Kitching (2000) developed an *in situ* hybridization using biotin-labelled oligodeoxynucleotides and tyramide signal amplification for sensitive detection of foot-and-mouth disease virus in infected cells. The tyramide signal amplification detection enhances by at least 100-fold compared to the sensitivity of *in situ* hybridization.

The genome segments of infectious bursal disease viruses (IBDV) in the bursa of Fabricius from experimentally infected chicken or field samples were detected by tissue print hybridization (TPH) with subsequent reverse transcriptase-PCR (Liu, 2000). Bursae were imprinted onto nylon membrane and then hybridized with a cloned digoxygenin-labelled cDNA probe. Tissue prints on nylon membrane were readily distinguished from control prints by colour development and differences in signal intensity.

Katari *et al.* (2000) developed two probes for differentiation of IBDV strains of varied virulence by dot blot hybridization. The *VP2* gene probe was able to detect both field and vaccine strains of IBDV under high as well as low stringency, while the *VP1* gene probe could differentiate under high stringency field isolates from vaccine strains. *In situ* hybridization was used in a pathogenesis study of three vaccine pathotypes (Delaware variant A, D78 and Bursa Vac) of IBDV (Sellers *et al.*, 2001). With an antisense VP2 gene probe, viral nucleic acid was detected in bursa from both D78 and Bursa Vac-infected chicken at 24, 48, 72 and 120 hour post-infection (HPI). However, viral RNA was detected only in the Delaware variant A-infected birds at 72 HPI.

Diagnosis of Bovine Herpesvirus Infection using Biotinylated cDNA Probes

BHV-1 is an important disease of bovines that can cause a variety of disease syndromes including infectious bovine rhinotracheitis, conjunctivitis, encephalitis, vulvovaginitis/balanoposthitis and abortion. Latent infection often follows viral infection. Biotinylated nucleic acid hybridization probes have been developed to detect BHV-1 nucleic acid in cells by *in situ* hybridization and in tissue homogenate and nasal exudates by blot hybridization.

A series of HBV-1 probes have been prepared by *Hind* III digestion of BHV-1 DNA and fragments cloned into pBR 322. One probe, pCB-31, proved to be the most sensitive and specific for *in situ* hybridization. Another probe, a pCB 2 probe (9.6 kb *Hind* III fragment), proved to be the most sensitive for blot hybridization. Probes were labelled by nick translation with biotin or radionucleotide (Dorman *et al.*, 1985; Dunn *et al.*, 1986).

In the case of *in situ* hybridization, the glass slide with smear is treated with 0.01% Triton X-100 and protease (0.4 mg/ml in 0/05 M Tris. HCl, pH 7.6 and 5 mM EDTA). The smear is incubated with hybridization solution with biotinylated probe. The slides are incubated at 80°C for about 5 minutes, at 40°C for 5 minutes and then at 37°C overnight. After hybridization, the slides are washed in 50% formamide in 2 × SSC at room temperature. Hybridized biotin-labelled probes are localized by HRP staining. The slide is then treated with goat anti-biotin IgG for 1 hour at 37°C. After wash, it is treated with biotinylated rabbit anti-goat serum and incubated at 37°C for 1 hour and again washed. Then the slides are treated with HRP conjugate and colour developed with 0.2 mg of diamino benzidine/ml in 2 × SSC containing 0.03% H_2O_2/ml. After wash, the slide is counterstained with Harris haematoxylin. Nuclear staining is seen in positive cases.

In the case of dot/slot hybridization method, either whole cells or nuclear DNA after extraction is applied to a nitrocellulose membrane (NCM). If whole cells are applied to NCM, then it is subjected to chemical treatment to release the DNA from the cells. NCM is then subjected to prehybridization and hybridization with labelled probes.

These experiments have demonstrated that molecular hybridization, using biotin-labelled DNA probes is a sensitive method for detection of BHV-1 nucleic acid in infected cells and tissues. This method can detect as little as 10 pg of purified viral DNA. Viral DNA could be detected in nasal swab specimens which had been stored dry for over one year.

Diagnosis of Enterotoxigenic *E. coli* Infections

Enterotoxigenic *E. coli* (ETEC) is a major cause of diarrhoeal illness in humans and in animals throughout the world. Diagnosis of ETEC diarrhoea has been complicated by the difficulty in differentiating enterotoxigenic strains from normal flora. ETEC produce both heat labile (LT) and heat stable (ST I and ST II) toxins. Recently, genes coding for these toxins have been cloned, labelled with ^{32}P and used to detect toxic genes coding for A (1.3 kb) and B (0.85 kb) subunits of porcine LT, porcine ST (PST, 157 bp) and human ST (HST, 510 bp) and were used to genotype 20 strains of *E. coli*. Probes were labelled with Bio-11-dUTP or ^{32}P, by nick translation or random priming (Mosley *et al.*, 1982; Biolkowska-Hobrzanska, 1987).

^{32}P-labelled hybrids were detected by autoradiography; biotinylated probes were detected by streptavidin/biotin-conjugated polyalkaline phosphatase/ NBT-BCIP system. About 160 pg of target DNA was detected with biotin system; however, the ^{32}P-labelled probes were approximately eightfold more sensitive.

Diagnosis of Sickle Cell Anaemia

Prenatal diagnosis of many of these genetic disorders is not possible with the analysis of foetal DNA being isolated from cells in amniotic fluid. Linkage analysis of DNA polymorphism has been used to identify foetuses at risk of sickle cell anaemia and thalassemias (Geever *et al.*, 1981). For example, a variant *Hpa* I restriction site is frequently associated with the sickle gene in the genome of black Americans. Greater specificity can be introduced into the diagnosis of sickle cell anaemia by the use of specific restriction enzymes that recognize the DNA sequences involved in the sickle mutation, CCTGAGG to CCTGTGG. For example, the restriction enzyme *Dde* I, which recognizes CTNAG (where N is any nucleotide) has been used to detect sickle mutation (Chang and Kan, 1982). A culture of amniotic fluid is necessary to obtain sufficient DNA for analysis. An alternate enzyme *Mst* II has an additional two-base specificity (CCTBAGG) which results in larger cleavage products. Thus, its use increases the sensitivity of the assay and precludes the need for culture of amniotic cells (Orkin *et al.*, 1982).

A probe, 1.8 kb *Bam* HI fragment that contains the sickle cell mutation site inserted into the *Pst* I site of pBR 322, was labelled with biotin-11-dUTP by nick translation. DNA was extracted from blood, digested with *Mst* II, resulting fragments separated on gel electrophoresis and transferred to NCM. The probe was denatured, hybridized and the presence of hybrids were detected using streptavidin, biotinylated alkaline phosphatase polymer and ANT-BCIP. The respective hybridized restriction fragments were revealed as purple bands.

The *Mst* II enzyme cleaves the DNA at the unmutated recognition site, the 5´ end of the normal β-globin gene is located on the 1.14-kb fragment. In contrast, since the recognization site is not present in the β^s gene, the 5´ end of the gene is a 1.34-kb fragment. The two fragments are readily separated and identified by the probe.

DNA Fingerprinting for Genotype Analysis

Analysis of restriction fragment length polymorphisms (RFLPs) is widely used in genetic studies, linkage studies of inherited disorders and in forensic medicine (Jeffreys *et al.*, 1985; Donis-Keller *et al.*, 1986). DNA fingerprinting is produced upon RFLPs associated with hypervariable minisatellite or variable number of tandem repeat (VNTR) regions present in the human genome. A typical RFLP patterns obtained with DNA samples from individuals of different origin is shown in Figure 5.8. DNA fingerprints can be obtained from µl quantities of blood or semen and from specimens dried for up to 4 years (Jeffreys, 1986). The probability that fingerprints of individuals will be identical is less than 10^{-20}, and the DNA fingerprints are stably inherited.

M13 bacteriophage hybridizes with hypervariable sequences in human and animal DNA. Labelled M13, instead of cloned human probes, can be used in DNA fingerprint analysis. DNA was obtained from blood cells, digested with *Hae* III, and the resultant fragments were separated in agarose gels. After transfer to nitrocellulose, filters were probed using a 280-bp M13 *Hae* III *Cla* I restriction fragment labelled with ^{32}P. Filters were then exposed to Kodak X-ray films (Vassart *et al.*, 1987). Each bovine pattern, all from animals of the same breed, was distinct. The presence of highly polymorphic patterns with DNA of equine, murine and canine origin were also reported.

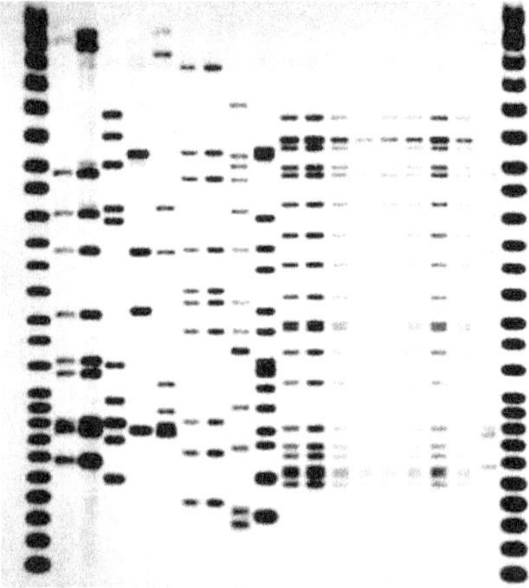

Figure 5.8   RFLP patterns of individuals of different orign

DNA fingerprinting is proving to be a significant tool with applications in paternity testing, forensic medicine and the more conventional genetic analysis. The potential for application of this technique for genotyping and for pedigree analysis of animal species would seem to be enormous.

REVIEW QUESTIONS

1. What are nucleic acid probes? How are they labelled?
2. What are the uses of nucleic acid probes?
3. What is meant by hybridization? Elaborate the hybridization strategies?
4. Discuss in detail the different steps involved in the hybridization method.
5. What are the different types of hybridization methods and their advantages?
6. Write an essay on DNA arrays and their usefulness.

REFERENCES

Ambrosio, R.E., Visser, E.S., Kockhoren, Y. and Kocan, K.M. (1998). "Hybridization of DNA probes to *A. marginale* isolates from different sources and detection in *Dermacenter andersoni* ticks." *Onderstepoort. J. Vet. Res.* 55: 227–229.

Baxi, M.K., Baxi, S., Clavijo, A., Burton, K.M. and Deregt, D. (2006). "Microarray-based detection and typing of foot-and-mouth disease virus." *Vet. J.* 172: 473–481.

Biolkowska-Hobrzanska, H. (1987). "Detection of enterotoxigenic. *E. coli* by dot blot hybridization with biotinylated DNA probes." *J. Clin. Microbiol.* 25: 338–343.

Caskey, C.T. (1987). "Disease diagnosis by recombinant DNA methods." *Science.* 236: 1223–1224.

Carnieli Jr., P., Ventura, A.M. and Durigon, E.L. (2006). "Digoxigenin-labelled probe for rabies virus nucleoprotein gene detection." *Rev. Soc. Bras. Med. Trop.* 39:159–162.

Chee, M., Yang, R., Hubbell, E., Berno, A., Huang, X.C., Stern, D., Winkler, J., Lockhart, D.J., Morris, M.S. and Fodor, S.P. (1996). "Assessing genetic information with high density DNA arrays." *Science.* 274: 610–613.

Dangler, C.A., Dunn, S.J., Squire, K.R.E., Scott, J.L. and Osburn, B.I. (1988). "Repaid identification of bluetongue virus by nucleic acid hybridization in solution." *J. Virol. Meth.* 20: 353–365.

Donis-Keller, A.D., Barker, R., Knowlton, J., Schumm and Braman, J. (1986). "Application of RFLP probes to genetic mapping and clinical diagnosis in humans." In: *Current Communications in Molecular Biology.* (ed.). Lerman, L.S. Cold Spring Harbour Laboratory, New York.

Dorman, M., Blair, C., Collins, J. and Beaty, B. (1985). "Detection of bovine herpes virus 1 DNA immobilized on nitrocellulose by hybridization with biotinylated DNA probes." *J. Clin. Microbiol.* 22: 990–995.

Dunn, D.C., Blair, C.D., Ward, D.C. and Beaty, B.J. (1986). "Detection of bovine herpesvirus-specific nucleic acids in *in situ* hybridization with biotinylated DNA probes." *Amer. J. Vet. Res.* 47: 740–746.

Gebbort, C.J., Ward, G.E. and Murtaugh, M.P. (1989). "Species-specific cloned DNA probes for the identification of *Campylobacter hyointestinalis*." *J. Clin. Microbiol.* 27: 2717–2723.

Geever, C.F., Wilson, L.B., Nallaseth, F.S., Milner, P.F., Bittner, M. and Wilson, J.T. (1981). "Direct identification of sickle cell anaemia by blot hybridization." *Proc. Natl. Acad. Sci.* 78: 5081–5085.

Ha, S.K., Jung, K., Choi, C., Ha, Y., Song, H.C., Lim, J.H., Kim, S.H. and Chae, C. (2005). "Development of *in situ* hybridization for the detection of *Mycoplasma haemosuis*

(*Eperythrozoon suis*) in formalin-fixed, paraffin wax-embedded tissues from experimentally infected splenectomized pigs." *J. Comp. Pathol.* 133: 294–297.

Haase, A.T., Brahic, M. and Stowring, L. (1984). "Detection of viral nucleic acids by *in situ* hybridization." *Meth. Virol.* 7: 189–224.

Hill, W.E. (1989). *DNA probes in infectious diseases.* (ed.). Tenover, F.C. CRC Press. p. 43.

Howley, P.M., Israel,M.A., Kew, M.F. and Martin, M.A. (1979). "A rapid method for detecting the mapping homology between heterologous DNAs." *J. Biol. Chem.* 254: 4876–4883.

Jeffreys, A.J., Wilson, V. and Theirn, S.L. (1985). "Hypervariable minisatellite regions in human DNA." *Nature.* 314: 67–73.

Jeffreys, A.J. (1986). "Hypervariable DNA and genetic fingerprints." In: *Current Communications in Molecular Biology.* (ed.). Lerman, L.S. Cold Spring Harbour Laboratory, New York. pp. 57–72.

Joens, L.A. and Morquer, R. (1988). Proc. of 10th Cong. Int. Pig Vety. School Brazil, p. 120.

Jung, K., Chae, C. (2004). "*In situ* hybridization for the detection of *Haemophilus parasuis* in naturally infected pigs." *J. Comp. Pathol.* 130: 294–298.

Katari, R.S., Tiwari, A.K., Butchaiah, G. and Kararia, J.M. (2000). "Development of probes for differentiation of infectious bursal disease virus strains of various virulence by dot-blot hybridization." *Acta. Virol.* 44: 259–263.

Kiehn, T.E. and Edwards, F.F. (1987). "Rapid identification using a specific DNA probe of *Mycobacterium avium* complex from patients with acquired immunodeficiency syndrome." *J. Clin. Microbiol.* 25: 1551–1552.

Kim, O. (2004). "Optimization of *in situ* hybridization assay using non-radioactive DNA probes for the detection of canine herpesvirus (CHV) in paraffin-embedded sections." *J. Vet. Sci.* 5: 71–73.

Krell, P.J., Sales, T. and Johnson, R.P. (1988). "Mapping of porcine parvovirus DNA and development of a diagnostic DNA probe." *Vet. Microbiol.* 17: 29–43.

Li, Y., Liu, D., Cao, B., Han, W., Liu, Y., Liu, F., Guo, X., Bastin, D.A., Feng, L., Wang, L. (2006). "Development of a serotype-specific DNA microarray for identification of some *Shigella* and pathogenic *Escherichia coli* strains." *J. Clin. Microbiol.* 44: 4376–4383.

Linne, T. (1987). "Diagnosis of pseudorabies virus infection in pigs with specific DNA probes." *Res. Vet. Sci.* 43: 150.

Liu, H.J. (2000). "Tissue print hybridization and reverse transcriptase PCR in the detection of infectious bursal disease viruses in bursal tissues." *Res. Vet. Sci.* 68: 99–101.

McDougall, J.K., Myersen, D. and Beckmann, A.M. (1986). "Detection of viral DNA and RNA by *in situ* hybridization." *J. Histochem. Cytochem.* 34: 33–38.

McGarrity, G.I. and Kotani, H. (1986). "Detection of cell culture *Mycoplasma* by a genetic probe." *Expt. Cell Res.* 163: 273–278.

Mc Laughlin, G.L., Edlind, T.D. and Ihler, G.M. (1986). "Detection of *Babesia bovis* using DNA hybridization." *J. Protozool.* 33: 125–128.

Molad, T., Mazuz, M.L., Fleiderovitz, L., Fish, L., Savitsky, I., Krigel, Y., Leibovitz, B. Molloy, J., Jongejan, F., Shkap, V. (2006). "Molecular and serological detection of *A. centrale-* and *A. marginale-*infected cattle grazing within an endemic area." *Vet. Microbiol.* 113: 55–62.

Mondal, S.P. and Cardona, C.J. (2003). "Characterization of infectious bronchitis virus isolates by slot blot hybridization." *Avian Dis.* 47: 725–730.

Mosley, S.L., Echeverriq, P., Seriwatana, J., Trirpat, J.C., Chaicumpa, W., Sakuldaipeara, T. and Falkow, S. (1982). "Identification of enterotoxigenic *E. coli* by colony hybridization using three enterotoxin gene probes." *J. Infect. Dis.* 145: 863–869.

Mulchay, L. (1986). "DNA probes: an overview." *Biochemistry.* Nov. 14–19.

Nho, W.G., Sur, J.H., Doster, A.R. and Kim, S.B. (1997). "Detection of canine parvovirus in naturally infected dogs with enteritis and myocarditis by *in situ* hybridization." *J. Vet. Diagn. Invest.* 9: 255–260.

O'Brien, R., Flynn, O., Costello, E., O'Grady, D. and Rogers, M. (2000). "Identification of a novel DNA probe for strain typing *Mycobacterium bovis* by restriction fragment length polymorphism analysis." *J. Clin. Microbiol.* 38: 1723–1730.

Orkin, S.H., Little, P.F.R., Kazarian, H.H. and Boehm, C.D. (1982). "Improved detection of the sickle mutation by DNA analysis." *New Engl. J. Med.* 307: 32–36.

Ramadass, P., Marshall, R.B. and Jarvis, B.D. (1990). "Species differentiation of *Leptospira interrogans* serovar *hardjo* strain *Hardjobovis* from Hardjoprajitno by DNA slot blot hybridization." *Res.Vet. Sci.* 49: 194–197.

Ramadass, P., Jarvis, B.D. Corner, R.J. Penny, D. and Marshall, R.B. (1992). "Genetic characterization of pathogenic *Leptospira* species by DNA hybridization." *Int. J. Syst. Bacteriol.* 42: 215–219.

Ravindran, R., Rao, J.R. and Mishra, A.K. (2006). "Detection of *Babesia bigemina* DNA in ticks by DNA hybridization using a nonradioactive probe generated by arbitrary PCR." *Vet. Parasitol.* 141:181–185.

Rock, D.L., Beam, S.L. and Mayfield, J.E. (1987). "Mapping bovine herpesvirus type 1 latency-related DNA in bigeminal ganglia of latently infected rabbits." *J. Virol.* 61: 3827–3831.

Rossi, M.S., Sadir, A.M., Schudel, A.A. and Palma, E.L. (1988). "Detection of foot and mouth disease virus with DNA probes in bovine oesophageal fluids." *Arch. Virol.* 99: 67–74.

Sachse, K., Hotzel, H., Slickers, P. and Ehricht R. (2006). "The use of DNA microarray technology for detection and genetic characterization of chlamydiae." *Dev. Biol. (Basel).* 126: 203–210.

Sellers, H.S., Villegas, P.N., El-Attrache, J., Kapczynski, D.R. and Brown, C.C. (2001). "Detection of infectious bursal disease virus in experimentally infected chickens by *in situ* hybridization." *Avian Dis.* 45: 26–33.

St. Amand, A.L., Frank, D.N., De Groote, M.A. and Pace, N.R. (2005). "Use of specific rRNA oligonucleotide probes for microscopic detection of *Mycobacterium avium* complex organisms in tissue." *J. Clin. Microbiol.* 43: 1505–1514.

Takahashi, Y., Kishida, M., Yamamoto, S. and Fukunaga, M. (1999). "Repetitive sequence of *Leptospira interrogans* serovar *iceterhaemorrhagiae* strain *icter.* No. 1: a sensitive probe for demonstration of *Leptospira interrogans* strains." *Microbiol. Immunol.* 43: 669–678.

Terpstra, W.J., Schoone, G.J. and Ter Schegget, J. (1986). "Detection of leptospiral DNA with ^{32}P-and biotin-labeled probes." *J. Med. Microbiol.* 22: 23–28.

Vassart, G., Georges, M., Monsieur, R., Brocas, H., Lequarre, A.S. and Christopher, D. (1987). "A sequence in M13 phage detects hypervariable minisatellites in human and animal DNA." *Science.* 235: 683–684.

Wang, X.W., Zhang, L., Jin, L.Q., Jin, M., Shen, Z.Q., An, S, Chao, F.H. and Li, J.W. (2007). "Development and application of an oligonucleotide microarray for the detection of food-borne bacterial pathogens." *Appl. Microbiol. Biotechnol.* May 10.

Zhang, Z. and Kitching, P. (2000). "A sensitive method for the detection of foot and mouth disease virus by *in situ* hybridization using biotin-labelled oligodeoxynucleotides and tyramide signal amplification." *J. Virol. Methods.* 88: 187–192.

6

GENE CLONING

INTRODUCTION

Remarkable developments have occurred in techniques for manipulating prokaryotic as well as eukaryotic DNA. This has allowed breakage of a DNA molecule at two desired places to isolate a specific DNA segment and then insert it into another DNA molecule at a desired position. The product obtained is called **recombinant DNA** and the technique is called **genetic engineering**. With this technique, we can isolate and clone single copy of a gene or a DNA segment into an indefinite number of copies, all identical. This became possible because vectors like plasmids and phages reproduce in a host (e.g. *E. coli*) in their usual manner even after insertion of a foreign DNA, so that the inserted DNA will also replicate faithfully, with the parent DNA. This technique is called gene cloning. Using this technique, genes can be isolated, cloned and characterized, and hence this technique has led to significant progress in all areas of molecular biology.

A variety of vectors have been developed which not only allow multiplication, but can also be manipulated in such a way that the inserted gene may express in the host.

CLONING VECTORS FOR RECOMBINANT DNA

Cloning vectors are small DNA vehicles designed to transport a foreign DNA fragment. DNA molecules are cloned for a variety of purposes including isolating novel genes, safeguarding DNA samples, facilitating sequencing, generating probes, and expressing recombinant protein in one or more host organisms. The DNA to be cloned can be produced from a number of sources including PCR, restriction digesting an existing vector, cDNA synthesis, RT-PCR, or genomic DNA preparations.

Few of the cloning vectors which are used in recombinant DNA technology include plasmids, bacteriophages, cosmids, phagemids and viruses. An ideal cloning vector should have the following characteristics:

1. It should have its own replicon and thereby be capable of autonomous replication in the host cell.

2. It should carry one or more selectable marker function, in order to permit the recognition of cells carrying the parental form of the vector or a recombinant between the vector and foreign DNA sequences.

3. It should have a cloning site, a region containing unique restriction enzyme cleavage site into which a foreign DNA can be inserted without interference in plasmid's ability to replicate or to confer suitable phenotype.

Plasmids as Vectors

Plasmids are autonomous elements, whose genomes exist in the cell as extrachromosomal units. They are self replicating circular duplex DNA molecules which are maintained in a characteristic number of copies in a bacterial cell, yeast cell or in eukaryotic cells. The cloning site is usually located in the middle of a selectable marker. The insertion of the cloned insert alters the associated phenotype. The circular plasmid DNA which is used as a vector can be cleaved at one site with the help of a restriction enzyme to give a linear DNA molecule. A foreign DNA segment can now be inserted by joining the ends of a linearized plasmid DNA to the two ends of a foreign DNA, thus regenerating a recombinant DNA molecule that can now be separated by gel electrophoresis on the basis of its size. Selection of recombinant DNA is facilitated by the resistance genes, which the plasmid may carry against one or more antibiotics. If a plasmid has two such genes conferring resistance against two antibiotics and if a foreign DNA insertion site lies within one of these two genes, then the chimeric vector loses resistance against one antibiotic. In such a situation, the parent vector in bacterial cells can be selected by resistance against two antibiotics and the recombinant DNA can be selected by retention of resistance against only one of the two antibiotics.

One of the standard cloning plasmid vectors widely used in the gene cloning experiments is pBR 322 (derived from *E. coli* plasmid ColE1), which is a 4362 bp DNA and was derived by several alterations of the earlier cloning vector. pBR 322 has been named after Bolivar and Rodriguez, who prepared this vector. (Figure 6.1). It has genes for resistance against two antibiotics (tetracycline and ampicillin), an origin of replication and a variety of restriction sites for cloning of restriction fragments obtained through cleavage with a specific restriction enzyme.

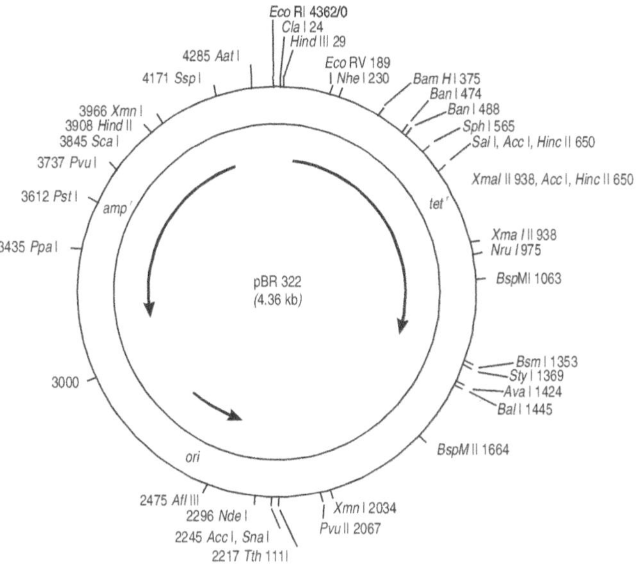

Figure 6.1 pBR 322 cloning vector

Another series of plasmids that are used as cloning vectors belong to pUC series (named after the place of their initial preparation, i.e., University of California). These plasmids are 2700 bp long and possess, (i) an ampicillin resistance gene; (ii) an origin of replication derived from pBR 322 and (iii) the *lacZ* gene derived from *E. coli*. Within the *lac* region is also found a polylinker sequence having unique restriction sites (identical to those found in phage M13). When DNA fragments are cloned in this region of pUC, the *lac* gene is inactivated. These plasmids, when transferred into an appropriate *E. coli* strain, which is *lac⁻* (e.g. JM103, JM109), and grown in the presence of isopropyl thiogalactoside (IPTG, which behaves like lactose, and induces the synthesis of β-galactosidase enzyme) and X-gal (substrate for the enzyme), will give rise to white or clear colonies. On the other hand, pUC having no inserts and transferred into bacteria will have an active *lacZ* gene and therefore will produce blue colonies, thus permitting identification of colonies having pUC vector with DNA segment. The cloning vectors belonging to pUC family are available in pairs with reversed orders of restriction sites relative to *lacZ* promoter, pUC 8 and pUC 9 make one such pair. Other similar pairs include pUC 12 and pUC 13, pUC 18 and pUC 19 (Figure 6.2) or pUC 118 and pUC 119.

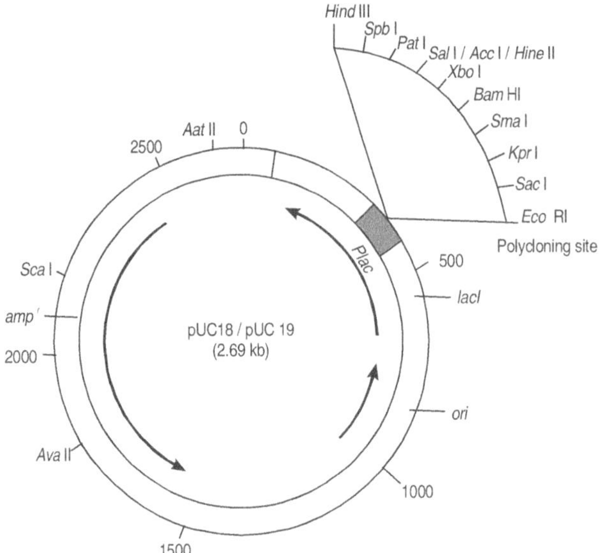

Figure 6.2 pUC 18 cloning vector

In pBR 322, the DNA is inserted at a site located within one of the two genes for resistance against antibiotics so that it will inactivate one of the two resistance genes. The insert bearing plasmids can be selected by their ability to grow in a medium containing both the antibodies. The plasmid carrying no insert on the other hand, will be able to grow in media containing one or both the antibiotics. In this manner, the presence of *lacZ* gene in pUC and resistance genes against ampicillin and tetracycline in pBR 322 allow selection of *E. coli* colonies transferred with plasmids carrying the desired foreign cloned DNA segment.

Yeast Plasmid Vectors

These are special vectors used for introducing DNA segments into yeast cells, a eukaryotic system that has been used extensively for developing genetically engineered yeast cells. Some examples of the yeast plasmid vectors are listed in the following section.

    i. **YEp or Yeast episomal plasmids**   These carry 2 μm circles of DNA sequence including the origin of replication and *rep* genes. One example is YEp13 (Figure 6.3). This yeast cloning vector is a **shuttle vector**. This has a selectable *LEU 2* gene. YEp13 also includes the entire pBR 322 sequence and can therefore replicate and be selected for both yeast and *E. coli*. Shuttle vectors are those which can replicate in the cells of more than one organism (e.g. in *E.coli* and yeast).

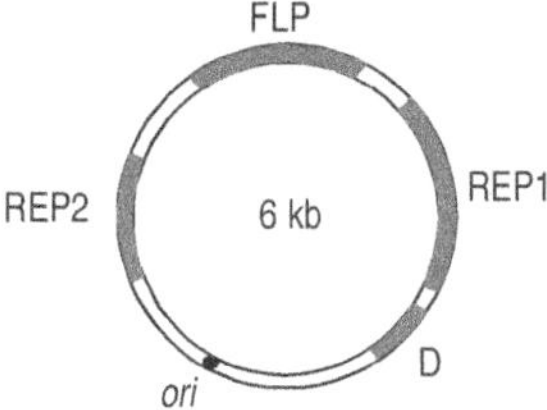

Figure 6.3 Yeast episomal plasmid

ii. **YIp or Yeast integrative plasmid** This allows transformation by crossing over and has no replication of origin. These plasmids will be stable in the yeast only on integration into the yeast chromosome. One example is YIp5 which is essentially a pBR vector with an inserted *URA3* gene (Figure 6.4a). This gene codes for orotidine-5-phosphate decarboxylase (an enzyme that catalyses one of the steps in the biosynthesis pathway for pyrimidine nucleotides) and is used as a selectable marker. This plasmid cannot replicate as a plasmid, but depends on yeast chromosomal DNA.

iii. **YRp or Yeast replicating plasmids** These carry autonomously replicating sequence; this sequence is very common with many yeast genes, and stable transformation can be achieved by crossing over. One example is YRp7, which is able to multiply as an independent plasmid. They carry origin of replication and are made of pBR 322 and yeast gene *TRP1* (Figure 6.4b). This gene is involved in tryptophan biosynthesis.

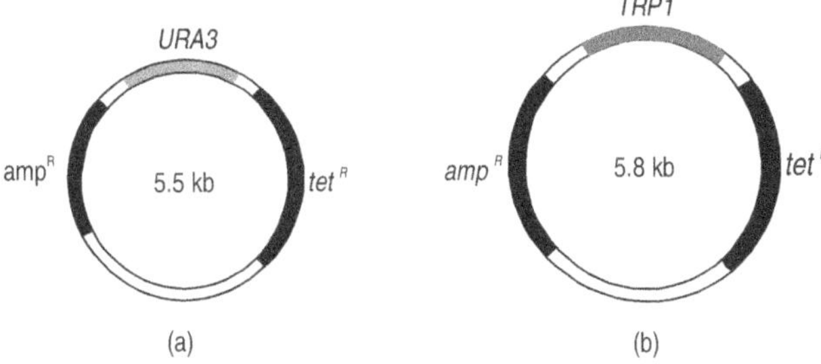

Figure 6.4 (a) Yeast integrative plasmid (YIp5); (b) Yeast replicative plasmid (YRp5)

iv. **YCp or Yeast centromere plasmids** These function as true chromosomes and segregate during mitosis and meiosis. This plasmid has *GEN* gene.

v. **pYAC or Yeast artificial chromosome vector** This carries centromere and telomere sequences and therefore can be used to obtain artificial chromosomes. One example is pYAC3, which is essentially a pBR 322 into which a number of yeast genes have been inserted. Two of these genes, *URA3* and *TRP1*, are seen as selectable markers for YIp5 and YRp7 respectively. The DNA fragment that carries *TRP1* gene also contains an origin of replication, but in pYAC3, this fragment is extended further to include the sequence called *CEN4*, which is the DNA from the centromere region of chromosome 4. The other component, the telomeres are provided by two sequences called *TEL*. This plasmid also contains *SUP4*, which is a selectable marker into which new DNA is inserted during the cloning experiment (Figure 6.5).

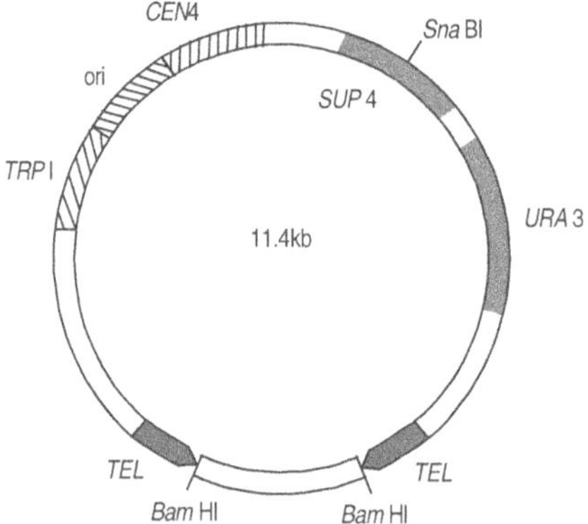

Figure 6.5 Yeast artificial chromosome (pYAC3)

For cloning, the vector is first restricted with a combination of *Bam* HI and *Sna* BI, cutting the molecule into three fragments. The *Bam* HI fragment is removed, leaving two arms, each bounded by one *TEL* sequence and one *Sna* BI site. The DNA to be cloned, which must have blunt ends (*Sna* BI, a blunt end cutter, recognizes TACGTA) is ligated between the two arms producing the artificial chromosome.

Bacteriophages as Vectors

Bacteriophages provide another source of cloning vectors. Since usually a phage has a linear DNA molecule, a single break will generate two fragments, which are later joined together with foreign DNA to generate a chimeric phage particle. The chimeric phage can be isolated after allowing it to infect bacteria and collecting progeny particles after a lytic cycle.

The lambda phage is a typical example of head and tail phage (Figure 6.6). The DNA is contained in the polyhedral head structure and the tail serves to attach the phage to the bacterial surface and to inject the DNA into the cell. The lambda DNA molecule is 49 kb in size and occurs both in

linear and circular forms. The linear molecule contains single-stranded complementary termini 12 nucleotide in length. Soon after entering a host bacterium, the cohesive termini associate by base-paring to form a circular molecule and the nicks are sealed by the host DNA ligase to generate a closed circular DNA molecule. An electron micrograph of a bacteriophage is shown in Figure 6.7.

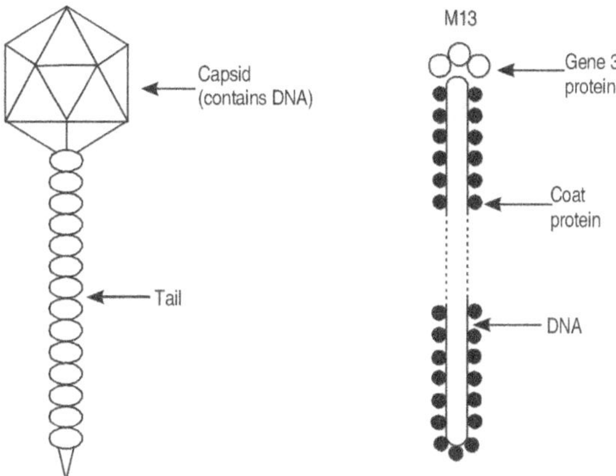

Figure 6.6   Phage and lambda cloning vectors

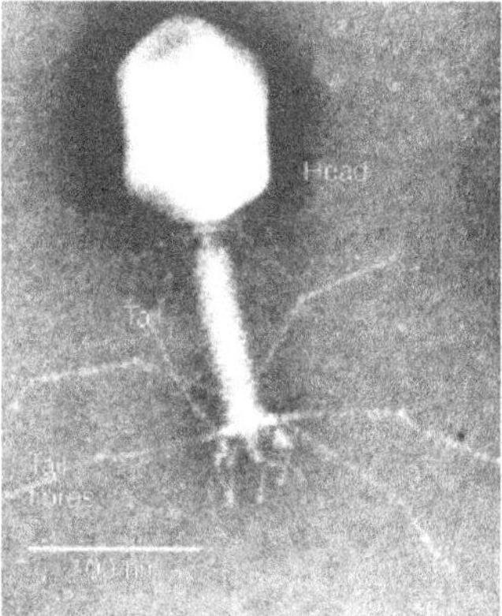

Figure 6.7   Electron micrograph of a bacteriophage

λ-phage vectors include λ-gt 10, λ-gt 11, EMBL3, EMBL4 and Charon. These vectors allow cloning of larger DNA fragments. Phage λ vectors permit cloning of segments up to 20–25 kb long and cosmid vectors accommodate up to 45 kb long. These λ vectors however, are easier and more efficient for making genomic and cDNA libraries.

λ-gt 10 and λ-11 are modified λ phages designed to clone cDNA fragments. The major difference between these two vectors is that λ-gt 11 is an expression vector, where inserted DNA is expressed as β-galactosidase fusion protein. λ-gt 10 is a 43-kb dsDNA for cloning fragments that are only 7 kb in length. The insertion of DNA inactivates cI^+ (repressor) gene generating cI^- bacteriophage. Non-recombinant λ-gt 10 is cI^+ and forms cloudy plaques on appropriate *E. coli* host, while recombinant cI^- λ-gt 10 forms clear plaques permitting screening of recombinant plaques.

λ-gt 11 is a 43.7-kb ds λ phage designed to clone DNA segments, which are less than 6 kb in length (usually for cDNA). Foreign DNA can be expressed as β-galactosidase fusion protein. Recombinant λ-gt 11 become gal^-, while non-recombinant λ-gt 11 remains gal^+, so that an appropriate *E. coli* host with recombinant phage (gal^-) will form white or clear colonies and that with non-recombinant phage (gal^+) will form blue colonies permitting screening in the presence of IPTG and X-gal.

Insertion vector   In these vectors a large segment of the non-essential region has been deleted and the two arms are ligated together. An insertion vector possesses at least one unique restriction site into which new DNA can be inserted. Two popular insertion vectors are: (i) λ-gt 10 which carries up to 8 kb of new DNA inserted into a unique *Eco*RI site located in the *cI* gene. Insertional inactivation of this gene means that recombinants are distinguished as clear rather than turbid plaques. (ii) λZAPII which can carry up to 10 kb new DNA into any of the six restriction sites with a polylinker which inactivates the *lacZ* gene carried by the vector. Recombinants give clear, rather than blue plaques on X-gal agar.

Replacement vectors   A replacement vector has two recognition sites for the restriction endonucleases used for cloning. These sites flank a segment of DNA that is replaced by the DNA to be cloned. Often, the replacement fragment (or stuffer fragment) carries additional restriction sites that can be used to cut it up into small pieces. Replacement vectors are generally designed to carry larger pieces of DNA than insertion vectors can handle. The recombinant selection is on the basis of size with non-recombinant vectors being too small to be packaged into the lambda phage heads. λEMBL 4 carries up to 20 kb of inserted DNA by replacing a segment flanked by a pair of *Eco* RI, *Bam* HI and *Sal* I sites.

Retriever vectors are another class of vectors, which are used to retrieve specific genes from the normal chromosome of an organism like yeast through recombination. They will multiply both in *E. coli* and yeast. Retriever vectors are very useful in isolation of genes from yeast for further molecular studies including sequencing.

EMBL 3 and EMBL 4 (replacement vectors) are the two vectors that are designed so that a central non-essential part of 44 kb long phage can be replaced by foreign DNA.

Cleavage of the phage with an appropriate enzyme generates 3 fragments (left arm, right arm and a central fragment called stuffer). The central fragment representing 40% of the phage genome is non-essential for propagation of the phage and can be replaced by foreign DNA that may be as long as 20–23 kb. Non-recombinant DNA resulting from the fusion of left arm and right arm will give a DNA that is too small for packaging, so there is automatic selection against non-recombinant phages. The two vectors have polylinkers with reverse order of restriction sites with respect to each other.

Charon 34 and Charon 35 differ from each other only in their central fragments and will accept fragments of 9–20 kb size. They have more extensive range of restriction targets within their polylinkers than do EMBL 3 and EMBL 4.

M13 phage is a filamentous bacteriophage of *E. coli* and contains a 7.2 kb long single-stranded circular DNA. M13 phage has been variously modified to give rise to M13 mp series of cloning vectors which can be used for cloning of a wide variety of DNA fragments and have the unique advantage of generating large quantities of DNA molecules that carry the sequence of one strand of the foreign DNA. Such single-stranded DNAs are the templates of choice for

- DNA sequencing by the dideoxy chain termination method

- generating DNA probes for hybridization that are radiolabelled in only one strand and

- site-directed mutagenesis using synthetic oligonucleotide.

Cloning vectors of M13 mp series have a *lacZ* gene that complements gal host (e.g. JM 103 or JM 104) giving rise to blue colonies. But if a foreign DNA segment is inserted at one of the polylinker sites associated with *lacZ* gene, it inactivates *lacZ* gene and no complementation is possible. Thus, on transformation, only white or clear plaques are obtained, permitting selection of recombinant M13 mp plaques. Once the foreign DNA is cloned in M13 vector, commercially available oligonucleotide primers are used for copying the insert in the presence of dideoxynucleotides, so that fragments of different sizes with known termini are produced, permitting the determination of the sequence. Reverse order of the restriction sites in polylinker is present in a pair of vectors like M13 mp8 and M13 mp9 permitting sequencing from both the ends of dsDNA molecule.

Cosmid vectors are modified plasmids that carry the DNA segments (*cos* sequences) required for packaging DNA into bacteriophage λ particles. Because cosmids carry on origin of replication (Col EI) and a drug resistance marker (amp), cosmid vectors can be introduced into *E. coli* by standard transformation procedures and propagated as plasmids. The advantage of the use of cosmids for cloning is that its efficiency is high enough to produce a complete genomic library of 10^6–10^7 clones from a mere 1 mg of insert DNA. The disadvantage, however is its inability to accept more than 40–50 kb of DNA.

Phagemids used as vectors are prepared artificially by combining features of phages with plasmids, as the name suggests. One such phagemid, which is commonly used in molecular biology laboratories is pBluescript II KS, which is derived from pUC 19 and is 2961 bp long. The KS designation indicates the orientation of polylinker, such that the transcription of *lacZ* gene proceeds from restriction site for *Kpn* I to that for *Sac* I (Figure 6.8). This phagemid has the following features:

1. a multi cloning site (MCS) flanked by T3 and T7 promoters to be read in opposite directions on two strands,

2. an inducible *lac* promoter (*lac* I), upstream of *lacZ* region, which complements with *E. coli* (lacZ⁻) and provides the facility for selection of chimeric vector DNA (recombinant vector) using the criteria of white colonies (as against blue colonies obtained if no foreign DNA is inserted)

3. f1 (+) and f1 (–) origins of replication derived from a filamentous phage for recovery of sense (+) and antisense (–) strands of *lacZ* gene, when host is co-infected with a helper phage

4. an origin of replication (ColEI *ori*) derived from plasmid, and used in the absence of helper phage

5. a gene having ampicillin resistance for antibiotic selection.

The **bacteriophage P1** cloning vector can allow cloning of 100 kb long DNA segments with an efficiency of 10^5 clones/mg of insert DNA. Therefore, in their capability, they fall between YACs and cosmids. The vector with insert DNA is amplified in *E. coli* and several micrograms of cloned DNA can be recovered from 5–10 ml of exponential phase of *E. coli* cells.

Few of plant viruses have been used as vectors for cloning. Cauliflower mosaic virus (CaMV), tobacco mosaic virus (TMV) and geminiviruses are 3 groups of plant viruses that have been used as vectors from cloning of DNA

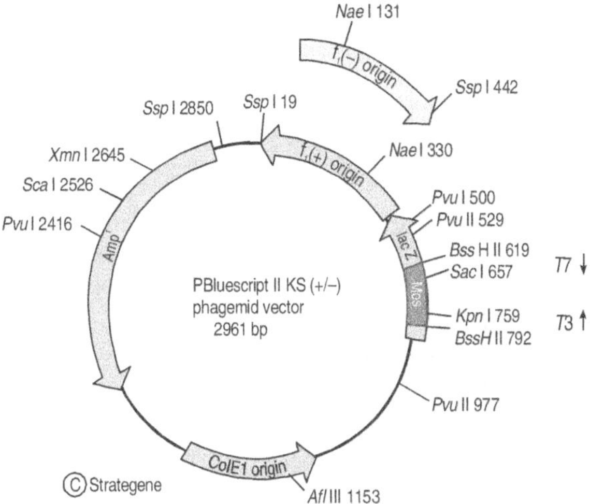

Figure 6.8 Phagemid vector

segments. A number of animal viruses are also used as vectors either for the delivery of nucleic acids into cultured cells followed by its integration with the host genome or for the amplification of high level expression of foreign gene, using promoters from virus genes.

Commercial Cloning and Expression Vectors

The TA Cloning® Kit is designed for cloning PCR products into bacteria directly from a PCR reaction without requiring modifying enzymes, purification, or restriction digestion. The TA Cloning® vector, pCR2.1, contains the *lacZ*-alpha complementation fragment used for blue-white colour screening, ampicillin, and kanamycin resistance genes for selection, as well as a versatile polylinker.

TOPO® Cloning is the most effective technology available for cloning DNA. It yields >95% recombinants* via a simple 5-minute, bench-top ligation. In TOPO® Cloning, the enzyme topoisomerase I is used in the place of DNA ligase, greatly increasing the success and speed of the cloning reaction. We have combined TOPO® Cloning with the TA Cloning® and Zero Background™ technologies to enable rapid, high-efficiency cloning of different types of DNA.

ViraPower™ lentiviral expression vectors   There are three pLenti-DEST™ Gateway® vectors (pLenti6/V5-DEST™, and pLenti4/V5-DEST™, pLenti6/UbC/V5-DEST™) and a single pLenti TOPO® Cloning vector (pLenti6/V5-D-TOPO®) that are available for reproducible, stable expression in the ViraPower™ Lentivirus Expression System.

pTYB11 is an *E. coli* cloning and expression vector (7414 bp) used in the IMPACT™ Protein Purification System which allows the over-expression of a target protein as a fusion to a self-cleavable affinity tag (1,2). It is an N-terminal fusion vector designed for in-frame insertion of a target gene into the polylinker, downstream of the intein tag (the *Sce* VMA intein/chitin binding domain, 55 kDa) (3,4). This allows the N-terminus of the target protein to be fused to the intein tag. The self-cleavage activity of the intein allows the release of the target protein from the chitin-bound intein tag, resulting in a single-column purification of the target protein. This vector can be used for expression and purification of a native target protein without any vector-derived residues.

RESTRICTION ENDONUCLEASES IN CLONING

The restriction enzymes have played a key role in the development of recombinant DNA technology. These enzymes have been found in nearly every bacterial species examined and are known to catalyse double-stranded breaks in DNA to yield restriction fragments. The designation **restriction endonuclease** derives from the natural function of these enzymes in restricting the growth of the virus that attack bacteria. The enzymes do this by binding to the viral DNA and cleaving it at highly specific sites within or adjacent to a particular sequence known as recognition sequence. The term **endonuclease** applies to sequence-specific nucleases that break nucleic acid chains somewhere in the interior, rather than at the ends, of the molecule. Nucleases that function by removing nucleotides from the ends of the molecule are called **exonucleases**.

A restriction endonuclease functions by "scanning" the length of a DNA molecule. Once it encounters its particular specific recognition sequence, it will bind to the DNA molecule and makes one cut in each of the two sugar–phosphate backbones of the double helix. The positions of these two cuts, both in relation to each other, and to the recognition sequence itself, are determined by the identity of the restriction endonuclease used to cleave the molecule in the first place. Different endonucleases yield different sets of cuts, but one endonuclease will always cut a particular base sequence the same way, no matter what DNA molecule it is acting on. Once the cuts have been made, the DNA molecule will break into fragments.

The site specificity is important since it enables the bacteria to defend its own DNA against attack by the restriction enzymes by methylating the corresponding sites of its own DNA. In almost all cases, a bacterium that makes a particular restriction endonuclease also synthesizes a companion DNA methyltransferase, which methylates the DNA target sequence for that restriction enzyme, thereby protecting it from cleavage. This combination of restriction endonuclease and methylase is referred to as a **restriction-modification system**. The observations suggesting the existence of restriction enzymes were made by W. Arber and co-workers in the 1960s.

These enzymes are classified into three groups. Type I endonucleases are large multimeric proteins that are capable of both cleaving and modifying DNA. Enzymes of this type require, ATP, Mg^{++} and a cofactor S-adenosyl methionine (SAM) for activity. Their recognition sites are complex and DNA cleavage occurs at non-specific sites. These enzymes contain 3 different types of subunits. Type II enzymes are smaller monomeric proteins that require only Mg^{++} for activity. These enzymes have only restriction activity; modification activity is carried by a separate enzyme. Unlike type I enzymes, this class cuts DNA in a predictable and consistent manner, at a site within or adjacent to the recognition sequence and is shown to produce DNA fragments with cohesive or blunt ends. Because of this site-specific nature of cleavage, type II restriction enzymes have become the primary tools in the analysis and restructuring of DNA. Type II enzymes are somewhat less complex. DNA cleavage occurs distal to the recognition site. These endonucleases contain only two types of subunits and require Mg^{++} and ATP for activity. It has both restriction and modification activities.

Well over 1200 different restriction enzymes have been identified to date and among them many are isochizomers, i.e., enzymes that recognize the same DNA sequences. Altogether, about 130 different specificities have been described. Different strains of individual species often contain different restriction enzymes; in *E. coli* for example, 141 restriction enzymes with 40 different specificities are known. In order to simplify the naming of these based on the name of the organisms from which the enzymes were isolated, the first initial of the genus and the first two initials of the species form the basic name. When the enzyme is present in a specific strain, these 3 italicized letters may be followed by a strain designation. The third portion of the name is reserved for a Roman numeral indicating the order of discovery of the enzyme in a particular strain. For example *Hae* III is purified from the strain *Haemophilus aegypticus* and is one of the three restriction enzymes present in the strain. Similarly, *Hinf* I is the first enzyme purified from the strain *Haemophilus influenzae* strain f´. The commonly used restriction enzymes are given in the Table 6.1. A typical DNA restriction enzyme digestion with

enzyme *Bam* HI is shown in Figure 6.9.

Table 6.1 Restriction enzymes, their origin and cutting sequences

Name of the enzyme	Source	Cutting sequence
Eco RI	*Escherichia coli*	G↓AATTC
Hind III	*Haemophilus influenzae*	A↓AGCTT
Hpa II	*Haemophilus parainfluenzae*	C↓CGG
Hae III	*Haemophilus aegypticus*	GG↓CC
Pvu I	*Proteus vulgaris*	CGAT↓CG
Bam HI	*Bacillus amyloliquefaciens*	G↓GATCC
Sal I	*Streptomyces albus*	G↓TCGAC
Pst I	*Providencia stuartii*	CTGCA↓G

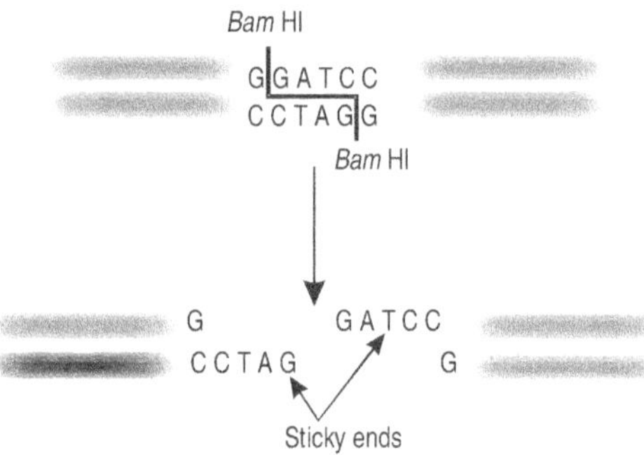

Figure 6.9 DNA restriction enzyme digestion

One unit of an enzyme is defined as the quantity required to cut 1 μg of DNA in one hour. Most restriction enzymes will function adequately at pH 7.4, but different enzymes vary in their requirements of ionic strength (provided usually by sodium chloride) and Mg^{++} concentration. It is also advisable to add a reducing agent such as dithiothreitol which will stabilize the enzyme and prevent its inactivation.

DNA restriction pattern obtained with digestion of different leptospiral serovar DNAs with restriction enzyme *Hind* III showing unique pattern for each serovar is shown in Figure 6.10.

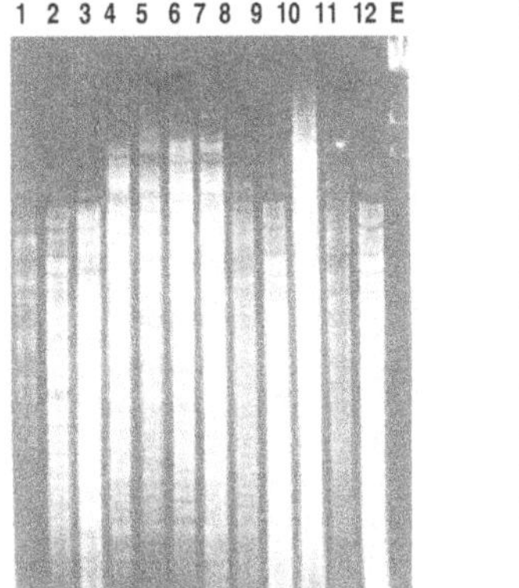
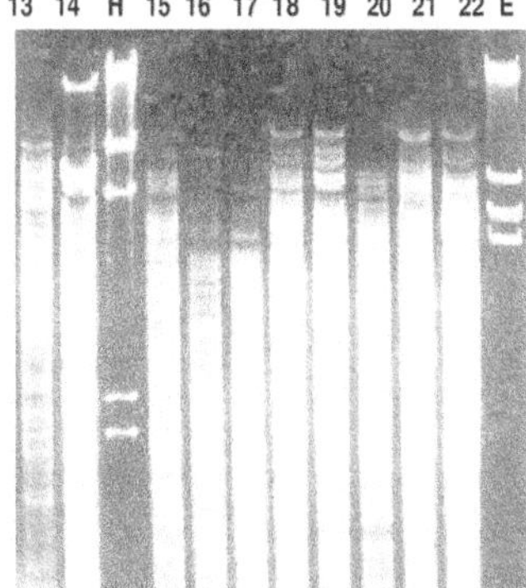

Figure 6.10 Restriction enzyme analysis of leptospiral DNA along with the genomic DNA of *Staphylococcus aureus and E.coli*

The recognition sites for type II restriction endonuclease are generally 4 (tetramers), 5 (pentamers); 6 (hexamers) or 8 (octamers) base pairs in length (Table 6.2). The length of restriction recognition sites varies: The enzymes *Eco* RI, *Sac* I and *Sst* I each recognize a 6 base-pair (bp) sequence of DNA, whereas *Not* I recognizes a sequence 8 bp in length, and the recognition site for *Sau*3 AI is only 4 bp in length. Length of the recognition sequence dictates how frequently the enzyme will cut in a random sequence of DNA. Enzymes with a 6 bp recognition site will cut, on average, every 4^6 or 4096 bp; a 4 bp recognition site will occur roughly every 256 bp.

Table 6.2 Different types of enzymes with their cutting sequences

Types of cutting	Enzymes	Cutting sequences
Tetramers (4 base cutters)	*Hae* III	GG ↓ CC
	Taq I	T ↓ CGA
	Hha I	GCG ↓ C
	Hpa II	CC ↓ GG
	Tha I	CG ↓ CG
Pentamers (5 base cutters)	*Hind* I	G ↓ ANTC
Hexamers (6 base cutters)	*Bam* HI	G ↓ GATCC
	Eco RI	G ↓ AATTC
	Hind III	A ↓ AGCTT
	Sma I	CCC ↓ GGG
	Dra I	TTT ↓ AAA
	Pst I	CTGCA ↓ G
Octamers (8 base cutters)	*Not* I	GC ↓ GGCCGC

The frequency of occurrence of a restriction site in a DNA can be estimated using the formula:

$$\text{Site frequency} = 1/4^n$$

where n = length of the restriction site sequence. For a tetrameric sequence, the frequency is $1/4^4$ bases or an average of 1 site for every 256 base pairs and for a hexameric sequence, the frequency is $1/4^6$ or approximately 1 site for every 4096 base pairs.

The recognition site for one enzyme may contain the restriction site for another. For example, note that a *Bam* HI recognition site contains the recognition site for *Sau*3 AI. Consequently, all *Bam* HI sites will cut with *Sau*3 AI. Similarly, one of the four possible *Xho* II sites will also be a recognition site for *Bam* HI and all four will cut with *Sau*3 AI.

The restriction enzymes most used in molecular biology labs cut within their recognition sites and generate one of three different types of ends. In the following illustrations, below, the recognition site is shown as shaded area and the cut sites indicated by triangles.

- **5′ overhangs** The enzyme cuts asymmetrically within the recognition site such that a short single-stranded segment extends from the 5′ ends. *Bam* HI cuts in this manner.

```
          ▼
5′–A–T–G–G–A–T–C–C–A–A–3′   Bam HI     –A–T–G          5′–G–A–T–C–C–A–A–
   | | | | | | | | | |      ───────►    | | |              | | |
3′–T–A–C–C–T–A–G–G–T–T–5′               –T–A–C–C–T–A–G–5′   G–T–T–
          ▲
```

- **3′ overhangs** Again, we see asymmetrical cutting within the recognition site, but the result is a single-stranded overhang from the two 3′ ends. *Kpn*I cuts in this manner.

```
          ▼
5′–G–A–G–G–T–A–C–C–C–T–3′   Kpn I      –G–A–G–G–T–A–G 3′       C–C–T–
   | | | | | | | | | |      ───────►    | | |                   | | |
3′–C–T–C–C–A–T–G–G–G–A–5′               –C–T–C        3′ C–A–T–G–G–G–A–
          ▲
```

- **Blunts** Enzymes that cut at precisely opposite sites in the two strands of DNA generate blunt ends without overhangs. *Sma* I is an example of an enzyme that generates blunt ends.

```
          ▼
5′–T–A–C–C–C–G–G–G–T–C–3′   Sma I      –T–A–C–C–C          –G–G–G–T–C–
   | | | | | | | | | |      ───────►    | | | | |            | | | | |
3′–A–T–G–G–G–C–C–C–A–G–5′               –A–T–G–G–G          –C–C–C–A–G–
          ▲
```

The 5´ or 3´ overhangs generated by enzymes that cut asymmetrically are called sticky ends or cohesive ends, because they will readily stick or anneal with their partner by base-pairing.

All restriction enzymes cleave the DNA substrates to form 5´ phosphate and 3´-hydroxyl termini on each strand. There are two types of restriction cleavages, viz., blunt end cutting as in *Sma* I (CCC↓GGG) or *Hae* III (GG↓CC) and cohesive end or sticky end cutting as in *Eco*RI (G↓AATTC).

CLONING STRATEGIES

The complexity of the cloning experiment depends on the overall aim of the work and the type of source material from which the nucleic acids will be isolated for cloning. Thus, a strategy to isolate and sequence a relatively small DNA fragment from *E. coli* will be different from a strategy to produce a recombinant protein in transgenic eukaryotic organisms.

Cloning strategies can be divided into two categories—those which involve the construction of **gene bank** or **gene library** and those require cloning of specific DNA fragments. For larger genomes of higher eukaryotes, it is necessary to clone the maximum number of different restriction fragments to obtain the gene of interest at a reasonable frequency. Thus, we obtain DNA sequences from large numbers of recombinants which contain a complete collection of nearly all DNA sequences in the entire genome. Such a collection of randomly cloned fragments that encompass the entire genome of a given species is called the **gene library**. Sometimes it is also called **gene bank**.

STEPS IN CLONING

The different stages of cloning involve:

1.   Isolation of DNA from the source organism;
2.   Cleavage of the vector and foreign DNA by restriction digestion;
3.   Combining the fragments of the foreign DNA with the vector to form recombinant DNA (rDNA) molecule (ligation);
4.   Introducing the rDNA into a host (transformation) and
5.   Screening hosts for a specific fragment of DNA.

The steps involved in gene cloning are depicted in Figure 6.11.

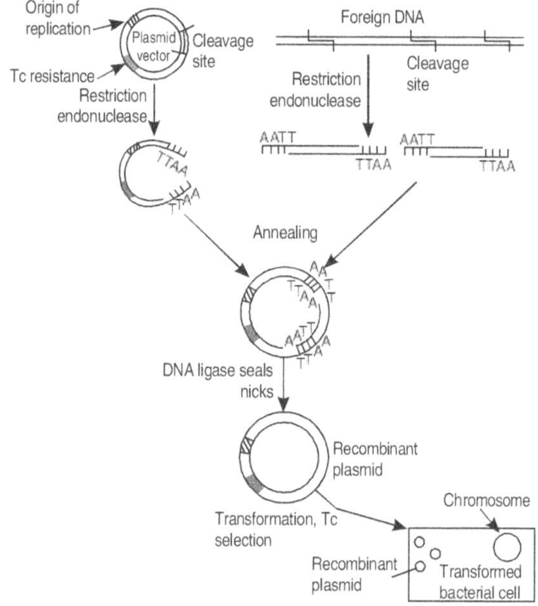

Figure 6.11 Steps in gene cloning

Preparation of DNA Fragments for Cloning

Cloning from either DNA or RNA, each has merit depending on the circumstances. Genomic DNA contains non-coding regions, such as introns, control regions and repetitive sequences. This can sometimes cause problems, particularly if the genome is large and the aim is to isolate a single copy gene. If the primary interest is the investigation of the overall structure of a particular gene and to examine the control sequences responsible for regulating gene expression, genomic DNA is the only alternative.

mRNA has two advantages over genomic DNA as a source material. It represents the genetic information that is being expressed by a particular cell type from which it is prepared. mRNA represents the coding sequence of the gene with any introns having been removed during RNA processing. Thus, production of recombinant protein is much more straightforward, if a clone of mRNA is available. If the lambda replacement vector such as EMBL4 is used, the maximum cloning capacity will be around 22 kb. Thus, fragments of this size must be available for the production of recombinants. In practice, a range of fragment sizes are used, often between 17 to 23 kb for a vector such as EMBL4. For a completely random library, the starting material could be from a high-mol. wt. DNA, and this should be fragmented by a totally random method.

Isolation of DNA in excess of 100 kb in length is desirable. Fragmentation of DNA is to be done by either mechanical shearing or by partial digestion with a restriction enzyme. Mechanical shearing by forcing DNA through a syringe and needle or by sonication will generate random fragments without cohesive termini. Thus, further filling or trimming in the ragged ends of the molecules will be required before the DNA can be joined to the vector, usually with linkers, adaptors or homopolymer tails.

Cloning from DNA   The genomic DNA used in gene cloning is predominantly chromosomal, which is large and fragile. The DNA has to be isolated and purified. The method of harvesting the genomic DNA varies extensively depending upon the source.

Cloning from mRNA   It is not possible to clone mRNA directly, so it has to be converted into a DNA before being inserted into a suitable vector. This is achieved using the enzyme reverse transcriptase to produce complementary DNA (cDNA). Method of cDNA synthesis utilizes the poly(A) tract at the 3´ end of the mRNA to bind an oligo(dT) primer, which provides the 3´-OH group required by reverse transcriptase. Given the four dNTPs, and suitable conditions, reverse transcriptase will synthesize a copy of the mRNA to produce a cDNA–mRNA hybrid. The mRNA can be removed by alkali hydrolysis and the single-stranded cDNA converted into ds cDNA by using a DNA polymerase. In this second strand synthesis, the priming 3´-OH is generated by short hairpin loop regions that form at the end of the ss cDNA. After second strand synthesis, the ds cDNA can be trimmed by S1 nuclease to give a flush-ended molecule, which can be cloned into a suitable vector.

Cloning cDNA in plasmid vectors   Although many workers prefer to clone cDNA in a bacteriophage vector system, plasmids are still often used, particularly when isolation of the desired cDNA sequence involves screening a relatively small numbers of clones. Joining the cDNA fragments to the vector is usually achieved by one of the 3 methods, viz., blunt-end ligation, use of linker molecule and homopolymer tailing.

Blunt end ligation   This refers to the joining of DNA molecules with blunt ends, using DNA ligase. In cDNA cloning, the blunt ends may arise as a consequence of the use of S1 nuclease, or by filling in the protruding ends with DNA polymerase. High concentrations of the participating DNAs must be used, so that the chances of two ends coming together are increased. One potential disadvantage of blunt-end ligation is that it may not generate restriction enzyme recognition sequences at the cloning site, thus hampering excision for the insert from the recombinant. This is usually not a major problem, as many vectors now have a series of restriction sites clustered around the cloning site. Thus,

DNA inserted by blunt-end ligation can often be excised by using one of the restriction sites in the cluster.

Use of linker molecule   The linkers are self-complementary oligomers that contain recognition sequence for a particular restriction enzyme. One such sequence is 5´-CCGAATTCGG-3´, which in ds form will contain the recognition sequence for *Eco* RI (GAATTC). Linkers are synthesized chemically and can be added to cDNA by blunt-end ligation. The cDNA/linker is then cleaved with the linker-specific restriction enzyme, thus generating sticky ends prior to cloning. This can pose problems if the cDNA contains sites for the restriction enzyme used to cleave the linker, but these may be overcome by using methylase to protect any internal recognition sites from digestion by the enzyme.

A second approach to cloning by addition of sequences to the ends of DNA molecules involves the use of adaptors. These are ss-non-complementary oligomers that may be used in conjunction with linkers. When annealed together, the linker-adaptor with one blunt end and one sticky end is produced, which can be added to the cDNA to provide sticky-end cloning without digestion of linkers.

Homopolymer tailing   In this technique, the enzyme terminal transferase is used to add homopolylinkers to dA, dT, or dC of a DNA molecule. Early experiments in recombinant production used dA tails on one molecule and dT tails on the other, although the technique is now most often used to clone, cDNA into the *Pst* I site of a plasmid vector by dG–dC tailing. Homopolymers have two main advantages over other methods of joining DNAs from different sources. Firstly, they provide longer regions for annealing DNAs together than, for example, cohesive termini produced by restriction enzyme digestion. A second advantage is specificity. As the vector and insert cDNA have different but complementary 'tails' there is little chance of self annealing, and the generation of biomolecular recombination is favoured over a wider range of effective concentrations than in the case of other annealing/ligation reactions.

In the homopolymer tailing, the vector is cut with *Pst* I and tailed by terminal transferase in the presence of dGTP. This produces dG tails. The insert DNA is tailed with dC in a similar way and the two can then be annealed. This generates the original *Pst*I site, which enables the insert to be cut out of the recombinant using this enzyme.

Restriction Enzyme Digestion

For preparation of rDNA, the genomic DNA and vector DNA are to be cleaved with the same restriction enzyme which yields the compatible sticky ends. Genomic DNA is cleaved randomly to yield large number of fragments, while the vector is cut only once in a predetermined location. The two sets of molecules are then combined and enzymatically coupled by DNA ligase (Figure 6.12). Complete restriction digestion of DNA will not be suitable for generating a representative library. Partial digest will produce a collection of fragments that are essentially random. This can be achieved by varying the enzyme concentration or the time of digestion. The degree of digestion of a DNA by a restriction endonuclease is measured by agarose gel electrophoresis. By comparing the digestion products to know size strands, an estimate of the average size of the digest can be determined.

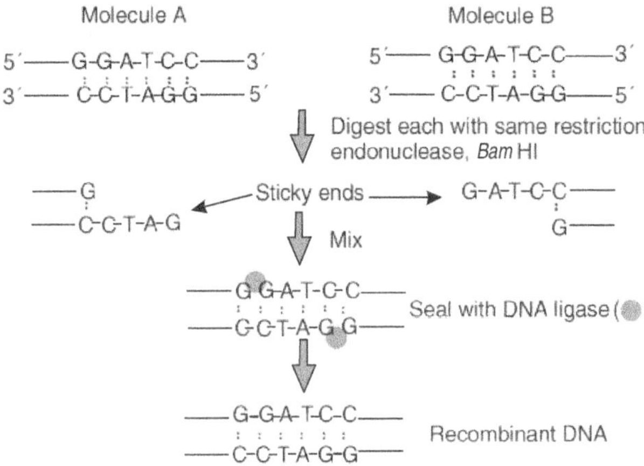

Figure 6.12   Steps involved in restriction endonucleus cleavage

Ligation of DNA

In order to generate a recombinant vector (plasmid), the plasmid DNA and the foreign DNA are exposed to an appropriate restriction enzyme, mixed together and treated with DNA ligase. The recombinant plasmid thus produced is also called as a **chimera**.

The construction of recombinant DNA is dependent on the ability of the DNA ligase to covalently seal nicks in the single-stranded DNA. When a restriction enzyme produces sticky ends, the cohesive ends will come together and will be held by hydrogen bonding between the complementary bases. The process in which two separate strands of nucleic acids interact to form a duplex molecule by interaction between complementary base pairs present in the individual strands, is often termed as **annealing**. The two molecules are then sealed and stabilized by a joining enzyme DNA ligase. This enzyme catalyses the formation of a phosphodiester bond between 3′-hydroxyl and 5′-phosphate of the adjacent nucleotide, establishing the structural continuity of the DNA strand.

DNA ligase plays a central role in the recombination, DNA replication and repair of the damaged DNA. DNA ligase from *E. coli* requires NAD+, while T4 DNA ligase requires ATP. T4 DNA ligase has the unique ability to join the blunt-end DNA fragments and because of this property, it is preferred for *in vitro* ligation of blunt ends of DNA fragments. Joining of cohesive ends produced by restriction endonuclease is accomplished by T4 DNA ligase and *E. coli* ligase. Cohesive end ligation proceeds 100 times faster than the blunt-end ligation.

To increase the efficiency of formation of rDNA molecule, alkaline phosphatase is used to remove the 5´ phosphate groups. Removal of the terminal 5´ phosphates on both ends of the vector suppresses self ligation and circularization of the plasmid DNA. However, a second molecule with a terminal 5´ phosphate can be ligated to a dephosphorylated vector. The resultant construct would have one nick on both strands, such a nicked molecule can be introduced into *E. coli* in which the nicks are repaired *in vivo* and the DNA is replicated.

Transformation

After inserting a foreign DNA into a vector of choice, the next step is the introduction of the recombinant vector in a host cell. The host cell should be easy to handle and propagate, should be available as a wide variety of genetically defined strains, and should accept the range of vectors. The bacterium *E. coli* fulfils these requirements and is commonly used in cloning protocols (Figure 6.13). Other organisms such as *Bacillus subtilis* and *Saccharomyces cerevisiae* are also used as hosts.

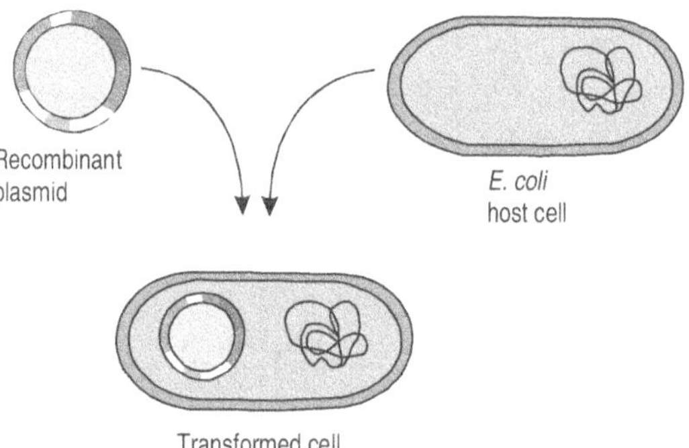

Figure 6.13 Bacterial transformation

The recombinant plasmid is introduced into an *E. coli* cell by the method of **transformation**. This process first requires the preparation of competent *E. coli* cells to take up the DNA. The natural transformation described by Griffith and Avery is an exceedingly rare event. However, in 1970 Mandel and Higa found that *E. coli* becomes markedly competent for transformation by foreign DNA when cells are suspended in cold calcium chloride solution and subjected to a brief heat shock at 42°C. Cells arrested in early to mid-log growth can be rendered more competent than cells in other stages of growth. Following uptake inside the *E. coli* cells, the plasmid replicates and expresses antibiotic resistance genes that allow selection of a transformed phenotype. The transformed bacterial cell is then grown on nutrient agar plates under optimum conditions for the colonies to develop. As each colony of cells is the progeny of a single cell, all cells have the same genetic make-up. This is called a **clone**. Besides transformation of *E. coli* with recombinant plasmid DNA, other methods include **transfection** with recombinant phage DNA and **transduction** with either recombinant phage or cosmid.

Identification, Selection, Screening and Analysis of Recombinants

Success of the cloning experiment depends on being able to identify the desired gene sequence among the many different recombinants that may be produced. Selection is where some sort of pressure (e.g. the presence of antibiotics) is applied during the growth of host cells containing recombinant DNA. The cells with the desired characteristics are therefore selected by their ability to survive. Screening is the procedure by which a population of viable cells is subjected to some sort of analysis that enables the desired sequences to be identified. Because only a small proportion of the large number of bacterial colonies or bacteriophage plaques being screened will contain the DNA sequence(s) of interest, screening requires methods that are highly sensitive and specific.

Selection for the presence of recombinant vector   The next step in gene cloning is to detect the inserted sequences, i.e., foreign DNA in the vector, or to determine whether it has acquired a foreign DNA or not. In the case of a plasmid vector, this can be recognized by altered antibiotic resistance. For instance, insertion of a foreign DNA fragment in *Pst* I site of pBR 322 inactivates the ampicillin resistance mechanism, but does not affect tetracycline resistance. In the case of phage vectors, advantage is taken of the fact that certain strains produce blue plaques when plated in the presence of chromogenic substrate (X-gal), but in the case of successful insertion in the gene region responsible for this colour change, colourless plaques are formed.

Selection for the presence of recombinant clones   After identifying the colonies in which recombinants have been formed, the next step is to screen and characterize recombinants which contain a specific DNA sequence. For selection and screening, several approaches are possible (Table 6.3).

Table 6.3   Methods used for identification, characterization and selection of recombinant colonies

Purpose	Methods
Selection of clones with a vector	(a) Antibiotic resistance
	(b) Plaque formation
Selection of clones with a recombinant vector	(a) Antibiotic resistance
	(b) X-gal test
Selection of recombinant clones with a specific DNA sequence	(a) Genetic
	(b) Nucleic acid hybridization
	(c) Recombinational

Genetic Methods

Genetic selection is a powerful tool since it can be applied to large populations. All useful vectors carry selectable genetic markers. Plasmid and cosmid vectors carry drug resistance, nutritional markers and for phages, plaque formation itself is a selection property. The genetic method depends on the inserted foreign DNA being expressed. The inserted gene may complement a metabolic defect in a strain of *E. coli* which has a requirement for a particular growth factor which an inserted DNA can synthesize. Transformants can then be selected by using an appropriate culture medium.

Genetic selection and screening rely on the expression (or non-expression) of certain traits. Usually, these traits are encoded by the vector. One of the simplest genetic selection methods involves the use of antibiotics to select for the presence of vector molecules. For example, the plasmid pBR 322 contains genes for ampicillin and tetracycline resistance. Thus, the presence of the plasmid in cells can be detected by plating potential transformants on an agar medium that contains either (or both) of these antibiotics. Only cells that have taken up the plasmid will be resistant and these cells will therefore grow in the presence of the antibiotic.

Insertional inactivation   The presence of cloned DNA fragments can be detected if the insert interrupts the coding sequence of a gene. This approach is known as insertional inactivation and can be used with any suitable genetic system. Three systems will be described to illustrate the use of the technique.

Antibiotic resistance can be used as an insertional inactivation system, if DNA fragments are cloned into a restriction site within an antibiotic-resistance gene. For example, cloning DNA into a *Pst* I site of pBR 322 (which lies within

the *amp*^r gene) interrupts the coding sequence of the gene and renders it
non-functional. Thus, cells that harbour a recombinant plasmid will be *Amp*^s
Tet^r. This can be used to identify recombinants as follows: if transformants
are plated firstly on to a tetracycline-containing medium, all cells that contain
the plasmid will survive, and form colonies. If a replica plate is then taken
and grown on ampicillin-containing medium, the recombinants (Amp^s Tet^r)
will not grow, but any non-recombinant transformants (Amp^r Tet^r) will. Thus,
recombinants are identified by their absence from the replica plate and can
be picked from an original plate and used for further analysis.

The **X-gal system** can also be used for screening cloned sequences. If a
DNA fragment is cloned into a functional β-galactosidase gene (e.g. into an
Eco RI site in Charon 16A), any recombinant will be genotypically lacZ⁻ and
therefore will not produce β-galactosidase in the presence of IPTG and X-gal.
Plaques formed by such phages will therefore remain colourless.
Non-recombinant phage will retain a functional *lacZ* gene, and therefore give
rise to blue plaques. This approach can also be used with the α-complementation
system; in the case, the inserted DNA inactivates the *lacZ* region in vectors
such as the M13 phage and pUC plasmid series. Thus, complementation will
not occur in recombinants, which will be phenotypically lacZ⁻ and will therefore
give rise to colourless plaques or colonies.

Plaque morphology can also be used as a screening method for certain
lambda vectors such as λ-gt 10, which contains the *cI* gene. This gene encodes
the cI repressor which is responsible for the formation of lysogens. Plaques derived
from *cI*⁺ vectors will be slightly turbid, due to the survival of some cells that
have become lysogens. If the *cI* gene is inactivated by cloning a fragment into a
restriction site within the gene, the plaques are clear and can be distinguished
from the turbid non-recombinants. This system can also be used as a selection
method.

Screening using nucleic acid hybridization   The power of nucleic acid
hybridization lies in the fact that complementary sequences will bind to each
other with a high degree of fidelity. When a suitable sequence has been selected,
it is labelled with ³²P. This produces a radioactive fragment of high specific
activity and can be used as an extremely sensitive screen for the gene of interest.
Alternatively, non-radioactive labelling methods may also be used.

Colonies or plaques are not suitable for direct screening, so a replica is
made on either a nitrocellulose membrane or nylon filters. This can be done
either by growing cells directly on the filter, or on an agar plate and by 'lifting'
replica from the plate (colonies or plaques). The DNA on the filter are

denatured. The probe is also denatured (usually by heating) and placed in a sealed plastic bag with the filters and incubated at a suitable temperature to allow hybrids to form with the homologous probe under standard conditions of incubation which is usually around 65–68°C. The duration of incubation may be up to 48 hours after hybridization; the filters are washed and allowed to dry. They are then exposed to X-ray film to produce an autoradiogram, which can be compared with the original plates to enable identification of the desired recombinant.

Immunological screening for expressed genes   An alternative method of screening with nucleic acid probes is to identify the protein products to a cloned gene by immunological techniques. The technique requires that the protein is expressed in recombinants and is most often used for screening cDNA expression libraries that have been constructed in vectors such as λ-gt 11. Instead of a nucleic acid probe, a specific antibody is used. The method used for immunological screening is similar to **plaque lift** screening with nucleic acid probe. Recombinant λ-gt 11 cDNA clones will express cloned sequences as β-galactosidase fusion proteins. Proteins can be picked up on a nitrocellulose filter and probed with an antibody. Detection can be carried out by a variety of methods, most of which use non-specific second binding molecule such as Protein A from bacteria, or a second antibody which attaches to the specifically bound primary antibody. Detection may be by a radioactive label (^{125}I-labelled protein A or second antibody) or by a non-radioactive method which produces a coloured product.

These methods use antigen–antibody reactions, if specific antibodies are available to the gene product (antigen) of the inserted DNA. Such immunochemical detection of clones synthesizing foreign protein has been found to be successful in cases where the inserted gene sequence is expressed. Antibodies adsorb strongly to plastic and if a plastic sheet is coated with a specific antibody and then gently applied to the surface of the lysed bacterial colonies on a replica plate, those colonies synthesizing a particular antigen will bind to the antibody. The immobilized antigen is then localized using a second antibody which is radiolabelled and then the plastic sheet is subjected to autoradiography. By referring back to the master plate, relevant colonies synthesizing a particular antigen can be identified.

Recombination   This method is based on homologous recombination in *E. coli* host. The probe sequence is inserted into a specially constructed plasmid vector. Foreign DNA in the form of genomic phage lambda libraries are propagated on recombination-proficient *E. coli*-containing plasmid with an inserted probe. Phage-carrying sequences homologous to the probe require an integrated copy of the plasmid by homologous recombination. This is screened and isolated under appropriate conditions.

EXPRESSION OF CLONED DNA IN *E. COLI* PLASMIDS

In order to achieve an efficient expression of a foreign gene in a bacterial cell, it is necessary to put that gene under the control of transcriptional and translational machinery of the host cell. For a eukaryotic gene to be transcribed within a prokaryotic cell, it is necessary to replace the eukaryotic promoter element with a prokaryotic promoter. Most plasmid expression vectors utilize either the promoter of the *lac* or *trp* operon from *E. coli* or the β-lactamase promoter from pBR 322.

The *lac* Promoter

There are a number of plasmid vectors which utilize segments of the *lac* operon including the regulatory region. These vectors can in principle be used as expression vectors. The *lac* operon has three genes, viz., *lacZ*, *lacY*, and *lacA* genes. When the cells are grown on rich media or minimal media with glucose, transcription of this operon is blocked by the *lac* repressor (produced by the neighbouring *lacI* gene) which binds to the operator upstream of *lacZ* and prevents RNA polymerase from binding to the promoter sequence. When these cells grow on media containing lactose or related analogs, the *lac* repressor no longer binds to the operator and the RNA polymerase synthesizes a single mRNA containing *lacZ*, *lacY* and *lacZ* encodes the enzyme β-galactosidase which cleaves lactose into glucose and galactose, which are utilized by the cells. The 'lac' operon is negatively regulated by *lac* repressor and can be repressed by the addition of an inducer such as isopropyl- β-D-thiogalactoside (IPTG) (Dickson *et al.*, 1975). The control region of the '*lac*' operon interacts with the repressor, the catabolite activator protein and RNA polymerase which contains a 203-base pair *Hae* III fragment. This 203-bp *Hae* III fragment has been isolated from *lac*-transducing phages and incorporated into plasmids to provide for the expression of foreign genes. The fragment contains the operator, promoter and the first eight codons of β-galactosidase.

The *trp* Promoter

A similar series of promoter element from the *trp* operon has also been incorporated into expression vector. The *trp* operon encodes five enzymes for the biosynthesis of tryptophan. It is under the negative control of a complex formed between the *trp* repressor and tryptophan, so that when the amino acid is in abundant supply, the genes responsible for its biosynthesis are repressed.

The *tac* Promoter

This is a hybrid that has now been constructed between elements of the *trp* and *lac* promoters (De Boer *et al.*, 1983). The first of these, *tac* I contains a '-35´ sequence from the *trp* promoter and the Pribnow box from the *lac* promoter. The second, *tac* II also has the *trp* –35 sequence, but the Pribnow box is a hybrid between the *trp* and *lac* promoters.

EXPRESSION OF CLONED GENES IN ANIMAL CELLS

The ability to transfer DNA into mammalian cells and so to complement a mutation has been known for over two decades. Using this technique, it is possible to transfer the thymidine kinase (*tk*) gene from herpes simplex virus (HSV) DNA into cells deficient for this enzyme. Such Tk cells survive because the normal pathway for the synthesis of dTTP is from dCDP and not from dTMP produced by the phosphorylation of thymidine by thymidine kinase. Tk cells do not survive in HAT medium which contains hypoxanthine and aminopterin that block the synthesis of dTTP from dCDP and thymidine. Tk cells that acquire the HSV *tk* gene can be selected by their growth in HAT medium.

SV-40 Vectors

For introduction of DNA into cells, there is a need to utilize a viral replicon together with selective systems in much the same way that is seen in bacterial cloning systems. The detailed knowledge of molecular biology of the small DNA tumour virus, simian virus 40 (SV-40), has made it a primary candidate for development as a vector for introducing foreign DNA sequences into cultured mammalian cells. The virus can have two kinds of interaction with its host cell, depending upon the species. In permissive monkey cells, one sees productive infective cycle resulting in cell death on the release of progeny virions. In non-permissive mouse or rat cells, a small proportion of cells become stably transformed and the viral DNA becomes integrated into the host cell chromosomes.

As with the case of bacteriophage lambda, there is a physical limit of the amount of DNA which can be packaged into the virus capsid.

The gene that codes the haemagglutinin (HA) glycoprotein of influenza virus has also been expressed in SV-40 vectors. The influenza HA becomes anchored to the surface membrane of infected cells. In cells that are expressing the *HA* gene cloned in SV-40, the protein becomes glycosylated and is transported to the membrane where it is biologically active and can agglutinate erythrocytes. SV-40 expression vectors contain promoter, splicing and polyadenylation sequences that prove useful for the expression of cDNA fragments.

Adenovirus Vectors

The human adenovirus genome can carry exogenous genes. Specifically, a number of natural recombinants between adenovirus DNA and SV-40 DNA were isolated during the production of adenovirus vaccine in monkey cells contaminated with SV-40. Adenovirus serotypes 2 and 5 have been the most extensively studied. The virion contains a linear 35-kb DNA molecule with inverted terminal repeats. The 5´ends of the DNA are covalently attached to a 55 kDa protein, which plays a critical role in DNA replication. The presence of protein does confer a higher degree of infectivity upon the DNA. Viral DNAs are selected that are resistant to restriction endonucleases. Variants of AD5 DNA lacking *Xha* I sites were selected by cleaving the DNA with *Xha* I and then relegating the fragments. Those fragments that are regenerated by ligation to intact viral genomes can be selected by virtue of their infectivity.

The use of adenovirus as a cloning vehicle is at an early stage. The recombinants that have so far been described contain the SV-40 large T-antigen gene, the expression of which can be selected in monkey cells.

Bovine Papilloma Virus Vectors

The papilloma viruses are another group of viruses whose potential as expression vectors has been recently recognized. The papilloma viruses are responsible for causing warts and other tumours which are usually benign. It is possible to use the DNA of bovine papilloma virus (BPV) to transform the growth properties of cultured cells. The transforming ability resides on a restriction fragment that comprises 69% of the BPV-1 genome (Lowy *et al.*, 1980). This fragment has been used as a vector to introduce the rat pre-proinsulin gene into mouse cells. Proinsulin is efficiently expressed and secreted into the medium. Efficient expression of the human growth hormone gene carried by BPV-1 DNA has also been reported (Pavlakis & Hamer, 1983). Zimm *et al.* (1983) cloned the β-interferon gene into a BPV vector containing a segment of the human globin gene that stimulates transformation and which has single receptor sites for *Bam* HI and *Sal* I fragments. They observed consistent levels of expression in independent cell lines, transformed by recombinants made with this vector. The β-interferon mRNA is induced to 400-fold when the cells are treated with poly(I)–poly(C). These recombinants have enabled two regulatory regions upstream of the gene to be identified.

Retroviral Vectors

These viruses have single-stranded RNA genomes containing basically 3 genes: *gag*, which encodes viral core proteins, *pol*, which encodes reverse transcriptase and *env*, which encodes the protein of the viral envelope. Such a virus will not transform cultured cells. Cellular transformation is brought about by another gene designated as *onc* for oncogene. Usually, the viral oncogene disrupts one or more of these three genes. Commonly *gag* and the *onc* genes are expressed as *gag–onc* fusion protein. The retroviral *onc* genes have cellular homologues which are transcribed in normal cases. It seems that the transducing agents and therefore their genomes could be manipulated *in vitro* for use as vectors. Furthermore, the virus normally has a very wide host-range.

Several groups have used retroviruses as vectors for the herpes simplex virus *tk* gene. Wei *et al.* (1981) inserted *tk* gene of herpes simplex virus into a cloned retrovirus genome. The insertion of foreign RNA into the cloned proviral RNA of a retrovirus can produce a defective viral genome. It is therefore necessary to use helper virus to propagate such recombinants.

Vaccinia Virus Vector

Vaccinia virus replicates and transcribes in the cell cytoplasm. It can therefore rely on the host cell machinery for these functions. Its genome is large (180 kb) and the foreign genes could be inserted into the vaccinia genome to produce viable recombinants. The viral genome encodes about 200 polypeptides. About half of these genes are expressed early; that is to say before the onset of viral DNA replication.

One easily characterized marker of early gene expression is the virally encoded thymidine kinase. The restriction fragment which carries this gene has been identified by translation assays. Since vaccinia virus DNA is not infectious, it is necessary first to infect cells with the *tk* virus and then attempt rescue by introducing restriction fragments by co-precipitation with calcium phosphate. Recombination occurs *in vivo* to produce wild type virus A plasmid carrying a vaccinia. *tk* gene interrupted by a segment of foreign DNA is introduced by co-precipitation with calcium phosphate into the cells infected with wild type vaccinia virus. The *tk* gene inactivated by the insertion of foreign DNA recombines with the wild type vaccinia genome and thus *tk* recombinants can be selected.

A viable hybrid vaccinia virus which contains the genes for the surface antigens of hepatitis B virus has been constructed (Smith *et al.*, 1983a). In this case, the hepatitis B virus gene was inserted into the cloned *tk* gene of vaccinia.

Recombination between the vaccinia genome and this plasmid effectively results in the insertion of the gene for the hepatitis surface antigen into the *tk* gene of vaccinia.

A similar approach has been used to insert the gene for influenza virus haemagglutinin into the vaccinia *tk* gene. The inoculation of hamster with this hybrid virus leads to the production of circulating antibodies which give protection against respiratory infection with influenza virus (Smith *et al.*, 1983b).

APPLICATIONS OF GENETIC ENGINEERING

The analysis of the molecular basis of life has been revolutionized by the recombinant DNA technology. Complex chromosomes are rapidly being mapped and dissected into units that can be manipulated and deciphered. Analysis of genes and cDNA can reveal the existence of a previously unknown protein which can be isolated and purified. New kinds of proteins could be created by altering genes in specific ways. Site-specific mutagenesis helps to understand how proteins fold, recognize other molecules, catalyse reactions and process information.

Protein engineering involves altering the structure of proteins via alteration of the gene sequence that has become possible due to the technique of mutagenesis *in vitro*. The desired effect might be the alteration of the catalytic activity of an enzyme by the modification of the residues around the active site, or an improvement in the stability of a protein used in industry or medicine. The amplification of genes by cloning has provided abundant quantities of DNA for sequencing.

Production of Recombinant Pharmaceuticals

A number of human disorders can be traced to the absence or malfunction of a protein normally synthesized in the body. Most of these disorders can be treated by supplying the patient with the correct version of the protein, but for this to be possible, the relevant protein must be available in relatively large amounts. If the defect can be corrected only by administering the human protein, then obtaining sufficient quantities will be a major problem unless donated blood can be used as the source.

Recombinant insulin   Insulin, synthesized by the β-cells of the islets of Langerhans in the pancreas, controls the level of glucose in the blood. An insulin deficiency manifests itself as diabetes mellitus, a complex of symptoms which may lead to death if untreated. The insulin used in treatment has traditionally been obtained from the pancreas of pigs and cows slaughtered for meat production. Although animal insulin is generally satisfactory, problems may arise in its use to treat human diabetes. Insulin is a relatively small protein, comprising two polypeptides, one of 21 amino acids (A chain) and one of 30 amino acids (B chain). In humans these are synthesized as a precursor called preproinsulin which contains the A and B segments linked by a third chain C and preceded by a leader sequence. The leader sequence is removed after translation and the C chain excised, leaving the A and B polypeptides linked to each other by two disulphide bonds. Artificial insulin genes are produced in *E. coli*. Two recombinant plasmids were constructed, one carrying the artificial gene for the A chain and one the gene for the B chain. In each case the artificial gene was ligated to a *lacZ* reading frame present in a pBR 322-type vector. The insulin genes were therefore under the control of the strong *lac* promoter and were expressed as fusion proteins. The purified A and B chains were then attached to each other by disulphide bond formation in the test tube.

Recombinant human growth hormones (hGH)   Somatostatin was the first human protein to be synthesized in *E. coli*. Being a very short protein only 14 amino acids in length, it was ideally suited for artifical gene synthesis. The strategy used was the same as that described for recombinant insulin, involving insertion of the artificial gene into a *lacZ* vector, synthesis of a fusion protein and cleavage with cyanogen bromide. Somatotropin presented a more difficult problem. This protein is 191 amino acids in length, equivalent to almost 600 bp. A combination of artificial gene synthesis and cDNA cloning was used to obtain a somatotropin-producing *E. coli* strain. mRNA was obtained from the pituitary, the gland that produces somatotropin in the human body, and a cDNA library prepared. The somatotropin cDNA turned out to have a unique site for the restriction endonuclease *Hae* III, which cuts the gene into two segments. The longer segment, consisting of codons 24 to 191, was retained for use in construction of the recombinant plasmid. The smaller segment was replaced by an artificial DNA molecule that reproduced the start of the somatotropin gene and provided the correct signals for translation in *E. coli*. The modified gene was then ligated into an expression vector carrying the *lac* promoter. Goeddel *et al.* (1979a) expressed the mature hGH in *E. coli*. They were able to isolate a specific restriction fragment of about 550 bp containing the coding sequence for amino acids 24–191 at COOH terminus of hGH. A synthetic DNA molecule was constructed corresponding to the first 24 amino acids of mature hGH, plus an AUG initiation codon. These two fragments were ligated and inserted into an expression vector so that the cloned DNA was under the control of a *lac* promoter. The resulting plasmid gave rise to the production of 2 mg of hGH per litre of culture, corresponding to nearly 200,000 molecules of hGH per cell.

Recombinant factor VIII An example of a recombinant pharmaceutical produced in eukaryotic cells in human factor VIII, a protein which plays a central role in blood clotting. The commonest form of haemophilia in humans results from an inability to synthesize factor VIII, leading to a breakdown in the blood clotting pathway and the well-known symptoms associated with the disease. The only way to treat haemophilia is by injection of purified factor VIII protein, obtained from human blood provided by the donors. Purification of factor VIII is a complex procedure and the treatment is very expensive. Recombinant factor VIII, free from contamination problems would be a significant achievement for biotechnology. Factor VIII gene is very large, over 186 kb in length, and is split into 26 exons and 25 introns. The mRNA codes for a large polypeptide (2351 amino acids) which undergoes a complex series of post-translational processing events, eventually resulting in a dimeric protein consisting of a large subunit, derived from the upstream region of the initial polypeptide, and a small subunit from the downstream segment. It has not been possible to synthesize an active version in *E. coli*.

The synthesis and purification of proteins from cloned genes is one of the most important aspects of genetic manipulation, particularly where valuable therapeutic proteins are concerned. Examples include insulin, human growth hormone, interferons, β-endorphin and factor VIII. Insulin is normally prepared from bovine or porcine pancreas. However, some people become allergic to these and may require human insulin, which is produced by genetic engineering.

Interferons Many types of human and animal cells produce, as a consequence of exposure to certain viruses or other inducing agents, glycoproteins known as interferons, which have a molecular weight of about 20,000. Interest in interferons first arose from their ability to render cells resistant to virus attack, and subsequently from the possibility that interferons may be useful as anti-cancer agents. However, supplies of interferons are severely limited; for example, two litres of human blood (one of the principle sources) are required to produce 1 μg of purified human leucocyte interferon. This made interferon an obvious target for gene cloning.

Goeddel *et al.* (1980) obtained a recombinant plasmid with an insert of about 1000 bp, which could be excised and cloned into another vector to place it under the control of a *trp* promoter. This gave a much higher level of expression (about 480,000 units per litre of culture). Knowledge of the DNA sequence of the 1000-bp fragment enabled a further series of manipulations to be carried out to obtain expression of mature leucocyte interferon at a high level. This involved removing the sequence coding for the signal precursor initiation ATG codon. Another DNA fragment was then inserted which carried the

E. coli trp operator/promoter region and ribosome-binding site. One clone obtained after these manipulations produced 2.5×10^8 units of interferons per litre of culture. The interferon produced was shown to have biological activity by its ability to protect squirrel monkeys against encephalomyocarditis virus.

Identification of Genes Responsible for Human and Animal Diseases

A second major area of medical research in which gene cloning is having an impact is in the identification and isolation of genes responsible for human diseases. A genetic or inherited disease is one that is caused by a defect in a specific gene, individuals carrying the defective gene being predisposed towards developing the disease at some stage of their lives. With some diseases like haemophilia, the gene is present on the X chromosome, so all males carrying the gene express the disease state, while females with one defective gene and one correct gene are healthy but can pass the disease to their male offspring. With some genetic diseases, the symptoms manifest themselves early in life, with others the disease may not be expressed until the individual is middle-aged or elderly. Cystic fibrosis is an example of the former, neurodegenerative diseases such as Alzheimer's and Huntington's are examples of the latter.

The breast cancer gene BRCA1　More than one in ten women develop breast cancer with over half of these dying from the disease. Women who carry the defective gene have a 90% probability of developing the disease during their lifetime. Identification of the gene responsible for breast cancer is clearly an urgent priority. The first task in isolation of any gene from human genome is to determine which chromosome carries the gene and to map, as accurately as possible, the position of the gene on this chromosome. This is usually carried out by linkage analysis, which involves comparing the inheritance pattern for the target gene with the inheritance patterns for genetic loci whose map positions are already known. Linkage analysis carried out in DNA samples from individuals from families predisposed to the genetic disease from families predisposed to the genetic disease possessed the same version of an RFLP called D17S74. This RFLP had previously been mapped to the long arm of chromosome 17. *BRCA1* must therefore also be located on the long arm of chromosome. The DNA segment for genes that might be *BRCA1* is identified by hybridization probes and Southern blotting.

The diagnosis of genetically based diseases has been revolutionized by the techniques of gene manipulation. A genetic disease is one that is caused by a specific defect in an individual gene. One important development is the use of restriction fragment length polymorphisms (RFLP) as genetic markers. RFLPs are differences in the lengths of specific restriction fragments generated when DNA is digested with a particular enzyme. RFLPs serve as a powerful tool in the analysis of genetic diseases such as sickle cell anaemia. Gene therapy is the cure of genetic diseases through insertion of an extra gene with the intention that the gene product will play a therapeutic role. Gene replacement therapy is the cure of similar diseases by replacing the abnormal genes with the normal one. Recombinant DNA is also providing highly specific diagnostic reagents such as DNA probes for the detection of genetic diseases, infections and cancer.

Luo *et al.* (1997) cloned and characterized the major outer membrane protein gene (*ompH*) of *Pasteurella multocida* X-73. The *ompH* gene encoding mature protein has been expressed in *E. coli* by using a regulatable expression system. Protection studies showed that protein OmpH was able to induce homologous protection in chicken.

A genomic library of *Mycoplasma synoviae* (MS) was generated using bacteriophage lambda gt-11 as a cloning and expression vector (Abdelmoumem *et al.*, 1999). Expression of the recombinant clones in *E. coli* yielded a predominant reactive fusion protein of 165 kD. Two clones (MS 2/28 and MS 2/12) that yielded inserts of different sizes were selected. The 2 MS DNA inserted were subcloned in a plasmid vector, labelled with digoxigenin and used as probes for the specific recognition of several MS strains. A high degree of conservation was demonstrated for the MS 2/12 and MS 2/28 genes in tested MS strains. In addition, neither DNA fragment recognized any other avian mycoplasma species, thus indicating their high specificity for MS.

A gene for increased serum survival, *iss*, has been found to be strongly correlated with *E. coli* associated with birds suffering from colibacillosis (Horne *et al.*, 2000). The *iss* gene and its protein product, Iss, have been suggested as potential targets for detection and control of avian colibacillosis.

A 97-kDa purified aminopeptidase N (PepN) of *Brucella melitensis* was previously identified to be immunogenic in humans. The *B. melitensis pepN* gene was cloned, expressed in *Escherichia coli* and purified by affinity chromatography (Contreras-Rodriguez *et al.*, 2006). The recombinant PepN (rPepN) exhibited the same biochemical properties, specificity and susceptibility to inhibitors as the native PepN. rPepN was evaluated as a diagnostic antigen in an indirect enzyme-linked immunosorbent assay (ELISA) using sera from patients with acute and chronic brucellosis. Higher sensitivity was obtained using rPepN compared with crude extract from *B. melitensis*. Anti-PepN sera did not exhibit serological cross-reaction to crude extracts from *Rhizobium tropici*, *Ochrobactrum anthropi*, *Yersinia enterocolitica* 09 or *E. coli* O157H7.

M. *bovis* BCG genomic DNA was extracted and purified, and the *mpb70* gene was amplified by PCR (Soliman *et al.*, 2004). The gene was then ligated to an expression vector, PQE. After transformation of the expression *E. coli* M15 host strain with the PQE plasmid, the expression was induced using 10 mM of IPTG. The diagnostic potential of the re-MPB70 was evaluated using sera from experimentally sensitized guinea pigs with different strains of mycobacteria (M. *bovis* BCG, M. *tuberculosis*, M. *kansasii* and M. *intracellulare*) using ELISA test. The results indicated the efficiency of MPB70 but not bovine PPD to discriminate between M. *bovis* sensitized guinea pigs and those sensitized with other mycobacterial strains at serum dilution of 1150. Fifty serum samples from tuberculin positive and 6 from tuberculin-negative cattle were used as well as 10 tuberculin-positive buffaloes. All positive animals were confirmed to be M. *bovis* infected by post-mortem analysis, bacteriological examination. ELISA results revealed that rMPB70 could recognize the tuberculin-positive infected animals at serum dilution of 1/50 and that it could diagnose tuberculosis in cattle as well as buffaloes.

Jimenez-Clavero *et al.* (1998) identified a gene coding for the capsid protein precursor of swine vesicular disease virus (SVDV) P1 from a Spanish isolate, cloned and expressed it in bacteria. The antigenicity and immunogenicity of the recombinant products were evaluated.

In order to identify the origin of an Indian isolate of bovine herpesvirus-I and its molecular relationship to known strains of BHV-1, a 680-bp region of the glycoprotein gene *gIV* was amplified by PCR, cloned and sequenced (Sivaraman *et al.*, 1999). Comparison of this sequence with the corresponding one of a European strain of BHV-1 revealed more than 99% nucleotide homology.

cDNA of a part of gene for major variable immunogenic region, VP1 of FMDV of four serotypes (A22, I, C and Asia 1), was amplified by PCR and cloned into an expression vector (Ratish *et al.*, 1999). The expression of the 16-kDa protein gene was induced with IPTG. This protein was found to be immunoreactive and useful in the FMD diagnosis.

Wu *et al.* (2003) expressed two immunogenic dominant epitopes of foot-and-mouth disease virus (FMDV) serotype O in tobacco plant using a vector based on a recombinant tobacco mosaic virus (TMV). The recombinant viruses TMVF11 and TMVF14 contained peptides of 11 and 14 amino acid residues respectively, from FMDV VP1 fused to the open reading frame of TMV coat protein (CP) gene between amino acid residues 154 and 155. TMVF11 and TMVF14 systemically infected tobacco plant and produced large quantities of stable progeny viral particles assembled with the modified CP subunits. Guinea pigs, mice and swine were used to test the protective effects of the recombinant viruses against FMDV infection. Most guinea pigs were protected against FMDV challenge after parenteral injection with TMVF11, TMVF14, or the mixture TMVF11/TMVF14, but not wtTMV. The TMVF11/TMVF14 mixture protected all animals when challenged with 150 guinea pig 50% infection dosage (GPID(50)) FMDV. The protective effect of TMVF11 was also demonstrated in swine, where preliminary tests showed that nine pigs immunized with TMVF11 in three experiments were protected against FMDV challenge with 20 minimal infecting dose (MID).

Peste des petits ruminants (PPR) is a viral disease of goats and sheep characterized by erosive stomatitis, enteritis, and pneumonia. Ismail *et al.* (1995) cloned and sequenced the cDNA of the nucleocapsid (N) gene of the Nigeria 75/1 strain of PPR virus (PPRV). A recombinant baculovirus that expressed the N protein in insect cells and larvae (*Spodoptera frugiperda*) was generated. The recombinant protein, characterized by western blot analysis, was shown to have a molecular weight of 58 kDa and was recognized by anti-PPRV antibodies. The recombinant protein was used successfully as a coating antigen in an enzyme-linked immunosorbent assay for the serological diagnosis of PPRV.

A cDNA from *Babesia gibsoni* merozoite mRNA, encoding a 50-kDa protein was cloned and designated the *P50* gene (Fukumoto *et al.*, 2001). Confocal laser microscopic analysis showed that the P50 protein was located on the surface of *B. gibsoni* merozoites. The recombinant P50 protein expressed by baculovirus in insect cells was used as the antigen in an ELISA, to differentiate between *B. gibsoni*-infected dog serum and *B. canis*-infected dog serum or non-infected dog serum. The results from this study demonstrated that recombinant P50 protein might be a useful diagnostic reagent for detection of antibodies to *B. gibsoni* in dogs.

McBride *et al.*, (2001) cloned an immunoreactive *Ehrlichia canis* surface protein gene of 1170 bp, which encodes a protein with a mass of 43 kDa (P43). Forty-two dogs exhibiting signs and/or haematologic abnormalities associated with canine ehrlichiosis were tested by indirect fluorescent antibody (IFA) and by recombinant western immunoblot. Among the 22 samples that were IFA positive for *E. canis*, 100% reacted with rP43. The results indicated that the rP43 is a sensitive and reliable serodiagnositc antigen for *E. canis* infection.

Recombinant Vaccines

Vaccines as recombinant proteins   The use of gene cloning in this field centres on the discovery that virus-specific antibodies are sometimes synthesized in response not only to the whole virus particle, but also to isolated components of the virus. This is particularly true of purified preparations of the proteins present in the virus coat. If the genes coding for the antigenic proteins of a particular virus could be identified and inserted into an expression vector, then the method described above for the synthesis of animal proteins could be employed in the production of recombinant proteins that might be used as vaccines. The vaccines would have the advantages that they would be free of intact virus particles and they could be obtained in large quantities. Unfortunately, this approach has not been entirely successful mainly because it has turned out that recombinant coat proteins often lack the full antigenic properties of the intact virus. The one notable success has been with hepatitis B virus, whose coat protein has been synthesized in *Saccharomyces cerevisiae*, using a vector based on the 2-μm plasmid. The protein was obtained in reasonably large quantities and when injected into monkeys, provided protection against hepatitis B.

Live recombinant vaccines   Recombinant vaccinia viruses are now used as live vaccines against other diseases. If a gene coding for a virus coat protein, for example the hepatitis B major surface antigen, is ligated into the vaccinia genome under control of a vaccinia promoter, then the gene will be expressed. After injection into the bloodstream, replication of the recombinant virus will result not only in new vaccinia particles, but also in significant quantities of the major surface antigen. Immunity against both smallpox and hepatitis B would result. Recombinant vaccinia viruses expressing a number of foreign genes have been constructed and shown to confer immunity against the relevant diseases in experimental animals. The possibility of broad-spectrum vaccines has been shown by the demonstration that a single recombinant vaccinia expressing the genes for influenza virus haemagglutinin, hepatitis B major surface antigen and herpes simplex virus glycoprotein confers immunity against each disease in monkeys.

By recombinant techniques, it is possible to develop vaccines based on select antigenic determinants of viruses. In addition, it is possible to construct recombinant vaccinia viruses that express a number of foreign genes, which may enable multiple vaccination to be achieved by single inoculation. Examples of animal diseases for which recombinant vaccines have been produced include foot-and-mouth disease virus, parvovirus, herpesvirus, rabies virus, etc.

Hardy *et al.* (1981) cloned a core antigen of hepatitis B virus in *Bacillus subtilis*. Subsequently, the HBsAg had been expressed in yeast and it was from this source that recombinant hepatitis B vaccine is produced (Valenzuela *et al.*, 1982; McAleer *et al.*, 1984). Another alternative approach is to insert antigen genes into derivatives of the vaccinia virus which can then be used as a live vaccine. For example, Smith *et al.* (1983) and Paoletti *et al.* (1984) constructed vaccinia recombinants containing the *HBsAg* gene and showed that vaccinated rabbits rapidly produced antibodies to HBsAg. Several different genes could be introduced into the same virus, thus creating a live vaccine which can confer immunity to several diseases simultaneously. For example, Perkus *et al.* (1985) produced a recombinant vaccinia strain carrying genes from hepatitis B, herpes simplex and influenza viruses and showed that rabbits immunized with this recombinant strain produced antibodies against all three viruses. The envelope gene of AIDS virus (HIV) has also been expressed by recombinant vaccinia virus (Chakrabarthi *et al.*, 1986) and vaccinia strains carrying a gene from the rabies virus have been produced as a possible method for oral vaccination of foxes in order to control the disease among wild animals (Blancou *et al.*, 1986).

The gene encoding MPB83 from *Mycobacterium bovis* Vallee111 chromosomal DNA was amplified by using polymerase chain reaction (PCR) technique, and the PCR product was approximately a 600-bp DNA segment (Jiang *et al.*, 2006). Using TA cloning technique, the PCR product was cloned into pGEM-T vector and subcloned into the expression vector pET28a(+), and expressed in the prokaryotic expression vector pET28a-83. Plasmid containing pET28a-83 was transformed into competence *Escherichia coli* BL21 (DE3). The expressed protein of 26 kDa was analysed using western blotting. The results indicated that the protein was of antigenic activity of M. *bovis*. The results were expected to lay foundation for further studies on the subunit vaccine and DNA vaccine of *MPB83* gene in their prevention against bovine tuberculosis.

The full-length open reading frame coding for a potentially immunogenic 35-kDa protein of *Mycobacterium avium paratuberculosis* was generated using polymerase chain reaction technology (Basagoudanavar *et al.*, 2004). The gene was inserted in-frame into *Escherichia coli* expression plasmid pQE32. The resulting recombinant plasmid pPMP35 was transformed into *E. coli* M15. Analysis of the *E. coli* induced with isopropyl-beta-D-thiogalactopyranoside revealed that the protein accumulated into the cytoplasm as insoluble inclusion bodies. The P35 reacted with a rabbit antiserum raised against a sonicate of M. *avium paratuberculosis*. The protein was also recognized by serum from a goat with clinical paratuberculosis. The results indicate that the 35-kDa protein of M. *avium paratuberculosis* is a membrane protein, having a role in the cellular immune response.

The expression of infectious bronchitis virus (IBV) S1 glycoprotein in potatoes and its immunogenicity in mice and chicken were investigated (Zhou *et al.*, 2003). Potato plants were genetically transformed with a cDNA construct encoding the IBV S1 glycoprotein with the *Agrobacterium* system.

Genomic DNA and mRNA analyses of the transformed plantlets confirmed the integration of the foreign cDNA into the potato genome, as well as its transcription. Mice and chicken vaccinated with the expressed IBV S1 glycoprotein produced antibodies that neutralized IBV infectivity. After three immunizations, vaccinated chicken were completely protected from virulent IBV infection. These results demonstrated that transgenic potatoes expressing IBV S1 glycoprotein can be used as a source of recombinant antigen for vaccine production.

The nucleoprotein (N) gene of rabies virus CTN strain was cloned, sequenced and expressed in *Escherichia coli* as a fusion with maltose-binding protein (MBP) (He *et al.*, 2006). The antigenicity of this recombinant MBP-N fusion protein was examined by western blotting and enzyme linked immunosorbent assay (ELISA). Subsequently, an indirect ELISA was developed to detect rabies specific antibody levels. Using sera from naive and vaccinated animals the ELISA results were compared with virus-neutralizing antibodies detected by a rapid fluorescent focus inhibition test (RFFIT). Neutralizing titres by RFFIT were found to correlate well with the OD values in the ELISA (r = 0.9436) and the sensitivity and specificity of the ELISA were shown to be 93.4 and 100% respectively. The data indicated that the recombinant MBP-N fusion protein can be expressed and isolated directly and may be useful as a safe and abundant source of antigen to monitor seropositivity in vaccinated canines.

Hu *et al.* (2006) described the construction and characterization of a replication-competent recombinant canine adenovirus type-2 expressing the rabies virus glycoprotein (SRV9 strain) by a different strategy from that reported previously, i.e., the recombinant genome carrying the glycoprotein cDNA was generated by a series of strictly gene cloning steps. Infectious recombinant virus was obtained by transfecting the recombinant genome into a canine kidney cell line, MDCK. This recombinant virus, CAV-E3delta-CGS, was subcutaneously injected into dogs. All vaccinated dogs produced effective neutralizing antibodies after one inoculation and a stronger anamnestic immune response was produced after booster injection. The immunized dogs could survive the challenge of 60,000 mouse LD50 CVS-24, which is lethal to all unimmunized dogs and was comparable to the conventional vaccines. The immunity lasted for months with a protective level of neutralizing antibody. This recombinant virus would be an alternative to the attenuated and the inactivated rabies vaccines and be prospective in immunizing dogs against rabies.

Gene Therapy

This is the name given to methods that aim to cure an inherited disease by providing the patient with a correct copy of the defective gene. There are two basic approaches to gene therapy: germ-line therapy and somatic cell therapy. In germ-line therapy, a fertilized egg is provided with a copy of the correct version of the relevant gene and reimplanted into the mother. If successful, the gene is present and expressed in all cells of the resulting individual. Germ-line therapy is usually carried out by microinjection of DNA into the isolated egg cell and theoretically could be used to treat any inherited disease.

Somatic cell therapy involves manipulation of ordinary cells, usually ones which can be removed from the organism, transfected, and then placed back in the body. The technique has most promise for inherited blood disease (e.g. haemophilia, thalassemia) with genes being introduced into stem cells from the bone marrow, which give rise to all the specialized cell types in the blood. The strategy is to prepare a bone extract containing several billion cells, transfect these with a retrovirus-based vector, and then reimplant the cells. Subsequent replication and differentiation of transfectants leads to the added gene being present in all the mature blood cells. The advantage of a retrovirus vector is that this type of vehicle has an extremely high transfection frequency, enabling a large proportion of the stem cells in a bone marrow extract to receive the new gene. Somatic cell therapy also has potential in the treatment of lung diseases such as cystic fibrosis.

Transgenic animals are generated by incorporation of genes into eggs or early embryos. Transgenics could be used for a variety of purposes. The study of embryological development has been extended and there is a scope for the manipulation of farm animals by incorporation of desirable traits via transgenesis. The use of whole organism for the production of recombinant protein is also possible. Examples of protein production in transgenics include the expression of human tissue plasminogen activator in transgenic mice and of human blood coagulation factor IX in transgenic sheep which are secreted into the milk of lactating animals.

REVIEW QUESTIONS

1. What are the different cloning vectors used in gene cloning technique? List their merits and demerits.

2. What are restriction endonucleases, how are they useful in gene cloning? What are the types of restriction enzymes?

3. What are the steps involved in gene cloning methods?

4. Explain in detail how genetic engineering could be used for improving the

health and productivity of animals and humans.

REFERENCES

Abdelmoumem, B.B., Roy, R.S. and Brousseau, R. (1999). "Cloning of *Mycoplasma synoviae* genes encoding specific antigens and their use as species-specific DNA probes." *J. Vet. Diagn. Invest.* 11: 162–169.

Basagoudanavar, S.H., Goswami, P.P., Tiwari, V., Pandey, A.K. and Singh, N. (2004). "Heterologous expression of a gene encoding a 35 kDa protein of *Mycobacterium avium paratuberculosis* in *Escherichia coli.*" *Vet. Res. Commun.* 28: 209–224.

Contreras-Rodriguez A., Seleem, M., Schurig, G.G., Sriranganathan, N., Boyle, S.M. and Lopez-Merino, A. (2006). "Cloning, expression and characterization of immunogenic aminopeptidase N from *Brucella melitensis.*" *FEMS Immunol. Med. Microbiol.* 48:252–256.

De Boer, H.A., Comstock, L.J. and Vasser, M., (1983). "The 'lac' promoter': a functional hybrid derived from the 'trp' and 'lac' promoters." *Proc. Natl. Acad. Sci. USA.* 76: 106–110.

Dickson, R.C., Abelson, J., Barns, W.H. and Reznikoff, W.S. (1975). "Genetic regulation, the 'lac' control region." *Science.* 187: 27–35.

Fukumoto, S., Xuan, X., Nishikawa, Y., Inoue, N., Igarashi, I., Nagasawa, H., Fujisaki, K. and Mikami, T. (2001). "Identification and expression of a 50-kDa surface antigen of *Babesia gibsoni* and evaluation of its diagnostic potential in an ELISA." *J. Clin. Microbiol.* 36: 2603–2609.

Glover, D.M. (1984). (ed.). *Gene Cloning: The Mechanics of DNA Manipulation.* Chapman & Hall, London & New York.

Gupta, P.K. (1994). (ed.). *Elements of Biotechnology.* Rastogi & Company, India.

He, Y., Gao, D. and Zhang, M. (2006). "Expression of the nucleoprotein gene of rabies virus for use as a diagnostic reagent." *J. Virol. Methods.* 138: 147–151.

Horne, S.M., Pfaff-McDonough, S.J., Giddings, C.W. and Nolan, L.K. (2000). "Cloning and sequencing of the *iss* gene from a virulent avian *E. coli.*" *Avian Dis.* 44: 179–184.

Hu, R., Zhang, S., Fooks, A.R., Yuan, H., Liu, Y., Li, H., Tu, C., Xia, X. and Xiao, Y. (2006). "Prevention of rabies virus infection in dogs by a recombinant canine adenovirus type-2 encoding the rabies virus glycoprotein." *Microbes Infect.* 8: 1090–1097.

Ismail, T.M., Yamanaka, M.K., Saliki, J.T., el-Kholy, A., Mebus, C. and Yilma, T. (1995). "Cloning and expression of the nucleoprotein of peste des petits ruminants virus in baculovirus for use in serological diagnosis." *Virology.* 208: 776–778.

Jiang, X.Y., Wang, C.F., Zhang, P.J. and He, Z.Y. (2006). "Cloning and expression of *Mycobacterium bovis* secreted protein MPB83 in *Escherichia coli*." *J. Biochem. Mol. Biol.* 39:22–25.

Jimenez-Clavero, M.A., Escribano-Romero, E., Sanchez-Vizcaino, J.M. and Ley, V. (1998). "Molecular cloning, expression and immunological analysis of the capsid precursor polypeptide (P1) from swine vesicular disease virus." *Virus Res.* 57: 163–170.

Lowy, D.R. Doveretzky I.D. and Shober, R. (1980). "*In vitro* tumerogenic transformation by a defined subgenomic fragment of bovine papilloma virus." *Nature.* 287: 72–74.

Luo, Y., Glisson, J.R., Jackwood, M.W., Hancock, R.E., Bains, M., Cheng, I.H. and Wang, C. (1997). "Cloning and characterization of the major outer membrane protein gene (*oompH*) of *Pateurella multocide* X-73." *J. Bacteriol.* 179: 7856–64.

McBride, J.W., Corstvet, R.E., Breitschwerdt, E.B. and Walker, D.H. (2001). "Immunodiagnosis of *Ehrlichia canis* infection with recombinant proteins." *J. Clin. Microbiol.* 36: 315–322.

Nicholl, D.S.T. (1994). (ed.). *An Introduction to Genetic Engineering.* Cambridge University Press, U.K.

Pavlakis, G.N. and Hamer, D.H. (1983). "Regulation of a metallothionein-growth hormone hybrid gene in bovine papilloma virus." *Proc. Natl. Acad. Sci. USA.* 80: 397–401.

Ratish, G., Viswanathan, S. and Suryanarayana, V. V. (1999). "C-terminal region of VP1 of selected foot and mouth disease virus serotypes: expression *E. coli* and affinity purification." *Acta Virol.* 43: 205–211.

Sivaraman, K.R., Sreekumar, E. and Rasool, T.J. (1999). "Cloning and sequencing of truncated gIV glycoprotein gene of an Indian isolate of bovine herpesvirus 1." *Acta Virol.* 43: 387–389.

Smith, G.L., Mackett, M. and Moss, B. (1983a). "Infectious vaccinia virus recombinants that express hepatitis B virus surface antigen." *Nature.* 305: 490–495.

Smith, G.L., Murphy, M. and Moss, B. (1983b). "Construction and characterization of an infectious vaccinia virus recombinant that expresses the influenza haemagglutinin gene and induce resistance to influenza virus infection in hamsters." *Proc. Natl. Acad. Sci. USA.* 80: 7155–7159.

Soliman, Y. A., Ghonim, M., Nasr, E.A., Mousse, I., Ebiary, E.A. and Selim, S.A. (2004). "Cloning and expression of recombinant MPB70 protein antigen from *Mycobacterium bovis* BCG for diagnosis of tuberculosis." *J. Immunol.* 11: 21–29.

Wei, C.M., Gibson, M., Spear, P.G. and Scolnick, E.M. (1981). "Construction and isolation of a transmissible retrovirus containing the '*src*' gene of Harvey Murine Sarcoma virus and the thymidine kinase gene of herpes simplex virus type I." *J. Virol.* 36: 935–944.

Wu, L., Jiang, L., Zhou, Z., Fan, J., Zhang, Q., Zhu, H., Han, Q. and Xu, Z. (2003). "Expression of foot-and-mouth disease virus epitopes in tobacco by a tobacco mosaic virus-based vector." *Vaccine.* 21: 4390–4398.

Zimm, K., Di, D., Maio and Maniatis, T. (1983) "Identification of two distinct regulatory regions adjacent to the human β-interferon gene." *Cell.* 34: 865–879.

Zhou, J.Y., Wu, J.X., Cheng, L.Q., Zheng, X.J., Gong, H., Shang, S.B. and Zhou, E.M. (2003). "Expression of immunogenic S1 glycoprotein of infectious bronchitis virus in transgenic potatoes." *J. Virol.* 77: 9090–9093.

7

MANIPULATION OF REPRODUCTION

INTRODUCTION

Artificial insemination has contributed considerably to the improvement of cattle industry. In recent years, tremendous progress has been made in the manipulation of reproduction in cattle using techniques like control of oestrous cycle. The embryo transfer technique seems to be the technique of choice to increase the number of offsprings from excellent parent stock. The non-surgical embryo recovery and transfer into cattle has allowed transfer of this technology from the laboratory to the farm so that increased number of offsprings from a superior cow is possible. The embryo transfer technology in cattle includes techniques like superovulation and breeding of donors, embryo collection and evaluation, selection of recipients and transfer of embryos.

EMBRYO TRANSFER (ET) TECHNOLOGY

The first successful transfer of fertilized rabbit eggs was reported in 1931 by Walter-Heape at Cambridge (UK). In farm animals embryo transfer was successfully done for the first time in 1934 in sheep and goats (Warwick *et al.*, 1934). The first offspring after transfer of embryo in cattle and swine were reported in 1951 (Willet *et al.*, 1951).

Embryo transfer can be used to produce a large number of offspring per female in a shorter time than possible with normal reproduction. Since a producer's lifespan covers only 6 or 7 generations of cows, the chances of obtaining a super bull from a superior cow are limited. By the use of superovulation and ET, the number of opportunities is increased. In beef cattle, replacements for both sexes can be selected on their own performance prior to reproductive age. A highly selective ET scheme could double the rate of genetic improvement. Embryo transfer allows dairy producers to increase the number of offspring from cows believed to be genetically superior. Each cow can be selectively bred to three or four superior bulls in a given year.

Embryo transfer is being done for a number of purposes which include the following:

1. To increase the number of offspring, either male or female, from a genetically superior female.

2. For genetic testing of bulls for inherited defects. Mating a bull to six or eight of his superovulated daughters will exhibit the recessive genes a bull carries more accurately and in much less time than by testing the population at large.

3. To decrease variability in research subjects, for the study of the physiology, pathology and immunology of reproduction or other traits.

4. For long-term storage by freezing.

5. For disease control. Many diseases present in the mother will not be transmitted by the embryo.

6. For the treatment of infertility. In general, reproductive traits are lowly heritable. However, producers should consider possible decreases in fertility of subsequent generations before using infertile donors extensively.

7. For rapid genetic change within a small population (e.g. from grade to pure-bred herd).

Compared to *Bos taurus* breeds, *Bos indicus* breeds of cattle present several differences in reproductive physiology (Baruselli *et al.*, 2006). Follicular diameter at deviation and at the time of ovulatory capability are smaller in *B. indicus* breeds. Furthermore, *B. indicus* breeds have a greater sensitivity to gonadotropins, a shorter duration of oestrus, and more often express oestrus during the night. These differences must be considered when setting up embryo transfer programs for *B. indicus* cattle. In recent studies, we evaluated follicular dynamics and superovulatory responses in *B. indicus* donors with the objective of implementing fixed-time AI protocols in superstimulated donors. Protocols using oestradiol and progesterone/progestrogen-releasing devices to control follicular wave emergence were as efficacious as in *B. taurus* cattle, allowing the initiation of superstimulatory treatments (with lower dosages of FSH than in *B. taurus* donors) at a self-appointed time. Furthermore, results presented herein indicate that delaying the removal of progesterone/progestrogen-releasing devices, combined with the administration of GnRH or pLH 12 hours after the last FSH injection, resulted in synchronous ovulations, permitting the application of fixed-time AI of donors without the necessity of oestrus detection and without compromising the results.

Embryo transfer includes a sequence of various steps which include:

1. selection of donors,

2. induction of superovulation,

3. embryos collection,

4. evaluation of embryos,

5. selection of recipients and

6. transfer of embryos. Until now, embryo transfer techniques have been extensively used in cattle and more sporadically in sheep, goats and pigs.

Selection of Donor

Regularly cycling cows or heifers can successfully respond to superovulatory treatment and be used for embryo transfer. The following selection criteria for donors will ensure a good probability of producing embryos of high quality.

1. Ages between 3 and 10 years

2. Being free of genetic diseases and conformational abnormalities

3. Exhibiting regular oestrous cycle

4. Showing superior production traits of economic importance

5. Being free from infectious diseases

6. Showing previous sound reproductive performances including no more than two inseminations per conception

Induction of Superovulation

Superovulation means induction of multiple ovulation by application of exogenous hormones (PMSG—pregnant mare serum gonadotrophin; FSH—follicle stimulating hormone; HMG—human menopausal gonadotrophin) in the early follicular or the luteal phase of the oestrous cycle, in order to collect a large number of fertilized eggs. Most frequently PMSG or FSH is used. A single injection of PMSG (2000 to 3000 IU) is sufficient as the substance has a long half-life time. In contrast, multiple injections of FSH (35–50 mg) are required to induce superovulatory responses (twice-daily injections for over four days). HMG (1000–1500 IU) also requires multiple injections; however, the drug is more expensive and is only rarely used.

Superovulatory response, embryo quality and fertility after treatment with different gonadotrophins in native Mertolengo cattle were studied (Lopes da Costa *et al.*, 2001). In one experiment, superovulatory response (SR), embryo quality and plasma progesterone (P4) levels between donors treated with equine chorionic gonadotrophin (eCG) vs those treated with FSH, were tested during progestogenic implantation. Significantly more viable embryos were yielded by FSH-treated donors, than by those treated with eCG. Fertilization rates were significantly higher in bred than in inseminated donors. Plasma P4 levels were significantly different (higher) between responder and non-responder donors on the day of embryo recovery.

Forty-eight to sixty hours after beginning the gonadotrophin treatment, prostaglandins (PGF2a) are administered to induce oestrus. Inseminations are performed at oestrus and the embryos are recovered 6–8 days after insemination. On an average, only one-third of the superovulated bovine donors show a good ovarian response.

Embryo Collection

In bovines, embryos are recovered by non-surgical methods. Special catheters are introduced via the cervix into the uterine horn and embryos are flushed with 250 to 300 ml of flushing medium (phosphate buffered solution with 1–2% foetal calf serum). In the cow, the embryos usually enter the uterus on day four after oestrus. Non-surgical techniques avoid the potential damage of the reproductive tract by adhesions, are repeatable and do not require elaborate facilities. Embryos are collected by non-surgical methods 6–8 days after the onset of oestrus. Prior to day 5, the embryos may be in the oviduct and after day eight they may be hatched (escape from the zone pellucida), difficult to visualize and are fragile. At day 7, bovine embryos are about 150–190 μ in diameter, are still within the zona pellucida and are at late morula or blastocyst stage of development.

In the case of other farm animals like sheep, pigs and goats, embryo recovery is performed by surgical methods. The abdomen of the donors are opened by a midline cut and embryos are recovered by flushing the oviduct and/or uterine horns. Embryos are collected 4–5 days after insemination or mating.

Evaluation of Embryos

After recovering ova and embryos from the flushing medium, their further development capacity has to be evaluated. Morphological criteria in embryo evaluation are shape, colour, number and compactness of cells, size of the perivitelline space, number and size of vesicles and the status of the zona pellucida. Embryos are classified into 4 groups.

1. Excellent embryos which are in the appropriate developmental stage with a perfect morphology.
2. Good embryos which are in the appropriate stage of development with slight morphological deviations.
3. Degenerated and/or retarded embryos.
4. Unfertilized ova.

Selection of Recipients

The normal physiological and health conditions, the reproductive status, lack of any reproductive disorders, compatibility of the donor with respect to the size of the foetus and oestrous synchronization are of considerable importance. In bovine, it has been proved that heifers and young cows are best suited as recipients. The oestrus of donors and recipients should be synchronized within 24 hours.

Spell *et al.* (2001) studied the effects of corpus luteum (CL) characteristics, progesterone concentration, donor–recipient synchrony, embryo quality and type and development stage on pregnancy rates after embryo transfer. Of the 526 recipients presented for embryo transfer, 122 received a fresh embryo and 326 received a frozen embryo. Pregnancy rates were greater with fresh embryos (83%) than frozen-thawed embryos (69%). Pregnancy rates were not affected by embryo grade, embryo stage and donor–recipient synchrony of the palpated integrity of the CL. There was a significant, positive simple correlation between CL diameter or luteal tissue volume and plasma progesterone concentration.

Transfer of Embryos

Special catheters are available to perform non-surgical transfer in cattle. It is important to transfer the embryos into the tip of the uterine horn without damaging the endometrium. In cattle, after superovulatory treatment, 8–15 follicles are detected at oestrus and about 6–11 ovulations on the day of embryo recovery. The average number of embryos recovered is 5–10 representing 60–70% of the number of corpora lutea. Pregnancy rate after transfer of freshly collected embryos is 55–65% and of frozen/thawed embryos is 50–60%. An average of 5–7 embryos per successful flushing can be recovered in dairy cattle. About 60–70% of these embryos are considered to be capable of further development. Statistically 2.5 to 3 calves per successful flushing can be produced.

The oestrous cycle of the donor and recipient should be closely synchronized if transferred embryos are to survive. Pregnancy rates are generally reduced unless the embryo is placed in the lumen of the uterine horn on the same side of the corpus luteum.

Cryopreservation of Embryos

A positive breakthrough in embryo transfer at a larger scale was due to the development of cryopreservation method. Modifications and adaptations of different procedures have made it more feasible to freeze and store bovine embryos at –196°C for varying time intervals. An optimal method for bovine embryos includes a one-step addition of 1.4 M glycerol as a cryoprotectant, a 20-min. equilibration period, 0.25 ml straws as embryo containers, slow cooling (0.3°C to 0.1°C/min.) down to –35°C and subsequent plunging into liquid nitrogen (–196°C). Embryos are thawed by placing the straw directly into warm water. Glycerol is removed using sucrose in two steps. Using this procedure, the percentage of morphologically intact embryos obtained was 95.3% and the pregnancy rate after non-surgical transfer was 51%.

Embryo Splitting

Mammalian eggs are more tough and can sustain extensive damage and still develop normally. Early cleavage-stage embryos are favoured for manipulation to produce identicals. Embryos from the late morulae up to late blastocyst stage can be collected easily by non-surgical procedures, and after splitting into two halves, they can be returned directly to recipients non-surgically.

Embryos to be bisected are collected non-surgically 6–7 days after oestrus. After this time embryos are compact morulae or blastocysts. After collection, the embryos are washed in modified PBS supplemented with 20% heat-inactivated FCS and classified morphologically. Only embryos of good quality should be selected for splitting. The embryo of good quality is fixed on the holding pipette and the zona pellucida is cut by a vertical motion of the microknife (Figure 7.1). The crack in the zona pellucida is opened by moving the glass needle and the hook, the embryo is either split inside the zona or after moving it out of the zona. Bisection can be done by cutting with the knife or by separating the two halves with the glass needle. One half remains in and will be put back into the original zona. The second half is transferred to a surrogate zona (made by preparing a zona of a degenerated embryo). Bisection of bovine embryos results in two genetically identical demi-embryos). The demi-embryos can be cultured for several hours or transferred immediately after splitting. The transfer can be done in one of the following ways:

1. each half in one recipient into the ipsilateral horn (unilateral transfer);
2. both halves in one recipient into both horns (bilateral transfer) and
3. both halves in one recipient into the ipsilateral horn.

The second and third alternative are similar or identical to naturally arising twin pregnancies. The first alternative will be used in commercial programmes, because it will result in the highest number of calves per collected embryos.

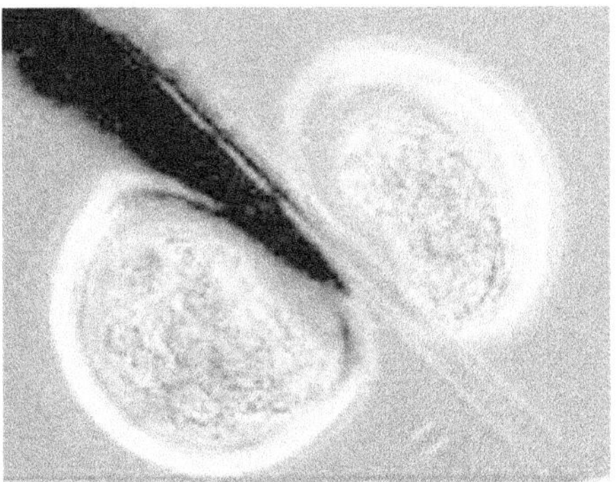

Figure 7.1   Embryo splitting

Improvements in embryo micromanipulation techniques led to the use of embryo bisection technology in commercial embryo transfer programmes, and made possible the direct genetic analysis of preimplantation bovine embryos by biopsy. For example, aspiration and microsection allow bovine embryo sexing by detection of male-specific Y-chromosome in a sample of embryonic cells. Pregnancy rates achieved with fresh bisected or biopsied embryos (50–60%) were similar to the fresh intact embryos (55–61%). The PCR protocol used for embryo sexing showed 92% to 94% of efficiency and 90 and 100% of accuracy (Lopes *et al.*, 2001). These results demonstrate that these procedures are suitable for use in field conditions.

Cryopreservation of microsurgically split bovine embryos allow the production of identical twins of different ages. Survival rates are significantly lower, only about 25% of frozen/thawed embryos will survive.

EMBRYO SEXING

Several methods can be used for sexing animal embryos at present, including: karyotyping, H-Y antigen analysis, X-chromosome-linked enzymatic activity test, Y-specific probe hybridization and polymerase chain reaction method. The sex of an individual is determined at fertilization and depends on whether the X-bearing ovum is fertilized by an X or Y-bearing sperm. There are several methods for diagnosing the sex of embryos.

1.   Sex chromosome identification
2.   Demonstration of H-Y antigen

3. Assay of sex-linked enzyme

4. Blotting DNA probes

Park *et al.* (2001) developed a rapid and reliable PCR method for the sexing of 8–16-cell stage bovine embryos. The BOV97M and bovine 1.715 satellite DNA sequences were selected for amplification of male- and bovine-specific DNA, respectively. In consecutive and multiplex PCR, the first 10 PCR cycles were done with male-specific primer followed by an additional 23 cycles with bovine-specific primer. This PCR method was applied successfully using groups of 8, 4, 2 and 1 blastomeres dissociated from the embryos, and the sexing efficiency was 100, 96.3, 94.3 and 92.1%, respectively. This method proved to be a rapid (within 2 hours) and effective PCR method for the sexing of 8–16-cell stage bovine embryos using a single blastomere.

Sex Chromosomal Analysis

Oocysts fertilized with X-bearing spermatozoa develop into phenotypic females and those fertilized by Y-bearing sperm into males. Karyotypes of each of the domestic animal differ in the number of chromosomes and each have morphologically characteristic differences in X and Y chromosomes. The X chromosomes contain approximately 4% more DNA than the Y chromosomes.

Sex chromosomal analysis is done using cells in the metaphase. Cells have to be removed from an embryo to be sexed and cultured *in vitro* until they reach mitosis. Sex diagnosis was possible in 68% of embryos. Splitting of bovine morulae and blastocysts allow an experimental approach to the cytogenetic sexing of embryos collected on days 6 to 7. The procedure starts with splitting of embryo. If the two parts do not look exactly the same, the good half is used for transfer or freezing, the poor half serves for cytogenetic study. Poor halves are cultured in medium with colcemid for 4 hours, cells are fixed, stained and examined under the microscope for identification of sex chromosomes. After sexing the first half, the second half can be transferred or frozen for any desired period.

The karyotyping method principally relies on cytogenetic analysis against blastomeres of an embryo and determines sex type of the embryo on basis of the occurrence of X or Y sex chromosomes. The karyotyping method had been developed by Gardner and Edwards (1968) who observed the blastocysts of rabbits and determined the femininity of rabbit embryo based on the fact that there is an inactivated X chromosome, or so-called Barr body, existing in that cell. However, according to King (1984), this method is not suitable for sexing embryos of common domestic animals, since, in somatic cells of common domestic animals, observable Barr body is less than 50%, and further, the cytoplasm of embryos of common domestic animals including embryos of pork, beef, sheep and goat, are not as clear as rabbit embryos, that observation of Barr body in those embryos is difficult. At present, karyotyping done on embryos of domestic animals consists usually of treating embryo firstly with cholchicine to induce its blastomere at metaphase of cell division and verifying whether there is Y chromosome in the karyotype of the embryo according to the shape of the chromosome. However, sexing embryo based on karyotyping involving treatment with cholchicine has several disadvantages in that it is a cumbersome and time-consuming procedure. Also the identifiability of the cell is low, and in particular, with a single blastomere, the probability to obtain an identifiable karyotype is even smaller.

The disadvantages of the karyotyping method for sexing are the following:

1. only 60% accuracy;
2. embryo viability may be reduced to some degree;
3. time-consuming procedure;
4. requires well-trained personnel for cytokaryology typing and
5. additional expense for splitting.

Demonstration of H-Y antigen

A new approach for identification of embryonic sex utilizes immunological detection of a male-specific protein, referred to as either histocampatibility-Y (H-Y) antigen or serologically detectable male (SDM) antigen. Many workers suggest that these two factors (viz., H-Y and SDM antigens) are the same or had parts of the same molecule.

H-Y antigen is present at the 8–16-cell stage of mammalian embryos. The detection of the H-Y antigen by antibodies may be an effective method for "sex selection" prior to transfer of embryos. Zona pellucida-free embryos are incubated in diluted H-Y antiserum in the presence of guinea pig complement and scored for lysis. Surviving embryos were karyotyped and transferred to show the genetical sex. Highly specific monoclonal antibodies have been prepared against H-Y antigen which can be used in the fluorescent system (Kohler and Milstein, 1975). The advantages of sex selection using MAbs against the H-Y antigen are the following. 1) accuracy is about 85%; 2) the test needs about 2 hours and does not affect the viability of embryos and 3) the test can be applied by inexperienced people.

Enzyme immunoassay eliminates the need for the fluorescent microscope. The principle is based on the reaction of an enzyme (carried by monoclonal H-Y antibody) with a substrate (added to solution) to form a coloured product signifying the presence of H-Y antigen (Wachtel, 1984).

Metabolic Activity of X-linked Enzymes

This is another method to identify the sex of embryos by enzymes that are associated with the number of active X-chromosomes. In mice, levels of X-linked enzymes such as glucose 6-phosphate dehydrogenase, hypoxanthine guanine phosphoribosyl transferase, phosphoglycerate kinase and α-galactosidase have been shown to be dependent upon the number of X-chromosomes. Female embryos containing two functional X-chromosomes theoretically produce greater levels of enzymes intrinsic to X chromosome expression.

One of the two X-chromosomes of a normal female is inactivated in early embryonic development. Prior to X-inactivation, the levels of X-linked enzymes are known to be dependent on the number of X-chromosomes. Such enzymes are glucose 6-phosphate dehydrogenase (G6PD), phosphoglycerate kinase, β-galactosidase and hypoxanthine guanine phosphoribosyl transferase. Williams (1986) devised an effective method of visual colorimetric *in situ* assay system for the X-linked enzyme G6PD in predicting the sex of mouse embryos.

DNA-based Methods

Sex-determination of embryos by means of Y-specific nucleotides is based on two molecular biological principles, the first one involving the use of Y-specific DNA sequence as probe to hybridize genomic DNA extracts derived from the embryo while the second one involving designing suitable primers based on Y-specific nucleotide sequences and carrying out *in vitro* amplification of gene sequence by means of PCR. Because of its high sensitivity, high accuracy and high efficiency, polymerase chain reaction method will become the most potentially practical technique among various methods.

DNA probes   DNA probes have been used for sexing embryos. Foetal sex can be determined with DNA samples of as little as 20 ng using a repetitive Y probe pS4 (Disteche *et al.*, 1984). A cDNA sequence (probe) is produced, which binds to a gene present only on the Y-chromosome. This probe can be labelled to identify the presence of Y-chromosome in cell preparations. In another study, DNA sample was prepared from blastomeres of bovine embryos and dot-blotted with Y-specific radiolabelled probe. Leonard *et al.* (1987) identified the sex of blastocyst-stage bovine embryos. Within 30 hours a total of 150 biopsies were evaluated. Definite results were obtained in 57% of these biopsies.

Fluorescence in situ hybridization (FISH)   This is a sensitive technique for molecular diagnosis of chromosomes on single cells and can be applied to sex determination of embryos (Lee *et al.*, 2004). The objective has been to develop an accurate and reliable bovine Y chromosome-specific DNA probe in order to sex biopsied blastomeres derived from IVF bovine embryos by FISH. Bovine Y chromosome-specific PCR product derived from BtY2 sequences was labelled with biotin-16-dUTP (BtY2-L1 probe), and FISH was performed on karyoplasts of biopsied blastomeres and matched demi-embryos. Our FISH signal was clearly detected in nuclei of blastomeres of male embryos. FISH analysis of bovine embryos gave high reliability (96%) between biopsied blastomeres and matched demi-embryos. These results indicated that the BtY2-L1 bovine Y chromosome-specific FISH probe was an effective probe for bovine embryo sexing, and the FISH technique of probe detection could improve the efficiency and reliability.

Polymerase chain reaction   In order to apply PCR to sex determination of embryos of domestic animals, a nucleotide sequence specific for one gene should be recognized at first, as for example SRY gene associated with testis determining factor (Sinclair *et al.*, 1990), certain Y-specific repetitive sequences (Nakahori *et al.*, 1986), ZFX, ZFY genes homologous to X and Y chromosomes (zinc finger X chromosome, zinc finger Y chromosome) (Pollevick *et al.*, 1992) and so on which are known to have sex-specific nucleotide sequences. However, in those PCR-based sexing methods described above, each primer pair can recognize sex-specific gene fragments derived from only one sex chromosome whereas it cannot detect simultaneously fragments derived from both of X and Y chromosomes so that an internal control primer derived from intrinsic gene must be added in the reaction. The existence of two groups of primers in the amplification may result in the formation of primer dimer, thus reducing the sensitivity of the method. Against this defect, the invention designs a group of primers homologous to both X and Y chromosomes, which can amplify gene fragments of different lengths simultaneously from X chromosome and from Y chromosome so that sex type of early cow embryo before implantation can be determined straight away from results of electrophoresis.

A simple and rapid method was developed for sex determination of buffalo embryos produced *in vitro* amplifying Y-chromosome-specific repeat sequences by polymerase chain reaction (PCR) (Manna *et al.*, 2003). Buffalo oocytes collected from slaughtered animals were matured, fertilized and cultured *in vitro* for 7 days. On day 7, embryos were evaluated and divided into six groups according to the developmental stage (2, 4, 8, 16 cell stage, morulae and blastocyst). Each embryo was stored singly in phosphate-buffered saline at −20°C until PCR was carried out. Two different methods of extraction of DNA were compared: a standard procedure (ST), using a normal extraction by phenol-chloroform, isoamyl alcohol and final precipitation in absolute ethanol, and a direct procedure (DT), using a commercial kit (Qiaquik–Qiagen mini blood). A pair of bovine satellite primers and two pairs of different bovine Y-chromosome-specific primers (BRY4.a and BRY.1) were used in the PCR assay on embryos and on whole blood samples collected from male and female adult buffaloes, used as control. The trial was carried out on 359 embryos (193 for ST and 166 for DT). When DNA samples from blood were amplified, the sex determined by PCR always corresponded to the anatomical sex. Both extraction protocols recovered sufficient quantities of target DNA at all developmental stages, but the time required for the ST (24 h) limits its use in embryo sexing and supports the use of commercial extraction kits (5 h). Figure 7.2 shows the use of PCR array for sexing of embryos. Double bands indicate male embryos and single bands indicate female embryos.

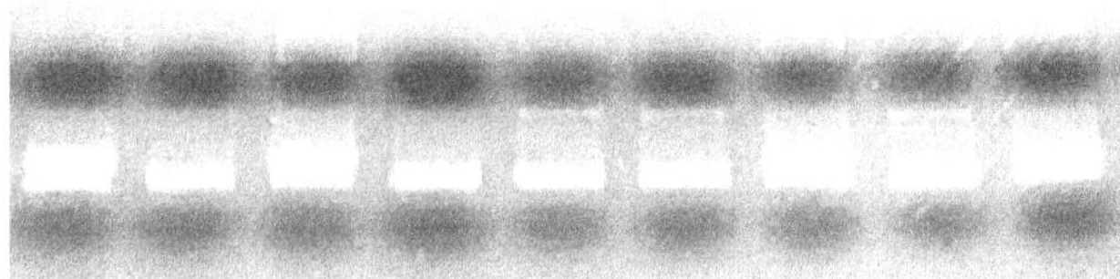

Figure 7.2   Sexing of embryo by polymerase chain reaction. Double bands indicate male embryos and single band indicates female embryos

Alves *et al.* (2006) used the random amplified polymorphic (RAPD) DNA technique to search for male-specific sequences present as monomorphic markers in genomic DNA from zebu and taurine bulls. A male-specific RAPD band was found to be present and highly conserved in both subspecies, as demonstrated by Southern blotting, fluorescent *in situ* hybridization (FISH) and DNA sequencing.

Loop-mediated isothermal amplification (LAMP) was used to identify a Y chromosome-specific sequence in water buffalo and to establish an efficient procedure for embryo sexing by LAMP (Hirayama *et al.*, 2006). Sexing by LAMP was performed using primers for swamp buffalo repeat Y-associated 2 gene. A 12S rRNA was also amplified by LAMP as a control reaction in both male and female. The minimal amount of the template DNA required for LAMP appeared to be 0.1–10 pg. Embryo sexing was also performed using blastomeres from interspecific nuclear transfer embryos. The sex determined by LAMP for blastomeres corresponded with the sex of nuclear donor cells in analyses using four or five blastomeres as templates. The LAMP reaction required only about 45 minutes and the total time for embryo sexing, including DNA extraction, was about 1 hour. In conclusion, the present procedure without thermal cycling and electrophoresis was reliable and applicable for water buffalo embryos.

Mara *et al.* (2004) developed a fast and easy duplex polymerase chain reaction (PCR) method, for sex determination of ovine *in vitro* embryos, produced prior to implantation. They tested 107 samples of autosomal cells (oviductal sheep cells and male lamb fibroblasts), divided into three groups for each sex according to the number of cells employed (30, 5, 2, respectively). The PCR utilized two different sets of primers: the first pair recognized a bovine Y-chromosome-specific sequence (SRY), that showed 100% homology with the corresponding sequence of the ovine Y-chromosome and is amplified in males only. The second pair recognized the bovine 1.715 satellite DNA (SAT) which was amplified in all ovine samples but, when submitted to the GenBank database did not show homology with any of the reported ovine sequences. Eight lambs were born and the sex, as determined by PCR, corresponded to the anatomical sex in seven (87.5% accuracy). These results confirmed that this method could be applied in ovine breeding programs to manipulate sex ratio of offspring.

IN VITRO FERTILIZATION

In the last two decades, rapid progress has been made in the development of new technologies of reproduction for genetic improvement in farm animals. The recombination of artificial insemination and embryo transfer has completely transformed the breeding scheme in bovines. Recent developments in the oocyte maturation and *in vitro* fertilization in farm animals is likely to offer a new dimension to research and development for future application in the improvement of animal reproduction.

In vitro fertilization (IVF) is a procedure that involves retrieving eggs (oocytes) and sperms from the bodies of the male and female and placing them together in a laboratory dish to facilitate fertilization. Fertilized eggs are then allowed to develop *in vitro* and after several days are transferred into a female's reproductive tract where implantation and embryo development can occur. The sequence of events in *in vitro* fertilization is illustrated in Figure 7.3.

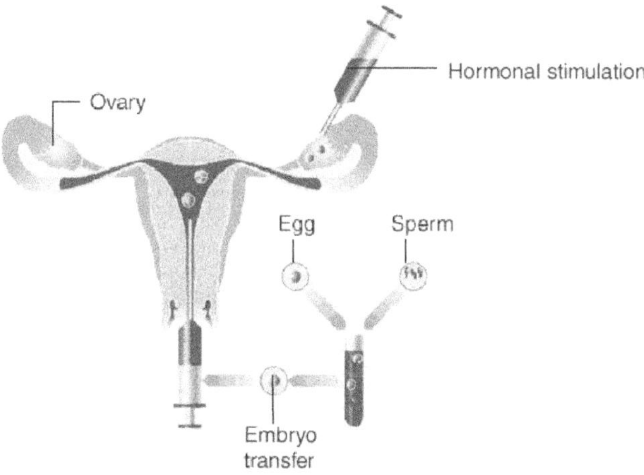

Figure 7.3 *In vitro* fertilization

Today, there are applications for the use of IVF in both humans and animals. IVF is only one of a number of innovative techniques available worldwide for infertile couples to achieve a family. As couples exhaust all clinical and surgical treatments for their infertility problems, they turn to the rapidly growing group of new technologies to bring children into the world. As a whole, these new technologies are called Advanced Reproductive Technologies or ART. Artificial insemination or AI is one of the most frequently used ART techniques. AI involves the transfer of male sperm directly into the reproductive tract of the female. This technology works best when the infertility problem is identified as the male's inability to produce and/or deliver viable sperm into the female's reproductive tract. A choice may be made to employ the use of IVF in cases when the female has blocked fallopian tubes or other problems that could prevent the successful release or fertilization of an oocyte.

Animal research is being done continuously to improve the techniques involved in IVF. This technology is also being used to benefit animals directly. IVF has been used in livestock with the intent of improving and expanding genetics. A genetically superior cow, for example, that for some reason becomes infertile due to old age, disease of the reproductive tract, or cystic ovaries, can produce calves through IVF procedures. IVF can also be performed with oocytes collected from animals dying of a terminal illness like cancer or even from cows that have died suddenly due to an accident.

Birth of live offspring following *in vitro* fertilization plus embryo transfer has been reported in several experimental animals. In humans, satisfactory results of *in vitro* fertilization and transfer used to treat infertility have been reported from several countries since 1978. However, there have been very few reports of successful *in vitro* fertilization in farm animals. In 1982, the first *in vitro* fertilized calf was born in the United States and since then more calves were born by this technique in other countries.

The basic steps involved are:

1. Collection of oocytes or eggs from a donor female.
2. Fertilization of the oocyte in a laboratory dish.
3. Culture or development of the embryo in a laboratory dish.
4. Transfer of the embryo back into a female recipient.

The major differences in the application of IVF technique when comparing humans and animals lie in the fact that in cattle the oocytes are collected at an immature stage and are matured *in vitro* (in a laboratory dish). Also, historically cattle embryos have been allowed to develop to about the day seven stage while human embryos are transferred sooner.

Preparation and Collection of the Oocyst

The duration of the capacity of the ovulated oocyte to be fertilized in the oviduct is estimated to be 12–24 hours in the cow. The time for the penetration of spermatozoa into the reproductive tract is estimated as 3.5 hours after ovulation in the cow. To induce superovulation for *in vitro* fertilization experiments, gonadotrophin is usually used in animals. The combination of the pregnant mare's serum gonadotrophin (PMSG) or follicle stimulating hormone (FSH) both in combination with prostaglandin (PG) has been used to induce superovulation. The advantages of this method are that many oocysts are retrieved at the same time and the timing of ovulation can be adjusted. Usually follicular oocysts are collected from the Graafian follicle.

The immature oocyte is collected from the ovary and cultured *in vitro* to induce maturation. Cow ovaries are collected at the slaughterhouse and maintained in PBS at 30–32°C for transportation to the laboratory. Following aspiration of follicles, the oocytes are then observed under a dissecting microscope. The oocytes are graded according to their morphology, washed twice with HEPES [N-(2-hydroxyl ethyl) piperazine N´-(2-ethane sulphonic acid) and once with BMOC-3 (Branchat and Olivhant medium for oocyte collection)]. After washing, 5–10 oocytes are transferred to 0.2 ml of BMOC-3 under paraffin oil at 37°C with 5% CO_2 in 95% air for 12–32 hours.

Using complex media containing serum and somatic cells, pregnancy rates following transfer of single, day 7 advanced-stage blastocysts approached 60%, while pregnancy rates of morulae or day 8 blastocysts were substantially lower (Hasler, 2000). The different stages of oocyte development are shown in Figure 7.4. Pregnancies resulting from *in vitro* derived embryos were characterized by the following features: sex ratio skewed in favour of males, increased spontaneous abortion rate throughout gestation, reduced intensity of labour in recipients and increase in birth weights, dystocias, cell mortality and foetal abnormalities.

In vitro maturation (IVM)   The oocytes are then placed in a culture-medium-filled dish and allowed a period of about 24 hours to reach maturity. During this time, the cumulus cells remain in contact with the oocyte in what is called a cumulus–oocyte complex (COC). The dish is incubated at about 38.5°C. (normal cow body temperature)

In vitro fertilization (IVF)   After twenty-four hours in IVM, the matured COCs are washed, moved into new dishes with a special IVF culture medium, and sperms are added at a concentration that is optimal for sperm capacitation and fertilization. These dishes are then incubated for 18 hours.

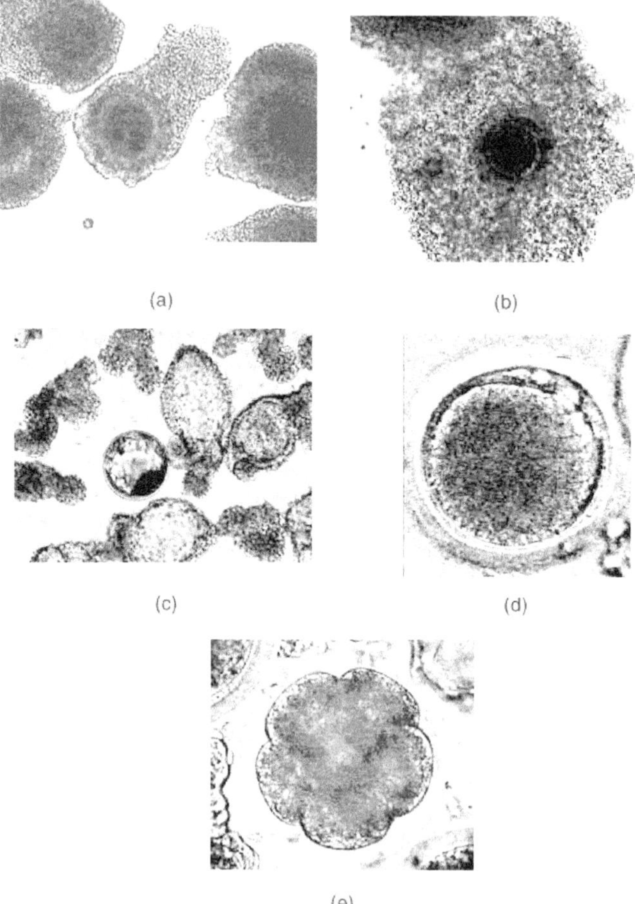

Figure 7.4 Different stages of cumulus cells, blastocyst and embryo. a. Cumulus cells; b. Expanded cumulus cells; c. Blastocyst; d. Embryo; e. Morula

In vitro culture (IVC) After approximately 18 hours, the IVF dishes are checked closely to ensure hyperactivation of sperm. Any presumptive zygotes are then washed and removed from IVF medium and placed in IVC culture medium. This medium contains uterine tubal cells (UTC) used to enhance growth and is formulated for optimal embryo development. The dishes are incubated for about seven days. They are checked after the first two days to verify cleavage.

Experiments were carried out to develop and improve *in vitro* culture systems for IVM–IVF prepubertal goat oocytes. Cumulus–oocyte complexes (COC) were obtained by slicing ovaries from slaughtered prepubertal goats (Izquierdo *et al.*, 1999).

In the experiment 1, co-culture systems using oviductal epithelial cells (OEC) and cumulus cells (CC) both of caprine and bovine, were compared; in experiment 2 the presence of serum and/or OEC, in experiment 3 four culture media (TCM 199, Ham's F10, CZB and SOF) for co-culture with OEC and in experiment 4 conditioned medium with OEC were compared. In experiment 1, the percentage of morulae plus blastocysts was higher in culture with OEC, in both caprine and bovines (15.1 and 14.8%, respectively) than with CC (4.1 and 6.7%, respectively). In experiment 3, this percentage was higher using OEC with TCM-199 compared to CZB medium (21.3 and 12.3%, respectively) and in the experiment 4, the results were 3.7, 11.2 and 21.3% for TCM-199 without cells, conditioned medium and co-culture with OEC, respectively.

Laparoscopic ovum pick-up (LOPU) is a convenient methodology by which oocytes can be recovered and used either for *in vitro* production of zygotes or as a source of cytoplasts in nuclear transfer (NT) procedures (Baldassarre *et al.*, 2002). The pregnancy and transgenesis rates achieved with IVM/IVF of LOPU-sourced oocytes followed by subsequent DNA microinjection of zygotes are similar to the rates obtained when using *in vivo*-produced oocytes or zygotes. Similarly, pregnancy rates and kids born by using LOPU-sourced and *in vitro* matured oocytes as recipient cytoplasts in NT programmes are comparable with those reported by others using *in vivo*-matured oocytes collected by oviduct flushing. The use of LOPU allows for improved control over the stage of maturation/ development of the oocytes and produced zygotes, a less invasive means of recovery, thereby allowing for repeated usage of the oocyte donor animals and the ability to source the oocytes from live animals of known health status. LOPU followed by IVM/IVF has demonstrated great potential for the early propagation of valuable animals, in particular, transgenic animals.

Preparation of the Spermatozoa

The capacitated spermatozoa is essential for *in vitro* fertilization. After capacitation, the acrosome reaction in the head of the spermatozoa allows the release of enzymes for penetration into the zona pellucida of the oocytes. The capacitation process is completed during the movement of the spermatozoa from the oviduct to the uterus. Capacitation involves the removal, probably by the enzyme action, of macromolecular material located on the surface of the spermatozoa. Re-exposure of the spermatozoa to seminal plasma leads to loss of capacitation and the process may well-involve the restoration of the surface layer or **decapacitation factor** which normally coats the spermatozoa as they pass through the male reproductive tract. Experimental procedures to induce the capacitation or acrosome reaction of the spermatozoa involve *in vitro* incubation with follicular fluid and serum.

Embryo Transfer (ET)

On about day seven after fertilization, the embryos (at the blastocyst stage) are transferred non-surgically to recipient cows. Recipient cows are synchronized to ensure timing of embryo placement postovulation. Care is taken to place embryos in the uterine horn on the same side as the corpus luteum where ovulation occurred to facilitate recognition of pregnancy.

In vitro Fertilization in Farm Animals

Among farm animals, *in vitro* fertilization has been achieved in the cow, pig and goat; but the number of newborn animals was limited. The oocytes are recovered from ovarian follicles or from oviducts near the time of *in vitro* ovulation following treatment of donors with PMSG and PG. For *in vitro* capacitation, semen is incubated, treated with high ionic strength medium and subsequently incubated in defined medium prior to insemination of the oocytes.

Lambert *et al.* (1986) produced six calves from a combination of *in vitro* fertilization and incubation in the rabbit oviduct before transfer. Laparoscopy was used to recover *in vivo* matured follicular oocytes from superovulated heifers. The oocytes enclosed in the expanded cumulus were then mixed *in vitro* for fertilization with fresh semen in a single medium and used at a concentration of 1×10^6 sperms/ml. Following the fertilization period (16–20 hours), the embryos were cultured *in vitro* up to 2 or 8-cell stage. Following *in vitro* development, the 2 or 8-cell embryos were either transformed into the cow oviduct or the rabbit oviduct. The embryo (8-cell stage to blastocysts) obtained after incubation in the rabbit were transformed to the uterus of virgin heifers either surgically or non-surgically. Numabe *et al.* (2000) evaluated the difference in birth weight and gestation length between Japanese Black calves obtained from transfer of bovine embryos produced *in vitro* (IV) and those developed *in vivo* (IV). Both the IVD and IVP embryos were transferred non-surgically to Holstein recipients on day 7 + 1 of oestrous cycle. Results indicated that IVP calves had heavier birth weights than IVD calves, but that the average gestation length of IVP calves was not always longer than that of IVD calves. Further, the birth rate of heavier calves and the incidence of stillbirth and perinatal mortality up to 48 hour post partum in IVP calves were greater than those in IVD calves. The transfer of *in vitro* fertilized embryos is now practicable, but the laboratory methods of fertilization still need development.

The recent developments in IVP technology in cattle can be adapted to embryo production in small ruminants to overcome limitations exhibited by surgical procedures on preserving the reproductive potential of donors and the efficiency of embryo production (Cox and Alfaro, 2007). The aim was to assess the current procedures used in cattle for the production of IVP embryos in goats and sheep based on oocytes supplied by laparoscopic-aided ovum pick-up (LOPU). Sexually matured goat and sheep donors were treated during the breeding season with FSH and subjected to laparoscopic-guided follicular puncture under general anaesthesia. The collected cumulus–oocyte complexes were matured in medium 199 and fertilized by frozen, thawed spermatozoa using Talp medium supplemented with heparin and oestrus-sheep serum. Cleaved ova were either cultured in sheep *in vitro* fertilization medium plus amino acids or transferred to sheep oviducts. Blastocyst rate, hatching rate and development rate up to term were used as markers of embryo function. The results are comparable to those obtained in small ruminants and in bovines suggesting that requirements for embryo production and development are similar.

Experiments were conducted to evaluate *in vitro* fertilization (IVF) of *in vitro* matured (IVM) bitch oocytes using dog spermatozoa frozen in three different extenders (De los Reyes *et al.*, 2006). Sperm-rich fraction from eight ejaculates of five dogs was frozen in each one of three egg yolk Tris extenders with additional materials including (i) 1.4 g citric acid and 0.8 g glucose; (ii) 0.7 g citric acid and 3.5 g glucose; or (iii) 1.4 g citric acid and 0.8 g fructose (all with 5% glycerol in 100 mL milliQ water). Thawed sperms were co-incubated with IVM bitch oocytes for 6 hours. Oocytes were fixed and evaluated under an epifluorescence microscope; penetrated oocytes were defined as those having sperm heads in the perivitelline space or in the oocyte cytoplasm. Higher penetration rates ($P < 0.05$) were obtained in oocytes cultured with spermatozoa frozen in extenders B and C than those frozen in extender A (33.1, 34.2 and 26.4%, respectively).

Lim *et al.* (2007) developed a defined culture medium that supported improved *in vitro* bovine embryo development and calving rate after embryo transfer (ET). *In vitro*-matured bovine oocytes from abattoir-derived ovaries from Korean native HanWoo cattle were fertilized with frozen-thawed spermatozoa and embryos were cultured in two-step culture media. In Experiment 1, embryos were cultured in media supplemented with 8 mg/mL BSA, or 0.1 mg/mL PVA and 8 mg/mL BSA + 2.77 mM myo-inositol or 0.1mg/mL PVA + 2.77 mM myo-inositol. Although defined culture media containing PVA supported lower developmental competence compared to undefined media (containing BSA; 8% versus 34%, respectively), defined culture media containing 2.77 mM myo-inositol increased rates of blastocyst formation up to 28%. In experiment 2, the effect of energy substrate (1.5 mM glucose or 1.2 mM phosphate) in PVA-myo-inositol-defined culture medium on *in vitro* embryo development was investigated. Defined culture media containing PVA, myo-inositol and phosphate supported better embryo development to blastocysts compared to medium supplemented with both glucose and phosphate (43% versus 31%). In experiment 3, the effect of epidermal growth factor (EGF) in PVA+myo-inositol-phosphate two-step culture medium on *in vitro* embryo development was investigated. Among 0, 1, 10 and 100 ng/mL EGF concentrations, the maximal effect was observed with 10 ng/mL EGF (52% blastocyst formation). In experiment 4, total cell number and calving rate were compared between defined PVA-myo-inositol-phosphate-EGF medium and undefined medium containing BSA, glucose and phosphate. No differences in total cell number of blastocysts obtained from the two groups were observed, however, the rate of viable offspring production was increased using the defined culture medium, compared to the undefined culture medium. In experiment 5, the relative abundance of mRNA transcripts [interferon-tau (If-tau), glucose transporter-1 (glut-1) and insulin-like growth factor 2 receptor (Igf2r)] were analysed in blastocysts derived from undefined or defined culture media. Gene expression of If-tau, glut-1 was significantly increased in defined culture medium compared to undefined medium. In conclusion, chemically defined culture media without BSA or FBS improved developmental competence of *in vitro*-cultured bovine embryos and delivery of viable calves after ET.

In vitro fertilization (IVF) is a feasible way to utilize sex-sorted sperm to produce offspring of a predetermined sex in the livestock industry. The objective of the present was to examine the effects of various factors on bovine IVF and to systematically improve the efficiency of IVF production using sex-sorted sperm (Xu *et al.*, 2006). Both the bulls and the sorting contributed to the variability among differential development rates of embryos fertilized by sexed sperm. As few as 600 sorted sperm were used to fertilize an oocyte, resulting in blastocyst development of 33.2%. Post-warming of vitrified sexed IVF embryos resulted in high morphological survival (96.3%) and hatching (84.4%) rates, similar to those fertilized by non-sexed sperm (93.1 and 80.6%, respectively). A 40.9% pregnancy rate was established following the transfer of 3,627 vitrified, sexed embryos into synchronized recipients. This was not different from the rates with non-sexed IVF (41.9%, $n = 481$), or *in vivo*-produced (53.1%, $n = 192$) embryos. Of 458 calves born, 442 (96.5%) were female and 99.6% appeared normal. These technologies (sperm sexing-IVF-vitrification-embryo transfer) provide farmers, as well as the livestock industry, with a valuable option for herd expansion and heifer replacement programmes. In summary, calves were produced using embryos fertilized by sex-sorted sperm *in vitro* and cryopreserved by rapid cooling vitrification.

REVIEW QUESTIONS

1. How could embryo transfer technology be effectively used for improvement of livestock? What are the advantages and disadvantages of this method?

2. What are the steps involved in embryo transfer technology?

3. What are the methods available for sexing of embryos?

4. What do you understand by *in vitro* fertilization? What are the steps involved in the method?

REFERENCES

Alves, B.C., Mayer, M.G., Taber, A.P., Egito, A.A., Fagundes, V., McElreavey, K. and Moreira-Filho, C.A. (2006). "Molecular characterization of a bovine Y-specific DNA sequence conserved in taurine and zebu breeds." *DNA Seq.* 17: 199–202.

Baldassarre, H., Wang, B., Kafidi, N., Keefer, C., Lazaris, A. and Karatzas, C.N. (2002). "Advances in the production and propagation of transgenic goats using laparoscopic ovum pick-up and *in vitro* embryo production techniques." *Theriogenology.* 57: 275–284.

Baruselli, P.S., de Sa Filho, M.F., Martins, C.M., Nasser, L.F., Nogueira, M.F., Barros, C.M. and Bo, G.A. (2006). "Superovulation and embryo transfer in *Bos indicus* cattle." *Theriogenology.* 65:77–88.

Cox, J.F. and Alfaro, V. (2007). "*In vitro* fertilization and development of OPU derived goat and sheep oocytes." *Reprod. Domest. Anim.* 42:83–87.

De los Reyes, M., Carrion, R. and Barros, C. (2006). "*In vitro* fertilization of *in vitro* matured canine oocytes using frozen-thawed dog semen." *Theriogenology.* 66:1682–1684.

Disteche, C., Luthy, D., Haslam, D.B. and Hoar, D. (1984). "Prenatal identification of a deleted Y chromosome by cytogenetics and a Y-specific repetitive DNA probe." *Human Genet.* 67: 222–224.

Gardner, R.L. and Edwards, R.G. (1968). "Control of the sex ratio at full term in the rabbit by transferring sexed blastocysts." *Nature.* 218: 346–349.

Hesler, J.F. (2000). "*In vitro* production of cattle embryos: problems with pregnancies and parturition." *Human Reprod.* 5: 47–58.

Hirayama, H., Kageyama, S., Takahashi, Y., Moriyasu, S., Sawai, K., Onoe, S., Watanabe, K., Kojiya, S., Notomi, T. and Minamihashi, A. (2006). "Rapid sexing of water buffalo (*Bubalus bubalis*) embryos using loop-mediated isothermal amplification." *Theriogenology.* 66:1249–1256.

Izquierdo, D., Villamediana P. and Paramio, M.T. (1999). "Effect of culture media on embryo development from prepubertal goat IVM-IVF oocytes." *Theriogenology.* 5247–861.

King. (1984). *J. Reprod. Fert.* 98: 335–340.

Kohler, G and Milstein, C. (1975). "Continuous culture of fused cells secreting antibody to predetermined specificity." *Nature.* 256: 495–497.

Lambert, R.D., Sirad, M.A., Bernard, C., Beland, R. Rioux, J.E., Leclere, P., Menard, D.P. and Bedoya, M. (1986). "*In vitro* fertilization of ovine oocytes matured *in vitro* and collected at laparoscopy." *Theriogenology.* 25: 117–133.

Lee, J.H., Park, J.H., Lee, S.H., Park, C.S. and Jin, D.I. (2004). "Sexing using single blastomere derived from IVF bovine embryos by fluorescence *in situ* hybridization (FISH)." *Theriogenology.* 62:1452–1458.

Leonard, M., Kirszenbauwm, N., Continot, C., Chesne, P., Heyman, Y., Stinnakre, M.G., Bishop, C., Detouis, C., Vaiman M. and Fellous, M. (1987). "Sexing bovine embryo using Y-chromosome specific DNA probe." *Theriogenology.* 27: 248.

Lim, K.T., Jang, G., Ko, K.H., Lee, W.W., Park, H.J., Kim, J.J., Lee, S.H., Hwang, W.S., Lee, B.C. and Kang, S.K. (2007). "Improved *in vitro* bovine embryo development and increased efficiency in producing viable calves using defined media." *Theriogenology.* 67:293–302.

Lopes, R.F., Forell, F., Oliveria, A.T. and Rodrigues, J.L. (2001). "Splitting and biopsy for bovine embryo sexing under field conditions." *Theriogenology.* 56: 1383–1392.

Lopes da Costa, L., Chagas, J., Silva, E. and Robalo Silva, J. (2001) "Superovulatory response, embryo quality and fertility after treatment with different gonadotrophins in native cattle." *Theriogenology.* 56: 65–67.

Manna, L., Neglia, G., Marino, M., Gasparrini, B., Di Palo, L. and Zicarelli, R. (2003). "Sex determination of buffalo embryos (*Bubalus bubalis*) by polymerase chain reaction." *Zygote.* 11: 17–22.

Mara, L., Pilichi, S., Sanna, A., Accardo, C., Chessa, B., Chessa, F., Dattena, M., Bomboi, G. and Cappai, P. (2004). "Sexing of *in vitro* produced ovine embryos by duplex PCR." *Mol. Reprod. Dev.* 69:35–42.

Nakahori, Y., Mitani, K., Yamada, M. and Nakagone, Y. (1986). "A human Y-chromosome specific repeated DNA family (DYZ1) consists of a tandem array of pentanucleotides." *Nucleic Acids Res.* 14: 7569–7580.

Numabe, T., Oikawa, T., Kikuchi, T. and Horiuchi, T. (2000). "Birth weight and birth rate of heavy calves conceived by transfer of *in vitro* or *in vivo* produced bovine embryos." *Anim. Reprod. Sci.* 64: 13–20.

Park, J.H., Lee, J.H., Choi, K.M., Joung, S.Y., Kim, J.Y., Chung, G.M., Jin, D.I. and Im, K.S. (2001). "Rapid sexing of preimplanatation bovine embryo using consecutive and multiplex PCR with biopsied single blastomere." *Theriogenology.* 55: 1843–1853.

Pollevick, G.D., Giambiagi, S., Mancardi, S., de Luca, L., Burrone, O., Frasch, A.C.C. and Ugalde, R.A. (1992). "Sex determination of bovine embryo by restriction fragment length polymorphisms of PCR amplifed ZFX/ZFY loci." *Biotechniques.* 10: 805–807.

Spell, A.R., Beal, W.E. Corah, L.R. and Lamb, G.C. (2001). "Evaluating recipient and embryo factors that affect pregnancy rates of embryo transfer in beef cattle." *Theriogenology.* 56: 287–297.

Wachtel, S.S. (1984). "H-Y Antigen in the study of sex determination and control of sex ratio." *Theriogenology.* 21: 18–28.

Warwick, B.L., Berry, R.O. and Horlacher, WR. (1934). "Results of mating rams to Angora female goats." Proc. 27th Ann. Meet. Am. Soc. Anim. Prod. 225–227.

Willet, E.L., Black, W.G., Casida, L.E. Stone, W.H. and Buckner, P.J. (1951). "Successful transplantation of a fertilized bovine ovum." *Science.* 113: 247.

Williams, T.J. (1986). "A technique for sexing mouse embryo by a visual colorimetric assay of the X-linked enzyme, glucose-6-phosphate dehydrogenase." *Theriogenology.* 25: 733–739.

Xu, J., Guo, Z., Su, L., Nedambale, T.L., Zhang, J., Schenk, J., Moreno, J.F., Dinnyes, A., Ji, W., Tian, X.C., Yang, X. and Du. F. (2006). "Developmental potential of vitrified Holstein cattle embryos fertilized *in vitro* with sex-sorted sperm." *J. Dairy Sci.* 89:2510–2518.

8

TRANSGENIC ANIIMALS

INTRODUCTION

The genetic alteration of animals by man has been the driving force behind the domestication of wild animals. The animal breeding programmes carried out since the last century have become much more consistent in pursuing the selective, purposeful and noticeable alteration of individual traits. The extreme development of certain traits is the result of formulating distinct breeding aims. The improvement of milk yields in dairy cows may serve as an example. While the annual milk yield per cow was approximately 1000 kg in the middle of the 19th century, it reached more than 7000 kg in some populations at the end of the 1980s with herd averages of over 10,000 kgs.

The difficult problem in animal breeding is with the complex genotypes responsible for the development of desired phenotypic traits. Frequently, information about the genetic causes of certain phenotypic characters can be obtained only by analysing an extensive body of data obtained from related animals (pedigree analysis and estimation of breeding values). The most important question regarding phylogenetically determined characters is the distinction between genetic and environmental influences for the evaluation of the cause-and-effect relationships between genotypes and phenotypes.

During the last few years, new methods have been developed which, unlike conventional breeding techniques, allow the genetic composition of organisms to be modified directly and selectively rather than indirectly by estimating the breeding values on the basis of phenotypes in order to establish ranking order for selection. In principle, these new techniques of genetic manipulation can be categorized into those which treat the entire genome as a compact unit (genome manipulation; e.g. nuclear transfer) and those in which individual genes are manipulated (gene manipulations; e.g. gene transfer). Further, these techniques can be differentiated into those allowing manipulation of somatic cells (e.g. somatic gene therapy) and those directed at altering the germ line of animals. The latter techniques give rise to transgenic lines of animals characterized by the stable transmission of the genetic modification.

The application of gene transfer techniques has furnished new insights in the developmental biology and the principle underlying tissue-specific gene expression. Gene transfer allows the development of new production system for pharmaceutically important proteins. As far as animal production is concerned, gene transfer appears to be a promising technique for improving disease resistance, performance of the quality of animal products by modifying, for example, metabolic pathways, and hormone status.

Recently, progress has been achieved in reproductive biotechnology in swine with special reference to *in vitro* production of embryos, generation of identical multiples and transgenic pigs useful for xenotransplanation (Niemann and Rath, 2001). *In vitro* production (*in vitro* maturation, *in vitro* fertilization and *in vitro* culture) of viable porcine embryos is possible, although with much lower success rates than in cattle. The main problems are insufficient cytoplasmic maturation of porcine oocytes. The generation of the first piglets from somatic cell nuclear transfer has been achieved. DNA microinjection into pronuclei of porcine zygotes has reliably resulted in the generation of transgenic pigs, which have special importance for the production of valuable pharmaceutical proteins in milk and xenotransplantation. It has been demonstrated that by expression of human complement regulatory proteins in transgenic pigs, the hyper-acute rejection response occurring after xenotransplantation can be overcome in a clinically relevant manner.

TRANSFER OF GENES

The term "transgenic" was used for the first time by Gordon and Ruddle (1981) to describe animals harbouring new genes within their genome. This term is now generally applied to the characterization of certain variants of species whose genome has been altered by the transfer of genes. Transgenic animals and genetically modified organisms (GMOs) are organisms with a segment of foreign DNA incorporated into their genome, or with any modification introduced artificially in their genome sequence.

Escherichia coli was the first genetically modified bacteria (Cohen *et al.*, 1973). Since then, the technology of genetic manipulation of organisms has had a remarkable progress with a variety of bacteria, fungi, protists, plants and animals as GMOs. Regarding transgenic animals, the first genetically modified mouse was obtained by Jaenisch and collaborators 3 decades ago (Jaenisch *et al.*, 1975).

Even today, the mouse is the most important model for genetic studies in mammals. In 2002 the sequencing of the mouse genome was finished (Waterston *et al.*, 2002), providing valuable knowledge for genetic engineering in livestock animals, such as cattle, pig, sheep and goats (Table 8.1).

Table 8.1   Transgenic landmarks and reproductive technology in livestock animals (cattle, goats, sheep and swine)

Important transgenic landmarks	Reference
Transgenic pig and sheep	Hammer *et al.* (1985)
Embryonic cloning by nuclear transfer in sheep	Willadsen (1986)
Transgenic dairy cattle	Krimpenfort *et al.* (1991)
Transgenic sheep producing altered milk	Wright *et al.* (1991)
Transgenic pigs resistant to viral infection	Muller *et al.* (1992)
Pig expressing inhibitor of human complement system	Fodor *et al.* (1994)
Somatic cloning by nuclear transfer in sheep (Dolly)	Wilmut *et al.* (1997)
Transgenic livestock production as a model of human disease	Petters *et al.* (1997)
Transgenic cattle produced by nuclear transfer	Cibelli *et al.* (1998)
Transgenic sheep produced by gene targeting	McCreath *et al.* (2000)
"Ecologically correct" transgenic pig	Golovan *et al.* (2001)
Production of biopolymer fibre from transgenic cells	Lazaris *et al.* (2002)
Calf with human artificial chromosome	Kuroiwa *et al.* (2002)
Transgenic cattle producing altered milk protein compounds	Brophy *et al.* (2003)
Complete gene inactivation in pigs	Phelphs *et al.* (2003)
Sequential inactivation of 2 bovine genes	Kuroiwa *et al.* (2004)
Transgenic cow resistant to bacterial infection (mastitis)	Wall *et al.* (2005)

The first successfully reported foreign DNA transfer to a mammal with germ-line transmission was obtained by using a retrovirus (Jaenisch *et al.*, 1975). The retrovirus is an efficient transgene-delivery vehicle and it has been used to infect the bovine embryo when injected into the perivitelline space, between the surface of the fertilized egg and the zona pellucida (Chan *et al.*, 1998). However, DNA transfer by a retrovirus has some limitations, such as the following:

1. preferential integration of retroviruses into dividing cells;

2. recognition of specific target cells; and

3. high probability of chimeric animal production, which is incapable of transferring the transgene to the next generation.

Studies are continued with the use of lentivirus vectors (family Retroviridae) in order to overcome some of these limitations, since lentiviruses do not require dividing cells for their integration and have a wide spectrum of target host cells. The sperm-mediated DNA transfer in rabbits, during *in vitro* fertilization, was the pioneer experiment to produce transgenic animals (Brackett *et al.*, 1971). However, the generated animals were generally mosaic for the transgenes, so the genes were not always expressed in the second generation. The transformed sperms have been used for *in vitro* fertilization (Maione *et al.*, 1998).

The production of transgenic animals by embryonic stem (ES) cells, generates chimeric embryos composed of 2 distinct cell lines, and 1 of them carries the desired gene(s). ES cell lines are pluripotent cells derived from the inner cell mass of the blastocysts, and are capable of producing all tissues of an individual, including the germinative tissue. This gene transfer method involves the injection of transgenic ES cells into expanded blastocysts, and was first utilized in mice. This method is feasible in all animals in which ES cells can be manipulated and transfected *in vitro* (Robertson *et al.*, 1986). Mice were the first mammals from which ES cells were isolated and cultured (Evans and Kaufman, 1981; Martin, 1981). Some of the advantages of ES cells in the production of transgenic animals include the fact that ES cell lines can be targeted through homologous DNA recombination, which can be screened and selected for the incorporation of the foreign DNA before the transgenic embryo production.

Another method for introducing foreign genes into animals is by direct pronuclear injection, where the gene of interest is directly injected into one of the pronuclei of a zygote (Figure 8.1). This technique was first developed in mice (Gordon and Ruddle, 1981; Palmiter *et al.*, 1982) and is still the main choice in production of transgenic rodents. Subsequently, the pronuclear injection was responsible for the production of transgenic livestock species (including rabbits, sheep and swine) with the rat and human growth hormone gene (Hammer *et al.*, 1985), as well as the first transgenic bovine (Krimpenfort *et al.*, 1991) with microinjected DNA.

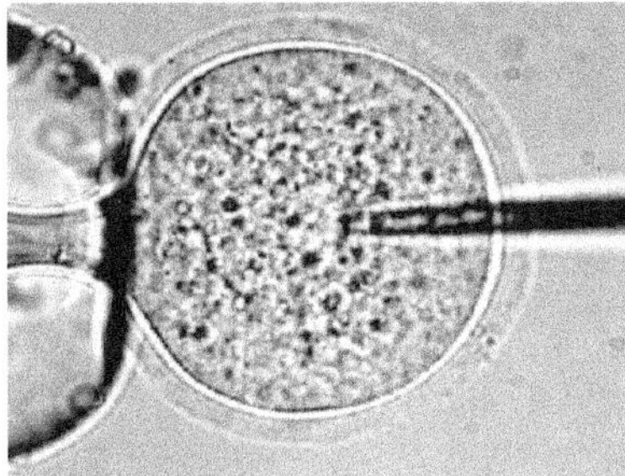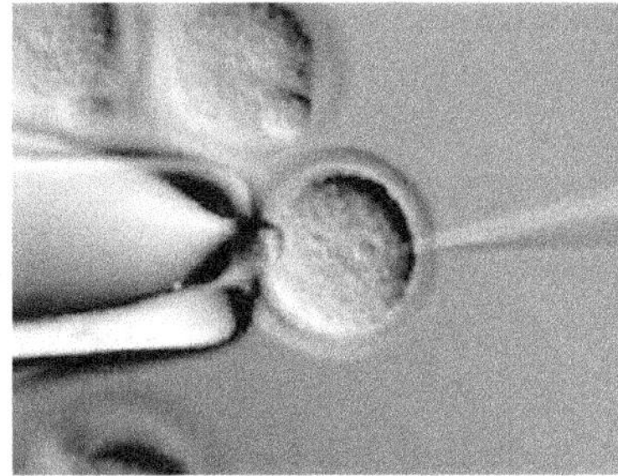

Figure 8.1 Microinjection

Recently, the first successful nuclear transfer (NT) using somatic cells from the mammary gland has lead to the birth of the famed clone Dolly (Wilmut *et al.*, 1997). In this method, DNA from the MII oocytes is removed (enucleation), leaving only the cytoplasm (the cytoplast). Following enucleation, a donor nucleus which can be almost from any cell of the body, is injected into the perivitelline space and fused to the cytoplast by electrofusion. After fusion, the zygote clone is activated by either chemical or mechanical stimulation in order to initiate embryo development (Campbell *et al.*, 1996; Wilmut *et al.*, 1997).

The NT associated with *in vitro* cell culture manipulation resulted in the first livestock animal (sheep) expressing human clotting Factor IX (Schnieke *et al.*, 1997). Afterwards, 5 transgenic bovine clones resistant to antibiotic G418 were successfully produced by NT (Cibelli *et al.*, 1998). Since then, many other transgenic animals have been generated by this technique (Keefer, 2004; Kues and Niemann, 2004).

This process is known as cloning, since a series of identical individuals can be generated from a single DNA donor cell line. The greatest impact of NT on transgenic livestock production is in the possibility to cultivate primary cells *in vitro* during a long lifespan without loss of viability (Kasinathan *et al.*, 2001), and to execute the genetic manipulation before NT in order to produce transgenic embryos with the same rate as that of non-transgenic ones (Iguma *et al.*, 2005).

The first animal containing experimentally introduced foreign DNA was obtained by microinjection of SV-40 DNA into the blastocoel of mice (Jaenisch and Mintz, 1974). Apart from retroviral injection of mouse embryos, another technique, i.e., the direct microinjection of DNA into the pronuclei, has been used extensively for the production of transgenic animals. A recombinant plasmid composed of segments of herpes simplex virus and SV-40 viral DNA inserted into the bacterial plasmid pBR 322 has been microinjected into the pronuclei of fertilized mouse oocytes by Gordon *et al.* (1980). Two of 78 mice contained the foreign DNA sequences, demonstrating that genes can be introduced into the mouse genome by direct injection of DNA into the nuclei of early embryos.

Microinjection

One aim in the generation of transgenic animals is to guarantee that all somatic cells and germ line in particular, will carry the foreign gene construct. It is therefore mandatory that the transfer of foreign genes is carried out as early as possible during the development of an animal. This usually means that the foreign genes are introduced into fertilized oocytes or into 2-cell-stage embryos at the latest. This could be achieved by direct microinjection of DNA solutions into the pronuclei and zygote of embryos. The steps involved in the generation of transgenic mammals consist of the following:

1. Cloning of the gene construct
2. Preparation of the DNA solution to be used for microinjection
3. Preparation of oocytes and embryos
4. Microinjection of the DNA into the pronuclei
5. Transfer of injected embryos into suitable foster animals
6. Detection of the transgenes in the newly born animals

Any DNA fragment, e.g. chemically synthesized DNA, cloned DNA or fragments of chromosomes can be microinjected and will be integrated into the host genome with more or less the same frequency. It has been observed that linearized DNA molecules integrate approximately five-fold better than circular molecules. Recombinant plasmids or cosmids containing the gene construct used for the establishment of transgenic animals are isolated as supercoiled DNA molecules by standard procedures from bacterial cultures. They are subsequently cleaved by suitable restriction enzymes to obtain the insert. A solution of approximately 1 µg/ml will contain several hundred copies of the construct/picolitre (pl). This concentration is optimal for obtaining the highest integration frequencies.

Preparation of embryos   Between the 8th and 12th day after the onset of oestrus, the cows receive 2000 to 3000 IU of pregnant mare serum gonadotrophin (PMSG). Two days after the PMSG injection, the donor animals receive 2 ml of prostaglandin F2α to induce luteolysis of the corpus luteum. The oestrus was observed approximately 48 hours later and was exploited for artificial insemination. A second insemination follows 12 hours later. The most suitable time for the recovery of the embryos is 78 to 82 hours after the prostaglandin treatment. Approximately 12 oocytes can be collected per donor animal and about half of them are suitable for microinjection.

DNA microinjection The embryo to be injected is held by holding the pipette under reduced pressure and positioned so that the pronuclei are visible. The injection pipette has a tip diameter of 1 mm and is filled with DNA solution at the tip. For DNA injection, the tip is inserted carefully through the zona pellucida and the cell membrane until the tip is positioned within the pronucleus (Figure 8.1). The volume of the pronucleus increases by approximately 50% if 1–2 pl of DNA solution is injected. This visible swelling of the pronucleus is the only indication for a successful injection of the DNA solution into the nucleus. All injected embryos are transferred to culture medium and are kept at 37°C until they are used for transfer into a recipient animal.

Sato *et al.* (2002) attempted to transfect testicular spermatozoa with plasmid DNA by direct injection into testes to obtain transgenic animals (this technique was thus termed "testis-mediated gene transfer"). When injected males were mated with superovulated females 2 and 3 days after injection, high efficiencies (more than 50%) of gene transmission were achieved in the mid-gestational F0 foetuses. This indicated that plasmid DNA introduced into a testis is rapidly transported to the epididymis and then incorporated by epididymal spermatozoa.

Embryo transfer The embryo will usually be transferred to a synchronized recipient animal after a short period of *in vitro* culture (up to several hours). Morphologically defective embryos are discarded and not used for the embryo transfer. The oestrous cycle of the recipient animals must be synchronized with that of the animals used as embryo donors so that ovaries, oviducts and uterus will be in a condition that will guarantee the physiological development of the transferred embryos (Figure 8.2). Synchronization is achieved with prostaglandins

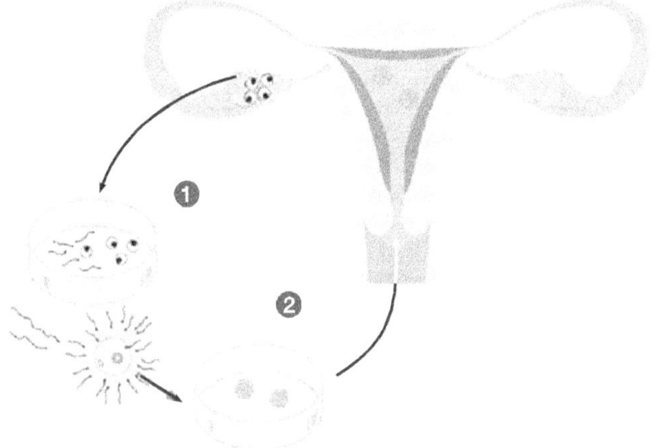

Figure 8.2 Embryo transfer

in cattle. The direct transfer of microinjected oocytes into the oviducts of ultimate recipients will always require surgery.

Tissue or blood samples of calves at the day of birth can be obtained for preparation of DNA to obtain data about the integration of the transferred gene construct. The expression of phenotypes distinguishing transgenic and non-transgenic animals will of course depend on the specific construct used. Since transgenes are normally not expressed in all individuals, unequivocal proof of the presence of an integrated gene construct must be obtained by analysis of the genomic DNA. Isolated DNAs are cut with restriction enzymes and Southern-blotted for analysis of detection of gene construct. King and Wall (1988) demonstrated for the first time the detection of transgenes in pre-implantation embryos by polymerase chain reaction (PCR). If the site of integration of a transgene is to be obtained at the chromosomal level, the method of choice is *in situ* hybridization of metaphase chromosomes.

Use of Embryonic Stem Cells

Undetermined pluripotent cells lines may be obtained and established either directly from embryos or indirectly from teratocarcinomas. The former are known as ES cells (embryonic stem cells) and the latter as EC cells (embryonic carcinoma cells). The first visible sign of differentiation of mammalian embryos is the formation of the blastocyst in which trophoectodermal cells develop into embryonic cell lines. Only a few cells constitute the so-called inner cell mass (ICM) which give rise to all organs of the growing foetus including the germ cells. The pluripotent stem cells are an essential constituent of teratocarcinoma.

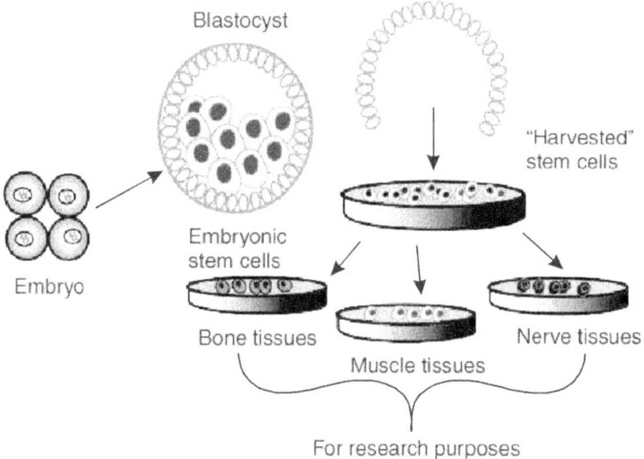

Figure 8.3 Embryonic stem cells

These malignant tumours of ectodermal origin develop in the gonads of male and female mammals and consist of differentiated and undifferentiated cells comprising a variety of different tissues including muscles, brain, skin, glands, blood, bone, hair and teeth and so on in each stage of development. Ectopic transplantation into the kidney or testis of syngenic mice of 5–6-day-old embryo, or of cells derived from the embryo of 12-day-old mouse foetuses will result in the development of teratomas or teratocarcinomas. These tumours can be propagated by subcutaneous or intraperitoneal transplantation into the mice. Production of different tissues from embryonic stem cells is shown in Figure 8.3.

Production of chimeras   EC cells lose their tumorigenic potential once they are injected into blastocysts and grow in the embryonal environment. The cells may then differentiate into normal cells participating in the development of all organs and this gives rise to chimeric animals. Chimeras may also be obtained by the aggregation of stem cells with the morula stages of embryos.

Gene transfer via stem cell transformation   The embryonic stem cells are used as vehicles for introduction of novel genes. A number of techniques are available which allow the genome of ES cells to be altered *in vitro*. They include the transfer of chromosomes, cell hybridization, microcell fusion, DNA transfer by the calcium phosphate precipitation techniques, infection with retroviral vectors, DNA microinjection and also electroporation. In the mouse, it has been shown that the gene constructs introduced into embryonic stem cells are expressed in the somatic tissues of the resulting mouse chimeras.

Gene transfer mediated by the use of transformed stem cell lines has several advantages. On the one hand, the integration and possibly the expression of the transgene can also be monitored in the resulting cell lines. On the other hand, the transfer can be effected by manipulation of morula and blastocyst stages. Since these stages can be isolated non-surgically, particularly in cattle, this procedure would greatly facilitate gene transfer programmes. The definite advantage of gene transfer involving embryonic stem cells is the possibility of direct manipulation of individual genes by homologous recombination.

RETROVIRAL VECTORS

Retroviruses are viruses that have single-stranded RNA genome which is replicated via a double-stranded DNA and stably integrated into a cellular DNA (provirus). Most retroviruses do not destroy their host cells and may even bestow selective growth advantages upon them. The proviruses whose genes are transcribed and translated by the cellular machinery are also stably inherited together with flanking DNA sequences. Retroviruses possess a number of features which suggest their use as gene transfer vehicle. These viruses can infect cells on their own and are efficiently integrated into the cellular genome. The retroviral genes are also efficiently expressed inside the cells. Many retroviruses already contain cellular sequences (oncogenes) and may be considered natural vectors allowing the transfer of non-viral and cellular DNA sequences.

The advantages of using retroviral vectors as vehicles for the gene transfer into cells are, the relatively high efficiency of infection, a single-copy integration, and the ease with which gene transfer in suitable target cells can be effected. A disadvantage of these vectors is the fact that the upper size limit of genes that can be incorporated into a retroviral genome is in the range of 6–8 kb.

The most frequently used retrovirus vectors are based on murine leukaemia virus (C type retroviruses). These vectors consists of two components, i.e., the vector construct and a packaging cell line. The genes encoding the structural proteins, *gag*, *pol* and *env* have been removed from the vector construct (proviral DNA) and replaced by other genes such as a β-galactosidase genes (*gal*) or the neomycin resistance gene (*neo*^{R}). Since the structural proteins are essential for the replication cycle of retroviruses, suitable packaging cell line must be used which provides these proteins in transgenics.

The production of transgenic animals by retroviral vectors has many advantages. Most notable are the stable integration of the desired genes, the low copy numbers (usually one), and the co-linear integration into the host cell DNA with the heterologous gene always flanked by the LTR sequences. The transfer of foreign genes by means of retroviral vectors is not restricted to the gene transfer into the germ line during the prenatal development. Recombinant retrovirus vectors are particularly well-suited for the transformation of haemopoietic stem cells and hence may offer some possibilities for somatic gene therapy. These vectors can also be used for the investigation of insertion mutations and for tagging sites of chromosomes. From a particular point of view the greatest advantage of the retroviral system is the ease with which the gene transfer can be effected because, in principle, it only requires co-cultivation of the embryos (without zona pellucida) with the cells producing the vector.

It was shown for the first time in 1974 that the injection of SV-40 (simian virus 40) DNA into the blastocoel of murine blastocysts might yield adult animals with cells containing this DNA (Jaenisch, 1974). A general strategy for the efficient insertion of recombinant retroviral vector DNA into the mouse germ line is the infection of preimplantation mouse embryos. Hussar *et al.* (1985) integrated bacterial *neo*^{R} gene into the germ line mice of retrovirus-mediated gene transfer. Soriano *et al.* (1986) produced transgenic mouse lines by infection of preimplantation embryos with recombinant retroviruses containing the complete human β-globin gene under the control of its own promoter and the bacterial neomycin phosphotransferase gene under the control of a viral promoter. Stewart *et al.* (1987) have infected preimplantation mouse embryos with a retroviral vector containing the *v-myc* gene and a neo^R gene under the control of the thymidine kinase promoter. Of 104 embryos, 17 (17%) developed into mice, of which 6 (35%) animals were positive and 3 were used as founder animals which transmitted the transgene to approximately 8% of their offsprings.

One serious disadvantage of the generation of transgenic animals with retrovirus vectors appears to be the lack of expression of the transgenes in mice. It appears that some factors suppress the expression of these vectors in early embryos and also in embryonic stem cells and only allow a very limited expression in adult animals. This may be overcome by constructing vectors which contain the desired structural gene and in addition also a suitable external promoter sequence. Another disadvantage is rather the limited capacity of retroviruses, since 8 kb of the foreign DNA constitutes the upper limit of DNA that may be incorporated into such vectors.

There are three problems associated with the retrovirus-mediated gene transfer which are more or less specific for retroviruses. The first problem is concerned with the spread of the vector virus in infected organisms. This possibility can be ruled out in most instances, because the virus vector systems generally in use do not require the use of helper viruses. Secondly, it has been shown that integration of retrovirus into the host genome may lead to the activation of cellular oncogenes by viral transcription signals. The third problem concerns the probability of creating active forms of retroviruses by recombination of the defective retroviruses generally used as vectors with endogenous retroviral sequences of the host.

TRANSMISSION OF TRANSGENES

The transmission of the transgene requires that all or at least some of the germ line cells of the primary animals (those derived from the infected embryos) contain the transgene. The analysis of newly born transgenic animals and also attempts to establish transgenic offsprings have demonstrated that genetic mosaics may develop although the DNA had been injected into the pronuclei of fertilized oocytes. Genetic mosaics are animals that consists of cell lines which are derived from the same zygote, but possess different genotypes. In animals which have stably integrated the gene construct within their genomes and transmit it to their offsprings, the mode of inheritance is usually Mendelian because integration usually has occurred at a single site within a chromosome. By definition such animals are hemizygous transgenic. Fifty per cent of the offspring of such transgenic animals will contain the transgene.

In rare cases, one observes more than one integration site in a primary transgenic animal. In a transgenic animal with two independent integration sites, one may expect that 75% of the offspring will inherit a transgene with 25% of the animals inheriting either of the two transgenes and 25% inheriting both integration sites. If hemizygous transgenic animals are crossed, 50% of the resulting progeny will be hemizygous transgenic, 25% will be homozygous transgenic and 25% of the animals will not be transgenic.

The introduction of a certain gene construct into a breeding population will require more than a single transgenic animal. At least 5–10 primary transgenic animals will be needed to ensure that appropriate transgenic lines will be established with an acceptable probability.

NUCLEAR TRANSFER (NT)

Nuclear transfer is the reconstruction of mammalian embryos by transfer of a blastomere nucleus to an enucleated oocyte or zygote for the production of genetically identical individuals. This had advantages for research (i.e., multiplication of genetically valuable livestock). However, the number of offsprings that can be produced from a single embryo is limited both by the number of blastomeres (embryos at the 32–64-cell stage are the most widely used in farm animal species) and the limited efficiency of the nuclear transfer procedure. The ability to produce live offspring by nuclear transfer from cells that can be propagated and maintained in culture offers many advantages, including the production of many identical offsprings over an extended period and the ability to modify genetically or to select populations of cells of specific genotypes or phenotypes before embryo reconstruction. This objective has been achieved with the production of lambs using nuclei from cultured cells established from embryonic, foetal and adult material. In addition, lambs transgenic for human factor IX have been produced from foetal fibroblasts transfected and selected in culture (Eyestone and Campbell, 1999). Figure 8.4 shows a cloned lamb (Dolly) produced by cloning a mammary cell from an adult sheep as nuclear donor and an enucleate ovum as recipient.

Somatic cell nuclear transfer in sheep is a complex and intricate procedure with a low success rate (Ritchie, 2006). Despite advances in embryo culture and the production of specialized tools and equipment, success rate has remained poor. Oocytes come from an *in vitro* system, which may make the procedure more economical and flexible than before. Oocytes are produced from ovaries collected at an abattoir. The cumulus–oocyte complexes (COCs) are recovered from the ovary with hypodermic needle and syringe. Selected COCs are matured for 18 to 20 hours in medium that is supplemented with hormones. Cumulus

Figure 8.4 Cloned lamb (Dolly) by nuclear transfer

cells are stripped from the oocytes using hyaluronidase, and mature oocytes with polar bodies, selected for enucleation. These oocytes are treated with a cytoskeletal inhibitor to prevent lysis of the oocyte during enucleation and with a DNA-specific dye to visualize the chromosomes with the aid of ultraviolet light. These permit removal of the metaphase II chromosomes and polar body prior to reconstruction of the embryo. A diploid cell is injected under the zona pellucida of the enucleated oocyte, which can then be fused to the cytoplast using an electrical pulse. The fused embryos are activated and allowed to develop in culture to the morula or blastocyst stage, when they are surgically implanted into previously prepared synchronized, recipient animals.

Cloned bovine embryos were produced at the blastocyst stage (Malenko *et al.*, 2006). Prior to enucleation, oocytes were freed from the zona pellucida. Fibroblasts isolated from the bovine foetus were used as nuclear donors. Pairs of foetal fibroblasts and enucleated oocytes (cytoplasts) were glued in phytohaemagglutinin solution under a binocular microscope. The subsequent electrofusion of 39 foetal fibroblast–cytoplast pairs yielded 36 reconstructed one-cell embryos (92.3%). After culturing in synthetic oviduct fluid for 7.5 days, seven cloned embryos developed to the blastocyst stage (19.4%) and six blastocysts were considered fit for transplantation. The applied technique of bovine embryo growth allowed 31.1% zona-free oocytes parthenogenetically activated to reach the blastocyst stage.

The successful application of nuclear transfer (NT) technology in cats was demonstrated by the birth of domestic and non-domestic cloned kittens at a similar level of efficiency to that reported for other mammalian species (Gomez *et al.*, 2006). In cats, it has been demonstrated that either oocytes matured *in vivo* or *in vitro* can be used as donor cytoplasts. The length of *in vitro* oocyte maturation affects *in vitro* development of reconstructed embryos, and oocytes matured *in vitro* for shorter periods of time are the preferred source of donor cytoplasts. For NT, cat somatic cells can be synchronized into the G0/G1 phase of the cell cycle by using different methods of cell synchronization without affecting the frequency of *in vitro* development of cloned embryos. Interspecific NT has potential application for preserving endangered felids, as live offspring of male and female African wildcats (AWC, *Felis silvestris lybica*) have been born and pregnancies have been produced after transferring black-footed cat (*Felis nigripes*) cloned embryos into domestic cat (*Felis silvestris catus*) recipients. Also, successful *in vitro* embryo development to the blastocyst stage has been achieved after intergeneric NT of somatic cells of non-domestic felids into domestic cat oocytes, but no viable progeny have been obtained. Several live domestic and AWC cloned kittens have been born that are seemingly normal and healthy. It is important to continue evaluating these animals throughout their lives and to examine their capability for natural reproduction.

True clones—a group of two or more individuals with identical genetic make-up derived, by asexual reproduction, from a single common parent or ancestor—may be produced by embryo splitting or blastomere separation either artificially or as it occurs naturally in the production of identical twins. The term clone has been applied to animals produced by the technique of nuclear transfer. In this asexual process, nuclear genetic material is transferred from a donor cell (karyoplast) into a recipient cell (cytoplast) from which the genetic material has been removed. In farm animals, the cytoplast of choice is the matured oocytes (or unfertilized egg), thus the animals developing from this technique are not true clones as each cytoplast is often derived from a different animal. The resultant animals may therefore be more aptly described as genomic copies. Numerous factors affect the development of embryos reconstructed by nuclear transfer including, the nature of the cell and the culture method (Campbell, 1999).

Figure 8.5**a** shows transgenic calves with genes for increased milk protein; Figure 8.5**b** shows transgenic sheep with genes for production of α-1 antitrypsin in their milk; Figure 8.5**c** shows goat with genes for production of human lysozyme in their milk; and Figure 8.5**d** shows transgenic pigs with genes for production of omega-3 fatty acids in their milk. Figure 8.6 shows transgenic chicks with increased viability.

(a)

(b)

(c)

(d)

Figure 8.5 Transgenic cattle (a) with genes for increased milk protein, transgenic sheep (b) with human alpha-1 anti-trypsin in the milk, transgenic goat (c) with increased human lysozyme in the milk and transgenic pig (d) with increased omega-3-fatty acids in milk.

Figure 8.6 Transgenic chicks with increased viability

It has been notoriously difficult to successfully cryopreserve swine embryos, a task that has been even more difficult for *in vitro*-produced embryos. The first reproducible method of cryopreserving *in vivo*-produced swine embryos was after centrifugation and removal of the lipids. Li *et al.* (2006) reported the adaptation of a similar process that permitted the cryopreservation of *in vitro*-produced somatic cell nuclear transfer (SCNT) swine embryos. These embryos developed to the blastocyst stage and survived cryopreservation. Transfer of 163 cryopreserved SCNT embryos to two surrogates produced 10 piglets. Application of this technique might permit national and international movement of cloned transgenic swine embryos, storage until a suitable surrogate is available, or the long-term frozen storage of valuable genes.

APPLICATIONS OF TRANSGENIC MICE

Disease Models

There are two different approaches, viz., the directed and specific establishment of disease models and the chance observation made in transgenic mice or their progeny, which may then be used as disease models. Wagner *et al.* (1983) studied the insertion mutation in transgenic animals for the first time. Following the injection of a hGH gene they obtained 6 transgenic mouse lines of which only 4 could be used to produce homozygous transgenic mice by crossing between transgenic founder animals. The insertion of foreign DNA into the β-1 collagen gene has yielded a lethal mutation in which the foetuses degenerate approximately between day 12–15 of gestation (Jaenisch *et al.* 1983).

Stacey *et al.* (1988) introduced a well-known mutation into the pro-α collagen gene and used these mutated genes for the production of transgenic mice. They observed that the normal function of collagen was already disturbed if the level of expression of the mutated transgene was in the order of 10% of that of the endogenous gene. The intracellular accumulation of the transgene product induces not only the degradation of the mutated but also of the normal collagens, thus either causing a reduction in the level of collagen or disrupting the correct alignment of collagen and the formation of fibrils. The transfer of a mutated collagen gene into the embryo of normal mice therefore yielded a disease model for the perinatal osteogenesis imperfecta type II.

The expression of a transgene mostly depends on the regulatory sequences which have been used in the design of the gene construct. A classical example is the transgenic giant mice in which the expression of a growth hormone (GH) gene is driven by the metallothionein promoter (Palmiter *et al.*, 1982). Transgenic giant mice are useful models for the study of human growth deviations such as gigantism and acromegaly.

Wirak *et al.* (1991) devised a transgenic disease model for Alzheimer's disease. They produced transgenic mouse lines which expressed the human amyloid β-protein under the control of the human amyloid protein promoter in the dendrites of some but not all hippocampus neurons in a year-old mice.

Aggregates of the amyloid β-protein formed amyloid-like fibrils that were similar in appearance to those in the brains of patients suffering from Alzheimer's disease.

The activity of an individual gene may be neutralized by the transfer of a gene construct which contains a structural gene encoding an antisense RNA for the corresponding endogenous transcript. Intracellular hybridization of an antisense RNA with the mRNA (sense RNA) encoded by the gene under study will either efficiently inhibit translation or leads to the synthesis of functionally crippled protein fragments. If in a particular transgenic model the aim is to repress the growth of a certain cell population rather than the activity of an individual gene, this can be achieved by a process known as genetic ablation or amputation. Breitmann *et al.* (1987) demonstrated the feasibility of such an approach by constructing a fusion gene in which the gene encoding a chain of diphtheria toxin was coupled to the σ-2 crystalline gene promoter. Due to the toxic effect of the gene, it was possible to suppress in part and also completely the growth of a particular cell population of the lens, thus producing microphthalmia in the transgenic mice expressing the construct. Palmiter *et al.* (1987) used such an approach to suppress the growth of a particular cell population of the lens, thus producing microphthalmia in the transgenic mice expressing the construct.

There are many instances in which a suitable model can be established only if the function of an endogenous gene is inactivated effectively by the introduction of a mutation. Smithies *et al.* (1985) showed that homologous recombination of a foreign gene construct with an endogenous gene locus is indeed possible. They transferred carcinoma cells with a gene construct consisting of 4.6 kb of the β-globin locus, the *neo* and *supF* gene. In approximately 0.1% of all cases they were able to demonstrate that this construct had integrated into an endogenous β-globin gene by homologous recombination.

The next step in the development of this technique has been the use of homologous recombination for directed mutagenesis in embryonic stem cells of mice. The hypoxanthine phosphoribosyl transferase (HPRT) gene has been chosen as a target gene because it is located on the X-chromosome and a selection procedure for HPRT-negative mutants is available. Another reason for choosing this gene is the fact that HPRT-negative transgenic mice ought to be a valuable model for the Leish–Nyhan syndrome. It has been possible to mutate the endogenous HPRT gene in embryonic stem cells by gene targeting and also to correct any HPRT-negative stem cells by homologous recombination and to produce transgenic mice (Doetschman *et al.*, 1987). The results of these experiments have been very promising and have suggested that it might be possible to use these embryonic stem cells to produce chimeras and mice carrying a mutated gene at a desired gene locus. Since it is difficult to mutate both alleles of the gene on both homologous chromosomes, the resulting transgenic mice are chimeric and their F1 offspring is hemizygous. By crossing these homozygous transgenic mice, it is possible to obtain homozygous mutants in the next generation, if the mutated gene is not lethal or does not create other problems.

The microinjection of DNA construct can also be used to correct the deleterious effects of a deleted endogenous gene. Brinster *et al.* (1989) injected 5´ sequence of a functional *MHC* IIE gene into mouse cells carrying a 630-bp deletion in the endogenous locus. The injection of over 10,000 oocytes yielded 1800 live mice, 500 of which were transgenic.

Oncogenes

Gene transfer is particularly valuable for studying the effects of oncogene expression in animals. As proto-oncogenes are important for normal development, the transfer and expression of these genes may be lethal in transgenic mice. Both viral and cellular oncogenes have been used for generating transgenic mice. The detailed analysis of these mice may be helpful in elucidating at a molecular level the mechanisms of (proto) oncogene functions during normal and malignant development.

A variety of viral oncogenes, including the large T antigen gene of SV-40 virus, polyoma virus large and middle T genes, bovine papilloma virus, human hepatitis B virus and other viruses have been used for generating transgenic mice. Brinster *et al.* (1984) used a gene construct of SV-40 genes and the murine metallothionein promoter to produce transgenic mice. These mice developed papillomas and carcinomas of the choroid plexus. Further investigations demonstrated that a 72-bp element of the SV-40 promoter was of crucial importance for the generation of these tumours (Palmiter *et al.*, 1985). When promoters of insulin and crystalline genes were used, the resulting transgenic mice suffered from tumours of the pancreas and the eye lens, respectively.

Windle *et al.* (1980) reported that the expression of the viral SV-40 T antigen oncogene in the retina of transgenic mice produced heritable ocular tumours with histological, ultrastructural and immunohistochemical features identical to those of human retinoblastomas.

The transfer of bovine papilloma virus 1 (BPV-1) genome has yielded transgenic mice that developed typical picture of fibropapillomas of the skin. The tumour tissue contained extra-chromosomal copies of BPV-1 DNA, while normal tissues contained only integrated copies of the viral DNA. These BPV-1 transgenic mice are an excellent model for the study of the tissue specificity in the expression of BPV, the activity of the BPV oncogene and the genetic make-up leading to the development of neoplasias.

The human immunodeficiency virus (HIV) is a retrovirus whose genome contains at least 6 genes in addition to the normal *gag*, *pol* and *env* genes. These additional genes are important for the regulation of the expression of viral gene and also influence the replication of the viral genomes. The *tat* gene in particular is a potent transactivator *in vitro*. HIV/LRT-tat 2 transgenic mice develop skin alterations that resemble those of Kaposi sarcomas observed in AIDS patients (Vogel *et al.*, 1988). Transgenic mice which harbour the entire intact provirus did not develop HIV viraemia and remained healthy. Transgenic F1 offspring of one line, however, developed the syndrome and infectious viruses were recovered from spleen, lymph nodes and the skin of the animals.

The establishment of transgenic mice is also invaluable for the analysis of the immune system (Storb, 1987), in particular for studies of B- and T-cell development and the mechanisms of action of lymphokine. Transgenic mice, rabbits and pigs carrying the genes for the light and heavy chains of a mouse monoclonal antibody will yield up to 1 g MAb/l of serum in one transgenic pig line. Most T cells carry T-cell receptors (TCRs) made of α and β chains and possess a constant and variable region. In transgenic mice showing low level expression of a transgenic TCR β-chain, the same cells expressing the transgene also express the endogenous TCR chain. The introduction of a mutated TCR β-chain led to the arrest of thymocyte maturation with transgenic mice being completely deficient in functional α-β T cells.

Interleukins play a decisive role in mediating the growth and differentiation of leucocytes and also in eliciting immune response and inflammatory processes. The over-expression of a transgene encoding granulocyte–macrophage colony-stimulating factor (GM-CSF) leads to pathological alterations in the retina and causes blindness and also characteristic pathological consequences. Transgenic mice harbouring the human gene for IL-1, or the TAC-IL-2 receptor constitutively express IL-2 in the thymus, spleen, bone marrow, lungs, muscle and skin. Among other things, these animals show pronounced growth retardation and die prematurely.

The expression of IL-4 in transgenic animal impairs the maturation of T cells and reduces considerably the population of immature CDH/CDS-positive thymocytes and peripheral T cells, while the number of mature CD-8-positive thymocytes increase markedly.

APPLICATIONS OF TRANSGENIC LIVESTOCK

The emergence of the first transgenic mice expressing the rat growth hormone fused to the metallothionein promoter sequence, during the 1980s (Palmiter *et al.*, 1982), opened the possibility to use transgenesis as an instrument to increase meat production. Thus, the application of livestock transgenesis could promote the improvement of carcass composition, meat quality, milk production, wool quality, increased prolificacy and disease resistance, besides other economically important characteristics. Recently, genetically modified bovines were generated by introducing additional copies of genes CSN2 and CSN3, which encode bovine α- and β-casein, respectively (Brophy *et al.*, 2003). Over-expression of CSN2 and CSN3 resulted in an up to 20% increase in α-casein, and a 2-fold increase in α-casein levels in milk. These results demonstrated that it is possible to alter milk composition by the transgenic approach. Moreover, it can also increase the efficiency of cheese production, which is a billion-dollar market.

It is possible to alter by gene transfer only those traits (of economically important animals) that are caused by single or few genes. A pre-requisite for such interventions is a detailed knowledge of the underlying physiological processes at the molecular level because that allows suitable gene constructs to be developed and applied successfully. The growth which is one of the classical quantitative traits in animal breeding can be converted into a qualitative trait simply by the transfer of the major gene, the growth hormone gene, whose expression makes the synthesis of growth hormone independent of any regulatory feedback mechanisms. A similar effect has been described for the application of bovine somatotropin in dairy cows.

At present, research in transgenic animals of economic importance is focused on features such as growth, new metabolic pathways, quality of animal products, gene farming and disease resistance. There are four major fields of transgenic livestock applications: production of pharmaceuticals and biomolecules, livestock production, xenotransplants, and transgenic animals as a model for human diseases.

Growth and Carcass Composition

Growth is a very complex process which is influenced by the interaction of hormones and autocrine/paracrine factors, nutritional conditions and environmental factors set against a discrete genetic background. Among the genetically determined factors, the genes encoding proteins of the growth hormones cascade are of particular interest. This cascade is initiated in the hypothalamus and the pituitary, includes the liver, and eventually effects peripheral target organs.

The hypothalamus is subjected to regulation by a number of different factors, in particular the serum concentrations and growth affecting hormones such as growth hormone (GH), IGF-1 and others. Among other things, it is responsible for the circadian expression of the stimulatory hormone somatoliberin (GHRH, growth hormone releasing hormone), and the inhibiting hormone somatostatin (SRIF, somatotropin release inhibiting factor). GHRH is an oligopeptide of 43 amino acids which is synthesized in the nucleus arcuatus and the ventromedialis of the hypothalamus. Porcine and human proteins differ in three positions only, while rat and human hormones differ by 14 amino acid positions. SRIF consists of two different molecules of 14 and 28 amino acids in length, whose sequence has been found in different species. The hormone is synthesized not only in the hypothalamus but also in the pancreas and the gut.

Growth hormone (GH, somatotropic hormone, STH, somatotropin) is synthesized in the somatotropic cells of the anterior lobe of the pituitary and consists of 190 or 191 amino acids. Next to the growth hormone comes somatomedin C (IGF-1, insulin-like growth factor I) which is a mitogenic basic polypeptide of 70 amino acids with a molecular mass of 7.5 kDa. It is synthesized in the liver, and also in other organs including kidneys, lungs, heart, testes, mammary gland and epiphysis of bones. IGF-I acts with an endocrine hormone, but also exerts autoparacrine effects. The release of growth hormone depends on the concentration of GHRH and SRIF.

Growth hormone-transgenic mouse was produced for the first time in 1982 (Palmiter *et al.*, 1982). These mice showed a very pronounced growth increase with a four-fold increase in growth rates and a two-fold increase in the final body weight (Figure 8.7). Subsequently, growth hormone growth of other species and also other genes of the growth hormone cascade such as GHRH and IGH-1 have also been used to produce transgenic mice.

In contrast to the results obtained with growth hormone-transgenic mice, the administration of growth hormone in economically important animals did not show corresponding biological effects in other animals. Transgenic pigs and rabbits expressing the growth hormone gene did not show increase in growth rates (Hammer *et al.*, 1985).

Improvement of carcass quality as well as meat production is an important application of transgenesis in livestock. One of the first attempts in this field was the production of mice that express the rat growth hormone. Those mice showed a remarkable increase in growth rate and body weight (Palmiter *et al.*, 1982), and were the pioneer transgenic mice. However, transgenic pig expressing the growth hormone exhibited only a slight increase in growth and a high incidence of collateral effects such as gastric ulcers, arthritis, cardiomegaly, dermatitis, and renal disease (Pursel *et al.*, 1989).

Figure 8.7   Transgenic mouse (white) with inserted growth hormone gene along with normal mice (black)

Transgenic pigs bearing a human metallothionein promoter/porcine growth-hormone gene construct showed significant improvements in economically important traits such as growth rate, feed conversion and body fat/muscle ratio without the pathological phenotype known from previous growth hormone constructs (Nottle *et al.*, 1999; Pursel *et al.*, 1989). Similarly, pigs transgenic for the human insulin-like growth factor-I had ~30% larger loin mass, ~10% more carcass lean tissue and ~20% less total carcass fat (Pursel *et al.*, 1989). The commercialization of these pigs has been postponed due to the current lack of public acceptance of genetically modified foods. Recently, an important step towards the production of more healthful pork has been made by the creation of the first pigs transgenic for a spinach desaturase gene that produces increased amounts of non-saturated fatty acids. These pigs have a higher ratio of unsaturated to saturated fatty acids in striated muscle, which means more healthful meat since a diet rich in non-saturated fatty acids is known to be correlated with a reduced risk of stroke and coronary diseases (Niemann, 2004; Saeki *et al.*, 2004).

A transgenic pig harbouring an MLV-rGH gene construct and expressing a biologically active rat growth hormone did not show an increased weight gain in the main growth period between the 2nd and 6th month compared to control animals. It appeared, however, that the growth period of this pig had been prolonged so that at the age of 8 months, it was 28% heavier than a non-transgenic male sibling (Ebert *et al.*, 1988). Only 3 out of 11 transgenic sheep harbouring the bovine growth hormone under the control of a mouse transferring enhancer/promoter and 2 out of 4 animals harbouring the human growth hormone-releasing factor gene under the control of the murine albumin enhancer/promoter, expressed the corresponding genes (Rexroad *et al.*, 1991). An increase in growth performance has not been observed, probably because these animals showed pathological alterations and fertility disturbances and developed a pronounced diabetic condition.

GH and GHRH transgenic sheep show elevated levels of growth hormone although they do not display increased growth performance (Rexroad *et al.*, 1990). Although the growth performance of these animals were not altered, the transgenic sheep only had 5–7% body fat in contrast to control animals (25–30%).

Biochemical Pathways and Quality of Products

The alteration of biochemical and metabolic pathways is a very interesting approach to improve the productivity of economically important animals. The main interest in altering metabolic pathways lies in the introduction of biosynthetic processes for essential factors. It has been known for some time that the availability of the amino acid cysteine is a factor limiting the synthesis of sheep wool. Moreover, the serum level cannot be raised by simply adding cysteine to the food because it is degraded in the ruminant stomach of sheep. It would be of considerable advantage if transgenic sheep were capable of synthesizing these amino acids. This would require the introduction of genes encoding the enzymes serine transacetylase and *o*-acetyl serine sulphydrolase of *E. coli* for providing them with suitable regulatory sequences allowing their expression in mammalian cells of the gastric epithelium of such transgenic sheep that would be capable of synthesizing cysteine by utilizing the hydrogen sulphide present in the stomach.

$$\text{Serine} + \text{Acetyl-CoA} \xrightarrow{\text{Serine transacetylase}} o\text{-acetylserine} + \text{CoA-SH}$$

$$o\text{-Acetylserine} + \text{H}_2\text{S} \xrightarrow{o\text{-acetylserine sulphydrolase}} \text{Cysteine} + \text{Acetate}$$

Ward *et al.* (1990, 1991) linked these genes with the MT-promoters of sheep and with GH sequence at the 3´ regions and used this construct to produce transgenic mice. Surprisingly, they found that constructs free of introns are also expressed quite satisfactorily. After the transfer of a construct containing the coding sequences of the genes, the SV-40 late promoter and the SV-40 polyadenylation signal, a constitutive expression of both genes were observed in transgenic mice and also in sheep (Rogers, 1990). These findings suggested that the bacterial genes were transcribed and translated correctly in some sheep tissues.

Another example of the establishment of new metabolic pathways is the introduction of glyoxylate pathway which would allow sheep to synthesize glucose from acetate. This approach is of principal interest because ruminants require glucose as a source of energy for certain tissues such as the brain and the foetus. Glucose, therefore, has to be synthesized via gluconeogenesis from free fatty acids such as propionic acid and also from amino acids. Acetate is usually available in amounts exceeding those of metabolites required for gluconeogenesis. The isocitrate is synthesized from acetate and oxaloacetate with isocitrate lyase yielding glyoxylate and succinate. In each cycle, two molecules of acetate will yield one molecule of succinate. The presence of active isocitrate lyase and malate synthase has been demonstrated by using transformed cells and also transgenic mice.

The introduction of wool biosynthetic pathways is not a trivial task. The process requires a number of parameters and conditions to be observed and fulfilled, e.g. the availability of the corresponding enzymes in cells which require them and the necessity to utilize pre-existing cellular substrates and coenzymes.

Modifications of the milk composition (Table 8.2) could be brought about by the transfer of suitable gene constructs. A model prepared by Mercier (1986) suggests a reduction of the lactose content of milk. By establishing transgenic sheep or cattle containing or expressing a lactase gene under the control of an udder-specific promoter, it should be possible to cleave lactose into glucose and galactose. The milk of such animals should be tolerated by those patients suffering from lactose intolerance due to the lack of the enzyme lactase.

Table 8.2 Endogenous milk proteins of ruminants

Encoded proteins	**Concentration (μg/ml)**
Caseins (Ruminants)	
αS_1	10,000–12,000
αS_2	3,400–3,800
β	10,000–16,000
Kappa	3,900–4,600
Major whey proteins	
α-Lactalbumin	800–1,000
β-Lactalbumin	2,000–3,000
Whey acidic proteins (Rodents)	2000

The alterations of the milk composition per se could also be tried. The aim would be an alteration of cow's milk so that it would resemble more closely the composition of human milk. Such a product might be an advantage in feeding human babies and infants.

The genomic fragment carrying the recombinant WAP (principle milk protein) gene locus was cloned and the 8.5-kb long 5´ untranslated part of the *rWAP* gene to target the expression of human erythropoietin (hEPO) gene cloned from the human phage genomic library into the mammary gland of the mouse (Mikus *et al.*, 2001). The vectors carrying either the *hEPO* gene or the *rWAP-hEPO* hybrid gene were injected into the mouse ova, and 12 transgenic animals were identified by PCR. Recombinant hEPO was produced in the milk of a transgenic mouse female at a secretion level of 5.3 mIU/ml, as detected by ELISA.

Seven Friesian human lactoferrin (hLf)-transgenic primiparous dairy cows expressing recombinant hLf (rhLf) in their milk were produced (Hyvonen *et al.*, 2006). After calving, concentrations of rhLf and bovine LF (bLf) in the milk, somatic cell count and milk yield were determined. The concentration of rhLf was found to be constant, about 2.9 mg/mL, throughout the early lactation period of 3 months. The concentration of bLf in colostrum was higher after calving, but decreased rapidly during the first few days of lactation. The mean concentration of bLf was 0.15 mg/mL, but concentrations varied between cows from 0.07 mg/mL to 0.26 mg/mL. Based on that, it may be possible to improve the non-specific host defence mechanism in the mammary gland of dairy cows by enhancing the content of rhLf in the milk.

Maga *et al.* (2006) generated one line of transgenic goats as a model for the dairy cow designed to express human lysozyme in the mammary gland. The milk from 5 transgenic females of this line expressing human lysozyme in their milk at 270 microg/mL or 68% of the level found in human milk. Milk from transgenic animals had a lower somatic cell count, but the overall component composition of the milk and milk production were not different from controls. Milk from transgenic animals had a shorter rennet clotting time and increased curd strength. Milk of such nature may be of benefit to the producer by influencing udder health and milk processing.

Transgenic animal mammary gland bioreactors are used to produce recombinant proteins. A simple and efficient method was established to evaluate the functionality of animal mammary gland tissue-expressed cassettes (Li *et al.*, 2006). The gene transfer vector pGBC2LF was constructed, and the expression of human lactoferrin (LF) gene was controlled by the goat beta-casein gene 5´ flanking sequence. Transfection of exogenous DNA into rabbit spermatozoa was found to be efficient using 30 microg mL^{-1} DNA, DMSO at a final concentration of 3%, and a 3 : 1 ratio of linear-to-circular DNA, with 29 of 85 (34.1%) *in vitro*-fertilized embryos being transgenic. Using DMSO-sperm-mediated gene transfer (DMSO-SMGT), 89 rabbit offsprings were produced, with 46 of these (57.1%) being transgenic. As mammary gland bioreactor models, 17 of 21 (81%) transgenic female rabbits could express human LF protein in their glands. During lactation of the transgenic rabbits, the highest level of human LF protein expressed was 153 +/–31 microg mL^{-1}, and the mean expression level in all of the transgenic rabbits was 103 +/–20 microg mL^{-1} in the third week, declining gradually after this time. These results demonstrate that transgenic rabbits produced by DMSO-SMGT were able to express human LF protein in the correct tissue.

Production of Pharmaceuticals and Biomolecules

Due to the high cost, the production of transgenic animals such as pig, goat, sheep and cattle must bring an elevated profit in order to be a feasible economical investment. For this reason, the production of high-value pharmaceutical substances, which correspond to a billion-dollar market (Wall *et al.*, 1997; Miller, 2002), is actually the principal and most promising application for animal transgenesis. Despite the lower costs of producing biomolecules in microorganisms, like bacteria and yeast, these organisms do not properly execute several post-translational modifications, such as *N*-linked glycosylations, authentic *O*-linked glycosylations, and correct folding in order to produce a wide range of fully active human proteins. Therefore, many human polypeptides must be produced in mammalian cell systems to be recovered with their full activities. However, principally due to the low productive capacity, the price of human biomolecules produced *in vitro* by mammalian cell culture is extremely high. For this reason, many biotechnology companies have been focusing on the production of biopharmaceuticals at high concentrations in transgenic livestock bioreactors. Worldwide those companies are investing great effort in this promising technology.

Among them we may mention PPL Therapeutics (UK), GTC Biotherapeutics (USA), Hematech (USA), Genzyme (USA), ZymoGenetics (USA), Nexia Biotechnologies (Canada), Pharming (Netherlands), BioProtein Technologies (France), Avigenics (USA), Viragen (USA), and TranXenoGen (USA). Pharmaceutical products can be produced in a variety of biological fluids such as milk, urine, saliva, blood and seminal fluid (Dyck *et al.*, 2003), and their expression can be driven by tissue-specific promoters. Generation of transgenic mice expressing therapeutic heterologous proteins in urine, such as human recombinant erythropoietin (rhEPO) and human alpha1-antitrypsin (alpha1AT), using a uromodulin promoter has been demonstrated recently (Zbikowska *et al.*, 2002a; Zbikowska *et al.*, 2002b). Protein expression into the urine-based system has some advantages, including the expression in both sexes, low contaminant-protein content in urine, and being able to be harvested soon after birth and expressed throughout the life of the transgenic animal. However, the use of this system may be significantly more time- and cost-consuming than the mammary-gland-based system.

Due to its large-scale production, milk is the preferred vehicle to express proteins in transgenic animals. Table 8.3 lists biomolecules expressed in the mammary glands of livestock, which are at an advanced stage of clinical trials, and the respective holders of their commercial rights. The mammary gland can

Table 8.3   Human proteins secreted in the milk of transgenic livestock

Pharma-ceutical	Bioreactor species	Application/ treatment	Company	Reference
Antithrombin	Goat	Thrombosis, pulmonary embolism	GTC Biothera-peutics (USA)	Ebert *et al.* (1991) Denman *et al.* (1991)
tPA	Goat	Thrombosis	PPL Therapeutics (UK)	Ebert *et al.* (1994)
α-Antitrypsin	Sheep	Emphysema and cirrhosis	PPL Therapeutics (UK)	Wright *et al.* (1991)
Factor IX	Sheep	Haemophilia b	PPL Therapeutics (UK)	Schnieke *et al.* (1997)
Factor VIII	Sheep	Haemophilia a	PPL Therapeutics (UK)	Paleyanda *et al.* (1997)
Polyclonal antibodies	Cattle	Vaccines	Hematech (USA)	Kuroiwa *et al.* (2004)
Lactoferrin	Cattle	Bactericide	Pharming Group (NED)	van Berkel *et al.* (2002)
CI inhibitor	Rabbit	Hereditary angioedema	Pharming Group (NED)	van Doorn *et al.* (2005)
Calcitonin	Rabbit	Osteoporosis and hypercalcaemia	PPL Therapeutics (UK)	McKee *et al.* (1998)

Note: The Pharming Group Co. is the current holder of most patents developed by PPL Therapeutics.

express more than 2 g of heterologous recombinant proteins per litre of milk (Velander *et al.*, 1992; van Berkel *et al.*, 2002). Based on the assumption of average expression levels, daily milk volumes and purification efficiency, 5400 cows would be needed to produce the 100, 000 kg of human serum albumin that are required worldwide per year, 4500 ewes would be required for the production of 5000 kg of alpha 1-antitrypsin, 100 goats for 100 kg of monoclonal antibodies, 75 goats for 75 kg of antithrombin III and two sows to produce 2 kg of human clotting factor IX required per year (Rudolph, 1999).

Therefore, a small herd of transgenic livestock could supply the world demand for pharmaceuticals, which cannot be expressed by other systems such as bacteria or fungi, mainly due to the necessity of complex post-translational processing to ensure their proper function. Studies have reported the use of transgenic rabbits as bioreactors, using a rat whey acidic protein (WAP) promoter to produce the human growth hormone (hGH) in milk (Lipinski *et al.*, 2003).

Some biopharmaceutical products are well-advanced in clinical trials or are to be regulated by the US Food and Drug Administration (FDA). Such products include:

1. the production of antithrombin III in dairy goats (GTC Biotherapeutics), which is a potent inhibitor of the coagulation cascade used to treat patients with genetic heparin resistance and thrombosis

2. the production of human polyclonal antibodies in cows aiming at therapeutic application (Hematech)

3. the production of human serum albumin in cows (GTC Biotherapeutics), which is a blood derivate with a high market price

4. the production of α-antitrypsin in sheep (PPL) for treatment of lung emphysema, cirrhosis, and cystic fibrosis

5. the production of recombinant human factor IX protein in sheep (PPL) and human factor VIII in swine (Paleyanda *et al.*, 1997) for the treatment of haemophilia B and A, respectively

6. the production of human lactoferrin in cows (Pharming), which has a bacteriostatic, antiviral, and antifungal activity, and also acts as a natural stimulator of the innate immunological defence. Nevertheless, not all attempts to produce bioproducts in animal milk have been successful. Frequently the protein synthesized in milk causes side effects, such as for instance:

    ♦ the erythropoietin effect leading to problems in haematopoiesis and rabbit fertility (Massoud *et al.*, 1996);

- the premature cessation of lactation in the goat expressing the tissue plasminogen activator (tPA) (Ebert *et al.*, 1994); and

- low yields of biologically active recombinant human factor VIII produced in sheep (Niemann *et al.*, 1999).

An increase of milk production in pigs is also another goal to be reached, with an expected outcome of an improved value of the piglet per dam per year parameter. Due to the fact that enhanced α-lactalbumin synthesis is closely correlated with an increase in milk production, copies of the bovine α-lactalbumin genes were introduced in swine (Bleck *et al.*, 1998). The results of this work showed an increased milk production during the first 9 days of lactation in transgenic sows, as well as a significant increase in piglet growth rate in transgenic gilts (Noble *et al.*, 2002). This effect can easily mean a profit of millions of dollars for the pig industry. The implications of modifying milk properties by transgenesis go beyond just dairy products. Humanized bovine and goat milk can bring several benefits to human health. An example of this is the expression of human lactoferrin in milk. Antibacterial, antifungal and antiviral properties have been demonstrated for human lactoferrin (Soukka *et al.*, 1992; Hasegawa *et al.*, 1994; Nibbering *et al.*, 2001). Moreover, human lactoferrin stimulates the growth of the intestinal biota of healthy breast-fed infants, as well as promotes intestinal cell growth *in vitro* (Nuijens *et al.*, 1997). The production of recombinant human lactoferrin in milk of 4 transgenic cows at a concentration of up to 2 mgmL^{-1} is a promising step to improve infant innate defence (van Berkel *et al.*, 2002).

Antibody production in transgenic animals   Numerous monoclonal antibodies are being produced in the mammary gland of transgenic goats (Meade *et al.*, 1999). Cloned transgenic cattle produce a recombinant bispecific antibody in their blood (Grosse-Hovest *et al.*, 2004). Purified from serum, the antibody is stable and mediates target cell-restricted T-cell stimulation and tumour cell killing. An interesting new development is the generation of trans-chromosomal animals. A human artificial chromosome containing the complete sequences of the human immunoglobulin heavy and light chain loci was introduced into bovine fibroblasts, which were then used in nuclear transfer. Trans-chromosomal bovine offsprings were obtained that expressed human immunoglobulin in their blood. This system could be a significant step forward in the production of human therapeutic polyclonal antibodies (Kuroiwa *et al.*, 2002).

Gene Pharming

Gene 'pharming' entails the production of recombinant biologically active human proteins in the mammary glands of transgenic animals. This technology overcomes the limitations of conventional and recombinant production systems for pharmaceutical proteins (Meade *et al.*, 1999; Rudolph, 1999) and has advanced to the stage of commercial application (Dyck *et al.*, 2003; Ziomek, 1998). The mammary gland is the preferred production site, mainly because of the quantities of protein that can be produced in this organ using mammary gland-specific promoter elements and established methods for extraction and purification of that protein (Meade *et al.*, 1999; Rudolph, 1999). Products derived from the mammary gland of transgenic goats and sheep, such as antithrombin III (ATIII), α-antitrypsin or tissue plasminogen activator (tPA), have progressed to advanced clinical trials (Kues and Niemann, 2004). Phase III trials for a recombinant human ATIII product have been completed and an application has been filed for European Market Authorization. The product is employed for the treatment of heparin-resistant patients undergoing cardiopulmonary bypass procedures. At the same company that manufactures this product, 11 transgenic proteins have been expressed in the mammary gland of transgenic goats at more than one gram per litre.

Many different proteins are currently produced by recombinant DNA technology involving a large-scale expression of genes in bacteria, yeast, and in tissue culture. It has been suggested to use the mammary glands of transgenic animals for production of heterologous proteins like pharmaceuticals, diagnostics and food components. The expression of foreign proteins in the milk is a system which would allow the recovery of the product in a conventional way (i.e., by milking), without exerting any adverse effects on the animals. The efficiency with which mammary glands synthesize proteins is enormous. The concentration of endogenous milk proteins is in the order of 4–6%, depending on the species. By gene pharming if recombinant proteins are expressed in concentrations of 1–2 g/L, it would be sufficient to yield considerable amounts of important recombinant proteins. Milk is a pure and hygienic product. The purification of foreign proteins from milk should not create any unsurmountable problem.

The transfer of genomic clone encoding sheep β-lactoglobulin into mice has demonstrated for the first time that the mammary gland-specific expression of heterologous milk protein genes in transgenic animals may be very efficient. In 3 out of 5 female β-lac bLG transgenic mice, Simons *et al.* (1987) observed concentration of β-lactoglobulin of up to 23 mg/ml. This concentration was of 5 times the expression level observed in sheep's milk and it should be noted that this protein is not a normal constituent of mouse milk.

The whey acid protein (WAP) is an abundant milk protein in rodents but not present in milk of pigs and other farm animals. Pursel *et al.* (1990) transferred several hundred copies of 7-kb fragment containing the mouse *WAP* gene into five transgenic pigs. Mouse WAP was detected in the milk from all lactating female pigs at concentrations of 1 g/L.

Gordon *et al.* (1987) demonstrated for the first time that proteins that are usually not found in milk can also be expressed in the mammary gland of transgenic mice. By using the *WAP* promoter and the cDNA of human tissue plasminogen activator (tPA) with its cognate secretion signal sequence, they have been able to obtain transgenic mice whose milk contains up to 460 ng/ml of active tPA. The amount of tPA found in the milk of one line including the same construct was about 50 µg/ml as judged by an ELISA.

High-level expression in transgenic mouse milk has also been observed using a hybrid bovine αS_1 casein/human urokinase gene. Urokinase is normally synthesized in the kidneys. However, the transgenic mice secreted the active human enzyme in their milk at a concentration of 1–2 mg/ml (Meade *et al.*, 1990). Using a minigene containing the sheep β-lactoglobulin promoter fused to hα1, AT (anti-trypsin) sequences which comprise part of exon 1 and the remaining downstream introns and exons, excluding intron 1, transgenic sheep were generated. Two of the animals produced 1–5 g/L, while the third produced approximately 35 g of hα1 AT/L of milk (Wright *et al.*, 1991).

By using the rabbit β-casein promoter, Bohler *et al.* (1990) produced human interleukin-2 in the milk of transgenic rabbits. In the 4 female rabbits tested, expression levels reached only 50–430 ng/ml. Fourteen lines of transgenic rabbits were generated carrying 3 different hybrid gene constructs comprising bovine αS_1 casein gene regulatory sequences and the bovine prochymosin gene. The milk of female of 9 transgenic lines contained a concentration of up to 10 g/L of bovine prochymosin.

Even though the commercial application of these transgenic animals is quite a long way off, it can be expected that these systems in future will make an important contribution towards the production of proteins which are of medical importance or which are required for other purposes.

Disease Resistance

The susceptibility to infection has a polygenic hereditary basis. It is therefore not surprising that only a few cases are known in which a specific gene locus has been shown to be involved in disease resistance. A particular example is the well-studied systems of *Mx*1 gene product found in certain strains of mice which selectively protects against infections of influenza virus.

Gene transfer is of particular interest for the manipulation of disease resistance because it has been impossible so far to reduce the susceptibility to diseases in farm animals by conventional breeding programmes. Muller and Brem (1991) reviewed few molecular biological techniques to increase disease resistance in farm animals. Five different classes of mammalian genes are currently discussed as likely candidates for gene transfer experiments because these genes appear to be associated with the regulation of disease resistance.

1. MHC genes
2. T-cell receptor genes
3. Immunoglobulin genes
4. Genes encoding lymphokines
5. Specific disease resistance genes

One of the very few examples of a gene, specific for resistance against viral infections is the gene known as Mx gene (Lindenmann, 1962). Genetic analysis has revealed that the resistance of certain strains of mice against infection with influenza A and B viruses is caused by a single autosomal dominant $Mx1^+$ allele. The expression of this gene is controlled by type I interferons (IFN α/β). Mice which contain $Mx1^-$ allele are not protected against influenza virus infections. The transfection of the $Mx1^+$ allele into cells that are susceptible to influenza virus revealed that this gene is necessary and sufficient to protect against influenza virus infections. The mammals studied so far express at least two Mx-related genes. Several different Mx gene constructs have been integrated into the genome of pigs by gene transfer. The gene constructs contained the murine $Mx1$ cDNA driven by three different promoters, viz., the human metallothionein IIA promoters, the SV-40 promoter and the Cognate endogenous Mx promoter of mice.

Genetic immunization appears to be a realistic probability to protect animals against infectious agents by the application of gene transfer techniques. This approach essentially involves the expression of genes directing the synthesis of defined antibodies in order to induce a protective state. The results of a variety of studies have demonstrated that it is possible to transfer gene constructs encoding MAbs into mice and to express large amounts of the gene products (Brinster *et al.*, 1984). The most interesting aspect of this approach is the fact that these transgenic animals produce antibodies against specific antigens without even having been challenged with these antigens during their life.

Another interesting strategy to elicit protection against viral infections is the expression of antisense polynucleotides. The antisense polynucleotide can be used to suppress specifically the expression of the genes encoding the corresponding sense RNA. These antisense molecules either inhibit the translation of the mRNA or interrupt splicing processes or the transport of the mRNA into the cytoplasm. The first example of the establishment of transgenic animals expressing as RNA has been described by Erns *et al.* (1991) who produced transgenic rabbit expressing adenovirus H5 antisense RNA. About 90–98% of the cells in primary kidney cell cultures were resistant against infections with the virus.

Transgenic strategies to increase disease resistance Transgenic strategies to enhance disease resistance include the transfer of major histocompatibility-complex genes, T-cell-receptor genes, immunoglobulin genes, genes that affect lymphokines or specific disease-resistance genes (Müller and Brem, 1991). Transgenic constructs bearing the immunoglobulin-A (IgA) gene have been successfully introduced into pigs, sheep and mice in an attempt to increase resistance against infections (Lo *et al.*, 1991). The murine IgA gene was expressed in two transgenic pig lines, but only the light chains were detected and the IgA molecules showed only marginal binding to phosphorylcholine (Lo *et al.*, 1991). High levels of monoclonal murine antibodies with a high binding affinity for their specific antigen have been produced in transgenic pigs (Lo *et al.*, 1991). It has also proved possible to induce passive immunity against an economically important porcine disease in a transgenic mouse model (Castilla *et al.*, 1998). The transgenic mice secreted a recombinant antibody in milk that neutralized the corona virus responsible for transmissible gastroenteritis (TGEV) and conferred resistance against TGEV. Strong mammary-gland-specific expression was achieved for the entire duration of lactation.

The potential to increase disease resistance is another very important aspect of agricultural transgenic market. Mastitis is an inflammatory reaction of the mammary gland caused by a microbial infection. Five bacterial species are related to the bulk of bovine mastitis cases: *Staphylococcus aureus, Streptococcus uberis, Streptococcus dysgalactiae, Streptococcus galactiae* and *Escherichia coli*. However, *S. aureus* is responsible for the majority of the clinical cases and, because of its resistance to a variety of antibiotics, has been difficult to control (Kerr and Wellnitz, 2003). Lysostaphin is a potent peptidoglycan hydrolase naturally secreted by *Staphylococcus simulans*, and has a bactericidal effect against other staphylococci, such as *S. aureus*. The efficacy of lysostaphin protection against *S. aureus* conferred to the mammary gland has been demonstrated using transgenic mice that expressed lysostaphin in the mammary gland cells (Kerr *et al.*, 2001). Recently, transgenic cows secreting lysostaphin at concentrations up to 14 mg mL^{-1} in their milk (Wall *et al.*, 2005) were obtained. Those authors also infiltrated *S. aureus* into the mammary glands of 3 transgenic and 10 non-transgenic cows. Although none of the transgenic animals showed mastitis symptoms, all of the non-transgenic cows developed mastitis.

The most feared bovine disease in countries of the northern hemisphere is bovine spongiform encephalopathy (BSE), also known as mad cow disease. The cause of BSE probably lies in a mutation of the prion protein (PrPC), which is found in the outer surface of neurons. This prion disease is generated by a switch from the normal *PrPC* to a modified form of PrP (PrPSC). Healthy animals fed a diet supplement produced with animal by-products could develop BSE (Weissmann *et al.*, 2002). During the early 1990s transgenic mice were generated by knocking out the PrPC gene (Bueler *et al.*, 1992). The obtained mice presented normal development and were resistant to spongiform encephalopathy, even after inoculation with PrPSC prions (Bueler *et al.*, 1993). Afterward, knockout of the prion gene in transgenic sheep and cattle have opened a new perspective to create animals non-susceptible to spongiform encephalopathy (Denning *et al.*, 2001; Kuroiwa *et al.*, 2004). Animals with deleted *PrPC* were generated by multiple gene targeting events, which removed not only the *PrPC* gene, but also changed the entire bovine immunoglobulin gene loci for the human correspondent.

Brucellosis is an important disease worldwide, caused by *Brucella* bacteria, some of the world's major zoonotic pathogens. Brucellosis-resistant animals have been identified, and this resistance has been related to the presence of the *NRAMP1* variant gene in bovines. Recently, the *NRAMP1* gene was introduced into mice macrophage cell lines in order to evaluate its protection conferred during *Brucella* infection (Barthel *et al.*, 2001). The introduction of the *NRAMP1* allele offered new promising prospects for brucellosis resistance.

Two strains of transgenic mice were generated that secrete into their milk the 42-kDa C-terminal portion of *Plasmodium falciparum* merozoite surface protein 1 [MSP1 (42)]. One strain secreted an MSP1 (42) with an amino acid sequence homologous to that of the FVP parasite line, the other an MSP1 (42) where two putative N-linked glycosylation sites in the FVO sequence was removed. Both forms of MSP1 (42) were purified from whole milk, emulsified with Freund's adjuvant and used to vaccinate monkeys, before challenge with the homologous *P. falciparum* FVO parasite line. Vaccination with a positive control molecule, a glycosylated form of MSP1 (42) produced in the baculovirus expression system, successfully protected five of six monkeys. By contrast, vaccination with the glycosylated version of milk-derived MSP1 (42) conferred no protection compared with an adjuvant control. Vaccination with the non-glycosylated, milk derived MSP1 (42) successfully protected the monkeys with 4/5 animals able to control an otherwise lethal infection with *P. falciparum* compared with 1/7 control animals. This study demonstrated the potential for producing efficacious malarial vaccines in transgenic animals (Stowers *et al.*, 2002).

Transgenic approaches to increase disease resistance of the mammary gland The levels of the anti-microbial peptides, lysozyme and lactoferrin, in human milk is many times higher than in bovine milk. Transgenic expression of the human lysozyme gene in mice was associated with a significant reduction of bacteria and reduced the frequency of mammary gland infections (Maga *et al.*, 1995, Maga and Murray, 1995). Lactoferrin has bactericidal and bacteriostatic effects, in addition to being the main iron source in milk. These properties make an increase in lactoferrin levels in bovine transgenics a practical way to improve milk quality. Human lactoferrin has been expressed in the milk of transgenic mice and cattle at high levels (Krimpenfort *et al.*, 1991; Platenburg *et al.*, 1994) and is associated with an increased resistance against mammary gland diseases (van Berkel *et al.*, 2002). Lycostaphin has been shown to confer specific resistance against mastitis caused by *Staphylococcus aureus*. A recent report indicates that transgenic technology has been used to produce cows that express a lycostaphin gene construct in the mammary gland, thus making them mastitis-resistant (Wall *et al.*, 2005).

Environmentally Friendly Farm Animals

Phytase transgenic pigs have been developed to address the problem of manure-related environmental pollution. These pigs carry a bacterial phytase gene under the transcriptional control of a salivary-gland-specific promoter, which allows the pigs to digest plant phytate. Without the bacterial enzyme, phytate phosphorus passes undigested into manure and pollutes the environment. With the bacterial enzyme, faecal phosphorus output was reduced by up to 75% (Golovan *et al.*, 2001). Developers expect these environmentally friendly pigs to enter commercial production in Canada within the next few years.

Xenotransplant

The possibility of xenotransplants as an alternative source of organs has been investigated for decades. In 1963, a kidney transplant from a chimpanzee to a human was the first attempt of a xenotransplant (reviewed in Auchincloss and Sachs, 1998). Nowadays, studies have shown that the pig is the animal considered the best choice as an organ donor for humans. This fact is based on: (1) their organs being similar in size, anatomy and physiology to human organs; (2) pigs growing rapidly and being a prolific species; (3) the maintenance of high hygienic standards possible at a relatively low cost; (4) established transgenic techniques to modify immunogenicity of porcine cells and organs. The main immunologic obstacles to xenotransplants include the hyperacute rejection response (HAR), which occurs within seconds or minutes, the acute vascular rejection (AVR), which occurs within days, and the cellular and potentially chronic rejection, which occurs within weeks after the transplant (Auchincloss and Sachs, 1998). Several transgenic approaches have been developed to overcome the immunological response that activates the complement cascade, which is activated by the antigen–antibody complex and is responsible for the induction of HAR and AVR. The production of a transgenic pig expressing human proteins that inhibit the complement cascade such as hCD59 (Fodor *et al.*, 1994), hCD46 (Diamond *et al.*, 2001) and hDAF (Zaidi *et al.*, 1998), are examples of attempts to overcome this problem. In those experiments a survival rate of 23 days was reached when transplanting a transgenic pig heart into primates (Diamond *et al.*, 2001).

One of the major constraints to using pig organs for xenotransplantation is complement-dependant hyperacute rejection (HAR), which occurs within a few minutes following tissue transplantation between discordant species of animals. Hyper-acute rejection results in lyses of the endothelial cells in the vessels of the transplanted organs and ultimately in graft failure (Ye *et al.*, 1994). Human and old world monkeys have natural antibodies against α (1, 3)-galactosyl epitopes on pig cells. Therefore, the best way to circumvent HAR is to knock-out the function of the α-1, 3- galactosyltransferase gene in the pigs. Mice with an α-1, 3- galactosyltransferase gene inactivated through homologous recombination are viable, and when human serum bound to cells and tissues of these mice they showed substantially less xeno-antibody production than normal mice (Tearle *et al.*, 1996). One allele of the α-1, 3-galactosyltransferase locus in the pig was successfully knocked out through homologous recombination in somatic cells and subsequent nuclear transfer (Ramsoondar *et al.*, 2003). More recently, the second allele of α-1, 3-galactosyltransferase gene has also been inactivated by selection of a point mutation at the second base of exon 9, which resulted in inactivation of the gene. Four healthy piglets with double knock-out of α-1, 3-galactosyltransferase gene were produced by nuclear transfer (Phelps *et al.*, 2003).

Transgenic Animals as a Model for Human Diseases

Farm animals, such as pigs, sheep or even cattle, may be more appropriate models in which to study potential therapies for human diseases that require longer observation periods than those possible in mice, e.g. atherosclerosis, non-insulin-dependent diabetes, cystic fibrosis, cancer and neuro-degenerative disorders (Hansen and Khanna, 2004; Li and Engelhardt, 2003; Palmarini and Fan, 2001; Theuring *et al.*, 1997).

An important porcine model has been developed for the rare human eye disease retinitis pigmentosa (PR) (Petters *et al.*, 1997). Patients with PR suffer from night blindness early in life, a condition attributed to a loss of

photoreceptors. Transgenic pigs that express a mutated rhodopsin gene show great similarity to the human phenotype and effective treatments are being developed (Mahmoud *et al.*, 2003). The pig could be a useful model for studying defects of growth-hormone releasing hormone (GHRH), which are implicated in a variety of conditions such as Turner's syndrome, hypochrondroplasia, Crohn's disease, intrauterine growth retardation or renal insufficiency. Application of recombinant GHRH and its myogenic expression have been shown to alleviate these problems in a porcine model (Draghia-Akli *et al.*, 1999).

A diversity of neuro-degenerative human disorders such as the Gerstmann–Straussler–Scheinker syndrome, Creutzfeldt-Jakob disease or fatal familial insomnia were related to a defective prion protein (*PrPC*) gene. Deletions in the *PrPC* allele using mice or cattle models can be useful models to study related anomalies (Bueler *et al.*, 1992; Kuroiwa *et al.*, 2004), and also create resistant animals (Bueler *et al.*, 1993). Animal prion diseases include scrapie of sheep and goats, transmissible mink encephalopathy, chronic wasting disease of mule deer and elk, feline spongiform encephalopathy and BSE. The characteristics of these diseases are infectious and vacuolar degeneration of the gray matter neuropil. It was discovered by Prusiner in 1982 that the infectious agent was a protease-resistant protein, which was termed a prion (PrP). A prion has a molecular weight of 27–30 kDa and is encoded by a single copy gene, *Prnp*, in mammals. Prions exist in two major isoforms: the non-pathogenic or cellular form, designated PrPC, and the pathogenic or scrapie-inducing form, designated PrPSc. Prion diseases are caused by conversion of PrPC to PrPSc through infection or genetic mutation. Once a cell contains PrPSc, it appears to act as a conformational template by which PrPC is converted to a new molecule of PrPSc through protein–protein interactions (DeArmond and Bouzamondo, 2002). In the prion knock-out mice, homozygous null mice (*Prnp*0/0) fail to develop the characteristic clinical and neuropathological symptoms of scrapie after inoculation with mouse prions, and they do not propagate prion infectivity (Bueler *et al.*, 1994). Therefore, it is expected that removal of the *Prnp* gene from cattle and sheep will result in these animals being resistant to BSE and scrapie.

TRANSGENIC CHICKEN

Gene transfer in poultry is usually aimed at improving disease resistance or increasing food utilization and growth performance. The method of choice for gene transfer in poultry has been the use of retroviral vectors. Retroviruses are natural constituents of the genomes of birds and are also found as endogenous viruses. Salter *et al.* (1986) have injected the chicken syncytial strain of reticuloendotheliosis (CS-REV) and wild type and recombinant avian leucosis virus (ALV) near the blastoderm of inoculated fertilized embryos and CS-REV intra-abdominally at the day of hatch. A number of positive birds in the progeny were identified, suggesting that retroviral genetic information had been inserted into the germ line of chicken. Chicken containing the insert called *alv* 6 were highly resistant to infections by the pathogenic serogroup A of ALV.

REVIEW QUESTIONS

1. What are the steps involved in production of transgenic animal?
2. Explain different applications of transgenic livestock for human and animal health and production.
3. Write an essay on gene pharming.

REFERENCES

Brackett, B.G., Baranska, W., Sawicki, W. and Koprowski, H. (1971). "Uptake of heterologous genome by mammalian spermatozoa and its transfer to ova through fertilization." *Proc. Natl. Acad. Sci. USA.* 68: 353–357.

Brem, G., Brenig, B., Salmons, B., Wolf, E., Muller, M., Erfle, V., Gunzburg, H.W. and Dahme, E. (1991). "Unerwartete transgene expression eines gesaugespezifischen wachstumshormongenkonstruktes in den Bergmann-Giliazellen der Maus." *Tierarztl. Prax.* 19: 1–6.

Brietman, M.L., Calpoff, S., Rossant, J., Tsui, L.C., Globe, M., Maxwell, I.H. and Bernstein, A. (1987). "Genetic ablation: targeted expression of a toxin gene causes microphthalmia in transgenic mice." *Science.* 238: 1563–1565.

Brinster, R.L., Chen, H.Y., Messing, A., van Dyke, T., Levine, A.L. and Palmiter, R.D. (1984). "Transgenic mice harbouring SV-40 T-antigen genes develop characteristic brain tumours." *Cell.* 37: 367–379.

Brinster, R.L., Bruan, R.E., Lo, D., Avarbock, M., Oram, F., and Palmiter, R.D. (1989). "Targeted correction of a major histocompatibility class I Eα gene by DNA microinjected into mouse eggs." *Proc. Natl. Acad. Sci. USA.* 86: 7087–7091.

Brophy, B., Smolenski, G., Wheeler, T., Wells, D., L'Huillier, P. and Laible, G. (2003) "Cloned transgenic cattle produce milk with higher levels of β-casein and κ-casein." *Nature Biotechnol.* 21: 157–162.

Bohler, T.A., Bruy're, T., Went, D.F., Stranzinger, G. and Burki, K. (1990). "Rabbit β-casein promoter directs secretion of human interleukin-2 into the milk of transgenic rabbits." *Biotechnology.* 8: 140–143.

Bueler, H., Raeber, A., Sailer, A., Fischer, M., Aguzzi, A. and Weissmann, C. (1994). "High prion and PrPSc levels but delayed onset of disease in scrapie-inoculated mice heterozygous for a disrupted PrP gene." *Mol. Med.* 1:19–30.

Campbell, K.H. (1999). "Nuclear transfer in farm animal species." *Semin. Cell Dev. Biol.* 10: 245–252.

Campbell, K.H., McWhir, J., Ritchie, W. A. and Wilmut, I. (1996). "Sheep cloned by nuclear transfer from a cultured cell line." *Nature.* 380: 64–66.

Castilla, J., Pintado, B., Sola, I., Sánchez-Morgado, J.M. and Enjuanes, L. (1998). "Engineering passive immunity in transgenic mice secreting virus-neutralizing antibodies in milk." *Nature Biotechnol.* 16 (4): 349–354.

Chan, A.W., Homan, E.J., Ballou, L.U., Burns, J.C. and Bremel, R.D. (1998). "Transgenic cattle produced by reverse-transcribed gene transfer in oocytes." *Proc. Natl. Acad. Sci. USA.* 95: 14028–14033.

Cibelli, J.B., Stice, S.L., Golueke, P.J., Kane, J.J., Jerry, J. and Blackwell, C. *et al.* (1998). "Cloned transgenic calves produced from nonquiescent fetal fibroblasts." *Science.* 280: 1256–1258.

Cohen, S.N., Chang, A.C., Boyer, H.W. and Helling, R.B. (1973). "Construction of biologically functional bacterial plasmids *in vitro*." *Proc. Natl. Acad. Sci. USA.* 70: 3240–3244.

DeArmond, S.J. and Bouzamondo, E. (2002). "Fundamentals of prion biology and diseases." *Toxicology.* 181–182: 9–16.

Denman, J., Hayes, M.C., O'Day, T., Edmunds, C., Bartlett, S. and Hirani *et al.* (1991). "Transgenic expression of a variant of human tissue-type plasminogen activator in goat milk: purification and characterization of the recombinant enzyme." *Biotechnology.* 9: 839–843.

Doetschman, T., Greff, R.G., Maeda, N., Hooper, M.L., Melton, D.W., Thompson, S. and Smithies, O. (1987). "Targeted correction of a mutant HPRT gene in mouse embryonic stem cells." *Nature.* 330: 576–578.

Draghia-Akli, R., Fiorotto, M.L., Hill, L.A., Malone, P.B., Deaver, D.R. and Schwartz, R.J. (1999). "Myogenic expression of an injectable protease-resistant growth hormone-releasing hormone augments long-term growth in pigs." *Nature Biotechnol.* 17: 1179–1183.

Dyck, M.K., Lacroix, D., Pothier, F. and Sirard, M.A. (2003). "Making recombinant proteins in animals: different systems, different applications." *Trends Biotechnol.* 21: 394–399.

Ebert, K.M., Low, M.L., Overstrom, E.W., Buonomo, F.C., Baile, C.A., Lee, T.M., Mandel, G. and Goodmann, R.H. (1988). "A Moloney MLV-rat somatotropin fusion gene produces biologically active somatotropin in transgenic pig." *Mol. Endocrinol.* 2: 227–283.

Ebert, K.M., Selgrath, J.P., DiTullio, P., Denman, J., Smith, T.E., Memon, M.A. *et al.* (1991). "Transgenic production of a variant of human tissue-type plasminogen activator in goat milk: generation of transgenic goats and analysis of expression." *Biotechnology* (NY). 9: 835–838.

Ebert, K.M., DiTullio, P., Barry, C.A., Schindler, J.E., Ayres, S.L., Smith, T.E. *et al.* (1994). "Induction of human tissue plasminogen activator in the mammary gland of transgenic goats." *Biotechnology.* 12:699–702.

Ernst, L.K., Zakcharchenko, V.I., Suraeva, N.M., Ponomareva, T.I., Microshnichenko, O.I., Prokofev, M.I. and Tikchonenko, T.I. (1991). "Transgenic rabbits with antisense RNA gene targeted at adenovirus H5." *Theriogenology.* 35: 1257–1271.

Evans, M.J. and Kaufman, M.H. (1981). "Establishment in culture of pluripotential cells from mouse embryos." *Nature.* 292: 154–156.

Eyestone, W.H. and Campbell, K.H. (1999). "Nuclear transfer from somatic cells: applications in farm animal species." *J. Reprod. Fertil. Suppl.* 54: 489–497.

Fodor, W.L., Williams, B.L., Matis, L.A., Madri, J.A., Rollins, S.A., Knight, J.W. *et al.* (1994). "Expression of a functional human complement inhibitor in a transgenic pig as a model for the prevention of xenogeneic hyperacute organ rejection." *Proc. Natl. Acad. Sci. USA.* 91: 11153–11157.

Golovan, S.P., Meidinger, R.G., Ajakaiye, A., Cottrill, M., Wiederkehr, M.Z., Barney, D.J., Plante, C., Pollard, J.W., Fan, M.Z., Hayes, M.A., Laursen, J., Hjorth, J.P., Hacker, R.R., Phillips, J.P. and Forsberg, C.W. (2001). "Pigs expressing salivary phytase produce low-phosphorus manure." *Nature Biotechnol.* 19: 741–745.

Gomez, M.C., Pope, C.E. and Dresser, B.L. (2006). "Nuclear transfer in cats and its application." *Theriogenology.* 66:72–81.

Gordon, J.W., and Ruddle, F.H. (1981). "Integration and stable germ line transmission of genes injected into mouse pronuclei." *Science.* 214: 1244–1246.

Gordon, J.W., Scandgos, G.A., Plotkin, D.J., Barbosa, A. and Ruddle, F.H. (1980). "Genetic transformation of mouse embryo microinjection." *Proc. Natl. Acad. Sci. USA.* 77: 7380–7384.

Gordon, K., Lee, E., Vitale, J.A., Smith, A.E., Westphal, H. and Henninghausen, L. (1987). "Production of human tissue plasminogen activator in transgenic mouse milk." *Biotechnology.* 5: 1183–1187.

Grosse-Hovest, L., Muller, S., Minoia, R., Wolf, E., Zakhartchenko, V., Wenigerkind, H., Lassnig, C., Besenfelder, U., Müller, M., Lytton, S.D., Jung, G. and Brem, G. (2004). "Cloned transgenic farm animals produce a bispecific antibody for T cell-mediated tumor cell killing." *Proc. Natl Acad. Sci. USA.* 101: 6858–6863.

Hammer, R.E., Pursel, V.G., Rexroad, C.E., Wall, R.J., Bolt, D.J., Ebert, K.M., Palmiter, R.D. and Brinster, R.L. (1985). "Production of transgenic rabbits, sheep and pigs by microinjection." *Nature.* 315: 680–683.

Hansen, K. and Khanna, C. (2004). "Spontaneous and genetically engineered animal models: use in preclinical cancer drug development." *Eur. J. Cancer*, 40: 858–880.

Hussar, D., Balling., R., Kothary, R., Magli, M.C., Hozumi, N., Rossant, J. and Ernstein, A. (1985). "Insertion of a bacterial gene into the mouse germ line using an infectious retrovirus vector." *Proc. Natl. Acad. Sci. USA.* 82: 8587–8591.

Hyvonen, P., Suojala, L., Haaranen, J., von Wright, A. and Pyorala, S. (2006). "Human and bovine lactoferrins in the milk of recombinant human lactoferrin-transgenic dairy cows during lactation." *Biotechnol. J.* 1: 410–412.

Iguma, L.T., Lisauskas, S.F., Melo, E.O., Franco, M.M., Pivato, I. and Vianna, G.R. *et al.* (2005). "Development of bovine embryos reconstructed by nuclear transfer of transfected and non-transfected adult fibroblast cells." *Genet. Mol. Res.* 4: 55–66.

Jaenisch, R. (1974). "Infection of mouse blastocysts with SV-40 DNA: Normal development of infected embryos and resistence to SV-40 specific DNA sequences in the adult animals." *Cold Spring Harbor Symp.* 39: 375–380.

Jaenisch, R. and Mintz, B. (1974). "Simian virus 40 DNA sequences in DNA of healthy adult mice derived from preimplantation blastocysts injected with viral DNA." *Proc. Natl. Acad. Sci. USA.* 71: 1250–1254.

Jaenisch, R., Fan, H. and Croker, B. (1975). "Infection of preimplantation mouse embryos and of newborn mice with leukemia virus: tissue distribution of viral DNA and RNA and leukemogenesis in the adult animal." *Proc. Natl. Acad. Sci. USA.* 72: 4008–4012.

Jaenisch, R., Harbers, K., Schnieke, A., Lohler, J., Chumakov, I., Jahner, D., Grotkopp, D. and Hoffmann, E. (1983). "Germline integration of Moloney murine leukemia virus at the Mov 13 locus leads to recessive mutation and early embryonic death." *Cell.* 2: 209–216.

Karatzas, C.N. and Turner, J.D. (1997). "Toward altering milk composition by genetic manipulation: current status and challenges." *J. Dairy Sci.* 80: 2225–2232.

Kasinathan, P., Knott, J.G., Moreira, P.N., Burnside, A.S., Jerry, D.J. and Robl, J.M. (2001). "Effect of fibroblast donor cell age and cell cycle on development of bovine nuclear transfer embryos *in vitro*." *Biol. Reprod.* 64: 1487–1493.

Keefer, C.L. (2004). "Production of bioproducts through the use of transgenic animal models." *Anim. Reprod. Sci.* 82–83: 5–12.

King, D. and Wall, R.J. (1988). "Identification of specific gene sequences in preimplantation embryos by genome amplification: detection of a transgene." *Mol. Reprod. Dev.* 1: 57–62.

Krimpenfort, P., Rademakers, A., Eyestone, W., van der Schans, A., van den Broek, S., Kooiman, P., Kootwijk, E., Platenburg, G., Pieper, F., Strijker R. and de Boer, H. (1991). "Generation of transgenic dairy cattle using *in vitro* embryo production." *Biotechnology.* 9: 844–847.

Kues, W.A. and Niemann, H. (2004). "The contribution of farm animals to human health." *Trends Biotechnol.* 22: 286–294.

Kuroiwa, Y., Kasinathan, P., Choi, Y.J., Naeem, R., Tomizuka, K., Sullivan, E.J., Knott, J.G., Duteau, A., Goldsby, R.A., Osborne, B.A., Ishida, I. and Robl, J.M. (2002). "Cloned trans-chromosomic calves producing human immunoglobulin." *Nature Biotechnol.* 20: 889–894.

Kuroiwa, Y., Kasinathan, P., Matsushita, H., Sathiyaselan, J., Sullivan, E.J., Kakitani, M. *et al.* (2004). "Sequential targeting of the genes encoding immunoglobulin-M and prion protein in cattle." *Nat. Genet.* 36: 775–780.

Lazaris, A., Arcidiacono, S., Huang, Y., Zhou, J.F., Duguay, F. and Chretien, N. *et al.* (2002). "Spider silk fibers spun from soluble recombinant silk produced in mammalian cells." *Science.* 295: 472–476.

Li, Z. and Engelhardt, J.F. (2003). "Progress toward generating a ferret model of cystic fibrosis by somatic cell nuclear transfer." *Reprod. Biol. Endocrinol.* 1: 83.

Li, R., Lai, L., Wax, D., Hao, Y., Murphy, C.N., Rieke, A., Samuel, M., Linville, M.L., Korte, S.W., Evans, R.W., Turk, J.R., Kang, J.X., Witt, W.T., Dai, Y. and Prather, R.S. (2006). "Cloned transgenic swine via *in vitro* production and cryopreservation." *Biol. Reprod.* 75: 226–230.

Li, L., Shen, W., Min, L., Dong, H., Sun, Y. and Pan, Q. (2006). "Human lactoferrin transgenic rabbits produced efficiently using dimethylsulfoxide-sperm-mediated gene transfer." *Reprod. Fertil. Dev.* 18:689–95.

Lindermann, J. (1962). "Resistance in mice to mouse-adapated influenza A virus." *Virology.* 16: 203–204.

Lo, D., Pursel, V., Linton, P.J., Sandgren, E., Behringer, R., Rexroad, C., Palmiter, R.D. and Brinster, R.L. (1991). "Expression of mouse IgA by transgenic mice, pigs and sheep." *Eur. J. Immunol.* 21: 1001–1006.

Maga, E.A. and Murray, J.D. (1995). "Mammary gland expression of transgenes and the potential for altering the properties of milk." *Biotechnology.* 13: 1452–1457.

Maga, E.A., Anderson, G.B. and Murray, J.D. (1995). "The effects of mammary gland expression of human lysozyme on the properties of milk from transgenic mice." *J. Dairy Sci.* 78: 2645–2652.

Maga, E.A., Shoemaker, C.F., Rowe, J.D., Bondurant, R.H., Anderson, G.B. and Murray, J.D. (2006). "Production and processing of milk from transgenic goats expressing human lysozyme in the mammary gland." *J. Dairy Sci.* 89:518–524.

Mahmoud, T.H., McCuen, B.W., Hao, Y., Moon, S.J., Tatebayashi, M., Stinnett, S., Petters, R.M. and Wong, F. (2003). "Lensectomy and vitrectomy decrease the rate of photoreceptor loss in rhodopsin P347L transgenic pigs." *Graefes Arch. Clin. Exp. Ophtalmol.* 241: 298–308.

Maione, B., Lavitrano, M., Spadafora, C. and Kiessling, A.A. (1998). "Sperm-mediated gene transfer in mice." *Mol. Reprod. Dev.* 50: 406–409.

Malenko, G.P., Prokof'ev, M.I., Piniugina, M.V., Antipova, T.A., Mezina, M.N. and Bukreev, I.M. (2006). "Production of cloned bovine embryos by somatic cell transfer into enucleated zona-free oocytes." *Izv. Akad. Nauk. Ser. Biol.* (3): 284–91.

Martin, G.R. (1981). "Isolation of a pluripotent cell line from early mouse embryos cultured in medium conditioned by teratocarcinoma stem cells." *Proc. Natl. Acad. Sci. USA.* 78: 7634–7638.

McCreath, K.J., Howcroft, J., Campbell, K.H., Colman, A., Schnieke, A.E. and Kind. A.J. (2000). "Production of gene-targeted sheep by nuclear transfer from cultured somatic cells." *Nature.* 405: 1066–1069.

McKee, C., Gibson, A., Dalrymple, M., Emslie, L., Garner, I. and Cottingham, I. (1998). "Production of biologically active salmon calcitonin in the milk of transgenic rabbits." *Nat. Biotechnol.* 16: 647–651.

Meade, H., Gates, L., Lacey, E. and Lonberg, N. (1990). "Bovine α_1-casein gene sequences direct high level expression of active human urokinase in mouse milk." *Biochemistry.* 8: 443–446.

Meade, H.M., Echelard, Y., Ziomek, C.A., Young, M.W., Harvey, M., Cole, E.S., Groet, S. and Curling, J.M. (1999). "Expression of recombinant proteins in the milk of transgenic animals." In: *Gene Expression Systems: Using Nature for the Art of Expression.* Fernandez, J.M. and Hoeffler, J.P. (eds.)." Academic Press, San Diego. 399–427.

Miercier, J.C. (1986). "Genetic engineering applied to milk producing animals: Some expectations". In: *Exploiting New Technologies in Animal Breeding.* (eds.). Smith, C., King, J., MacKay, J.C. Oxford Press, Oxford.

Mikus, T., Maly, P., Poplstein, M., Landa, V., Trefil, P. and Lidicky, J. (2001). "Expression of human erythropoietin gene in the mammary gland of a transgenic mouse." *Folia Biol. (Praha).* 47: 187–195.

Muller, M. and Brem, G. (1991). "Disease resistance in farm animals." *Experientia.* 41: 923–934.

Muller, M., Brenig, B., Winnacker, E.L. and Brem, G. (1992). "Transgenic pigs carrying cDNA copies encoding the murine Mx1 protein which confers resistance to influenza virus infection." *Gene.* 121: 263–270.

Niemann, H. (2004). "Transgenic pigs expressing plant genes." *Proc. Natl. Acad. Sci. USA.* 101: 7211–7212.

Niemann, H. and Rath, D. (2001). "Progress in reproductive biotechnology in swine." *Theriogenology.* 56: 1291–2304.

Niemann, H., Halter, R., Carnwath, J.W., Herrmann, D., Lemme, E. and Paul, D. (1999). "Expression of human blood clotting factor VIII in the mammary gland of transgenic sheep." *Transgenic Res.* 8: 237–247.

Nottle, M.B., Nagashima, H., Verma, P.J., Du, Z.T., Grupen, C.G., MacIlfatrick, S.M., Ashman, R.J., Harding, M.P., Giannakis, C. Wigley, P.L., Lyons, I.G., Harrison, D.T., Luxford, B.G., Campbell, R.G., Crawford R.J. and Robins, A.J. (1999). "Production and analysis of transgenic pigs containing a metallothionein porcine growth hormone gene construct." In: *Transgenic Animals in Agriculture.* Murray, J.D., Anderson, G.B., Oberbauer, A.M. and McGloughlin, M.M. (eds). CABI Publishing, New York. 145–156.

Paleyanda, R.K., Velander, W.H., Lee, T.K., Scandella, D.H., Gwazdauskas, F.C. and Knight, J.W. (1997). "Transgenic pigs produce functional human factor VIII in milk." *Nat. Biotechnol.* 15: 971–975.

Palmarini, M. and Fan, H. (2001). "Retrovirus-induced ovine pulmonary adenocarcinoma: an animal model for lung cancer." *J. Natl. Cancer Inst.* 93: 1603–1614.

Palmiter, R.D., Brinster, R.L., Hammer, R.E., Trumbauer, M.E., Rosenfeld, M.G., Birnberg, N.C. and Evans, R.M. (1982). "Dramatic growth of mice that develop from eggs microinjected with metallothionein-growth fusion genes." *Nature*. 300: 611–615.

Palmiter, R.D., Chen, H.Y., Messing, A. and Brinster, R.L. (1985). "SV-40 enhancer and large T antigen are instrumental in development of choroid plexus tumours in transgenic mice." *Nature*. 316: 457–460.

Palmiter, R.D., Behringer, R.R., Quaife, C.J., Maxwell, F., Maxwell, I.H. and Brinster, R.L. (1987). "Cell lineage ablation in transgenic mice by cell-specific expression of a toxin gene." *Cell*. 50: 435–443.

Petters, R.M., Alexander, C.A., Wells, K.D., Collins, E.B., Sommer, J.R., Blanton, M.R., Rojas, G., Hao, Y., Flowers, W.L., Banin, E., Cideciyan, A.V., Jacobson, S.G. and Wong, F. (1997). "Genetically engineered large animal model for studying cone photoreceptor survival and degeneration in retinitis pigmentosa." *Nature Biotechnol*. 15: 965–970.

Phelps, C.J., Koike, C., Vaught, T.D., Boone, J., Wells, K.D., Chen, S.H., Ball, S., Specht, S.M., Polejaeva, I.A, Monahan, J.A. *et al.* (2003). "Production of alpha 1,3-galactosyltransferase-deficient pigs." *Science*. 299:411–414.

Platenburg, G.J., Kootwijk, E.A.P., Kooiman, P.M., Woloshuk, S.L., Nuijens, J.H., Krimpenfort, P.J.A., Pieper, F.R., de Boer, H.A. and Strijker, R. (1994). "Expression of human lactoferrin in milk of transgenic mice." *Transgenic Res*. 3: 99–108.

Prusiner, S.B. (1982). "Novel proteinaceous infectious particles cause scrapie." *Science*. 216:136–144.

Pursel, V.G., Pinkert, C.A., Miller, K.F., Bolt, D.J., Campbell, R.G., Palmiter, R.D., Brinster, R.L. and Hammer, R.E. (1989). "Genetic engineering of livestock." *Science*. 244: 1281–1288.

Pursel, V.G., Wall, R.J., Henninghausen, L., Pittius, C.W. and King, D. (1990). "Regulated expression of the mouse whey acidic protein gene in transgenic swine." *Theriogenology*. 33: 302.

Ramsoondar, J.J., Machaty, Z., Costa, C., Williams, B.L., Fodor, W.L., Bondioli, K.R. (2003) "Production of {alpha}1,3-galactosyltransferase-knockout cloned pigs expressing human {alpha}1,2-fucosylosyltransferase." *Biol. Reprod*. 2:2.

Rexroad, C.E., Hammer, R.E., Bohringer, R.R., Palmiter, R.D. and Brinster, R.L. (1990). "Insertion expression and physiology of growth regulating genes in ruminants." *J. Reprod. Fert. Suppl*. 41: 119–124.

Rexroad, C.E., Mayo, Bolt, D.J., Elsasser, T.H., Miller, K.F., Bohringer, R.R., Palmiter, R.D. and Brinster, R.L. (1991). "Transferin-and albumin-directed expression of growth-related peptides in transgenic sheep." *J. Amin. Sci.* 69: 2992–3004.

Ritchie, W.A. (2006). "Nuclear transfer in sheep." *Methods Mol. Biol.* 325: 11–23.

Robertson, E.A., Bradley, A., Kuehn, M., Evans, M. and (1986). "Germ-line transmission of genes introduced into cultured pluripotential cells by retroviral vector." *Nature.* 323: 445–448.

Rogers, G.E. (1990). "Improvement of wool production through genetic engineering." *Trends Biotech.* 8: 6–11.

Rudolph, N.S. (1999). "Biopharmaceutical production in transgenic livestock." *Trends Biotechnol.* 17: 367–374.

Saeki, K., Matsumoto, K., Kinoshita, M., Suzuki, I., Tasaka, Y., Kano, K., Taguchi, Y., Mikami, K., Hirabayashi, M., Kashiwazaki, N., Hosoi, Y., Murata, N. and Iritani, A. (2004). "Functional expression of a Delta12 fatty acid desaturase gene from spinach in transgenic pigs." *Proc. Natl. Acad. Sci. USA.* 101: 6361–6366.

Salter, D.W., Smith, E.J., Hughes, S.H., Wright, S.E., Fadly, A.M., Witter, R.L. and Crittenden, L.B. (1986). "Gene insertion into chicken germ line by retroviruses." *Poultry Sci.* 65: 1445–1458.

Sato, M., Ishikawa, A. and Kimura, M. (2002). "Direct injection of foreign DNA into mouse testis as a possible *in vivo* gene transfer system via epididymal spermatozoa." *Mo. Reprod. Dev.* 61: 49–56.

Schnieke, A.E., Kind, A.J., Ritchie, W.A., Mycock, K., Scott, A.R., Ritchie, M., Wilmut, I. *et al.* (1997). "Human factor IX transgenic sheep produced by transfer of nuclei from transfected fetal fibroblasts." *Science.* 278: 2130–2133.

Simons, J.P., McClenghan, M. and Clark, A.J. (1987). "Alteration of the quality of milk by expression of sheep β-lactoglobin in transgenic mice." *Nature.* 328: 530–532.

Smithies, O., Greeg, R.G., Boggs, S.S., Koralewski, M.A. and Kycherlapti, R.S. (1985). "Insertion of DNA sequences into the human chromosomal β-globin locus by homologous recombination." *Nature.* 317: 230–234.

Soriano, P., Cone, R.D., Mulligan, R.C. and Jaenisch, R. (1986). "Tissue specific and ectopic expression of genes introduced into transgenic mice by retroviruses." *Science.* 234: 1409–1413.

Stacey, A., Bateman, J., Choi, T., Mascara, T., Cole, W. and Laenisch, R. (1988). "Perinatal lethal osteogenesis imperfecta in transgenic mice bearing an engineered mutant pro-α_1 (1) collagen gene." *Nature.* 332: 131–136.

Stewart, C., Schuetze, S., Vanek, M. and Wager, E. (1987). "Expression of retroviral vectors in transgenic mice obtained by embryo infection." *EMBO J.* 6: 383–388.

Storb, U. (1987). "Immunoglobulin transgenic mice." *Ann. Rev. Immunol.* 5: 151–174.

Stowers, A.W., Lh Chen, L.H., Zhang, Y., Kennedy, M.C., Zou, L., Lambert, L., Rice, T.J., Kaslow, D.C., Saul, A., Long, C.A., Meade, H. and Miller, L.H. (2002). "A recombinant vaccine expressed in the milk of transgenic mice protects Aotus monkeys from a lethal challenge with *Plasmodium falciparum*." *Proc. Natl. Acad. Sci. USA.* 99: 339–344.

Tearle, R.G., Tange, M.J., Zannettino, Z.L., Katerelos, M., Shinkel, T.A., Van Denderen, B.J., Lonie, A.J., Lyons, I., Nottle, M.B., Cox, T. *et al.* (1996). "The alpha-1,3-galactosyltransferase knockout mouse. Implications for xenotransplantation." *Transplantation.* 61:13–19.

Theuring, F., Thunecke, M., Kosciessa, U. and Turner, T.D. (1997). "Transgenic animals as models of neurodegenerative disease in humans." *Trends Biotechnol.* 15: 320–325.

van Berkel, P.H., Welling, M.M., Geerts, M., van Veen, H.A., Ravensbergen, B., Salaheddine, M., Pauwels, E.K., Pieper, F., Nuijens, J.H. and Nibbering, P.H. (2002). "Large scale production of recombinant human lactoferrin in the milk of transgenic cows." *Nature Biotechnol.* 20: 484–487.

van Doorn, M.B., Burggraaf, J., van Dam, T., Eerenberg, A., Levi, M., Hack, C.E., *et al.* (2005). "A phase I study of recombinant human C1 inhibitor in asymptomatic patients with hereditary angioedema." *J. Allergy Clin. Immunol.* 116: 876–883.

Vogel, J., Hinriches, S.H., Reynolds, R.K., Luciw, P.A. and Jay, G. (1988). "The HIV 'tat' gene induces dermal lesions resembling Kaposi's sarcoma in transgenic mice." *Nature.* 335: 606–611.

Wagner, E.F., Covarrubians, L., Stewart, R.A. and Mintz, B. (1983). "Prenatal lethalities in mice homozygous for human growth hormone gene sequences as carrier for DNA uptake into cell." *Proc. Natl. Acad. Sci. USA.* 87: 3410–3414.

Wall, R.J., Powell, A., Paape, M.J., Kerr, D.E., Bannermann, D.D., Pursel, V.G., Wells, K.D., Talbot, N. and Hawk, H. (2005). "Genetically enhanced cows resist intramammary *Staphylococcus aureus* infection." *Nat. Biotechnol.* 23: 445–451.

Ward, K.A., Nancarrow, C.D., Byrne, C.R., Shanahan, C.M., Murray, J.D., Leish, Z., Townrow, C., Rigby, N.W., Wilson, B.W. and Hunt, C.L. (1990). "The potential of transgenic animals for improved agriculture productivity." *Rev. Sci. Tech. Off. Int. Epiz.* 9: 847–864.

Ward, K.A., Byrne, C.R., Wilson, B.W., Leish, Z., Rigby, N.W., Townrow, C.R., Hunt, C.L., Murray, J.D. and Nancarow, C.D. (1991). "The regulation of wool growth in transgenic animals." *Adv. Dermatol.* 1: 70–76.

Waterston, R.H., Lindblad-Toh, K., Birney, E., Rogers, J., Abril, J.F., Agarwal, P. *et al.* (2002). "Initial sequencing and comparative analysis of the mouse genome." *Nature.* 420: 520–562.

Wheeler, M.B., Bleck, G.T. and Donovan, S.M. (2001). "Transgenic alteration of sow milk to improve piglet growth and health." *Reproduction.* 58 (Suppl.): 313–324.

White, D. (1996). "Alteration of complement activity: a strategy for xenotransplantation." *Trends Biotechnol.* 14: 3–5.

Willadsen, S.M. (1986). "Nuclear transplantation in sheep embryos." *Nature.* 320: 63–65.

Wilmut, I., Schnieke, A.E., McWhir, J., Kind, A.J. and Campbell, K.H. (1997). "Viable offspring derived from fetal and adult mammalian cells." *Nature.* 385: 810–813.

Windle, J.J., Albert, D.M., O'Brein, J.M., Marcus, D.M., Disteche, C.M., Bernards, R. and Mellon, P.L. (1990). "Retinoblastoma in transgenic mice." *Nature.* 343: 665–669.

Wirak, D.O., Bayne, R., Kundel, C.A., Lee, A., Scangos, G.A., Trapp, B.D. and Unterback, A.J. (1991). "Regulatory region of human amyloid precursor protein (APP) gene promotes neuron-specific gene expression in the CNS of transgenic mice." *EMBO J.* 10: 289–296.

Wright, G., Carver, A., Cottom, D., Reeves, D., Scot, A., Simons, P. *et al.* (1991). "High level expression of active human alpha-1-antitrypsin in the milk of transgenic sheep." *Biotechnology* (NY). 9: 830–834.

Ye, Y., Niekrasz, M., Kosanke, S., Welsh, R., Jordan, H.E., Fox, J.C., Edwards, W.C. and Maxwell, D.K. (1994). "Cooper, the pig, as a potential organ donor for man. A study of potentially transferable disease from donor pig to recipient man." *Transplantation.* 57:694–703.

Ziomek, C.A. (1998). "Commercialization of proteins produced in the mammary gland." *Theriogenology.* 49: 139–144.

9

DNA FINGERPRINTING AND RFLP IN DOMESTIC ANIMALS

INTRODUCTION

The analysis of restriction fragment length polymorphisms (RFLPs) is one of the most efficient means of monitoring genetic diversity in domestic animals. The first results of RFLPs in pigs, sheep and cattle were published in 1985 (Chardon *et al.*, 1985). Since then, the number of RFLPs described in domestic animals have increased considerably. A higher degree of variability due to variable numbers of tandem repeats (VNTR) results in multiallelic RFLP patterns that can be seen in certain regions of repetitive DNA (Nakamura *et al.*, 1987). When such banding patterns are specific to a defined DNA locus, they are called as single locus VNTRs. Several independent DNA loci share sequence homology, thereby enabling the demonstration of many highly polymorphic DNA loci simultaneously. Such multilocus VNTRs (called 'DNA fingerprints' due to the high individual specificity) were studied methodically and applied by Jeffrey and co-workers (Jeffreys *et al.*, 1985a, b, c and 1986).

Jeffreys *et al.* (1985a) termed their repetitive sequences **"minisatellites"** which are characterized by tandemly repeated consensus sequence of 16–64 nucleotides. Some of these minisatellites comprise of **"families"** that are related to core sequences, characterized by a high degree of homology in their consensus sequences. These homologies allow the cloning of novel VNTR probes for screening genomic libraries, using known minisatellite sequences.

Some human VNTR probes can also be used in domestic animals because minisatellite sequences show not only intraspecies, but also substantial interspecies homologies. Probes 33.6 or 33.15 **(Jeffreys probes)** and related sequences are now known to display hypervariable banding patterns in different species. Initially, Jeffreys and Morton (1987) used both probes in dogs and cats;

subsequently, Georges *et al.* (1988) generated DNA fingerprints in cattle, horses, swine and dogs with a polymer of **"Jeffreys core sequence"**. Fingerprints in cattle were also obtained with the original Jeffreys probe 33.6. Jeffreys probes were also used to obtain fingerprints in different species of poultry (Hillel *et al.*, 1989).

APPLICATIONS OF DNA FINGERPRINTS IN DOMESTIC ANIMALS

Identification of individuals is important in the domestic animals, e.g. for the control of semen used in artificial insemination, the identification of cell lines and chimeras. DNA fingerprinting using skin samples, allows the diagnosis of zygosity even in species showing blood cell chimerism due to placental anastomoses. Paternity checks are performed in domestic animals in forensic cases (if paternity has been disputed by the owner of an animal) or for special breeding purposes, e.g. for calves born after embryo transfer.

DNA fingerprints also prove useful in investigating the phylogeny and genetic structuring of populations in particular cases. The control of inbred lines and determination of genetic relationships have already been realized by DNA fingerprints. Thus, oligonucleotide fingerprints can be used as a powerful tool in the analysis of genetic relationships. In poultry, the highly discriminating oligonucleotides $(CAC)_5$, $(CGAT)_4$ and $(GT)_8$ have enough discriminating potential to differentiate between individual chicken as well as of strains. Furthermore, the genetic make-up of inbred chicken can be evaluated by establishing fingerprints. Such data may be useful, e.g. for selecting lines or individuals in cross-breeding programmes.

DNA fingerprints can be utilized to establish genetic linkage maps. VNTR probes are expected to be a powerful tool to increase the number of available genetic markers. In human metaphase chromosomes, the $(CAC)_n/(GTG)_n$ loci coincide with the R bands and are therefore spread over the whole genome. This is an advantage for linkage analysis as compared to minisatellite lock, which are reported to be localized on telomeric region. In contrast to man, large pedigrees can be designed in domestic animals, enabling efficient linkage studies. These can be used to identify disadvantageous genes (defect gene) or to monitor advantageous genes. Even more complex traits can be investigated by linkage analysis using multilocus VNTRs.

Two minisatellite probes cloned from genomic DNA of bird by colony hybridization with probe 33.6 and 33.15 gave highly variable banding patterns in birds as well as in some mammals (Gyllenstein *et al.*, 1989). Different probes used in fingerprinting of domestic animals and birds are given in Table 9.1. Vassart *et al.* (1987) described the use of a DNA from bacteriophage M13 to obtain all multilocus VNTR patterns in humans and in cattle. A more detailed investigation of different cattle, pigs, horses and dogs were carried out by Georges *et al.* (1988). They found the M13 probe to be highly informative in all the species investigated including chicken (Kuhnlein *et al.*, 1989).

Table 9.1 Probes which have been used for characterization of various species of animals and birds

Species	Probes	References
Cat	33.6/33.15	Jeffreys and Morton, 1987
Dog	33.6/33.15	Jeffreys and Morton, 1987
	M13; pUC J	Georges *et al.* 1988
	pSP 64; 2.5 RI	
Swine	M13, pUC J, pS 3	Georges *et al.* 1988
Cattle	M13, pUC J	Georges *et al.* 1988
	33.6, pGBJ43a	Kashi, *et al.* 1990
Sheep	33.15, M13	Broad, 1988
Horse	M13, pUC J	Georges *et al.* 1988
	pSP 64, 2.5RI	
Poultry	33.6	Hillel *et al.* 1989
	M13	Kuhnlein *et al.* 1989

Bravi *et al.* (2004) developed a simple, quick assay in order to discriminate forensic samples among humans, and common domestic and livestock species of the Pampean region, Argentina. A mitochondrial cytochrome *b* fragment amplified with universal primers was separately digested with three restriction enzymes (*Alu I*, *Hae III*, and *Hinf I*) and the resulting fragments were resolved through electrophoresis in polyacrylamide gels. This PCR-RFLP method allowed the identification of the target species and to work on degraded samples. The assay was successfully applied in livestock robbery cases in Argentina, and may be useful when attempting a first assessment as to the specific status of a forensic evidence.

MINISATELLITE PROBES USED IN DOMESTIC ANIMALS

A murine probe (pSP 64; 2.5 RI) was used for fingerprints of domestic animals and found to be informative in dogs, swine and horses (Georges *et al.*, 1988). In cattle, this probe displayed Y-chromosome-specific hybridization patterns

(Parret *et al.*, 1990), thus providing a marker to monitor male genealogy. Human minisatellite probes using different oligonucleotides specific to known consensus sequences were used to obtain fingerprints in cattle (probe EFD 134.7), pig(probes pYNH24 and pYNA 23) and horses (probe EFD 134.7). Coppieters *et al.* (1990) discovered a pig minisatellite suitable for multilocus fingerprints of swine, horses and rabbits.

MICROSATELLITE PROBES

Hypervariable loci comprising of tandemly repeated short (bp) simple sequence motives, e.g. CAC CAC CAC... or GT GT GT... are termed as **simple tandem repeats** (str) or **microsatellites**. The first str used to obtain DNA fingerprint consisted of simple quadruplet repeat (sqr) sequences consisting of GATA/GACA repeats. Such sequences were initially found in the sex-specific satellite DNA of female colubrid snakes. Southern blot experiments using a probe from a genomic clone containing $(GATA)_n$ sequences derived from *Drosophila*, displayed hypervariable banding patterns in humans.

The allele nomenclature of six polymorphic short tandem repeat (STR) loci (PEZ3, PEZ6, PEZ8, PEZ10, FHC2161, and FHC2328) for canine genotyping (*Canis lupus familiaris*) was presented (Hellmann *et al.*, 2006). The nomenclature was based on the sequence data of the polymorphic region of the microsatellite markers as recommended by the DNA commission of the International Society of Forensic Haemogenetics (ISFH) in 1994 for human DNA typing. To cover commonly and rarely occurring alleles, a selection of homozygous and heterozygous animals were analysed and subjected to sequence studies. The alleles consisted of simple tri- and tetra-nucleotide repeat patterns as well as compound and highly complex repeat patterns. Several alleles revealing the same fragment size but different repeat structures were found. The allele designation described here was adopted to the number of repeats, including all variable regions within the amplified fragment. In a second step the most commonly occurring alleles were added to an allelic ladder for each marker allowing a reliable typing of all alleles differing in size. A total number of 142 unrelated dogs from surrounding municipal animal homes, private households, and canines in police duty were analysed. The data were added to a population database providing allele frequencies for each marker.

For identification of the offending dog involved in dog bites, Eichmann *et al.* (2004) evaluated canine STR typing of saliva traces on dog bite marks. The specificity of 15 canine-specific STRs was tested on human-canine DNA mixtures prior to an applied study in which 52 cases of dog bites were investigated. The first-aid wound bandages as well as swab samples from the surrounding area of the wound were used for DNA analyses. Generally, it was

possible to obtain a canine-specific STR profile from the dog's saliva left on the wound area, even when high background of human DNA was present (blood). Interestingly, it was found that canine STR typing was more successful when the bandages and swabs showed high amounts of human blood, i.e., when the dog bite was severe.

A forensic genotyping panel of 11 tetranucleotide STR loci from the domestic cat was characterized and evaluated for genetic individualization of cat tissues (Menotti-Raymond *et al.*, 2005). 49 candidate STR loci and their frequency assessment in domestic cat populations was examined. The STR loci (3–4 base pair repeat motifs), mapped in the cat genome relative to 579 coding loci and 255 STR loci, were well distributed across the 18 feline autosomes. All loci exhibit Mendelian inheritance in a multi-generation pedigree. Eleven loci that were unlinked and were highly heterozygous in cat breeds were selected for a forensic panel. Heterozygosity values obtained for the independent loci, ranged from 0.60–0.82, while the average cat breed heterozygosity obtained for the 11 locus panel was 0.71 (range of 0.57–0.83). A small sample set of outbred domestic cats displayed a heterozygosity of 0.86 for the 11-locus panel. The panel showed good potential for genetic individualization within outbred domestic cats. A multiplex protocol, designed for the co-amplification of the 11 loci and a gender-identifying locus, was species-specific and robust, generating a product profile with as little as 0.125 ng of genomic DNA.

OLIGONUCLEOTIDE FINGERPRINTING

Oligonucleotide fingerprints provide a powerful tool to demonstrate genetic variability in man, animal species and plants. Oligonucleotide $(TG)_{10}$ was reported to reveal highly variable banding patterns in horses, sheep and chicken (Kashi *et al.*, 1990). The use of oligonucleotide fingerprints provide some technical advantages that are important for practical individualization studies. The advantages of oligonucleotide fingerprints are:

1. Oligonucleotides are chemically synthesized, thus bypassing problems concerning alteration of repetitive sequences which may occur if highly repetitive sequences are cloned.

2. Hybridization directly in the gel is much faster and avoids loss of DNA as compared to Southern blot procedures.

3. The use of non-radioactively labelled oligonucleotides recently developed for hybridization directly in the gel, enables DNA fingerprints to be introduced in non-specialized laboratories also and

4. Reproducible fingerprints are obtained.

A synthetic polynucleotide $(TG)_n$ was used to get DNA fingerprints in horses to characterize different breeds (Ellegren *et al.*, 1992a). Similar probes were used for paternity testing in horses. PCR-based paternity testing systems have been developed. The microsatellites could easily be amplified from a single hair. A standardized paternity testing system based on PCR-analysed microsatellites is likely to be a future alternative to blood typing in horses (Ellegren *et al.*, 1992b).

DNA FINGERPRINTS FOR DETECTION OF MAJOR GENES

Several genes having major effect on quantitative traits in farm animal species have been detected in recent years. These include Booroola gene in sheep (Piper *et al.*, 1985), the double muscling gene in cattle (Hanset, 1986) and the porcine stress syndrome (PSS) gene in pigs (Webb *et al.*, 1982). All of these genes have been detected via their major effects on the phenotype and are of economic importance as well as provide useful scientific tools for the investigation of the genetic causes of phenotypic variations. Molecular genetic techniques have now shown that there is a wealth of genetic markers. Such polymorphisms are detected as variation in the length of fragments produced when DNA is cut using a restriction enzyme (RFLPs). In future, genetic maps based upon linked polymorphic marker loci are likely to be produced for the important farm animals. Quantitative trait loci (QTL) will be detected by their linkage to markers on the map.

The potential use of DNA fingerprints for the detection of major genes has been demonstrated by Georges *et al.* (1990) who identified a putative linkage between a fingerprint locus and the gene causing muscular hypertrophy in Belgium Blue cattle.

RESTRICTION FRAGMENT LENGTH POLYMORPHISMS (RFLP)

RFLPs are important tools for the identification of genetic variations. RFLPs occur as a result of DNA base changes due to deletion, insertion or rearrangements that either create, eliminate or translocate restriction enzyme cleavage sites. Gene deletion or insertion often leads to complete failure to produce the relevant protein. The types of DNA variation represent the molecular basis of RFLPs. In some cases, the RFLP has components of single-base RFLPs which are recognized by restriction endonuclease cleavage and detected by hybridization analysis. Jeffreys *et al.* (1985a) were the first to isolate probes that had highly polymorphic characters because of a variable number of tandem nucleic acid repeat sequences (VNTRs). VNTRs have proven clinically useful in the diagnosis of several genetic diseases and they are of general importance as markers in genome analysis. Furthermore, the use of RFLPs predominantly based on VNTRs has become applicable in forensic analysis as well as in animal population genetics. This method can be considered as a general means of genetically identifying individuals by DNA variation. The advantages of this approach are as follows; (a) The patterns obtained by hybridization with the probe are highly informative. The probability that the patterns of all individuals are identical has been calculated to be in the range of 10^{13}. (b) The test requires minimal amount of DNA. The classical hybridization techniques can now be supplemented by the PCR that allows amplification of the DNA material. In this way, DNA from the cells adhering to a single hair or even from a single sperm cell is sufficient for analysis. (c) The same oligonucleotides can be used to probe DNA from different species, because similar patterns of repetitive sequences do occur.

RFLPs in Cattle

RFLPs in cattle have been identified with cloned probe from several genes including fibronectin, the major histocompatibility complex, α and β-interferons, prolactin and thyroglobulin. The major histocompatibility complex (MHC) is one of the best investigated genomic DNA region. These regions containing several genes, which are in different ways involved in the immune response MHC genes have been associated with disease resistance in domestic animals. BHA-W26 of cattle was found to be associated with bovine leukaemia virus susceptibility. BoLA W6-positive animals are more likely to develop eye cancer. Beckmann *et al.* (1986) described RFLPs for the gene loci of growth hormone (bGH) and β-actin. For the bGH-locus, polymorphisms are shown after digestion with the enzyme *Bgl* II and *Bam* HI and hybridization with a bovine cDNA probe. Digestion with *Hind* III or *Eco* RI also yields informative patterns for the bGH locus.

RFLPs in cattle have been identified with cloned probes from several genes including γ-crystalline, fibronectin, growth hormone, the MHC, α and β interferons, prolactin and thyroglobulin (Thielmann *et al.*, 1989). These polymorphisms greatly enhance the pool of immunogenic, biochemical and molecular markers available in cattle for linkage analysis, testing of parentage and distinction of breeds (Figure 9.1).

RFLPs for the thyroglobulin locus (bTg) had also been studied. The hybridization probe consisted of two fragments of a bTg-cDNA called "right and left probe". Using this probe, they were able to differentiate variable patterns. Associations were found between RFLP patterns for bTg loci and recessive hereditary goitre.

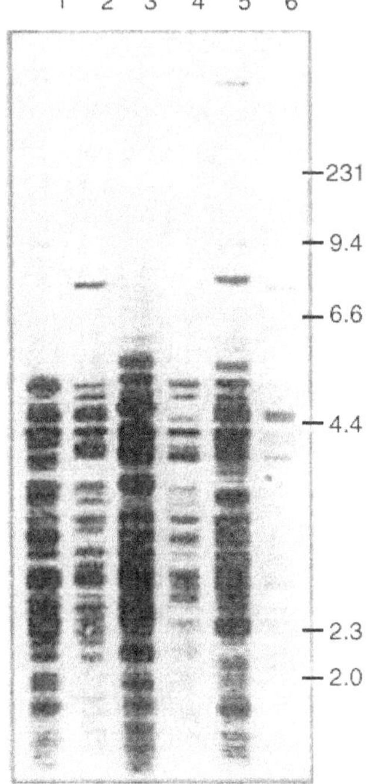

Figure 9.1 DNA fingerprints of cattle of different breeds with (CAC)$_5$ probe. DNA digested with *Hin f*1 enzyme and electrophoresed on a 0.7% agarose gel and hybridized with (CAC)$_5$ (Courtesy: Buitkamp *et al.,* 1991; *Electrophoresis*, 12: 169)

Restriction sites in the casein loci were found for α-S$_1$ and for κ-casein loci using *Pst* I and an α-S$_1$-cDNA probe. Rando *et al.* (1988) found RFLPs for the κ-casein after digestion with *Pst* I and hybridization with a cDNA probe. These DNA polymorphisms were correlated with the protein type A and B. RFLPs for α- and β-casein interferons (IFN α and IFN β) were reported by Atkinson *et al.* (1988). When they studied human cDNA for the IFN α and bovine cDNA for IFN α, they inferred for the IFN α a multiallelic polymorphism. The IFN β restriction pattern was interpreted as a result of allelic variants.

Lechniak *et al.* (1999) studied the relationship between the growth hormone gene RFLP polymorphism and bull sperm characteristics. The bGH genotypes were studied in AI bulls of dairy crossbred and beef breeds. The frequency of the Leu allele was 0.86 among dairy bulls and 0.38 in beef bulls (0.14 and 0.62 for the Val allele, respectively). Eight sperm characteristics and Day 60 non-return rates (NRR) were analysed. The three genotype groups (LL, VV and LV) and the effect of production type (dairy or beef) on sperm characteristics were considered. There was a tendency for the LL bulls to have a lower ejaculate volume and VV bulls higher NRR. Beef bulls were superior in sperm concentration and non-return rate, whereas dairy bulls excelled in individual fresh sperm motility.

Acyl-coenzyme A dehydrogenases (ACAD) are a family of nuclear-coded, mitochondrial flavoenzymes that catalyse the alpha- and beta-dehydrogenation of fatty acids. The eighth member of this family, ACAD8 catalyses the valine catabolism. In this study, the bovine ACAD8 full-length mRNA and genomic DNA sequence were obtained and its gene structure was determined through alignment of the genomic DNA sequence to the mRNA sequence (Li *et al.*, 2007). One A-G single nucleotide polymorphism (SNP) was revealed at nucleotide 13,408 (GenBank Accession No. DQ435445) in the bovine *ACAD8* gene by sequencing the polymerase chain reaction (PCR) products of 6 randomly selected individuals from the sample population. Different genotypes were determined by restriction fragment length polymorphism (RFLP). The association analysis of this SNP in bovine *ACAD8* with production traits in 178 unrelated steers from 5 breeds showed that it had a significant effect on the daily gain and the beef tenderness ($P < 0.05$). Cattle with the G allele grew more rapidly and the beef they produced was more tender than those with the A allele. Thus, this SNP of the bovine *ACAD8* gene can be used as an indicator to improve the growth rate and the beef tenderness.

Insulin-like growth factor-binding protein-3 (IGFBP-3) is a protein that binds the majority of insulin-like growth factors in circulation for regulation of its action on growth and metabolism of the animals. Animals belonging to Hariana, Holstein-Friesian (HF) and their crossbreds (HF × Hariana) were studied using polymerase chain reaction-restriction fragment length polymorphism and nucleotide sequencing of the *IGFBP-3* gene (Choudhary *et al.*, 2007). A 651-bp fragment of the *IGFBP-3* gene spanning over a part of exon 2, complete intron 2, exon 3 and a part of intron 3 was amplified and digested with *Hae* III restriction enzyme. Three patterns of restriction fragments were observed in HF and crossbred cattle revealing polymorphism in both the populations. The frequency of AA, AB and BB genotypes were 0.65, 0.32 and 0.03 in crossbreds and 0.29, 0.65 and 0.06 in HF, respectively. The allelic frequency of the A and B allele was 0.81 and 0.19 in crossbreds and 0.62 and 0.38 in HF cattle, respectively. Only one restriction pattern (AA genotype) was observed in all the animals of Hariana breed of *Bos indicus* showing the absence of polymorphism. Nucleotide sequencing revealed a C → A mutation in the intron 2 region of the *IGFBP-3* gene as the cause of the polymorphism. Animals of AB genotype showed higher birth weight and body weight than the animals possessing AA genotype.

Associations of DNA polymorphisms in growth hormone (GH) relative to growth and carcass characteristics in growing Brahman steers (N = 324 from 68 sires) were evaluated (Beauchemin *et al.*, 2006). Polymorphisms were an *Msp* I RFLP and a leucine/valine SNP in the *GH* gene as well as a *Hinf* I RFLP and a histidine/arginine SNP in transcriptional regulators of the GH gene, Pit-1 and Prop-1. Genotypic frequencies of the *GH* SNP, Pit-1 RFLP, and Prop-1 SNP were greater than 88% for one of the bi-allelic homozygous genotypes. Genotypic frequencies for the GH *Msp* I RFLP genotypes were more evenly distributed with frequencies of 0.43, 0.42, and 0.15 for the genotypes of +/+, +/–, and –/–, respectively. These polymorphisms within the GH gene and/or its transcriptional regulators do not appear to be informative predictors of growth and carcass characteristics in Brahman steers. This is partly due to the high level of homozygosity of genotypes. These findings do not eliminate the potential importance of these polymorphisms as predictors of growth and carcass traits in *Bos taurus* or *Bos taurus* × *Bos indicus* composite cattle.

In cattle, bovine leucocyte antigens (BoLAs) have been extensively used as markers for bovine diseases and immunological traits. Takeshima *et al.* (2007) developed a rapid, high-resolution sequence-based typing (SBT) system for BoLA-DQA1. They amplified 355 bp of BoLA-DQA1 by fully nested polymerase chain reaction (PCR) using the newly constructed primers and then performed direct sequencing of each product. Using this method, they investigated the locus in 51 animals whose BoLA haplotypes had been characterized at the Fifth International BoLA Workshop. They identified 15 distinct DQA1 alleles, and there was no conflict between the typing result of PCR-SBT and restriction fragment length polymorphism analysis. Together with the previously developed method for typing BoLA-DRB3, the PCR-SBT for BoLA-DQA1 clearly provided a useful tool for detailed class IIa haplotype analysis.

Wojdak-Maksymiec *et al.* (2006) investigated associations between combined defensin genotypes (CDGs) and somatic cell count (SCC) in Jersey cows. The study included a herd of 184 dairy Jersey cows from Wielkopolska region in Poland. Polymerase chain reaction-restriction fragment length polymorphism analysis with *Taq* I restrictase established the existence of 12 CDGs with a frequency of over 1%. The most frequent were A1A2B1B2C1C2 and A1A2B1B2C1- genotypes with a frequency of 56.9783% and 12.5% respectively. The highest SCC (transformed into a logarithmic scale) was found in the milk of cows with A1-B1-C1C2 genotype, whereas the lowest one in cows with A2-B1B2C1C2 genotype. Another aim of the project was to study the association between CDG and milk production traits, such as daily milk yield and fat and protein content. CDG was found to be a significant factor affecting daily milk yield and non-significant for fat and protein content. The highest daily milk yield was observed in cows with CDGs A1A2B1B2C2- and -A2-B1B2C1-, whereas the lowest one was characteristic of -A2-B1-C1C2 and A1A2B1-C1- animals. Fat content was found to be related to CDG genotype in the opposite way; the highest values were recorded in animals with -A2-B1B2C1- genotype, and the lowest in animals with -A2-B1-C1C2 genotype. Similar results were observed in protein content in milk, -A2-B1B2C1- being the highest content and -A2-B1-C1C2, the lowest content. The results confirm the hypothesis of using CDG as an SCC marker.

The use of polymorphic genes as detectable molecular markers is a promising alternative to the current methods of trait selection once these genes are proven to be associated with traits of interest in animals. The point mutations in exon IV of bovine β-lactoglobulin gene determine two allelic variants A and B. These variants were distinguished by polymerase chain reaction and restriction fragment length polymorphism (PCR-RFLP) analysis in two indigenous *Bos indicus* breeds, viz., Sahiwal and Tharparkar cattle (Rachagani *et al.*, 2006). DNA samples (228 in Sahiwal and 86 in Tharparkar) were analysed for allelic variants of β-lactoglobulin gene. Polymorphism was detected by digestion of PCR-amplified products with *Hae* III enzyme, and separation on 12% non-denaturing gels and resolved by silver staining. The allele *B* of β-lactoglobulin occurred at a higher frequency than the allele *A* in both Sahiwal and Tharparkar breeds. The genotypic frequencies of *AA*, *AB*, and *BB* in Sahiwal and Tharparkar breeds were 0.031, 0.276, 0.693 and 0.023, 0.733, 0.244, respectively. Frequencies of *A* and *B* alleles were 0.17 and 0.83, and 0.39 and 0.61 in Sahiwal and Tharparkar breeds, respectively. It was concluded that genotype frequencies of *AA* were the lowest compared to that of *BB* genotype in Sahiwal cattle while *AB* genotypes were more frequent in Tharparkar cattle. The frequency of *A* allele was found to be lower than that of *B* allele in both the breeds studied. These results further confirm that *Bos indicus* cat e are predominantly of β-lactoglobulin B type than *Bos taurus* breeds.

RFLPs in Sheep and Goats

DNA from sheep and goats were used for RFLP studies of globin genes. After hybridization with an ovine cDNA, polymorphic patterns for α-globin were found in sheep and goat DNA. A caprine cDNA of embryonic globin identified a biallelic polymorphism in sheep and goat being associated with β-A and α-B. Variable restriction patterns for the genes of haemoglobin A and B were detected with cDNA probe from rabbit, in order to detect variability in New Zealand sheep breeds (Romney and Merino). Montgomery and Hill (1988) screened these breeds with a human n-myc-oncogenic probe and demonstrated a polymorphic restriction site with the help of *Bg* l II.

Nine sheep breeds or strains, including 615 individuals were screened with forced PCR-RFLP method for the *FecB* gene to study the polymorphism and its effects on litter sizes, body weights and body sizes (Guan *et al.*, 2007). Results showed that the polymorphism frequencies of *FecB* gene were significantly imbalanced in these breeds or strains. The Hu sheep were all homozygous carriers (*BB*). In the Chinese Merino prolific meat strain, the genotype frequencies of *BB*, *B+* and *++* are 51%, 30% and 19%, respectively, whereas all the other flocks had only the wild-type (*++*) genotype. Results within Chinese Merino prolific meat strain showed that mean litter sizes of ewes with genotype *BB* and *B+* are 2.8 (+/–0.74) and 2.3 (+/–0.63) (P > 0.05), whereas *++* ewes had a litter size of only 1.2 (+/–0.68) (P < 0.01). At 90 days after birth, the body weights of *BB/B+* lambs were higher than that of *++* lambs (18.6 +/– 3.70 kg, 18.0 +/– 3.71 kg versus 15.6 +/– 2.22 kg, P < 0.05). In addition, the heart girth and chest width of *BB/B+* lambs were significantly longer than *++* lambs (P < 0.05). In addition to the additive effect on litter size, these findings show for the first time that the *FecB* gene had a positive effect on early postnatal body growth. Figure 9.2 shows the PCR-RFLP for identification of sheep and goat foetuses.

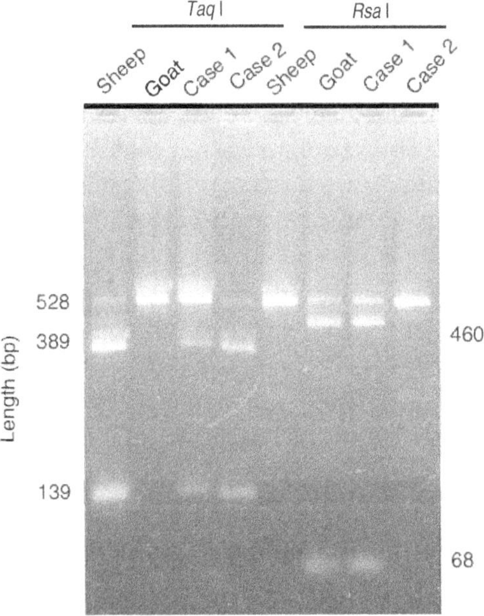

Figure 9.2   RFLP in sheep and goat. Case 1 was a foetus from a goat after a pregnancy of 93 days. Case 2 was sheep lamb suspected to have been sired by a goat.

Muramatsu *et al.* (1992) investigated restriction fragment length polymorphism (RFLP) on the *PrP* gene and the frequencies of RFLP patterns in 35 healthy Suffolk sheep randomly collected. According to the combinations of *PrP* encoding DNA fragments generated by restriction enzymes *Eco* RI and *Hind* III, the RFLP patterns were classified into six types and designated as types I to VI. The frequencies of these types were as follows: I, 8.6%; II, 11.4%; III, 17.6%; IV, 11.4%; V, 28.6%; and VI, 22.9%. In 10 sheep diagnosed as having natural scrapie, RFLP types, I, III, IV, and V were determined. To examine the correlation between the RFLP type and the occurrence of scrapie, the frequencies of RFLP types in sheep infected with natural scrapie were compared with those in healthy sheep. It was found that the frequency of type I in the sheep with natural scrapie was 70%, about eight times higher than that in randomly collected healthy sheep.

Animal fibres are highly valuable industrial products often adulterated during marketing. Currently, there is no precise method available to identify and differentiate the fibres. In this study, Subramanian *et al.* (2005) used PCR-RFLP technique to differentiate cashmere wool fibres derived from goat and sheep, respectively. The presence of DNA in animal hair shafts enabled the isolation of DNA from scoured cashmere wool fibres. The mitochondrial cytochrome *b* sequences of both species were amplified by PCR using primers designed from conserved regions. The polymorphism observed between the two species was detected by restricting the amplified product by endonucleases, viz., *Bam* HI and *Ssp* 1. The RFLP profile clearly distinguishes the cashmere wool fibres and this technique can also be exploited to test adulteration in animal fibres qualitatively.

Goat milk proteins show extensive qualitative and quantitative polymorphisms of α-S_1-caseins, controlled by at least seven autosomal alleles. Alleles α-S_1-$C_{n,A,B,C}$ are associated with a high α-S_1 casein content. RFLP was used to characterize α–S_1–C_n alleles. This procedure enables to identify alleles, i.e., α-S_1-$C_{n,A,B,C}$ which are associated with high level of protein synthesis. Goat milk protein α-S_1 casein shows extensive qualitative and quantitative polymorphisms.

RFLPs in Pigs

For MHC genes, polymeric patterns with 7–10 bands were found using human class 1 cDNA probe (Chardon *et al.*, 1985). While studying RFLPs and linkage for complement C4, C2 and factors B, Lie *et al.* (1987) found polymorphisms for C4 and B1 but not for C2. RFLPs in β-globin locus of the pigs were studied by a cDNA probe from rabbit and the enzyme *Hind* III was used. From the results of family studies 3 allelic forms of the gene were provided based on point mutation.

Polymorphisms to glucose phosphate isomerase gene locus were presented by Dav *et al.* (1987, 1988). In this study, *Pvu* II and *Sac* I were used with a porcine cDNA probe. This gene is of interest in animal breeding because there is a linkage to the Halothane locus and correlation to the porcine stress syndrome.

Kennes *et al.* (2001) characterized four polymorphisms in swine leptin (*LEP*) gene and evaluated for association with economically important production traits in Yorkshire, Landrace and Duroc pigs. Results showed that these polymorphisms were generally of low frequency. Two polymorphisms (A2845T and T3469C) might have been associated with feed intake and growth rate traits in Landrace pigs.

RFLPs in Horses

A synthetic polynucleotide $(TG)_n$ was used to get DNA fingerprint in horses to characterize different breeds (Ellegren *et al.*, 1992a). Similar probes were used for paternity testing in the horses. PCR-based paternity testing systems have been developed. The microsatellites could easily be amplified from a single hair. A standardized paternity testing system based on PCR-analysed microsatellites is likely to be a future alternative to blood typing horses (Ellegren *et al.*, 1992).

Alexander *et al.* (1987) described polymorphisms in class I and class II genes of the MHC. Human cDNA probes were used for hybridization. It was also found that in the horse genome, more than one gene is present for DR β and also for DQ α. RFLPs for globin genes using a cDNA probe from rabbits, were identified.

RFLPs in Poultry

In chicken, there is only little information on RFLPs. Polymorphism in the class I gene region were published by Andersson *et al.* (1987). They showed RFLPs by use of human cDNA probes.

Human minisatellite probe 33.6, cross-hybridizes to multiple variable loci in poultry DNA and produces highly informative DNA fingerprints. The average number of bands detected by probe 33.6 in the different birds analysed ranged between 17 bands in leghorns to 21 in ducks (Hillel *et al.*, 1989). These ranges are similar to the figures reported for humans, dogs, mice and wild birds but higher than in cats. The frequency of large fragments in poultry DNA fingerprints is higher than in human and other animals. Thus, in all poultry types examined (with exception of duck), over 40% of the bands detected were larger than 10 kb compared with 20% in human, 13% in dogs and none in cats. It is possible that the minisatellites in chicken might be located in clusters and therefore appear in each DNA fragment. The total number of detectable loci in broilers, using probe 33.6 was estimated as 62, of which 13 loci are available for use. Poultry fingerprints can be used for individual identification, linkage analysis and as an aid in breeding programmes, for either elimination of unwanted genome or for selection of desired one.

β-complex was identified as the MHC in chickens. There are 3 classes of highly polymorphic genes encoded in β-complex, designated β-F, β-L and β-G or class I, II and IV, respectively. The β-complex has been associated with resistance to many diseases in chicken including Marek's disease and fowl cholera and also with some important economic traits. Resistance to these two diseases are associated with β haplotype in the S_1 line (Chen and Lamont, 1992).

Alleles of the growth hormone (GH) and GH receptor (GHR) gene were analysed for association with juvenile body weight (HBWT), age at first egg (AFE), the hen-day rate (HDR) of egg production, egg specific gravity (SPG) and egg weight (EWT) in a strain of white leghorns (Feng *et al.*, 1997). The GH genotype was significantly associated with AFE as well as HDR. The GHR genotype had trends associated with HBW and HDR. Regression analysis revealed that HDR was associated negatively with AFE and positively with HBWT.

The chicken growth hormone (cGH) gene plays a crucial role in controlling growth and metabolism, leading to potential correlations between cGH polymorphisms and economic traits. Nie *et al.* (2005) screened DNA from four divergent chicken breeds for single nucleotide polymorphisms (SNPs) in the cGH gene using denaturing high-performance liquid chromatography and sequencing. A total of 46 SNPs were identified, of which 4 were in the 5´ untranslated region, 1 in the 3´ untranslated region, 5 in exons (two of which are non-synonymous), with the remaining 36 in introns. The nucleotide diversity in the cGH gene (theta = $2.7 \times 10(-3)$) was higher than that reported for other chicken genes, even within the same breeds. The association of five of these SNPs and their haplotypes with chicken growth and carcass traits were determined using polymerase chain reaction-restriction fragment length polymorphism analysis in an F2 resource population cross of two of the four chicken breeds (White, Recessive, Rock and Xinghua). This analysis shows that, among other correlations, G+1705A was significantly associated with body weight at all ages measured, shank length at three of four ages measured, and average daily gain within weeks 0 to 4. Thus, this cGH polymorphism, or another polymorphism that is in linkage disequilibrium with G+1705A, appears to correspond to a significant growth-related quantitative trait locus difference between the two breeds used to construct the resource population.

The insulin-like growth factor I (IFG-I) gene was screened for genetic variants associated with trait means and trait correlations (Nagaraja *et al.*, 2000). Analysis of an unselected randomly mated white leghorn population revealed a *Pst* I restriction fragment length polymorphism in the 5´ region of the gene which segregated at a frequency of 0.83 for the *Pst* I (+) allele (presence of a *Pst* I restriction site). A comparison of the three genotypic classes revealed that the *Pst* I (–/–) genotype was associated with a significantly lower egg weight, while the *Pst* I (+/–) genotype was significantly associated with a higher eggshell weight estimated from the egg weight and egg specific gravity.

Chicken growth hormone (cGH), a polypeptide hormone synthesized in and secreted by the pituitary gland, is involved in a wide variety of physiological functions such as growth, body composition, egg production, aging and reproduction, Ip *et al.* (2001) investigated the GH gene polymorphism in selected strains of native Chinese chickens. The authors reported that this polymorphism of the cGH gene might be useful for phylogenetic analysis, as well as in the design of breeding programmes.

RFLPs in Dogs and Cats

Human minisatellite probes 33.6 and 33.15 cross-hybridize to multiple dispersed autosomal loci in both cat and dog DNA to produce highly informative DNA fingerprints. The overall properties of these Southern blot patterns are similar to their human counterparts, except for the lower number of variability of DNA fragments detected in both cats and dogs. These DNA fingerprints can be used directly for animal identification and for example, the control of verification of semen used in artificial insemination. DNA fingerprints can be used to establish parentage. In a typical dog mating, each offspring will contain around 9 different maternal specific bands and a similar number of paternal specific fragments. In establishing paternity, for example, all paternal specific bands in an offspring should be present in the father. If paternity is incorrect, then approximately 5 offspring fragments should not be present in the putative father, producing multiple exclusions. DNA fingerprints therefore provide a useful method for establishing parent–offspring relationships as in human (Jeffreys and Morten, 1987).

DNA fingerprints with probes, 33, 15 and α–globin 3´ HVR were used to resolve 3 cases of disputed paternity in dogs. In a litter of Rhodesian Ridgebacks, 12 DNA bands were informative to establish paternity. In a second case of a litter of Afghan Hounds, 5 DNA bands established paternity. Lately, in a litter of Border Collies, 5 bands established paternity. Figure 9.3 shows RAPD profiles of individual animals representing four candidae species showing differences between different animals.

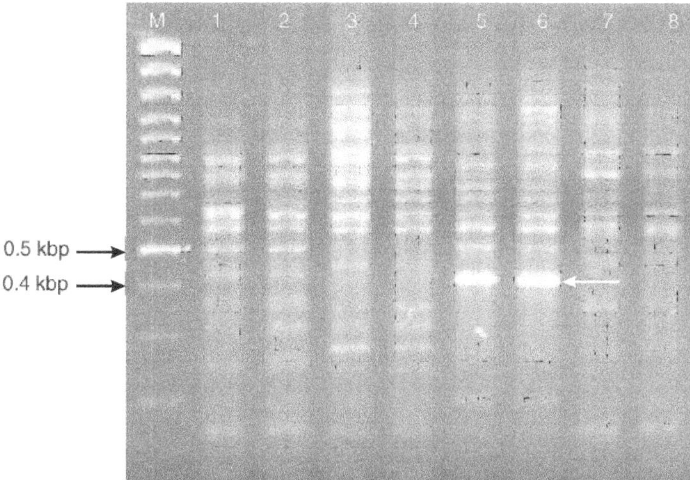

Figure 9.3 RAPD profiles of individual animals representing four Canidae species (amplified by primer M-9). Lanes: M=100-bp DNA Ladder plus (gene ruler TM); 1, 2= arctic fox; 3, 4= red fox; 5, 6 = chinese raccoon dog and 7, 8 = domestic dog.

(Courtesy: Stepniak et al. 2002. *J Appl. Genet.* 43: 489–499).

MAJOR HISTOCOMPATIBILITY COMPLEX (MHC) IN CATTLE

The MHC was discovered about 50 years ago as a major locus controlling the rejection of foreign tissue grafts (transplantation) in the mouse. The molecules encoded by the MHC therefore have been referred to as transplantation antigens. The MHC in man, designated HLA was discovered about 30 years ago and detected in lymphocytes. There are two classes of MHC molecules, denoted class I and class II, which are both integral cell-surface glycoproteins. The class I molecule is found on virtually all nucleated somatic cells. The class II molecules are primarily expressed in B lymphocytes and macrophages.

The MHC system in cattle is known as bovine lymphocyte antigens system (BoLA system). BoLA has been assigned to chromosome 23 in cattle (HLA to chromosome 6). At least 10 bovine genes which are comparable to the number of human class I genes have been isolated. The class II region harbours the genes encoding the α and β subunits of class II molecules. In bovine class II regions, DQ α, DQ β, DR β, DO β, DZ α and DY β genes were identified. In cattle, C4 and Z10H are the only genes which have been studied in class III region.

The antibody response to human serum albumin and to the polypeptide (T, G)-A-L in young bulls is reported to be MHC-linked. Similarly, the immune response to ovalbumin has also been found to be MHC-linked. Extensive studies have been carried out on the immunological function of BoLA in cell-mediated immune response of cattle to *Theileria parva*. *T. parva* is a protozoan parasite infecting bovine lymphocytes and causing an acute lymphoproliferative disease known as East Coast fever. It has been clearly shown that anti-theileria cytotoxic T cells are restricted by class I MHC molecules.

Lewin *et al.* (1988) reported that the subclinical progression of bovine leukaemia virus infection varies between animals carrying different BoLA types. BoLA polymorphism has been weakly associated with susceptibility to the level of infestation with the tick *Boophilus microplus*. Weak associations between the incidence of mastitis and BoLA-A (Solbu *et al.*, 1984) and BoLA-DQ polymorphisms have been reported.

REVIEW QUESTIONS

1. Write an essay on different applications of DNA fingerprinting in domestic animals?

2. What do you mean by RFLP? How is it useful to characterize different animals?

REFERENCES

Alexander, A.J., Baily, E. and Woodward, J.G. (1987). "Analysis of the equine lymphocyte antigen system by Southern blot hybridization." *Immunogenetics.* 25: 47–54.

Andersson, L., Lundberg, C., Rask, L., Gissel-Nielsen, B. and Simonsen, M. (1987). "Analysis of class II genes of the chicken MHC (B) by use of human DNA probes." *Immunogenetics.* 26: 79–84.

Atkinson, L.R., Loung, D.W. and Womack, J.E. (1988). "Somatic cell mapping and restriction fragment analysis of bovine firbonectin and γ-crystalline." *Cytogenet. Cell Genet.* 47: 155–159.

Beauchemin, V.R., Thomas, M.G., Franke, D.E. and Silver, G.A. (2006). "Evaluation of DNA polymorphisms involving growth hormone relative to growth and carcass characteristics in Brahman steers." *Genet. Mol. Res.* 5:438–447.

Beckman, J.S., Kashi, Y., Hallermann, E.M., Nave, A. and Soller, M. (1986). "Restriction fragment length polymorphism among Israeli Holstein Friesian dairy bulls." *Anim. Genet.* 17: 25–38.

Bravi, C.M., Liron, J.P., Mirol, P.M., Ripoli, M.V., Peral-Garcia, P. and Giovambattista, G. (2004). "A simple method for domestic animal identification in Argentina using PCR-RFLP analysis of cytochrome β gene." *Leg. Med. (Tokyo).* 6:246–51.

Broad, T. (1988). "Biotechnology. A major contributor to NZ's future growth." DSIR Report. Palmerston North, New Zealand.

Chardon, P.C. Renald, M., Kirszembaum, P.R., Culien, C., Geffrotin, D., Cohen and Vaiman, M. (1985). "Molecular genetic analysis of the major histocompatibility complex in pig families and ruminants". *J. Immunogenet.* 12: 139–149.

Chen, Y. and Lamont, S.J. (1992). "Major histocompatibility complex Class I. RFLP analysis in highly inbred chicken lines and lines selected for MHC and immunoglobulin production." *Poultry Sci.* 71: 999–1006.

Choudhary, V. Kumar, P., Bhattacharya, T.K., Bhushan, B., Sharma, A. and Shukla, A. (2007). "DNA polymorphism of insulin-like growth factor-binding protein-3 gene and its association with birth weight and body weight in cattle." *J. Anim. Breed. Genet.* 124:29–34.

Copplieters, W., van de Weghe, A., Depciker, A., Bouquest, Y. and van Zevergen, A. (1990). "A hypervariable pig DNA fragment." *Anim. Genet.* 21: 29–38.

Davies, W., Harbitz, I. and Hange, J.G. (1987). "A partial cDNA clone for porcine glucose phosphate isomerase: isolation, characterization and use in detection of restriction fragment length polymorphisms." *Anim. Genet.* 18: 283–240.

Davies, W., Harbitz, I., Fries, R., Stranzinger G. and Hauge, J.G. (1988). "Porcine malignant hyperthermia carrier detection and chromosomal assignment using a linked probe." *Anim. Genet.* 19: 203–212.

Eichmann, C., Berger, B., Reinhold, M., Lutz, M. and Parson, W. (2004). "Canine-specific STR typing of saliva traces on dog bite wounds." *Int. J. Legal Med.* 118:337–342.

Ellegren, H., Andersson, L., Johanson, M. and Sandberg, K. (1992a). "DNA fingerprinting in horses using a simple $(TG)_n$ probe and its application to population comparisons." *Anim. Genet.* 23: 133–142.

Ellegren, H., Johanson, M. Sandberg, K. and Andersson, L. (1992b). "Cloning of highly polymorphic microsatellites in the horse." *Anim. Genet.* 23: 133–142.

Feng, X.P., Kuhnlein, U., Aggrey, S.E., Gavora, J.S. and Zadworny, D. (1997). "Trait association of genetic markers in the growth hormone and the growth hormone receptor gene in a white leghorn strain." *Poult. Sci.* 76: 1770–1775.

Georges, M., Lequarre, A.S., Castelli, M., Hanset, R. and Vassart, G. (1988). "DNA fingerprints in domestic animals using four different minisatellite probes." *Cytogenet. Cell Genet.* 47: 127–131.

Georges, M., Lathrop, M., Hilbert, P., Marcotte, A., Schwers, A., Swillens, S., Vassart, G. and Hanset, R. (1990). "In the use of DNA fingerprints for linkage studies in cattle." *Genomics* 6: 461–474.

Grosclaude, F., Mahe, M.F., Brignon, G., Distasio, L. and Jeunet, R. (1987). "A Mendelian polymorphism underlying quantitative variations of goat α-S$_1$-casein." "*Evolution* 19: 399–412.

Guan, F., Liu, S.R., Shi, G.Q. and Yang, L.G. (2007). "Polymorphism of *FecB* gene in nine sheep breeds or strains and its effects on litter size, lamb growth and development." *Anim. Reprod. Sci.* 99:44–52.

Gyllenstein, V.B., Jakobsson, S., Temrin, H. and Wilson, W.C. (1989). "Nucleotide sequence and genomic organization of bird minisatellites." *Nucleic Acids Res*: 17: 2203–2214.

Hanset, R. (1986). In: *Exploiting New Technologies in Animal Breeding.* (eds.). Smith, C., King, J.W.B. and McKay, J. Oxford University Press, Oxford. pp. 71–80.

Hellmann, A.P., Rohleder, U., Eichmann, C., Pfeiffer, I., Parson, W. and Schleenbecker, U. (2006). "A proposal for standardization in forensic canine DNA typing: allele nomenclature of six canine-specific STR loci." *J. Forensic Sci.* 51:274–81.

Hillel, J., Plotzky, Y., Haberford, A., Lavi, U., Cahaher, A. and Jeffreys, A.J. (1989). "DNA fingerprints of poultry." *J. Anim. Genet.* 20: 145–155.

Ip, S.C., Zhang, X. and Leung, F.C. (2001). "Genomic growth hormone gene polymorphisms in native Chinese chickens." *Exp. Biol. Med.* (Maywood). 26: 458–462.

Jeffreys, A.J. and Morton, D.B. (1987). "DNA fingerprints of dogs and cats." *Anim. Genet.* 18: 1–15.

Jeffreys, A.J., Wilson, V. and Thein, S.L. (1985a). "Hypervariable 'minisatellite' regions in human DNA." *Nature.* 314: 67–73.

Jeffreys, A.J., Wilson, V. and Thein, S.L. (1985b). "Individual specific 'fingerprints' of human DNA." *Nature.* 316: 76–79.

Jeffreys, A.J., Brooksfield, J.F.Y. and Semeonoff, R. (1985c). "Positive identification of an immigration test case using human DNA fingerprints." *Nature.* 317: 818–819.

Jeffreys, A.J., Brooksfield, J.F.Y. and Semenoff, R. (1988). "DNA fingerprint analysis in immigration test cases." *Nature.* 322: 290–291.

Kashi, Y., Tickochinsky, Y., Genislav, E., Iraqi, F., Nave, A. and Beckman, J.S. (1990). "Large restriction fragments containing poly-TG are highly polymorphic in a variety of vertebrates." *Nucleic Acid Res.* 18: 1129–1132.

Kennes, Y.M., Murphy, B.D. Pothier, F. and Palin, M.F. (2001). "Characterization of swine letpin (LEP) polymorphisms and their association with production traits." *Anim. Genet.* 32: 215–218.

Kuhnlein, U., Dawe, Y., Zadworny, C. and Gavora, J.S. (1989). "DNA fingerprinting: A tool for determining genetic distances between strains of poultry." *Theor. Appl. Genet.* 77: 669–672.

Lechniak, D., Machnik, G., Szydlowski, M. and Switonski, M. (1999). "Growth hormone gene polymorphism and reproductive performance of AI bulls." *Theriogenology*. 52: 1145–1152.

Lewin, H.A., Wu, M.C., Stewart, J.A. and Nolan, T.J. (1988). "Association between BoLA and subclinical bovine leukemia virus infection in a herd of Holstein Friesian cows." *Immunogenetics*. 27: 388–344.

Lie, W.R., Rothschild, M.F. and Warner, C.M. (1987). "Mapping of C_2 bf and C_4 genes to the swine major histocompatability complex (Swine leukocyte antigens)." *J. Immunol.* 139: 3388–3395.

Marner, C.M., Rothschild, M.F. and Lamont, S.J. (1988). (eds.). *"The Molecular Biology of the Major Histocompatibility Complex of Domestic Animal Species."* Iowa State University Press, Ames, USA.

Menotti-Raymond, M.A., David, V.A., Wachter, L.L. Butler, J.M. and O'Brien, S.J. (2005). "An STR forensic typing system for genetic individualization of domestic cat (*Felis catus*) samples." *J. Forensic Sci.* 50:1061–1070.

Montgomery, G.W. and Hill, D.G. (1988). "DNA variation in New Zealand Romney and Merion sheep." XXIst International Conference on Animal Blood Groups and Biochemial Polymorphisms. Turin. July. 4–8.

Muramatsu, Y., Tanaka, K., Horiuchi, M., Ishiguro, N., Shinagawa, M., Matsui, T. and T. Onodera. (1992). "A specific RFLP type associated with the occurrence of sheep scrapie in Japan." *Archives of Virology*. 127: 1–9.

Nagaraja, S.C., Aggrey, S.E., Yao, J., Zadworny, D., Fairfull, R.W. and Kuhnlein, U. (2000). "Trait association of a genetic marker near the IGF-I gene in egg-laying chickens." *J. Herald.* 91: 150–156.

Nakamura, Y., Leppert, M., O'Connell, P., Wolff, R., Holm, T., Culver, M., Martin, C., Fujimoto, E., Hoff, M., Kumlin, E. and White, R. (1987). "Variable number of tandem repeat (VNTR) markers for human gene mapping." *Science.* 235: 1616–1622.

Nie, Q., Sun, B., Zhang, D., Luo, C., Ishag, N.A., Lei, M., Yang, G. and Zhang, X. (2005). "High diversity of the chicken growth hormone gene and effects on growth and carcass traits." *J. Hered.* 96:698–703.

Parret, J., Shia, Y.C., Fries, R., Vassart, G. and Georges, M. (1990). "A polymorphic satellite sequence maps to the pericentric region of bovine Y chromosome." *Genomics*. 6: 482–490.

Piper, L.R., Bindon, B.M. and Davis, G.H. (1985). "The single gene inheritance of the higher litter size in the Booroola Merino." In: *Genetics of Reproduction in Sheep*. (eds.). Land, R.B., Robinson, D.W. and Butterworths. London. pp. 115–126.

Rachagani, S., Gupta, I.D., Gupta, N. and Gupta, S.C. (2006). "Genotyping of beta-lactoglobulin gene by PCR-RFLP in Sahiwal and Tharparkar cattle breeds." *BMC Genet.* 7: 31.

Rando, A., Di Gregorio, P. and Masina, P.M. (1988). "Identification of bovine κ-casein genotypes at the DNA level." *Anim. Gene.* 19: 51–54.

Subramanian, S., Karthik, T. and Vijayaraaghavan, N.N. (2005). "Single nucleotide polymorphism for animal fibre identification." *J. Biotechnol.* 116:153-8.

Takeshima, S., Miki, A., Kado, M. and Aida, Y. (2007). "Establishment of a sequence-based typing system for BoLA-DQA1 exon 2." *Tissue Antigens.* 69:189–199.

Thielmann, J.L., Skow, L.C. Baker, J.F. and Womack, J.E. (1989). "Restriction fragment length polymorphisms for growth hormone: Prolactin, ostenectin, α-crystalline, γ-crystalline, fibronectin and 21-steroid droxylase in cattle." *Ani. Genet.* 20: 257–266.

Vassart, G., Georges, M., Monsieru, R., Brocas, H., Lequarre, A.S. and Cristiphe, D. (1987). "A sequence of M13 phage detects hypervariable minisatellites in human and animal DNA." *Science.* 235: 683–684.

Webb, A.J., Carden, A.E., Smith, C. and Imlah, P. (1982). "Porcine stress syndrome in pig breeding." Proc. of the 12th World Congress Applied to Livestock Production, Madrid V. pp. 41–54.

Wojdak-Maksymiec, K., Kmiec, M. and Zukiewicz, A. (2006). "Associations between defensin polymorphism and somatic cell count in milk and milk utility traits in Jersey dairy cows." *J. Vet. Med. A Physiol. Pathol. Clin. Med.* 53:495–500.

10

BIOTECHNOLOGY IN ANIMAL PRODUCTION

MANIPULATION OF GROWTH

Growth means increase of cell number, cell size and/or deposition of substances within cells. Individual growth successively focuses on development of the neural system, bones, muscles and fat. Animal productions are anabolic processes requiring nutrients above maintenance level, which may be provided by feed intake and by body resources such as lipid stores, glycogen and unstable proteins.

Insulin

Insulin is an anabolic hormone and it enhances glucose uptake in muscle and adipose tissues, fat synthesis either by provision of glucose and/or acetate utilization in ruminants and protein synthesis by facilitating uptake of amino acids and inhibiting proteolysis in muscles. Insulin also regulates somatotropin (STH) receptors in liver, hence it may act presumably for release of somatomedins. Insulin exerts a short feedback with the blood sugar level. Feed intake provokes insulin release, restricted feeding—the simplest measure to avoid fattening—depresses it, while starvation produces diabetes-like symptoms. On the contrary, glucagons induce hyperglycaemia by glycogenolysis in the liver and gluconeogenesis. It is postprandially released concomitantly with insulin, obviously to prevent the insulin-induced hypoglycaemia.

Growth Hormone Releasing Hormone (GHRH), Somatotropin (STH), Growth Hormone (GH) and Somatomedins

Somatomedins are hormones produced in the liver when stimulated by STH. High levels of somatomedin have been found in faster growing breeds of pigs, sheep, cattle and chicken; in bulls a parallel increase in STH concentration was shown during pubertal growth between 8–11 months of age. Contrary to insulin, STH concentration is elevated during restricted feeding resulting in lipolysis and glucose mobilization. The STH supplementation will have a beneficial effect to compensate for shortage of feed.

The ability of recombinant bovine somatotropin (BST) to enhance milk production is well established in cows and in other dairy ruminants. In dairy ewes, milk yield has been increased by 20–30% following treatment with BST (Baldi, 1999), without affecting the gross composition or coagulating properties of milk.

The effect of a transgene encoding ovine growth hormone and regulated by a metallothionein promoter was examined in the progeny of 69 Merino ewes and 49 Poll Dorset ewes that were inseminated by rams heterozygous for the gene construct (Adams *et al.*, 2002). The presence of the transgene had no effect on the progeny from one of the three rams used, as evinced by a normal concentration and secretion pattern of growth hormone and normal growth rate and fatness. In progeny from the other two rams that bore an actively transcribed and translated copy of the transgene, the mean concentration of growth hormone in the plasma was twice that of controls, but the pulsatility of secretion was lost. These animals grew faster (P < 0.001) and were leaner (P < 0.001), but had a greater parasite faecal egg count (P < 0.001). The impact of the transgene differed between breeds with greater wool growth rate (P < 0.01) and live weight increase (P = 0.06) in Merino progeny compared with Poll Dorset cross. The results indicate that the production effects of genetic manipulation may depend on the age, the breed, and the sex of the animal.

GHRH administration stimulates insulin secretion *in vitro* and releases STH in cattle. Frequent administration of STH preparations have produced anabolic responses, accompanied by reduced adipose tissue. The human pancreatic hpGHRH has been demonstrated to release STH in cattle. Depression of STH concentration has been observed following an excess of dietary proteins and after administration of somatostatin which is physiologically released from the hypothalamus and the pancreas. When a recombinant porcine STH was given at a dose of 140 µg/kg bwt, there was a doubling increase in the area of M. *longissimus dorsi*. The fat content was decreased linearly to a minimum of 25% with increasing doses of STH (Etherton *et al.*, 1986).

The direct transfer of recombinant DNA to embryos is conceptually a powerful method for the manipulation of the genetic potential of domestic animals, but in practice the technology has yet to fulfil its promise. Modification of the growth hormone status of sheep by the use of the ovine growth hormone gene has been achieved (Ward and Brown, 1998). This was under the regulation of an ovine metallothionein promoter. This approach has increased the growth rate of sheep while maintaining the animals in good health.

Somatostatin (Somatotropin Release Inhibiting Factor, SRIF)

Somatostatin, isolated from the hypothalamus inhibits STH release from the pituitary and hence antagonise the effect of GHRH. It causes

- a reduction of STH levels;
- a reduction of insulin;
- a reduction in glucagon secretion and
- suppressive effect on absorption of several nutrients by the gut.

Immunization of cattle against somatostatin reveals a 17.6% increase of daily weight gain accompanied by a reduction of food intake by 13% (Lawrence *et al.*, 1986).

Thyroid Hormones

Thyroid hormones exert a direct effect on the cell nucleus; they are essential growth hormones for several tissues including the CNS and bones. Thyroid function-supporting feed additives are iodinated casein and thyroprotein. Some positive results have been reported in veal calves, pigs, chicken and lambs. In mature cattle, the depression of the thyroid function by thyrostatics (methyl- or propyl-thiouracil) resulted in remarkable improvement of body weight gain during a four-week period.

Reproductive Steroids
(Androgens, Oestrogens, Gestagens)

Steroid hormones exert their biological effects on the target tissues by binding to specific cytosolic hormone receptors. The hormone–receptor complex then interacts with DNA in the cell nucleus initiating increased RNA synthesis followed by an increased protein synthesis and specific hormonal effects.

Castration changes the growth and fattening performance entirely. Weight gain and feed conversion ratios are reduced. Growth rates in bulls, especially around 8–10 months of age, were positively correlated to the response in plasma testosterone to choriogonadotropin injection. Exogenously administered reproductive hormones act anabolically. The mode of action of androgens directly on muscle cells is due to their having specific receptors within the cytoplasm. A similar mode of action is suggested for oestrogens. Indirect anabolic effects are probably mediated by stimulation of STH.

Use of anabolics have brought about improvement in the range of 10–15% true anabolic weight gain. A positive effect on feed conversion and in general no adverse effect on meat quality were shown.

The anabolic agents have two possible modes of action:

1. they enter the muscle cells and have a direct action on protein synthesis and/or degradation;

2. they enter other endocrine organs such as hypothalamus, anterior pituitary, gonads, pancreas and thyroids and alter the synthesis, metabolism or secretion of other hormones. These hormones may then produce the anabolic effect in muscle and affect intermediary metabolism in other parts of the body including fat and liver.

The administration of anabolic agents has been found to bring about a change in the concentration of circulating endogenous hormones. For example, oestradiol injection increases the concentration of circulating plasma growth hormone.

It is well established that ovulation rate can also be influenced by exogenous hormone administration and by environmental factors such as nutrition (Hunter *et al.*, 2004). It has become apparent that these nutritional effects are mediated by a direct action at the level of the ovary, involving insulin, insulin-like growth factors (IGF) I and II and their binding proteins among other factors. These factors can also affect the quality of the oocyte and consequently embryo development and survival. Recently, the regulation of follicular angiogenesis has been shown to be important for the development of ovulatory follicles, particularly vascular endothelial growth factor (VEGF) which is produced primarily by the granulosa cells within the ovary and can be stimulated by gonadotrophins. Administration of VEGF has been shown to stimulate pre-antral follicular growth and increase the number of pre-ovulatory follicles. Manipulation of the angiogenic process may also provide new opportunities for regulating the quality and number of follicles that ovulate.

Ovarian follicular dynamics in *B. indicus* cattle is characterized by the occurrence of two, three or sometimes four waves of follicular development. While dominance is similar to that in *B. taurus* cattle, maximum diameters of the dominant follicle and CL are smaller than those reported in *B. taurus* and are probably due to a lower capacity for LH secretion than in *B. taurus*. Duration of oestrus is approximately 10 hours and the interval from oestrus to ovulation is about 27 hours. However, the variability in response to prostaglandin F2alpha (PGF) treatments and the difficulty for oestrus detection in *B. indicus* cattle have limited the widespread application of artificial insemination (AI) and emphasizes the need for treatments that control follicular development and ovulation (Bo *et al.* 2003). Follicular-wave development in *B. indicus* cattle can be controlled mechanically by ultrasound-guided follicle ablation, or hormonally by treatments with GnRH or oestradiol and progestogen/progesterone in combination. Treatments with GnRH plus PGF and a second GnRH (synchronization protocol known as Ovsynch) or oestradiol benzoate (known as GPE) have resulted in acceptable pregnancy rates after fixed-time AI (FTAI) in cycling cows, but results were lower in heifers and cows in post-partum anoestrus. Alternatively, treatments with oestradiol and progestogen/progesterone-releasing devices resulted in synchronous emergence of a new follicular wave, and a second oestradiol or GnRH treatment after device removal resulted in synchronous ovulation and acceptable pregnancy rates to FTAI. Furthermore, oestradiol and progesterone treatments combined with eCG (given at the time of device removal) increased pregnancy rates in suckled *B. indicus* cows and may be useful for the treatment of cows in post-partum anoestrus. In summary, exogenous control of luteal and follicular development facilitates the application of assisted reproductive technologies in *B. indicus* cattle by offering the possibility of planning AI programs without the necessity of oestrus detection and without sacrificing the overall results.

The wave-like patterns of ovarian follicular development in cattle can be manipulated by shortening the luteal phase with prostaglandin F2alpha (PGF), lengthening the period of follicle dominance with progesterone or curtailing follicle development with GnRH or oestradiol as 17 beta, benzoate or cypionate (Macmillan *et al.*, 2003). These hormones can also be used to synchronize ovulation allowing timed inseminations without detected oestrus. Progesterone, PGF, GnRH and oestradiol benzoate have each been used to increase conception rates in some situations, but their use has reduced them in others. For example, inseminations made within 96 hours of a single injection of PGF administered during the luteal phase were associated with increased conception rates in dairy cows whereas double injection protocols reduced conception rates. The three forms of oestradiol and GnRH have greater effects on follicular development following divergence and dominance than following wave emergence. This can mean that follicles of differing maturity will be present about 7 days later and can result in varied intervals to the onset of oestrus following a PGF injection. The consequent variation in ovulation time can be reduced by injecting GnRH or an oestradiol during pro-oestrus.

Programmed management of follicle growth, regression of the CL and induction of ovulation led to development of the ovulation synchronization (Ovsynch) program (Thatcher and Santos, 2007). Pre-synchronization of oestrous cycles followed 12 to 14 days later with the Ovsynch program increased pregnancy rates to timed inseminations. Initiation of the Ovsynch program on day 3 of the oestrous cycle reduced ovulation to GnRH and resulted in a smaller proportion of excellent and good quality embryos following timed insemination. The pregnancy rate to a timed insemination of Ovsynch was greater when cows were ovulated to the first injection of GnRH. Treatment with hCG at day 5 after insemination increased pregnancy rate in lactating dairy cows. Injection of bovine somatotropin at insemination increased pregnancy rate, conceptus length and interferon-tau content in uterine luminal flushings and altered endometrial gene expression at day 17 of pregnancy. During heat stress, timed embryo transfer increased pregnancy rate and using embryos cultured with IGF-I and transferred fresh resulted in a greater pregnancy rate. Induction of ovulation with oestradiol cypionate, as a component of a timed insemination programme, increased fertility. Manipulation of the oestrous cycle to improve follicle/oocyte competence and management of the post-ovulatory dialogue between embryonic and uterine tissues should enhance embryo development and survival.

Probiotics as Growth Promoters

The term "probiotic" originated from two Greek words meaning "for life" and contrasted with the term antibiotic which means "against life". It is defined as a live microbial feed supplement, which beneficially affects the host animals by improving its intestinal microbial balance. The live cultures and natural and beneficial bacilli can be introduced into animal gut in massive numbers to supplement those already present. The increasing level of beneficial organisms in the gut crowd out the existing pathogens and as they proliferate, ensure that there is no room for others to be established.

The probiotic preparations available in the market are generally composed of organisms of lactobacilli and/or streptococci species. Few may contain *Bifidobacterium* and yeast cultures. These products are viable, freeze-dried or microencapsulated which proliferate and establish in large numbers in the gastrointestinal tract particularly in the small intestine.

Probiotic preparations benefit the host by:

i.   having direct antagonistic effect against specific groups of undesirable or harmful organisms through production of antibacterial compounds, and eliminating or minimizing their competition for nutrients;

ii. altering the pattern of microbial metabolism in the gastrointestinal tract;

iii. stimulation of immunity and

iv. neutralization of enterotoxin formed by pathogenic organisms.

Probiotics help to favourably influence the growth rate and feed efficiency, and prevent intestinal infections. The use of bacterial cultures is increasingly becoming popular in poultry and pig feeds, while the use of yeast cultures is gaining importance in dairy cattle rations.

The normal bacterial flora in the alimentary tract of pigs and birds are species of *Lactobacillus, Streptococcus, E. coli* and *Clostridium welchii*, in the descending order of concentration. The lactobacilli are the most common ones which include. *L. acidophilus, L. bulgaricus, L. plantarum, L. casei* and *Streptococcus cremoris* and *S. lactis*. The yeast cultures used in dairy cattle ration is *Saccharomyces cerevisiae*.

Administration of probiotics to the pregnant and lactating mother increased the immunoprotective potential of breast milk, as assessed by the amount of anti-inflammatory transforming growth factor β-2 (TGF-β2) in the milk (Rautave *et al.*, 2002). The risk of developing atopic eczema during the first 2 years of life in infants whose mothers received probiotics was significantly reduced in comparison with controls.

Ideal characteristics of a probiotic

1. It should be non-pathogenic to animals and man.
2. It should be gram-positive since gram-positive organisms are more resistant to destruction by digestive enzymes.
3. It should have high tolerance for bile and low pH.
4. It should be capable of producing lactic acid.
5. It should easily proliferate under *in vivo* and *in vitro* conditions.
6. It should have high survival rate after passing through manufacturing processes.
7. It should possess high stability at room temperature and long life when mixed with feed.

Mode of action of probiotics The normal gut microflora has a supportive role in disease protection and digestion of food. However, stressful conditions such as introduction of artificial rearing, castration, weaning, parturition, change of diet and many more, increase gut pH, which favour growth of pathogenic organisms. The probiotics bring about its beneficial effects through one of the following mechanisms.

1. Neutralization of toxin
2. Suppression of
    i. viable numbers of specific bacteria

> ii. antibacterial agents and
>
> iii. competition for adhesion sites

3. alteration of microbial metabolism
4. stimulation of immunity

Uses of probotics   Use of probitics in cattle ration has improved weight gain and better feed efficiency. The most striking observation made so far is with 'Yea-Secc' a yeast culture containing *Saccharomyces cerevisiae*. In different studies, increase in milk yield from 3.5 to 8.4% has been observed. In addition, the milk production was more economical with a 13% reduction in feed cost (Gunther, 1989; Huber *et al.,* 1989). The yeast culture was found to influence rumen metabolism by altering the microbial fermentation rates, the protein turnover of the microbes, the hydrogen ion exchange and the metabolism of easily soluble carbohydrates. Thus, there is an improvement in the digestibility of structural carbohydrates thereby increasing the energy supply to the microbes, inhibition of methane build-up, stimulating the formation of propionic acid, and increased flow of proteins of microbial origin resulting in better feed conversion for milk yield.

Diarrhoea is a common occurrence in neonatal calves. Several veterinary probiotics claiming to prevent or treat calf diarrhoea are available, but have not been well studied. Ewaschuk *et al.* (2004) assessed the capability of *Lactobacillus rhamnosus* strain GG (LGG) to maintain viability in the gastrointestinal tract of calves with diarrhoea. They also determined whether LGG can be administered in an oral rehydration solution (ORS) without compromising the efficacy of the ORS or the viability of LGG, and whether LGG produces D-lactate or not. To investigate the intestinal survival of LGG, 15 calves were randomized into 3 groups and LGG was administered orally with their morning milk feeding on 3 consecutive days at a low (LD), medium (MD), or high (HD) dosage. Faecal samples were collected on days 0 (control), 1, 2, 3, 5, and 7 and incubated for 72 hours on deMan, Rogosa, Sharpe agar. Twenty-four hours after the first feeding, LGG was recovered from 1 out of 5 calves in the LD group, 4 out of 5 calves in the MD group, and 5 out of 5 calves in the HD group. To determine if LGG caused the glucose levels in the ORS to drop below effective levels, 1.5 L of the ORS was incubated with LGG for 2 hours at 37°C and the glucose concentration was measured every 20 minutes using a glucose meter. This ORS was further incubated for 10 hours and aliquots analysed by high performance liquid chromatography to determine if D-lactate was produced by LGG. Glucose concentrations did not change over the 2 hours of incubation, and no D-lactate was produced after 48 hours. The LGG maintained viability in ORS. Therefore, this study demonstrated that LGG survives intestinal transit in the young calf, produces no D-lactate, and can be administered in an ORS.

Younts-Dahl *et al.* (2005) evaluated the effects of three doses of *Lactobacillus acidophilus* strain NP51 and a combination treatment of strains NP51 and NP45 on prevalence of *Escherichia coli* O157 in cattle. Three hundred steers were assigned randomly to 60 pens (five steers per pen) and received one of five treatments: (i) control, no added direct-fed microbial; (ii) HNP51, high dose of NP51 at 10^9 cfu per steer daily; (iii) MNP51, NP51 at 10^8 cfu per steer daily; (iv) LNP51, low dose of NP51 at 10^7 cfu per steer daily; and (v) NP51+45, NP51 at 10^9 cfu per steer daily and NP45 at 10^6 cfu per steer daily. All direct-fed microbial treatments included *Propionibacterium freudenreichii* at 10^9 cfu per steer. Individual rectal faecal samples were collected on arrival and every 28 days throughout the feeding period. Faecal and hide samples were collected on the day of harvest. Samples were analysed for the presence of *E. coli* O157 using immunomagnetic separation methods. Cattle receiving HNP51, MNP51, and LNP51 had a lower prevalence (P < 0.01) of *E. coli* O157 throughout the feeding period compared with the controls, and the dose response for NP51 was a linear decrease in prevalence with increasing dose (P < 0.01). No decrease in prevalence for cattle receiving the combination NP51+45 was detected compared with controls (P = 0.15). *E. coli* O157 prevalence averaged across collection times were 23.9, 10.5, 9.9, 6.8, and 17.3% for cattle in the control, LNP51, MNP51, HNP51, and NP51+45 groups, respectively. Least squares mean estimates of faecal prevalence at harvest of *E. coli* O157 were 31.7, 12.5, 17.4, 8.2, and 41.6% among cattle in the control, LNP51, MNP51, HNP51, and NP51+45 groups, respectively. The greatest decrease in *E. coli* O157 carriage was achieved using NP51 at 10^9 cfu per steer.

A 2-year study was conducted during the summer months (May to September) to test the effectiveness of feeding *Lactobacillus acidophilus* strain NP51 on the proportion of cattle shedding *Escherichia coli* O157:H7 in the faeces and to evaluate the effect of the treatment on finishing performance (Peterson *et al.*, 2007). Steers (n = 448) were assigned randomly to pens, and pens of cattle were assigned randomly to NP51 supplementation or no supplementation (control). NP51 products were mixed with water and applied as the feed was mixed daily in treatment-designated trucks at the rate of 10^9 cfu per steer. Faecal samples were collected (n = 3,360) from the rectum from each animal every 3 weeks, and *E. coli* O157:H7 was isolated by standard procedures, using selective enrichment, immunomagnetic separation, and PCR confirmation. The outcome variable was the recovery of *E. coli* O157:H7 from faeces, and was modelled using logistic regression accounting for year, repeated measures of pens of cattle, and block. No significant differences were detected for gain, intakes, or feed efficiency of control or NP51-fed steers. The probability for cattle to shed *E. coli* O157:H7 varied significantly between 2002 and 2003 (P = 0.004). In 2002 and 2003, the probability for NP51-treated steers to shed *E. coli* O157:H7 over the test periods was 13 and 21%, respectively, compared with 21 and 28% among controls. Over the 2 years, NP51-treated steers were 35% less likely to shed *E. coli* O157: H7 than were steers in untreated pens (odds ratio = 0.58, P = 0.008). This study is consistent with previous reports that feeding NP51 is effective in reducing *E. coli* O157:H7 faecal shedding in feedlot cattle.

In an experiment with 33 first-litter sows from day 90 of pregnancy to day 28 of lactation, the influence of a probiotic supplementation on weight performance, feed intake, litter sizes, litter weights, health status and microbiological profile was tested (Bohmer *et al.*, 2006). *Enterococcus faecium* DSM 7134 was supplemented in a concentration of 5×10^8 cfu/kg feed to the gestation and lactation diets of gilts. The supplemented sows showed a significant higher improvement of feed intake (4.16 vs. 3.71 kg/day), litter size (9.2 vs. 7.7 piglets) and weight performance. The average live weight of the probiotic sows at day 28 of lactation was 11 kg higher than of the controls. The bacterial counts/g faeces (lactobacilli, gram-positive anaerobes, gram-negative anaerobes, *Escherichia coli* and enterococci) and the incidence of adhesive and haemolytic *E. coli* organisms revealed no significant differences between the sows of the two groups or their piglets. While the litter size cannot necessarily be assumed as a primary effect of the probiotic supplementation, the significantly better feed intake and weight performance might be partly due to the probiotic use and can prevent "starvation sterility" of young sows after their first litter caused by reduced feed intake during lactation with high mobilization of body tissue accompanied by lack of energy.

This placebo-controlled double-blind study was conducted to evaluate effects of *Enterococcus faecium* DSM 10663 NCIMB 10415 (EcF) orally given from birth to weaning on diarrhoea and performance of piglets (Zeyner and Boldt, 2006). At the first 3 days post-natum (p.n.), piglets from 54 [verum group (VG)] and 60 [placebo group (PG)] sows got 1 g of a gel directly per mouth by a dosing device. Gel for the VG contained 2.8×10^9 colony forming units (cfu) EcF/g. From day 4 p.n. until weaning (24 +/– 3.2 days p.n.) a liquid additive was given that was administered twice a day 1.26×10^9 cfu EcF to each VG piglet. In case of diarrhoea, an electrolyte solution was used which provided daily 2.9 and 5.8 (week 1 and >/ = 2, respectively) $\times 10^8$ cfu EcF per VG piglet. Diarrhoea scores were defined as follows: (i) no diarrhoea; (ii) piglets developed diarrhoea, but were vital and (iii) piglets suffered from diarrhoea and additionally looked pale, developed rough coat, showed slackening of the

flank and lethargy. Counts of viable born, stillborn and weaned piglets were normal and not different between groups (p > 0.05). Placebo group vs. VG piglets suffered more frequently from diarrhoea (40.0 vs. 14.8%, p < 0.05). Duration of diarrhoea was not affected by feeding EcF (2.2 +/– 0.81 days, p > 0.05). Diarrhoea score was lower in VG vs. PG (1.2 vs. 1.5 +/– 0.54, p < 0.05) and the daily weight gain (DWG) was higher by 17 g/day (p < 0.05). Results suggest that the daily oral supplementation of EcF from birth to weaning reduces the portion of piglets suffering from diarrhoea. This may improve performance, as the higher DWG indicates.

Probiotics are microorganisms that are added to food to exert beneficial effects on the host. Biagi *et al.* (2007) studied the *in vitro* and *in vivo* evaluation of the effect of *Lactobacillus animalis* LA4 (isolated from the faeces of a healthy adult dog) on composition and metabolism of dog intestinal microflora. When added to dog faecal cultures, LA4 reduced enterococci and increased lactobacilli counts throughout the study, whereas *C. perfringens* counts were significantly reduced at 24 hours. After 8 hours of incubation, LA4 reduced ammonia and increased lactic acid concentrations. For the *in vivo* study, nine adult dogs received the freeze-dried preparation of *L. animalis* LA4 for 10 days. On day 11, faecal lactobacilli were higher than at trial start (6.99logcfu/g versus 3.35logcfu/g of faeces) and faecal enterococci showed a trend towards a numerical reduction (P=0.08). The results showed that LA4 was able to survive gastrointestinal passage and transitorily colonize the dog intestine where, based on the *in vitro* results, it could positively influence composition and metabolism of the intestinal microflora. These results suggested that *L. animalis* LA4 can be considered as a potential probiotic for dogs.

Probiotics are extensively used in poultry industry. In a broiler trial, birds receiving probiotics supplementation (100 mg/kg feed) gained 1268 g as against 1204 g in control group at the end of 8 weeks. There was also a marginal improvement in feed efficiency (2.3 vs 2.26). In a short-term feeding trial with layers, there was a 5.6% increase in egg production; a reduction in the incidence of thin-shelled eggs (6.7% vs 19.7%); a significant reduction in serum cholesterol (114.3 mg/dl vs 176.5 mg/dl) and a significant reduction in yolk cholesterol (11.28 mg vs. 14.69 mg) (Mohan, 1991).

Sixty chicken were randomly divided into two groups to determine the effect of oral administration of probiotics on the intestinal mucosal immune response and ultrastructure of caecal tonsils. The first group (control) was fed with a basic diet without antibiotics or probiotics. The second group was fed with the same diet as the control, except that they received drinking water with probiotics (4×10^9 cfu per chicken and day) from post-hatch to day 3 of age. The probiotic preparation was composed of *Bacillus subtilis* Bs964, *Candida utilis* BKM-Y74 and *Lactobacillus acidophilus* LH1F. Intestinal fluid, Peyer's patch and caecal tonsils were taken at day 1, 4, 7, 10 and 18 after administration of probiotics. The results were as follows:

i.   Compared to the control, probiotics enhanced the content of the following items: immunoglobulin (Ig)A in the intestinal fluid at day 7 ($p < 0.01$), the IgG-forming cells at day 10 ($p < 0.05$), IgM-forming cells in the Peyer's patch at day 7 ($p < 0.05$), IgA-forming cells at day 7–10 ($p < 0.05$), IgG-forming cells at day 7 ($p < 0.05$) and IgM-forming cells in cecal tonsils diffuse area at day 4–7 ($p < 0.05$).

ii. T lymphocytes in caecal tonsils were enhanced at day 7 ($p < 0.01$) after orally fed with probiotics.

iii. The density of microvilli and length of caecal tonsils increased after probiotics were administrated at day 3. With chicken ageing, the efficiency of probiotics would decrease.

These results suggested that probiotics enhance intestinal mucosal immunity of chicken at the early age.

Probiotics have proved to have potential to improve broiler growth, help to reduce cholesterol in eggs, and milk yield in cattle. Vero cytotoxin (VT)-producing *Escherichia coli* (VTEC), such as *E. coli* O157:H7, are emerging food-borne pathogens worldwide. VTs are associated with haemorrhagic colitis and haemolytic uraemic syndrome in humans. Attachment of B subunit of VTs to its receptor, globotriaosylceramide (Gb3), at gut epithelium is the primary step and consequently, the A subunit of VTs inhibit protein synthesis in the target cells. Proinflammatory cytokines, such as tumour necrosis factor-α (TNF-α) and interleukin-1 β (IL-1β), up-regulate Gb3 expression, increase sensitivity of VTs, and enhance VT action in developing disease. Administration of culture supernatants of *Bifidobacterium longum* HY8001, intragastrically into mice, inhibited the action of VTs (Kim *et al.*, 2001). Cytokine TNF-α and IL-1β levels in sera and expression of their mRNA were decreased and expression of Gb3 in renal tubule epithelial cells were reduced in mice treated with *B. longum* HY8001 culture supernatants. The results indicate that soluble substances in *B. longum* Hy8001 culture supernatants may have inhibitory activity on the expression of Hb3, VT-Gb3 interaction or both.

Probiotics are living microorganisms that can effect the host in a beneficial manner. Lactobacillus GG alone or the combination of *Bifidobacterium bifidum* and *Streptococcus thermophilus*, is effective in the treatment of *Clostridium difficile*, as well as in preventing the frequency and severity of infectious acute diarrhoea in children (Madsen, 2001). The mechanism of action of probiotics may include receptor competition, effects on mucin secretion or probiotic immunomodulation of gut-associated lymphoid tissue.

MANIPULATION OF LACTATION

Lactation may be improved by manipulation of mammogenesis, lactogenesis and galactopoiesis. Each stage is regulated by a multifactorial hormone complex. Hormone treatment to induce mammogenesis artificially (i.e., without pregnancy) in dairy ruminants employ steroid hormones (oestrogens, gestagens and glucocorticoids). Somatotropin has been shown to enhance mammary development. Hormonal manipulation of lactogenesis and especially galactopoiesis focuses on somatotropic hormone (STH). STH regulates the partitioning of nutrients within the organisms preferentially to the mammary gland and mobilizes them by its lipolytic active precursors for milk synthesis.

Mammogenesis

Mammogenesis is defined as the growth and differentiation of the mammary gland to the stage prior to active secretion. Generally in cows, little or no lobule-alveolar development occurs before first pregnancy, but appears to be a continuous exponential process through gestation and terminates at the onset of lactation; in cows a gradual involution is observed that becomes more pronounced during the dry period. Mammary development is largely stimulated by hormones. Steroid and thyroid hormones and proteohormones of the prolactin and growth hormones are responsible for ductal development, whereas progesterone and prolactin regulate lobule-alveolar proliferation. Maximal growth, however, was achieved by additional application of glucocorticoids. Oestrogen–progesterone treatment stimulated alveolar development.

Prolactin is important for the formation of alveoli in cows, goats and ewes. In goats and sheep, foetal numbers, concentration of placental lactogen during late pregnancy and milk yield have been correlated. STH treatment of heifers resulted in increased mammary parenchyma.

Insulin-like growth factor I (IGF-I) is known to regulate mammary gland development. This regulation occurs through effects on both cell cycle progression and apoptosis. Hadsell *et al.* (2002) studied the IGF-I-dependent regulation of these processes by using transgenic and knockout mouse models that exhibited alterations in the IGF-I axis. Studies of transgenic mice that over-expressed IGF-I during pregnancy and lactation demonstrated that this growth factor slowed the apoptotic loss of mammary epithelial cells during the declining phase of lactation but had minimal effects during early lactation on milk composition or lactational capacity. In contrast, our analysis of early developmental processes in mammary tissue from mice carrying a targeted mutation in the IGF-I receptor gene suggested that IGF-dependent stimulation of cell cycle progression was more important to early mammary gland development than potential anti-apoptotic effects. With both models, the effects of perturbing the IGF-I axis were dependent on the physiological state of the animal. The diminished ductal development that occurred in response to loss of the IGF-I receptor was dramatically restored during pregnancy, whereas the ability of over-expressed IGF-I to protect mammary cells from apoptosis did not occur if the mammary gland was induced to undergo forced involution.

Lactogenesis and Galactopoiesis

Induction and maintenance of lactation are dependent on reproductive steroids, prolactin (PRL), thyroid hormones, STH, insulin, corticoids and possibly placental lactogen. In ruminants, PRL seems to be essential for lactogenesis, but once lactation is established, it is not a limiting factor for milk production. Thyroid hormones play an important role in lactational performance of cattle. T3 and T4 (T3-tri-iodothyronine, T4-thyroxin) play a central role in metabolic regulations. Increase in milk and milk fat yields after treating cows with T4 have been reported. STH has been found in higher concentration in high-yielding cows. Milk yield increased from 8–41% when treated with STH in cattle. Increase in milk production due to STH application were also observed in cows.

Glucocorticoids were used to improve lactation after artificially stimulated mammogenesis. Milk yields of more than 10 kg/day have been obtained by the use of glucocorticoids.

MANIPULATION OF WOOL GROWTH IN SHEEP

Wool production in sheep is essentially a protein output. Wool growth is dependent on the availability of S-amino acid and total amino acids available in the blood supply to the wool follicle. About half the cysteine derived from intestinal absorption in sheep is converted into wool. There are three major possibilities for increasing the availability of S-amino acids and hence the production of wool. These include:

1. manipulating the microbes of the microbial fermentation digestive system of sheep to improve microbial growth efficiency,

2. manipulating the microbes in the rumen to increase the utilization of dietary protein that escapes rumen fermentation and

3. using recombinant DNA processes to create animals that are able to synthesize S-amino acids which are the primary limiting amino acids for wool growth.

Manipulation of the Rumen Microbial Digestive System

To increase wool production, manipulation of the rumen microbial ecosystem must be primarily aimed at increasing the availability of essential amino acids to the animal. This can be achieved by

- increasing the efficiency of microbial growth in the rumen;

- increasing the amount of intestinally-digestible dietary protein that escapes rumen fermentation;

- increasing the digestibility of fibrous feed in the rumen which increases amino acids (and also volatile fatty acid production) and possibly encourages the growth of more efficient organisms;

- producing, by recombinant means, bacteria that synthesize proteins with high S-amino acid content or that secrete S-amino acids and

- decreasing the amino acid utilization for other metabolic functions including deamination to supply glucogenic precursors and substrates for oxidative metabolism.

Manipulating the rumen ecosystem to achieve maximum microbial and dietary protein outflow from the rumen is the major strategy when attempting to selectively increase wool growth in the mature animal. A continuous supply of fermentable carbohydrates, ammonia, peptides, amino acids and other nutrients are needed to promote efficient utilization of ATP for microbial protein yield. Increased microbial synthesis of proteins was observed in the rumen of sheep on diets with casein as the N-source compared to urea-based diets (Hume, 1970).

Wool growth is increased by extra amino acids absorbed from the small intestine. Manipulation that increase the rate of fermentative digestion will increase microbial protein relative to energy. These include:

1. treatment of the forage with ammonia which increase digestibility;
2. addition of small amount of readily digestible fibre (which stimulates digestion of the basal dietary fibres);
3. maintaining a protozoa-free rumen and
4. optimizing nutrient availability by balancing microbial nutrients in the rumen.

The ruminant DNA material can be inserted into sheep embryos at the 4–8-cell stage by micromanipulation. Scientists are attempting to introduce bacterial DNA fragments into the ova of sheep coding for the synthesis of cysteine, which is the primary limiting amino acid for wool growth. The biosynthetic pathway of cysteine synthesis is:

$$\text{Serine} + \text{Acetyl-CoA} \xrightarrow{\text{Serine transacetylase}} \text{O-Acetylserine} + \text{CoA}$$

$$\text{O-Acetylserine} + H_2O \xrightarrow{\text{O-acetylserine sulphydrolase}} \text{Cysteine} + \text{Acetate}$$

H_2S is generated from sulphur compounds in the rumen and the S^{2-} is not used in microbial growth but is absorbed and converted to sulphate in the rumen epithelium or eructed. Eructed gases are inhaled into the lungs before the gases are finally exhaled. Thus, the lungs and the rumen epithelium are the only tissues receiving quantities of H_2S and therefore the only ones likely to synthesize significant quantities of cysteine in transgenic animals with the necessary enzymes in its genome. When protein intake is high, the availability of non-essential amino acid serine may not be limiting; however, serine is a glucogenic amino acid and if a drain of serine into cysteine creates a serine deficiency, then the animal must be capable of increasing the synthesis of this amino acid. The carbon fragments have to arise from the glucogenic precursors, mainly propionate or other glucogenic amino acids.

If sheep are given diets deficient in bypass protein and/or glucogenic precursors, then this is likely to be a detrimental manipulation. The process could result in serine becoming rare and limiting.

REVIEW QUESTIONS

1. Write in detail how biotechnology could be used for manipulation of growth and production in animals.

2. What are probiotics? How are they useful in animal production?

REFERENCES

Adams, N.R., Briegel, J.R. and Ward, K.A. (2002). "The impact of a transgene for ovine growth hormone on the performance of two breeds of sheep." *J. Anim. Sci.* 80: 2325–2333.

Baldi, A. (1999). "Manipulation of milk production and quality by use of somatotropin in dairy ruminants other than cow." *Domest. Amin. Endocrinol.* 17: 131–137.

Biagi, G., Cipollini, I.A., Pompei, G., Zaghini, D. and Matteuzzi, D. (2007). "Effect of a *Lactobacillus animalis* strain on composition and metabolism of the intestinal microflora in adult dogs." *Vet. Microbiol.* March, 28.

Bó, G.A., Baruselli, P.S. and Martínez, M.F. (2003). "Pattern and manipulation of follicular development in *Bos indicus* cattle." *Anim. Reprod. Sci.* 78:307–326.

Böhmer, B.M., Kramer, W. and Roth-Maier, D.A. (2006). "Dietary probiotic supplementation and resulting effects on performance, health status, and microbial characteristics of primiparous sows." *J. Anim. Physiol. Anim. Nutr.* (Berl). 90:309–315.

Etherton, T.D., Evock, C.M., Chung, C.S., Walton, P.E., Sillence, M.N., Magri, K.A. and Ivy, E.E. (1986). "Stimulation of pig growth performance by long-term treatment with pituitary porcine growth hormone (pGH) and a recombinant pGH." *J. Ani. Sci.* 63: (Suppl. 1), Abst. 144. p. 219.

Ewaschuk, J.B., Naylor, J.M., Chirino-Trejo, M. and Zello, G.A. (2004). "*Lactobacillus rhamnosus* strain GG is a potential probiotic for calves." *Can. J. Vet. Res.* 68:249–53.

Gunther, K.D. (1989). In: *Biotechnology in the Feed Industry.* (ed.). Lyons, T.P. pp. 39–46.

Huber, J.T. *et al.* (1989). In: *Biotechnology in the Feed Industry.* (ed.). Lyons, T.P.

Hadsell, D.L., Bonnette, S.G. and Lee, A.V. (2002). "Genetic manipulation of the IGF-I axis to regulate mammary gland development and function." *J. Dairy Sci.* 85:365–377.

Hume, I.D. (1970). "Synthesis of microbial protein in the rumen III. The effect of dietary protein." *Aust. J. Agric. Res.* 21: 305.

Hunter, M.G., Robinson, R.S. Mann, G.E. and Webb, R. (2004). "Endocrine and paracrine control of follicular development and ovulation rate in farm species." *Anim. Reprod. Sci.* 82-83:461–477.

Kim, S.H., Yang, S.J., Koo, H.C., Bae, W.K., Kim, J.Y., Park, J.H., Back, Y.J. and Park, Y.H. (2001). "Inhibitor activity of *Bifidobacterium longum* HY8001 against Vero cytotoxin of *Escherichia coli* O157:H7." *J. Food Prot.* 64: 1667–1673.

Lawrence, M.E., Schelling, G.T., Byers, F.M. and Green, L.W. (1986). "Improvement of growth and feed efficiency in cattle by active immunization against somatostatin." *J. Ani. Sci.* 63: (Suppl. 1). p. 215.

Macmillan, K.L., Segwagwe, B.V. and Pino, C.S. (2003). "Associations between the manipulation of patterns of follicular development and fertility in cattle." *Anim. Reprod. Sci.* 78:327–344.

Madsen, K.L. (2001). "The use of probiotics in gastrointestinal disease." *Can. J. Gastroenterol.* 15: 817–822.

Mohan, B. (1991). M.V.Sc dissertation submitted to the TANUVAS, Madras.

Peterson, R.E., Klopfenstein, T.J., Erickson, G.E., Folmer, J., Hinkley, S., Moxley, R.A. and Smith, D.R. (2007). "Effect of *Lactobacillus acidophilus* strain NP51 on

Escherichia coli O157:H7 fecal shedding and finishing performance in beef feedlot cattle." *J. Food Prot.* 70: 287–291.

Rautava, S., Kalliomaki, M. and Isolauri, E. (2002). "Probiotics during pregnancy and breast feeding might confer immunomodulatory protection against atopic disease in the infant." *J. Allergy Clin. Immuno.* 109: 119–121.

Thatcher, W.W. and Santos, J.E. (2007). "Control of ovarian follicular and corpus luteum development for the synchronization of ovulation in cattle." *Soc. Reprod. Fertil. Suppl.* 64:69–81.

Ward, K.A. and Brown, B.W. (1998). "The production of transgenic domestic livestock: successes, failures, and the need for nuclear transfer." *Reprod. Fertil. Dev.* 10: 659–665.

Younts-Dahl, S.M., Osborn, G.D., Galyean, M.L., Rivera, J.D., Loneragan, G.H. and Brashears, M.M. (2005). "Reduction of *Escherichia coli* O157 in finishing beef cattle by various doses of *Lactobacillus acidophilus* in direct-fed microbials." *J. Food. Prot.* 68:6–10.

Yurong, Y., Ruiping, S., Shimin, Z. and Yibao, J. "Effect of probiotics on intestinal mucosal immunity and ultrastructure of cecal tonsils of chickens." *Arch. Anim. Nutr.* 59:237–46.

Zeyner, A. and Boldt, E. (2006). "Effects of a probiotic *Enterococcus faecium* strain supplemented from birth to weaning on diarrhoea patterns and performance of piglets." *J. Anim. Physiol. Anim. Nutr.* (Berl). 90:25–31.

11

MANIPULATION OF RUMEN MICROORGANISMS

INTRODUCTION

Genetic engineering of microorganisms to increase rumen productivity in developing countries is being investigated in recent times. Fibrous feeds and cereal crop residues of low digestibility comprise the major proportion of feed available to most ruminants under small holder conditions in developing countries. In this context, the primary objective of the manipulation of rumen organisms must be to increase the rate and extent of digestibility of these fibrous feeds as this is a primary limiting factor. Because of feeding of ruminants with the feed of low digestible fibrous feeds, productivity of these livestock has been extremely low typified by low growth rates, late maturity, low reproduction rates, low milk yield and stunted mature body size. The low productivity can be overcome to some extent by supplementation to ensure that:

i. the nutrients for the growth of ruminal organisms are balanced, ensuring an efficient fermentative digestion and

ii. small quantities of bypass nutrients are supplied to balance the nutrients that arise from fermentative digestion. This ensures an efficient utilization of nutrients by the animal.

A 5–10 units increase in digestibility of a fibrous forage can lead to 100% increase in productivity. Supplementation strategies, combined with technology for increasing forage digestibility could potentially increase the productivity of each ruminants by approximately 5-fold.

Understanding the factors controlling rumen microbial activity may allow scientists to modify the rumen ecology in order to create conditions that will optimize the use of poor quality feed by ruminant livestock. The fastest way to improve rumen function in an animal is to introduce digestion-enhancing bacterial species from other animals or to selectively increase populations of species that inhabit the rumen only at low levels.

Rumen ecology may also be modified by altering the function of bacteria using genetic engineering techniques. Many approaches have been taken in attempts to transform rumen bacteria by introducing a foreign gene or genes into the bacteria.

THE TARGET ORGANISMS

Fermentative digestion of forages occurs in ruminants in the well developed forestomach (the reticulo-rumen). This organ is maintained in a relatively stable continuous anaerobic culture of mixed population of protozoa, bacteria, fungi and bacteriophage. The rumen fluid pH is maintained at close to neutrality by continuous saliva which balances the production of acidic end products of fermentation. The short chain fatty acids are either absorbed directly or pass out of the rumen in the digesta to the intestines where they are absorbed.

The anaerobic bacteria are the major organisms in the rumen which help in fibre digestion. The information on their nutrition and biology is lacking. The widely distributed anaerobic fungi which grow on fibrous feed particles in the rumen have only recently been discovered. The fungi are mostly fibrolytic with the major substrates being cellulose and hemicellulose. Their major role may be weakening of feed particle structure as most species have radii which penetrate plant tissues and therefore tend to digest plant materials from inside, whereas bacteria attach to and attack damaged parts of plants and digestion occurs from outside towards the centre. As the rate of fibre breakdown in the rumen is often a primary limitation to intake of low digestibility feeds, the possibility of developing fungi with enhanced ability to grow or secrete cellulase enzyme could be very important. A number of species of ciliate protozoa occur in the rumen. They are targeted for elimination from the rumen. Improved digestibility and improved efficiency of microbial growth can be achieved in the rumen of sheep on straw-based diets from which protozoa can be eliminated. Bacteriophage can be found in large numbers in the rumen of sheep and cattle. The phage outnumbers bacteria as much as 10 : 1.

Prevotella ruminicola (formerly *Bacteroides ruminicola*) is one of the most numerous groups of rumen bacteria and is found in ruminants fed on many different diets. In silage-fed cattle from Sweden, these bacteria accounted for up to 60% of all isolates. These are strictly anaerobic and involved in the degradation and utilization of starch and of plant cell polysaccharides such as xylans and pectins. These bacteria are also believed to play an important role in the degradation of proteins and in the uptake and fermentation of peptides. *Ruminobacter amylophillus* accounts for about 15% of isolates from rumen and involved in starch degradation. They also possess significant proteolytic activity.

Fibrobacter succinogenes is one of the most widespread cellulolytic bacteria of the rumen. This accounted for around 20% of the isolates recovered from a cow fed wheat straw; in cows fed other diets, *F. succinogenes* accounted for about 5% of the isolates. *Succinivibrio dextrinosolvens* ferments dextrins in animals fed starch. In addition to their capacity to digest starch, some strains also possess enzyme which could enable products of plant cell wall breakdown to be used.

Anaerovibrio lipolytica has two key properties that probably reflect its ecological roles, the hydrolysis of lipids and utilization of lactate. *Ruminococcus flavefaciens* and *R. albus* are involved in plant fibre degradation in the rumen.

GENE AND PROTEIN STRUCTURE, AND EXPRESSION IN RUMEN BACTERIA

The vast majority of genes that have been cloned and sequenced from rumen bacteria are involved in plant polysaccharide degradation. This is the reflection of the unique ability of the rumen system to rapidly and effectively degrade plant fibre and of the central importance of plant fibre degradation in rumen nutrition. These genes are of interest in an applied sense, both for their potential applications in other systems, and for the possibility that a better understanding of enzyme systems and constraints on digestion will allow improvement of plant fibre digestion in the rumen.

Prevotella ruminicola are generally effective xylan degraders, and genes cloned from the group are largely involved in xylan degradation. While attempts to clone genes involved in xylan degradation from this group of organisms into *E. coli* have been successful, further analysis has generally demonstrated that transcription, translation and processing signals recognized in *E. coli* are not the same as those operating in the native host, suggesting that the genes cloned to date may represent a subset of the total gene complement that is serendipitously expressed in *E. coli*. *Clostridum longisporum* has endoglucanase activity.

The *Ruminoccus* spp. constitute one of the most effective groups of crystalline cellulose-degrading organisms in the rumen. A number of investigators have attempted to clone genes involved in cellulose or xylan degradation from these organisms. Large numbers of genes have been cloned in *E. coli* and sequenced, but as with *Prevotella ruminicola*, subsequent analysis has generally shown that regulatory signals, post-translational processing and protein stability differ in the native and host organisms. The genes of *Butyrivibrio fibrisolvens* are involved in polysaccharide degradation, the exception being genes for glutamine synthetase.

APPLICATION OF RECOMBINANT DNA TECHNOLOGY TO RUMEN ORGANISMS

The practicability of inserting novel genes into bacteria has been proven by the widespread use of biochemical products derived from genetically engineered microorganisms. The type of alteration that is most desirable is an increased ability by rumen organisms to digest coarse plant fibre, by the increased production of cellulolytic and xylanolytic enzymes. While some species of bacteria (e.g. *Ruminococcus albus*) possess both types of enzymes, other dominant rumen species produce one or the other (e.g. many strains of *B. succinogenes* are cellulolytic and are unable to digest xylan, while *B. ruminicola* has only xylanolytic activity (Hungate, 1966). By endowing the dominant populations of bacteria with digestive capability for multiple cell wall constituents, it may be possible to enhance fibre digestion by a sort of mixed populations that occur naturally.

Using familiar bacterial species, the transfer of functional genes is now a routine procedure. Many bacterial genes have inducible promoters and such switches may be useful for the timed induction of nutrient synthesis, antibiotic production, anti-helminthic synthesis, etc. The initial step in this process is to identify inducible genes in the ruminal anaerobes and isolate promoter sequences responsible for control functions.

The genetics of ruminal anaerobes and their involvement with epigenetic factors such as plasmids and bacteriophage are at a very early stage of investigation. A highly significant increase in cellulolytic activity of *Ruminococcus albus* by mutation and selection (Taya *et al.*, 1983) has been reported. The introduction of completely novel enzymes into dominant species of rumen bacteria and the addition of highly manipulatable control mechanisms to specific genes are two clear examples.

Gene libraries have been constructed and used to isolate the genes for cellulose enzymes (Tether *et al.*, 1984) which can be expressed as active molecules in the *E. coli* host cells. The occurrence of plasmids and bacteriophage in rumen bacteria has been demonstrated (Klieve *et al.*, 1986). It is clear, therefore that essential requirements for producing genetically altered rumen bacteria already exist.

The genetic manipulation of rumen microorganisms by altering the inherited characteristic is the most powerful potential tool for enhancing the rate and extent of digestion of fibrous feeds within the rumen. Its success depends on the identification of the specific genetic information, the development of appropriate genetic techniques and creation of new organisms with high probability of survival in the harsh environment of the rumen of animals fed on highly fibrous feeds.

There is ample experimental evidence that gene transfer among microorganisms can and does occur in nature, through transformation, transduction and conjugation. The rumen environment has several qualities that might be thought to encourage inter- and intraspecies gene transfer. As an ecosystem, it has very high population density or very diverse microorganisms. Studies have indicated that *E. coli* could exchange plasmids carrying antibiotic-resistance markers at a measurable rate in the rumen of starved animals. Fliegerova (1993) demonstrated the presence of plasmid-associated multi-resistance in strains of *E. coli* isolated from the rumens of calves. Similarly, antibiotic-resistant strains of *Streptococcus bovis* have been isolated from the rumen of calves (Jonecova *et al.*, 1993), and the occurrence of *in vitro* conjugative transfer of tetracycline resistance from rumen isolates of *S. bovis* to strains of *S. bovis* and *Enterococcus faecium* have been shown (Jonecova *et al.*, 1994). Scott and Flint (1995) demonstrated the low level transfer of a 60-kb plasmid conferring both ampicillin and tetracycline resistance among rumen *E. coli* strains under conditions that stimulate those in the rumen.

NON-GENETIC MANIPULATION OF RUMEN MICROBES

Rumen fermentation has to be optimized to improve feed utilization in ruminants. This is to maximize the digestion of lignocellulosic compounds and the use of non-protein nitrogen, while allowing other nutrients (i.e., proteins, starch and lipids) to escape rumen fermentation. There are three steps to optimize fermentation.

1. Supplying the nutritional requirements of rumen microflora to ensure maximum microbial activity and microbial growth yield.

2. Modifying microbial population (bacteria, protozoa and fungi) to select the species that are most apt to supply biomass and metabolites that the host animal requires.

3. Altering the physiology of the digestive tract (i.e., salivary secretion, rumen motility, rates of liquid-phase and particle-phase outflow).

The first step requires that the known quantitative and qualitative needs of the ruminal microbial populations be met. This can be done by controlling the diet (nature, amount, physical form of feedstuff, balancing nutrients). The other two ways require means and methods that will bring lasting and significant modifications. These two aspects correspond to the manipulation of the rumen microbes.

Addition to Antibiotics

The antibiotics used as feed additives in the ruminants are frequently presented as a model to study the modes of action in the manipulation of rumen function. Among the antibiotics, ionophors (monesin and lasalocid) have been studied more frequently. Their mode of action on cellular ion transport induces important changes in rumen microbes and consequently in the modification of rumen fermentation.

Ionophors are potent against gram-positive bacteria and some gram-negative bacteria which have a gram-positive-like cell wall structure such as *Ruminococcus albus*, *R. flavefaciens* and *Butyrivibrio fibrisolvens*. By contrast, *Bacteroides ruminicola* and *R. succinogenes* become resistant to this antibiotic after 2–4 days growth in the presence of monensin. *Selenomonas ruminantium* is not affected by adding this antibiotic. That is why the enzyme activity in the rumen is modified by an increase in the succinate dehydrogenase activity, which leads to higher succinate production and to the conservation activity, which leads to higher succinate production and to the conservation of succinate into propionate. Similarly, acetate kinase activity is greatly reduced (Wallace *et al.*, 1981), which expresses a drop in the number of acetate-producing microorganisms. These antibiotics also have a negative effect on the lactate producers, which are, for the most part, gram-positive bacteria, without inhibiting ruminal lactate utilizers which produce propionic acid. These are the main reasons for modifications observed on improvements in molar proportions of propionate and decreases in acetate or butyrate, when animals are fed monensin (Thivend and Jouany, 1983).

In vitro calcimycin induces a decrease in propionic acid to the advantage of acetic and especially butyric acids and favours methane production. *In vitro* aberixine has effects similar to monensin, from which it is derived, but contrary to this molecule, it has no negative effects as regards microbial synthesis.

Elimination of Rumen Ciliate Protozoa (Defaunation)

Protozoa accounts for up to 50% of the biomass of rumen microorganisms and their elimination causes modifications in the bacterial population and in organic matter digestion in the rumen. Defaunation increases the total number of bacteria in the rumen, because ciliates prey on bacteria and compete with them for nutrient utilization. The composition of microflora is modified towards more small gram-negative rods and selenomonas-like bacteria in defaunated animals.

Protozoa host methanogenic bacteria that are closely associated with their surface. These bacteria capture the hydrogen released by protozoa to convert it into methane by reaction with CO_2. The number of fungi in the rumen increases after defaunation. The strongest effects of defaunation were observed on cell-wall carbohydrate digestion. The microbial nitrogen yield is largely enhanced by defaunation and results in the net microbial growth and a depressive effect on the organic matter fermented in the rumen of treated animals.

The defaunation can be considered to be an advantage only for animals fed low protein diets, in which bacterial nitrogen constitutes the main supply of intestinal proteins. Defaunation can be used for saving dietary proteins from being degraded in the rumen in low-protein diets, but it lowers cell-wall degradation in roughage-poor diets. Under conditions of defaunation, new strains of cellulolytic bacteria or genetically modified rumen bacteria could be developed to improve cell-wall carbohydrate digestion.

A study was conducted to evaluate *in vivo* the role of rumen ciliate protozoa with respect to the methane-suppressing effect of coconut oil (Machmuller *et al.*, 2003). Three sheep were subjected to a 2×2 factorial design comprising two types of dietary lipids (50 g $\times$ kg^{-1} coconut oil vs. 50 g $\times$ kg^{-1} rumen-protected fat) and defaunation treatment (with vs. without). Due to the defaunation treatment, which reduced the rumen ciliate protozoa population by 94% on average, total tract fibre degradation was reduced but not the methane production. Feeding coconut oil significantly reduced daily methane release without negatively affecting the total tract nutrient digestion. Generally, the present data illustrate that a depression of the concentration of ciliate protozoa or methanogens in rumen fluid cannot be used as a reliable indicator for the success of a strategy to mitigate methane emission *in vivo*. The methane-suppressing effect of coconut oil seems to be mediated through a changed metabolic activity and/or composition of the rumen methanogenic population.

Addition of Fats

When large amounts (5–10% of dry matter) of non-protected triglycerides or free polyunsaturated fatty acids from oils are added to the diets, a large decrease in ciliate protozoa in the rumen is induced (Tamming *et al.*, 1983), which probably leads to modification of populations. Adding these substances also provokes a drop in organic matter digestion, especially in cell wall carbohydrates, ammonia and methane production and in the acetate/propionate ratio. The microbial nitrogen yield in the rumen is increased by adding oil rich in polyunsaturated fatty acids.

Addition of Protein Degradation Protectors

Chalupa (1977) showed that diphenyliodonium chemicals inhibit rumen acid degradation, which results in large amounts of amino acids in the rumen. Amino acid degradation may be prevented by inhibiting their intracellular transport in bacteria. These molecules also have an effect on rumen fermentation products. They bring on a decrease in methanogenesis, but without modifying the H_2 concentration in the medium.

A series of *in vitro* studies was conducted to determine the effects of adding a commercial enzyme product on the hydrolysis and fermentation of cellulose, xylan, and a mixture (1 : 1 wt/wt) of both (Colombatto *et al.*, 2003). The enzyme product (Liquicell 2500, Specialty Enzymes and Biochemicals, Fresno, CA) was derived from *Trichoderma reesei* and contained mainly xylanase and cellulase activities. Addition of enzyme (0.5, 2.55 and 5.1 microL/g of DM) in the absence of ruminal fluid increased (P < 0.001) the release of reducing sugars from xylan and the mixture after 20 hours of incubation at 20°C. Incubations with ruminal fluid showed that enzyme (0.5 and 2.55 microL/g of DM) increased (P < 0.05) the initial (up to 6 hours) xylanase, endoglucanase, and beta-D-glucosidase activities in the liquid fraction by an average of 85%. Xylanase and endoglucanase activities in the solid fraction also were increased (P < 0.05) by enzyme addition, indicating an increase in fibrolytic activity due to ruminal microbes. It was concluded that enzymes enhanced the fermentation of cellulose and xylan by a combination of pre- and post-incubation effects (i.e., an increase in the release of reducing sugars during the pretreatment phase and an increase in the hydrolytic activity of the liquid and solid fractions of the ruminal fluid), which was reflected in a higher rate of fermentation.

Addition of Buffer Substances

The incorporation of buffer substances into ruminant feeds (bicarbonate alone or mixed with magnesium or calcium bicarbonate) prevents a drop in pH with diets rich in easily fermentable sugars and limits the risk of acidosis. Massive intake of salts cause rise in osmotic pressure in the rumen and disturb development of microorganisms.

Addition of Bacteria or Rumen Content

Adding rumen content (in fresh or in freeze-dried form) can be useful if rumen metabolism has been impaired (e.g. acidosis, bloating, per os antibiotic treatment). For example, when animals are defaunated, it is advisable to introduce freeze-dried and thawed rumen contents, to provide all species of bacteria, which can develop in this new condition.

Oeztuerk *et al.* (2005) examined whether *Saccharomyces boulardii* had an effect on parameters of rumen microbial metabolism at different dosages and whether the yeast would be suitable as a probiotic agent for ruminants. To test whether the potential positive effects of *S. boulardii* could be attributed to the yeast's viability or to its content of nutrients, living and autoclaved yeasts were tested simultaneously. For this purpose, incubation trials were carried out using the long-term rumen simulation technique. Living and autoclaved yeasts were added to fermentation vessels at a concentration of 0.5 or 1.5 grams/day. The addition of living and autoclaved yeasts stimulated microbial metabolism, with no major differences between the treatments. It was concluded that ruminal microbes digested the supplied yeast of *S. boulardii* as an additional substrate and that *S. boulardii*, at least in ruminants, is utilized as a prebiotic rather than as a probiotic agent.

Addition of Branched-chain Volatile Fatty Acids

In vitro addition of branched chain volatile fatty acids (isobutyric and isovaleric acids) as 1% of the feedstuff increased both the apparent dry matter digestibility and microbial growth in the rumen (Russel and Sniffen, 1984). The positive effects on microbial nitrogen synthesis are greater when protein sources have a low ruminal degradability. In such conditions, branched-chain volatile fatty acids supply branched-chain carbon skeletons which are lacking in bacteria because amino acids (leucine and valine) are not released. *In vitro* trials conducted on dairy cows have shown that the maximal dose was around 89 g per animal per day, which amounts to 0.45% of the diet. Investigations on these type of probiotic compounds might be worthwhile, because as they are usually produced in the rumen and metabolized by bacteria.

Feed Distribution Method

When the nature of the ration is known, the efficiency of its utilization can be improved by modifying the conditions of its distribution so as to obtain regular fermentations in the rumen. Several frequent feeding lessened the drop in milk fat content generally observed in the dairy cow with restricted diets (Sutton *et al.*, 1985). With mixed diets containing up to 60% concentrates, the flow rate of microbial nitrogen in the intestine doubled when the feeding schedule was changed from 1 to 2 meals per day to continuous distribution (Tamminga, 1981).

REVIEW QUESTIONS

1. Enumerate various ruminal microflora. What are their role in ruminal digestion of fibres? How can they be manipulated to improve ruminal digestion?

REFERENCES

Chalupa, W. (1971). "Manipulating rumen fermentation." *J. Anim. Sci.* 45: 585–599.

Colombatto, D., Mould, F.L., Bhatt, M.K., Morgavi, D.P., Beauchemin, K.A. and Owen, E. "Influence of fibrolytic enzymes on the hydrolysis and fermentation of pure cellulose and xylan by mixed ruminal microorganisms *in vitro*." *J. Anim. Sci.* 81:1040–50.

Fliegerova, K. (1993). "Multiresistant strains of *E. coli* isolated from the rumen of young calves." *Folia Microbiol.* 38: 363–366.

Hungate, R.E. (1966). (ed.). *The Rumen and its Microbes.* Academic Press, New York.

Jonecova, Z., Marekova, M. and V. Kmet. (1993). "The occurrence of antibiotic resistant strains of streptococci in the ingesta of calves." *Vet. Med.* 38: 360–372.

Jonecova, Z., Marekova, M. and V. Kmet. (1994). "Conjugative transfer of tetracycline resistance in rumen streptococcal strains." *Folia Microbiol.* 39: 83–86.

Klieve, A.V., Hudman, J.F. and Bauchop, T. (1986). "Bacteriophages in the rumen contents of sheep and cattle." *Aust. Microbiol.* 7: 150.

Machmüller, A., Soliva, C.R. and Kreuzer, M. (2003). "Effect of coconut oil and defaunation treatment on methanogenesis in sheep." *Reprod. Nutr. Dev.* 43:41–55.

Oeztuerk, H., Schroeder, B., Beyerbach, M. and Breves, G. (2005). "Influence of living and autoclaved yeasts of *Saccharomyces boulardii* on *in vitro* ruminal microbial metabolism." *J. Dairy Sci.* 88:2594–2600.

Russell, J.B. and Sniffen, C.J. (1984). "Effects of C4 and C5 volatile fatty acids on the growth of mixed rumen bacteria *in vitro*." *J. Dairy Sci.* 67: 987–994.

Scott, K.P. and Flint, H.J. (1995). "Transfer of plasmids between strains of *E. coli* under rumen conditions." *J. Appl. Bacteriol.* 78: 189–193.

Sutton, J.D., Broster, W.H. Napper, D.J. and Siviter, J.W. (1985). "Feeding frequency for lactating cows: effects on digestion, milk production and energy utilization" *Br. J. Nutr.* 53: 117–130.

Tamminga, S. (1981). "Nitrogen and amino acid metabolism in dairy cows." Thesis, Wageingen, Holland. p. 143.

Tamminga, S., Van Vuuren, A.M., Van Der Koelen, C.J., Kattab, H.M. and Van Gils, L.G.M. (1983). "Further studies on the effects of fat supplementation of concentrates fed to lactating animals and site of digestion of dietary components." *Neth. J. Agri. Sci.* 31. 249–258.

Taya, M., Ohmiya, K. Kobayashi, T. and Shimizu, S. (1983). "Enhancement of cellular digestion by mutants from an anaerobe, *Ruminococcus albus*." *J. Ferment. Technol.* 61: 197–199.

Teather, R.M., Erfle, J.D., Crosby, W.L., Collier, B. and Thomas, D.Y. (1984). Cloning and expression in *E. coli* of cellulose genes from the rumen aerobe *Bacteroides succinogenes*. Abstract of the Annual Meeting of Amer. Soc. Microbiol., Washington, D.C.p. 194.

Thivend, P. and Jouany, J.P. (1985). "Influence of lasalocid sodium on rumen fermentation and digestion in sheep." *Reprod. Nutr. Develop.* 23: 817–828.

Wallace, R.J., Czerkawski, J.W. and Breckenridge, G. (1981). "Effect of monensin on the fermentation of basal rations in the rumen simulation technique (Rusitec)." *Br. J. Nutr.* 46: 131–148.

12

POLYMERASE CHAIN REACTION

INTRODUCTION

The polymerase chain reaction (PCR) is an *in vitro* technique which allows the amplification of a specific deoxyribonucleic acid (DNA) region that lies between two regions of known DNA sequences. Before going into PCR technique, we should know about the DNA molecule. In its native state, DNA exists as a double helix. This helix comprises of two single strands of DNA running anti-parallel to each other and held together non-covalently by hydrogen bonds. The hydrogen bonds form between the complementary bases, i.e., adenine (A) with thymine (T) and guanine (G) with cytosine (C). The bases are each attached to a sugar molecule, deoxyribose and each sugar molecule is joined to the adjacent sugar molecule via a phosphate group. One unit of a DNA comprises a phosphate group, a sugar and a base and is known as nucleotide; sugar and a base is known as nucleoside. The numbers, for the sugar carbons each have a 'prime', e.g. 5´ and 3´. It is the 5´ and 3´ carbons of adjacent sugars that are linked via phosphate groups, hence the single-stranded DNA and the related molecule, ribonucleic acid (RNA) will have a 5´and a 3´end. The 5´ end of one strand of double-stranded DNA is complementary to the 3´ end of the other strand. RNA has a base uracil (U) instead of thymine (T), and has a hydroxyl group at its sugar moiety (ribose) at the 2´ position. In RNA, uracil (U) can base-pair with adenine (A).

The PCR amplification of DNA is achieved by using oligonucleotide primers, also known as amplimers. These are short, single-stranded DNA molecules which are complementary to a defined sequence of DNA template. The primers are extensions of a single-stranded DNA (template) denatured by a DNA polymerase, in the presence of deoxynucleoside triphosphates (dNTPs) under suitable reaction conditions. This results in the synthesis of new DNA strands complementary to the template strands.

These strands exist at this stage as double-stranded DNA molecules. Strand synthesis can be repeated by heat denaturation of the dsDNA, annealing of primers by cooling the mixture and primer extension by DNA polymerase at a temperature suitable for the enzyme reaction. Each repetition of strand synthesis comprises a cycle of amplification. Each new DNA strand synthesized becomes template for any further cycle of amplification and so the amplified target DNA sequence is selectively amplified, cycle after cycle.

The first extension products result from DNA synthesis on the original template and these do not have a distinct length as the DNA polymerase will continue to synthesize new DNA until it either stops or is interrupted by the start of the next cycle. The second cycle extension products are also of intermediate length; however, at the 3rd cycle, fragments of "target" sequence are synthesized which are of defined length corresponding to the position of the primers on the original template. From the 4th cycle onwards, the target sequence is amplified exponentially (Figure 12.1). This amplification, as a final number of copies of the target sequence, is expressed by the formula, $(2^n - 2n)X$, where

n = number of cycles

$2n$ = first product obtained after cycle 1 and second product obtained after cycle 2 with undefined length

X = number of copies of the original template

Potentially after 20 cycles of PCR, there will be 2^{20}- fold amplification, assuming 100% efficiency during each cycle. However, in practice, only 20–30% efficiency is achieved in PCR methods.

PCR was invented by Kary Mullis in 1983 while working for Cetus Corporation in California. In 1993, the Nobel Prize for Chemistry was awarded to Dr. Mullis for having invented PCR.

The original PCR protocols used Klenow fragment of *E. coli* DNA polymerase I to catalyse the oligonucleotide extension. However, this enzyme is thermally inactivated during the denaturation step of a PCR cycle and so the researchers had to add a fresh aliquot of enzyme at each cycle, to the amplification process. In later years, a thermostable polymerase like *Taq* polymerase was discovered for use in PCR methods.

Nucleic acid amplification techniques are most useful for the detection and characterization of viruses for which cell culture and serological methods are difficult, expensive or unavailable. Especially, the very high level of sensitivity provided by DNA amplification makes PCR the method of choice to detect viral DNA directly in clinical samples. The most obvious advantage of PCR for virus diagnosis is the selective amplification of extremely small number of viral DNA or RNA molecules in clinical samples to amounts sufficient for detection by simple methods. The high sensitivity of PCR makes this technique especially appropriate for diagnosis of viral infections where viral antigens or virus-specific antibodies cannot be detected and the presence of the viral genome is the only evidence of infection. This is particularly true for latent virus infections including herpesvirus, retrovirus or papilloma virus infections.

PCR is most useful to detect "non-cultivable" viruses such as certain enteric adenoviruses, papilloma viruses, astrovirus or rotaviruses, that are difficult, impossible and/or tedious to cultivate or viruses that grow without visible CPE such as respiratory syncytial virus, mucosal disease virus, coronavirus or certain gamma herpesviruses.

PCR can be used for the rapid detection of those pathogens whose *in vitro* cultivation is difficult, time-consuming or unavailable. RFLP patterns using PCR-amplified DNA is an excellent method for bacterial typing and has already been used for the identification of the bacterial strains involved in human food-borne outbreaks (Hill, 1996). Parasitic infestations will probably be the last field of veterinary clinical diagnosis to incorporate PCR techniques.

STANDARD PCR

PCR is typically done in a 50 or 100 µl volume and in addition to the sample DNA, it contains PCR buffer (50 mM, KCl, 10 mM tris. HC1, pH 8.3 at room temperature, 1.5 µM $MgCl_2$) 100 µ/ml gelatin, 0.25 µM of each primer, 200 µM of each deoxynucleoside triphosphate (dATP, dCTP, dGTP and dTTP) and 2.5 units of *Taq* polymerase, 25 µl of mineral oil is added at the top of the reaction mixture to prevent evaporation during denaturation. The amplification can be carried out in a thermal cycler. The machine can be programmed to go for a specified number of cycles. Usually 20–30 cycles are performed for each amplification. Each cycle consists of 3 segments: denaturation of target DNA at 94°C for 1–2 minutes, annealing of primers to target DNA at 55°C for 1–2 minutes and synthesis of new strands at 72°C for 1–2 minutes.

Template DNA

The sample may be from a variety of tissues and cell types. The samples may be single-or double-stranded DNA or RNA.

If the starting material is RNA, the first strand cDNA is prepared using reverse transcriptase. If high molecular weight genomic DNAs are being used, then amplification is improved by digesting DNA with rare-cutting restriction enzyme (e.g. *Not* I or *Sfi* I).

In many instances, enough of the DNA from a small number of tissue culture cells is made accessible to PCR mainly by the lysis of the cells during the heat-denaturation step. Methods of DNA purification from animal cells often involve the use of detergents to solubilize cell components and proteolytic enzymes to digest away proteins. This procedure is usually followed by extraction with organic solvents to remove residual protein and membrane components, followed by precipitation of nucleic acids by ethanol to remove traces of the organic solvents.

The extraction of DNA from fresh tissues is achieved by incubating the sample with proteinase K or simply by boiling in water. Proteinase K incubations are time-consuming, but result in high yield and good quality DNA. More rapid DNA extraction is achieved by boiling the tissue in sterile distilled water for about 15 minutes. This approach provides DNA of sufficiently good quality for most PCR applications, although the yield and quality of DNA are usually lower than for proteinase K extraction.

Low concentration of DNA is required for optimal PCR because both the primers and dNTPs should be in excess; an overabundance of template DNA will form annealing of the template sequence, rather than their annealing to the primer pair and will also increase the chances of forming non-specific products. Typically, a single copy gene can be amplified sufficiently in 30 cycles from less than 0.2 μg of DNA.

Primers

Primers are designed to be exactly complementary to the template DNA. The primers used in PCR are between 20–30 nucleotides in length. The primers, should if possible be made with an approximately equal number of each of the four bases, avoiding regions of unusual sequences such as stretches of polypurines or polypyrimidines. Primers should not be complementary to each other, otherwise, they will form **primer-dimer**. Primer-dimer is the amplification artifact often observed in the PCR product, especially when cycles of amplifications of a sample containing very few initial copies of template are performed. It is a double-stranded fragment whose length is very close to the sum of the two primers and appears to occur when one primer is extended by the polymerase over the other primer.

The primer concentration should not be greater than 1 μM. Higher primer concentration promotes mispriming, the formation of primer or the generation of non-specific products, thereby reducing the yield of the desired product.

Primer design and optimization A simple set of rules for the design of PCR primers includes the following.

1. Primer size should range from 15–30 bases in length.
2. Base composition should be 50–60% of guanine + cytosine.
3. Long runs with more than three or four of the same base should be avoided.
4. Primers should not have secondary structures (e.g. Hairpin loops).
5. Ideally, primers should not contain sequences that are complementary to each other.
6. Palindromic sequences should be avoided.
7. Primer melting temperature (T_m) between 55–80°C is preferred.

The melting temperature (T_m) for a particular primer can be calculated using the formula.

$$T_m = [(\text{no. of A + T}) \times 2°C + (\text{no. of G + C}) \times 4°C].$$

Many laboratories use annealing temperatures of 3–5°C below the T_m value.

The annealing temperature used in the PCR is dependent on both the length and composition of the primers. Ideally, the annealing temperature should be between 1–5°C lower than the lowest T_m value. Too low an annealing temperature results in non-specific annealing and therefore non-specific amplification. Too high an annealing temperature leads to a reduced yield of product.

Deoxynucleoside Triphosphates (dNTPs)

High purity dNTPs are supplied by several manufacturers either as four individual stock solutions or as a mixture of all four dNTPs. PCR is normally performed with dNTP concentration around 100 μM. However, the PCR optimally performed with dNTPs depends on (a) the $MgCl_2$ concentration, (b) the reaction stringency, (c) the primer concentration, (d) the length of the amplified product and (e) the number of cycles of PCR. Usually, each dNTP is used at a concentration between 50 μM and 200 μM. Higher concentration encourages misincorporation by the DNA polymerase. Concentration of 50 μM and 200 mM of each dNTP is sufficient to synthesize 6.5 μg and 25 μg of DNA, respectively.

PCR Buffers and $MgCl_2$

There are several buffers available for PCR and the most commonly used buffer has the following components in a 10 times concentration and must be diluted 1 : 10 (v/v) prior to use.

100 mM tris. HCl, pH 8.3 at room temperature

500 mM KCl

15 mM MgCl$_2$

0.1% (w/v) gelatin

The appropriate concentration of magnesium ions in the reaction buffer is also important for maximal *Taq* polymerase activity and because the dNTPs bind magnesium ions, the reaction mixture must contain an excess magnesium ions. As a rule of thumb, the magnesium concentration in the reaction mixture is generally 0.5 to 2.5 mM greater than the concentration of dNTPs. This concentration of magnesium ions also influences the efficiency of primer to template annealing. As a result, it is possible to modify the magnesium concentration rather than the annealing temperature to regulate primer specificity.

The MgCl$_2$ concentration in the final reaction mixture can be varied usually within the range of 0.5 to 5.0 mM. Mg^{2+} ions form a soluble complex with dNTPs which is essential for dNTP incorporation; they also stimulate the polymerase activity and increase the T_m of the double-stranded DNA and primer/template interaction. The concentration of MgCl$_2$ can have a dramatic effect on the specificity and yield in PCR (optimal concentration of MgCl$_2$ is between 1.0 to 1.5 mM for most reactions). Generally, insufficient Mg^{2+} leads to low yields and excess Mg^{2+} will result in the accumulation of non-specific products.

Salts like KCl and NaCl may help to facilitate primer annealing, but concentration of 50 mM will inhibit *Taq* polymerase activity. Detergents such as Tween 20, Triton X-100 or Nonidet P-40 and/or extra protein (e.g. gelatin or bovine serum albumin) may also be added to the reaction buffer. The addition of these reagents helps to prevent precipitation of the hydrophobic *Taq* polymerase in aqueous solutions.

PCR Enzymes

Initially, DNA polymerases such as Klenow fragment of *E. coli* DNA polymerase I and T4 DNA polymerase have been used which are heat-labile and hence fresh aliquots of enzymes have to be added at the start of each PCR cycle. In contrast, the thermostable enzyme like *Taq* DNA polymerase can be added in a single addition at the beginning of the amplification process without further addition during the reaction. The most commonly used DNA polymerase is *Taq* DNA polymerase, which is isolated from a bacterium found in hot springs known as *Thermus aquaticus*. This works optimally at 72°C and over the pH range of 8.0 to 7.5, adding ~100 nucleotides/sec. to the primer under these conditions.

One drawback of *Taq* polymerase is its lack of 3′ to 5′ exonuclease (proof reading) activity, which can lead to the misincorporation of nucleotides (~1 in 9000 bases). Table 12.1 lists some of the polymerases commonly in use in PCR.

Table 12.1 Thermostable DNA polymerases and their sources

DNA polymerase	Natural/recombinant	Source
Taq	Natural	*Thermus aquaticus*
Amplitaq[R]	Recombinant	*T. aquaticus*
HotTub[TM]	Natural	*Thermus flavus*
Vent[TM]	Recombinant	*Thermococcus litoralis*
Tth	Recombinant	*Thermus thermophilus*
Pfu	Natural	*Pyrococcus furiosus*
UITma[TM]	Recombinant	*Thermotoga maritima*

DNA polymerases are enzymes which catalyse the synthesis of long polynucleotide chains from monomer deoxynucleoside triphosphate using one of the original parental strands as template for the synthesis of a new complementary strand. DNA synthesis always proceeds in a 5′ to 3′ direction, since polymerization is always from the 5′ α-phosphate to the 3′ terminal hydroxyl group of the growing DNA strand. DNA polymerase, unlike RNA polymerase, requires a short DNA segment or primer, to anneal to a complementary sequence and strand synthesis. The dNTPs attach to the free 3′ hydroxyl group of the primer and form a strand complementary to the template strand.

Taq/Amplitaq[R] DNA polymerase   The thermostable DNA polymerase from *Thermus aquaticus* (*Taq*) has been the most extensively used enzymes in PCR. *T. aquaticus* was first isolated from a hot spring in Yellowstone National Park. *Taq* DNA polymerase has an optimal extension rate (polymerization rate) of 35–100 nucleotides per second at 70–80°C which is the optimum temperature range for the enzyme. Processivity, which is the average number of nucleotides incorporated before the enzyme dissociates from the DNA template, is relatively high for *Taq* DNA polymerase.

Taq DNA polymerase and Amplitaq[R] have a 5′ to 3′ exonuclease activity which removes nucleotides ahead of the growing chain. *Taq* DNA polymerase has been cloned and a modified version expressed in *E. coli* is Amplitaq[R] (Perkin Elmer, Cetus). Since Amplitaq[R] is a recombinant enzyme, the purity and reproducibility of this enzyme is higher than for *Taq* polymerase. However, PCR of bacterial targets contain DNA sequences homologous to those found in

E. coli (which is the host used for expression and production of Amplitaq^R). It may therefore be preferable to use *Taq* DNA polymerase rather than Amplitaq^R to avoid potential contamination of the PCR with DNA from the enzyme itself.

Vent™ DNA polymerase   Vent™ DNA polymerase was first isolated from *Thermococcus litoralis*, which is a thermophilic archaebacterium found on ocean floors at temperatures of up to 98°C. The gene encoding this enzyme has been cloned and expressed in *E. coli* (New England Biolabs). This polymerase is more thermostable than *Taq* DNA polymerase and is capable of extending primers to give products up to 8–13 kb in length. Vent™ polymerase possesses a 3´ to 5´ exonuclease activity that is responsible for the high level of fidelity, which is 5 to 15-fold greater than that of *Taq* DNA polymerase.

Pfu DNA polymerase   This polymerase isolated from the hyperthermophilic marine archaebacterium *Pyrococcus furiosus*, possesses both 5´ to 3´ DNA polymerase activity and 3´ to 5´ exonuclease proof-reading activity. The fidelity of DNA synthesis is 12-fold higher than that of *Taq* DNA polymerase.

Pfu DNA polymerase is available commercially from Stratagene, as a genetically engineered mutant of cloned *Pfu* DNA polymerase. When setting up reaction with this enzyme, it is essential to add the enzyme last (as for Vent™ DNA polymerase), since in the absence of dNTPs the 3´ to 5´ exonuclease activity of the enzyme results in the degradation of templates and primers.

Tth DNA polymerase   Five DNA polymerases have been isolated from *Thermus thermophilus* (*T*th) and two forms have been shown to be efficient in PCR. Recombinant *T*th DNA polymerase is a thermostable polymerase obtained by the expression in *E. coli* of a modified form of the DNA polymerase gene from *T. thermophilus*. In the presence of manganese, at a temperature around 70°C, this polymerase is used to reverse-transcribe RNA efficiently. Subsequent PCR can be performed in the same tube using the intrinsic DNA polymerase activity simply by chelation of manganese cation and the addition of magnesium. The use of *T*th DNA polymerase for RNA-PCR, therefore, has the advantage over conventional RNA (RT) PCR where a reverse transcriptase is used to synthesize cDNA and then a second reaction is required with thermostable DNA polymerase for the PCR step.

The thermostability and the reverse transcriptase (RT) activity of *T*th DNA polymerase is useful in amplifying DNA from RNA templates that contain G-C-rich sequences or secondary structures since the elevated temperatures serve to denature the template RNA.

UlTma™ DNA polymerase   UlTma DNA polymerase is encoded by a recombinant modified form of the *Thermotoga maritima* DNA polymerase gene. *T. maritima* is hyperthermophilic and was first isolated from geothermally heated marine sediments in Italy. UITma DNA polymerase exhibits greater thermostability than Amplitaq DNA polymerase. This enzyme is not associated with 5´ to 3´ exonuclease activity, but does have an inherent 3´ to 5´ exonuclease proof-reading activity.

Inhibitors and Enhancers of PCR

Many different types of biological specimens are used for PCR, including animal tissues and body fluids, bacterial samples, forensic and archaeological materials and plant tissues. Human DNA is typically obtained from peripheral blood cells, urine, faecal samples, cells, smears, hair roots, semen, cerebrospinal fluid, biopsy material, amniotic fluid, placenta and chorionic villus. In addition to these sources, DNA from animals and birds may be extracted from tail sections and feather roots, respectively. Many of these crude preparations contain inhibitory substances which have not been identified. Heparin, a commonly used anti-coagulant, should not be used for collection of blood if PCR is to be performed on the extracted DNA, since heparin is a potent PCR inhibitor. Other substances in the blood, probably porphyrin compounds which are also strong inhibitors of PCR, can be eliminated from the DNA preparation by lysis of the red blood cells and centrifugation to pellet the white cells.

Extraction of DNA from different sources routinely used detergents for cell lysis and denaturation. Often non-ionic detergents (e.g. Triton X-100, Tween-20, Nonidet P40) are used and these generally do not inhibit PCR at concentrations of up to 5%. On the other hand, ionic detergents such as sodium dodecyl sulphate (SDS) can only be tolerated at extremely low concentration and therefore should be removed by phenol extraction and used at the level of 1–2% (v/v) and SDS inhibits *Taq* DNA polymerase at concentrations higher than 0.01% (w/v). The inhibitory effects of low concentration of SDS (e.g. 0.01%) can be removed by certain non-toxic detergents (e.g. 0.5% Tween-20 and Nonidet P40).

Many DNA extraction procedures are performed in the presence of proteinase K or pronase that will digest denatured proteins. *Taq* DNA polymerase is susceptible to protease digestion and so proteinase K must be removed or inactivated. Thermal denaturation at 95°C is sufficient to achieve this. Heat treatment is usually followed by phenol extraction which will also denature the proteinase K.

Many researches have reported substances that can be added to PCR to increase the efficiency or specificity. Formamide at a concentration of 5%, dimethyl sulphoxide (DMSO) at a concentration of 10%, polyethylene glycol 6000 at a concentration of 50–15%, glycerol at a concentration of 10–15% and Tween-20 at 0.25% are found to enhance PCR amplification.

Steps in PCR

The components of PCR are mixed in reaction tubes. The mixture is covered with mineral oil to avoid evaporation and the tubes are placed in a thermal cycler-PCR machine. The thermal block is controlled by a microprocessor in order to increase and to decrease the temperature rapidly in accordance with a preselected programme. PCR commences with an extended initial denaturation step (usually 5 minutes at 94°C), which ensures that there is complete strand separation of the template DNA. The reaction mixture is then cycled usually between 20–35 times. A typical temperature programme would be:

Denaturation at 94°C for 20–30 seconds

Primer annealing at ~50°C for 20–60 seconds

Extension at 72°C for 30–60 secconds (one min./kb)

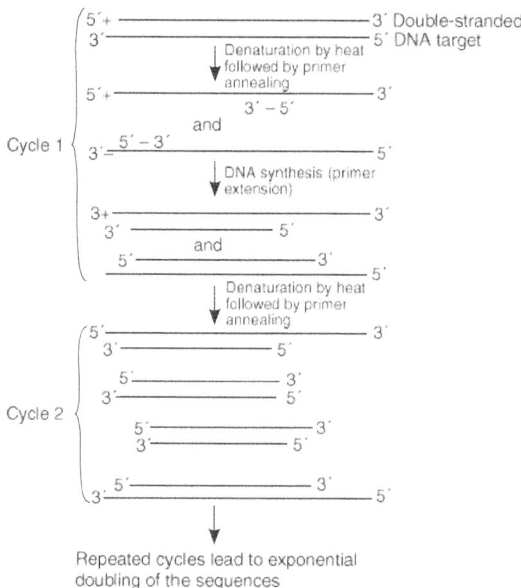

Figure 12.1 PCR thermal cycles of denaturation of target DNA, annealing of primers and synthesis of new strand

Cycling concludes with a final extension step at 72°C, usually for 5 minutes, assuming that all template DNA is in double-stranded conformation. The PCR is run by repeated cycles of heating and cooling (Figure 12.1). Each cycle consists of 3 steps; denaturing double-stranded (ds) DNA; annealing of primers to the single-stranded (ss) DNA template and extending nascent strands using *Taq* DNA polymerase enzyme.

Denaturation   The reaction mixtures are heated to approximately 90–95°C. At this temperature, the dsDNA is denatured, i.e., the ds molecules are separated to form single strands. This step is needed for annealing of the primer to the template DNA.

Primer annealing   The temperature of the thermal block is programmed to a predetermined value between 37°C and 70°C depending on the type of the primer, to allow specific annealing between the primers and the single strands of the target DNA. A suitable temperature is selected that allows only specific annealing (by base-pairing) between exact matches of primer and target sequences and does not allow non-specific annealing to other sequences. The optimal annealing temperature mainly depends on the length and on the guanine/cytosine (GC) content of the primers.

DNA synthesis (Primer extension)   The temperature is increased to between 67°C and 72°C. At this temperature, *Taq* DNA polymerase extends the primers by using each ss target as a template for the construction of a new complementary strand. This results in duplication of both original DNA strands. One PCR cycle is completed within 3–5 min.

Repeating the cycle   On completion of the first cycle, the thermal block is heated up again and begins the second round of amplification. In this cycle, the double strands will separate and serve yet again as templates for new DNA synthesis. Accordingly, each PCR cycle results in the duplication of the DNA target. If the number of cycles is n, then the amplification results in 2^n exponential increase in DNA. By performing 25–30 cycles, a 33.6 million-fold amplification of the target can theoretically be achieved within 3 hours. However, due to decrease of enzyme activity and other factors, one could expect more than approximately 10^6-fold amplification of the target in a single PCR assay. Figure 12.2 shows a PCR thermal cycler.

Figure 12.2   A PCR thermal cycler with two blocks with independent controls

Detection and Analysis of PCR Products

The PCR products or amplicons consist of a fragment (or fragments) of DNA which is normally of a length defined by the boundaries of the PCR primers. PCR products are generally less than 10 kb in length. Many techniques can be used to detect and confirm the identity of the amplified products. Normally, the simplest or most commonly used method is electrophoresis of an aliquot of the PCR product on an agarose or polyacrylamide gel and visualization by staining with ethidium bromide, which is a fluorescent dye that intercalates into the DNA. After staining, ultraviolet transillumination allows visualization of the DNA in the gel.

Detection and confirmation of identity of a PCR product can be performed by Southern blot hybridization or dot blot hybridization and detection with radioactive or non-radioactive specific probes. The Southern blot techniques involve capillary blotting of the agarose or polyacrylamide gel onto a nitrocellulose or nylon membrane. The filter is hybridized with a probe labelled with ^{32}P or non-radioactive label like digoxigenin label.

The PCR product can also be digested with restriction enzyme and the product is run on agarose gel for further characterization.

MODIFICATIONS OF PCR

There are several variations to the basic PCR technique, which allow varied applications of this technique.

Inverse PCR (Chromosome Crawling)

In this method, the amplification of those DNA sequences, which are away from the primers and not those which are flanked by the primers are done. For instance, if the border sequences of a DNA are not known and those of a vector are known, then the sequence to be amplified may be cloned in the vector and the border sequence of the vector may be used with primers in such a way that the polymerization proceeds in an inverse direction, i.e., away from the vector sequence flanked by the primers and towards the DNA sequence of inserted segment. Similarly, if the gene sequence is known, one can use its border sequences as primers, for an inverse PCR, to amplify the sequences flanking this gene, e.g. the regulatory sequences.

In this technique, the primers are designed so that their 3´ends face away from each other. Carrying out PCR will not be expected to amplify the target sequence, since the extension products of either primer will not contain sequences complementary to each other. However, if the linear DNA molecule containing the target and primer sequences is first circularized by restriction enzyme digestion followed by ligation, the target can be amplified by PCR.

This clever modification of the basic PCR technique allows amplification of unknown DNA sequences to one or the other or both sides of a known DNA segment (depending upon which restriction enzyme sites are chosen) with a result, it resembles chromosome walking by traditional cloning methods.

Anchored PCR

In the basic PCR technique and in inverse PCR, one has to use two primers representing the sequences lying at the ends of sequences to be amplified. But sometimes, we may have knowledge about the sequence at only one of the two ends of the DNA sequences to be amplified. In such cases, anchored PCR may be used, which will utilize only one primer instead of two. In this technique, due to the use of one primer, only one strand will be copied first, after which a poly(G) tail will be attached at the end of the newly synthesized strand. This newly synthesized strand with poly(G) tail at its 3´ end will then become template for the daughter strand synthesis utilizing an anchor primer with which a poly(C) sequence is linked to complement with poly(G) of the template. In the next cycle, both the original primer and anchored primer will be used for gene amplification. There are two anchored PCR methods for DNA isolation: rapid amplification of cDNA ends (RACE-PCR) and ligation-anchored PCR.

RACE-PCR   This is a method by which the 3´ and 5´ ends of a cDNA are amplified using a small stretch of known sequence within a gene. This small stretch can be derived by back translation from the amino acid sequence of the protein or can be from DNA sequence homology with other members of the gene family. The procedure was made more powerful by the use of nested primers, which reduce non-specific amplification and ensure the production of relatively pure specific product. For 3´ RACE, a reverse transcriptase primer that consists an oligo(dT) sequence linked to an adapter sequence (which may be long enough to permit binding of two nested primers or shorter for single primer binding) is used to prime the first strand cDNA synthesis. Primary amplification is then performed with a gene-specific primer (GSP-1) and the outer primer R1. A small fraction of the first amplification is then used for secondary amplification that is nested gene-specific primer (GSP-2) and inner primer R2.

For 5´ RACE, cDNA is prepared with either a specific primer or random hexamers and then the newly synthesized cDNA strand has homopolymeric tail (e.g. A residues) added to its 3´ end using terminal transferase. The complementary reverse transcription anchor primer is then used to generate second strand cDNA. This dsDNA can then serve as the template for PCR reaction as described for 3´ RACE.

Ligation-anchored PCR   This is a simple, efficient and sensitive technique for the amplification of cDNAs where the 5´ end has unknown sequence. In this method, T4 RNA ligase is used to covalently link the anchor oligonucleotide to the first strand of cDNA. Amplification is then carried out with one primer specific for a sequence within the gene of interest and one primer specific for the anchor. The anchor oligonucleotide must be 5´ phosphorylated. This is necessary for ligation of the cDNA. The 3´ end of the anchor oligonucleotide is blocked, for example, by the addition of a dideoxynucleotide using terminal deoxynucleotidyl transferase or by the incorporation of amino acid group during oligonucleotide synthesis. This 3´ modification is required to prevent the ligation of more than one anchor to the cDNA.

Reverse Transcriptase PCR

The RT-PCR protocol requires two separate major reactions, a reverse transcriptase step, followed by PCR amplification. Recently, the *Tth* DNA polymerase, isolated from *Thermus thermophilus,* is found to have the properties of both reverse transcription and DNA amplification. This enzyme, in the presence of manganese can reverse-transcribe RNA. Because the *Tth* DNA polymerase can utilize both DNA and RNA templates, the whole procedure can be carried out in a single tube.

For mRNA that contains a poly(A) tract at the 3´ end, oligo(dT), random hexamers or a gene-specific primer can be used to prime cDNA synthesis. RT-PCR is a highly sensitive tool in the study of gene expression at the RNA level and in particular, in the quantitation of mRNA or viral RNA levels. This technique, also known as message amplification phenotyping (MAPPing), permits simultaneous analysis of a large number of mRNAs from small numbers of cells. RT-PCR can also be used as a first step in preparing a cDNA library by PCR of all of the mRNAs in a sample of cellular DNA.

Reverse transcriptase is usually used to synthesize first-strand cDNA from RNA. Reverse transcriptase can be purified from several sources, e.g. avian myeloblastosis virus (AMV) and Moloney murine leukaemia virus (MMLV). AMVRT is an RNA-dependent DNA polymerase that uses single-stranded RNA as a template and can synthesize a complementary DNA (cDNA) in the 5´ to 3´ direction if the primer is present. In addition, it has DNA polymerase activity and this enzyme also exhibits ribonuclease H activity, which is specific for RNA : DNA heteroduplex molecules.

MMLV-RT acts in the same way as AMV-RT; however, it lacks DNA endonuclease activity and has lower RNase H activity. Thus, it has a greater chance of producing full-length copies of large mRNA species. Figure 12.3 shows RT-PCR for the detection of Newcastle disease virus *F* gene showing 254 bp amplicon in positive cases.

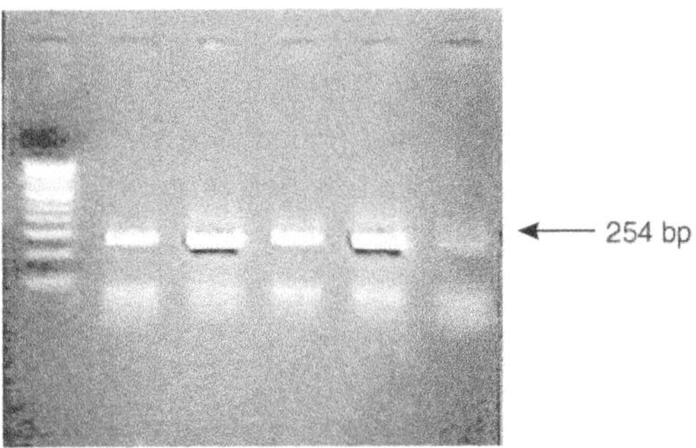

Figure 12.3 RT-PCR for Newcastle disease virus *F* gene showing 254-bp amplicon in positive cases

RT-PCR with primers targeted to the 3´ terminal portion of the nucleoprotein gene (N) was used with neck-skin samples of nine patients who had rabies for detection of rabies amplicons (Macedo *et al.*, 2006). Six of the eight post-mortem samples were found to be positive for rabies by RT-PCR, and one of the two samples collected ante-mortem was positive with this same technique. Results were confirmed by DNA sequencing. It was concluded that RT-PCR applied to neck-skin biopsies could allow early diagnosis and lead to more effective rabies treatment.

Two sets of specific primers for a one-step RT-PCR and a nested PCR, targeting a 640-bp fragment and a 297-bp fragment, respectively, were selected from the highly conserved region of the nucleocapsid protein (*NP*) gene of CDV (Shin *et al.*, 2004). These methods were used to amplify a part of canine distemper virus *NP* gene of a CDV vaccinal strain and samples of urine, blood, nasal discharge and saliva from 29 dogs suspected of suffering CD. The expected 640-bp fragment of the *NP* gene was detected in 11/22 (50%) blood, 10/20 (50%) urine, 5/25 (20%) saliva and 6/27 (22.2%) nasal swab samples by one-step RT-PCR, whereas the nested PCR amplified an expected 297-bp fragment of the *NP* gene in 18/22 (81.8%) blood, 15/20 (75%) urine, 14/25 (56%) saliva and 19/27 (70.3%) nasal swab samples. It was found that the nested PCR detected CDV in blood, urine, nasal swab and saliva more frequently than did the one-step RT-PCR. Therefore, it was concluded that this assay could be useful for ante-mortem diagnosis of CDV infections in dogs. Saito *et al.* (2005) evaluated the use of the RT-PCR in urine samples to diagnose canine distemper virus in dogs with progressive neurological disease. A fragment of the nucleoprotein gene of CDV was amplified from the urine of 22 distemper dogs. RT-PCR of urine samples was more sensitive than serum and leucocytes and at least as sensitive as CSF to screen for distemper in dogs with neurological signs.

Nested PCR

The sensitivity and specificity of PCR can be increased by using nested PCR (nPCR). In nPCR, two separate amplifications are used. The first uses a set of primers that yields a large product, which is then used as a template for the second amplification. The second set of primers anneal to sequences within the initial product producing a second smaller product. The primers for the second round of amplification are either both different from the first set or both located within the amplified DNA region. If only one of the second round primers is located within the amplified region and is used together with one of the first round primers, it is termed as **semi-nested PCR**. Nested PCR increases the specificity of the reaction because formation of the final product depends upon the binding of two separate sets of primers, which may preclude the need for verification of the PCR product by blotting, restriction digestion or sequencing. The second set of primers also serves to verify the specificity of the first product. A typical nested PCR reaction is shown in Figure 12.4.

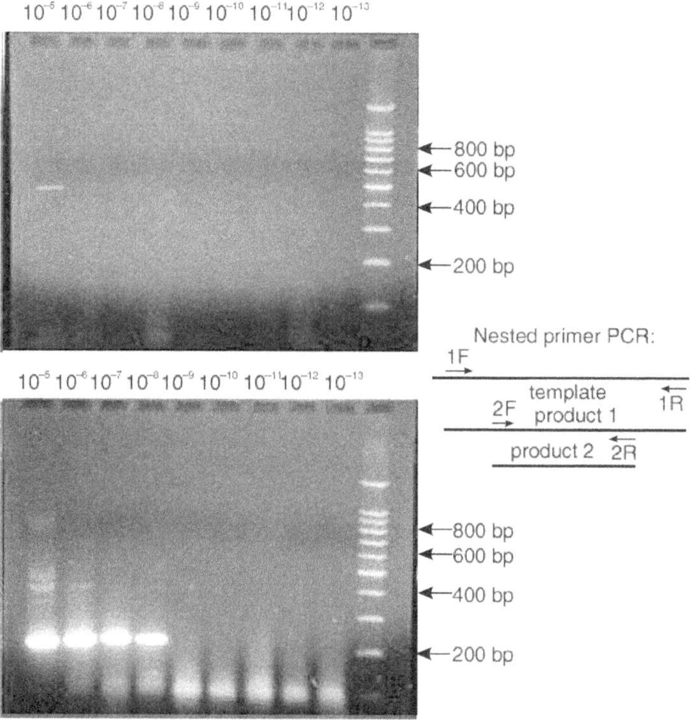

Figure 12.4 Nested PCR. The outer primers were able to detect the viral DNA down to a 10^{-5} dilution (top figure). In contrast, the full nested PCR system detected the viral DNA down to a 10^{-8} dilution.

Nested PCR primers are ones that are internal to the first primer pair. The larger fragment produced by the first round of PCR is used as the template for the second PCR. Nested PCR can also be performed with one of the first primer pair and a single nested primer. The sensitivity and specificity of both DNA and RNA amplification can be dramatically increased by using the nested PCR method. The specificity is particularly enhanced because this technique almost always eliminates any spurious non-specific amplification products. This is because, after the first round of PCR the non-specific products are unlikely to be sufficiently complementary to the nested primers to be able to serve as template for further amplification. Thus the desired target sequence is preferentially amplified. However, the increased risk of contamination is a drawback of this extreme sensitivity.

Asymmetric PCR

In this method, the addition of one primer in vast excess over the other during the PCR results in the generation of an excess of one amplified strand relative to the other, i.e., an asymmetric amplification. The single-stranded product is usually used for dideoxy sequencing. Asymmetric PCR can be performed directly on any given template, although reproducible high yields of single-stranded products are often difficult to obtain. Alternatively, it is more common to isolate a double-stranded PCR product by conventional symmetrical methods and re-amplify this by asymmetric PCR, using only one primer for approximately 20 cycles. The single-stranded product increases arithmetically with each cycle.

Dideoxy DNA sequencing of asymmetric PCR products can be performed with either the original PCR primer or by using a complementary sequence internal to the asymmetric PCR product. Asymmetric PCR sequencing is particularly useful for repetitive sequencing of a particular region where the asymmetric PCR conditions can be standardized.

Multiplex PCR

In multiplex PCR (mPCR), two or more primer pairs specific for different targets are included in the same amplification reaction. The major advantages of mPCR are conservation of reagents and template and reduction in preparation and analysis time required to identify multiple target sites in one assay, as opposed to running separate analysis for each target. However, the system requires careful optimization of the PCR conditions to ensure that one PCR reaction is not dominant over the other. Furthermore, the products obtained in each reaction should also be of different sizes to enable their visualization on agarose gel. The co-amplification of two or more products in a single reaction is dependent on the compatibility of the PCR primers used in the reaction. All the primers in the reaction must have similar melting temperature, so they anneal to and dissociate from complementary DNA sequences at approximately the same temperature; allowing each amplification to proceed at the selected temperature. Each amplification proceeds independently of the other and each specific amplification product is synthesized in an unencumbered way.

mPCR assay has been used for detection of *Campylobacter coli* and *C. jejuni* (Harmon *et al.*, 1997), *Clostridium perfringens* toxins (Meer and Songer, 1997), enterohaemorrhagic *E. coli* 0157:H7 toxic genes (Fratamico *et al.*, 2000), *Brucella abortus* and *Mycobacterium bovis* in cattle (Sreevatsan *et al.*, 2000), *Mycoplasma* sp. in small ruminants (Greco *et al.*, 2001) and *Staphylococcus aureus* enterotoxins (Mehrotra *et al.*, 2000).

A multiplex PCR was developed for rapid identification of *Mycobacterium bovis* in clinical isolates of both veterinary and human origin (Cobos-Marin *et al.*, 2003). They were also able to differentiate *M. bovis* from other members of *M. tuberculosis* complex using mPCR. A multiplex nested PCR was developed for the detection of *Brucella canis* and *L. interrogans* in the semen and cryoprotective agent (CPA) (Kim *et al.*, 2006). The results demonstrated the high sensitivity and simplicity of this technique in the detection of these organisms in canine semen.

A multiplex PCR was used for diagnosis of canine parvovirus infection and leptospirosis in dogs (Ramadass and Latha, 2005). Faecal samples for CPV infections and blood samples for leptospiral infection were prepared separately and PCR was carried out in a single tube with primers specific for CPV and *Leptospira*, which produced 536-bp and 285-bp amplicons for CPV and *Leptospira* respectively (Figure 12.5). In situations where concurrent infections with CPV and leptospirosis are suspected, this multiplex PCR would be useful.

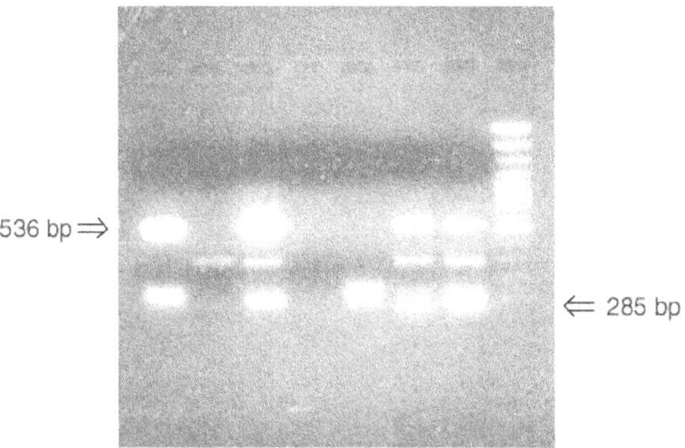

Figure 12.5 Multiplex PCR patterns of *Leptospira* (285 bp) and CPV (536 bp)

A multiplex polymerase chain reaction (PCR) assay was developed for the rapid detection of *Brucella ovis, Actinobacillus seminis* and *Histophilus somni* in fresh ram semen samples (Saunders *et al.*, 2007). The multiplex assay was based on the single PCR assays published for the detection of *A. seminis* and *B. ovis*, and the forward primer published for the detection of *H. somni*; an alternative reverse primer for *H. somni* was designed in this study. The multiplex PCR was far more successful in the detection of *H. somni* (45/295) than culture (23/295). *A. seminis* was also detected in more semen samples by multiplex PCR (29/295) than culture (13/295) and *B. ovis* was detected in three samples using both PCR and culture. This PCR could be used as a complementary test, or alternative to culture of ram semen and other biological samples for the detection *B. ovis, H. somni* and *A. seminis*.

Multiplex polymerase chain reaction (PCR) method was developed for differential detection of turkey coronavirus (TCoV), infectious bronchitis coronavirus (IBV), and bovine coronavirus (BCoV) (Loa *et al.*, 2006). Primers were designed from conserved or variable regions of nucleocapsid (N) or spike (S) protein gene among TCoV, IBV, and BCoV and used in the same PCR reaction. Reverse transcription followed by the PCR reaction was used to amplify a portion of *N* or *S* gene of the corresponding coronaviruses. The PCR products were detected on agarose gel stained with ethidium bromide. Two PCR products, a 356-bp band corresponding to *N* gene and a 727-bp band corresponding to *S* gene, were obtained for TCoV isolates. In contrast, one PCR product of 356 bp corresponding to a fragment of *N* gene was obtained for IBV strains and one PCR product of 568 bp corresponding to a fragment of *S* gene was obtained for BCoV. There were no PCR products with the same primers for Newcastle disease virus, Marek's disease virus, turkey pox virus, pigeon pox virus, fowl pox virus, reovirus, infectious bursal disease virus, enterovirus, astrovirus, *Salmonella enterica, Escherichia coli,* and *Mycoplasma gallisepticum.* These results indicated that the multiplex PCR is a rapid, sensitive, and specific method for differential detection of TCoV, IBV, and BCoV in a single PCR reaction.

A multiplex RT-PCR was used for simultaneous detection and differentiation of North American serotypes of bluetongue (BT) virus and epizootic hemorrhagic disease virus (EHDV) in cell culture and clinical samples (Aradaib *et al.*, 2003). Two pairs of primers (B1 and B4) and (E1 and E4) were designed to hybridize to non-structural protein 1 (NS1) genomes of BTV-11 and EHDV-1, respectively. The BTV primers generated a 790-bp product, while EHDV primers produced a 387-bp product. Two pairs of nested primers (B2 and B3) and (E2 and E3), internal to the annealing sites of primers (B1 and B4) and primers (E1 and E4), produced a 520-bp specific BTV and a 224-bp specific EHDV-PCR product from BTV and EHDV-first amplification products, respectively. These nested amplifications increased the sensitivity of the PCR assay and confirmed the specificity of the first amplified EHDV or BTV-PCR products.

A nested multiplex PCR was used for genotyping of rotavirus of neonatal calves isolated from Tamil Nadu (Saravanan *et al.*, 2006). Of 209 diarrhoeic faecal samples tested, nine samples were found positive for group A rotavirus by PAGE with silver staining. These positive samples were subjected to nested multiplex PCR using type-specific primers of common genotypes P(11), G10 and P(5), G6 to identify their prevalence. Three out of nine samples were found to be P(1), G10 genotype and none of the samples were of P(5), G6 genotype.

Arbitrarily Primed PCR

This method is also called as random amplified polymorphic DNA fingerprinting, is based on the amplification of genomic DNA with a single primer selected from an arbitrary nucleotide sequence. We use a short oligonucleotide primer to produce random, but reproducible sets of amplified DNA fragments. The multiple products resulting from RAPD analysis are then separated according to size by conventional agarose gel electrophoresis and the DNA banding patterns of different isolates can then be compared. This technique has proven to be very useful for subtyping diverse bacterial species.

This method has been used for characterization of *Brucella* sp. (Techerneva *et al.*, 2000). *Leptospira serovars* (Pershina *et al.*, 1999; Ramadass *et al.*, 2002), *Mycobacterium avium* complex strains from pigs and humans (Ramasoota *et al.*, 2001), *Mycoplasma bovis* isolates (Butler *et al.*, 2001), *Pasteurella multocida* isolates of pigs (Zucker *et al.*, 1986). A typical AP-PCR pattern is shown in Figure 12.6 differentiating different serovars of leptospires.

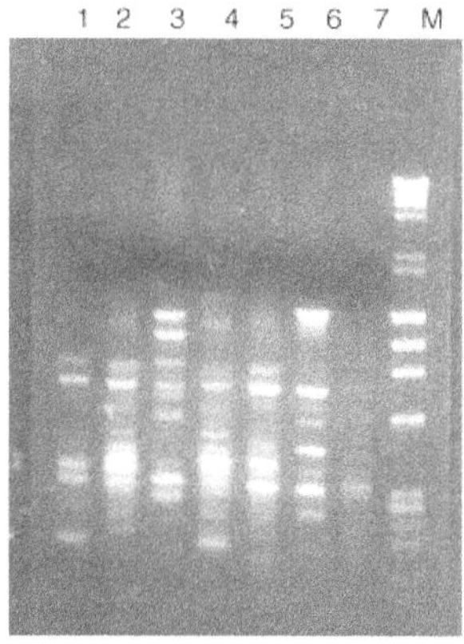

Figure 12.6 AP-PCR for Leptospires Lanes 1 to 7 show AP-PCR patterns of different leptospiral serovars Lane M—1 Kb DNA ladder

Quantitative PCR

To quantify template DNA in clinical specimens, approaches were made to develop quantitative PCR assays (Ferre, 1992). One approach involves the use of a competitive template that is largely identical to, but somehow discriminates the target sequence to be quantified (Piatak *et al.,* 1993). Known quantities of these competitive templates are introduced into the specimen with the unknown amount of the target DNA. When the amounts of target sequence and competitor templates are equivalent, equal amounts of the respective PCR products will accumulate. These can be quantified by a variety of techniques and the amount of target sequence present in the original specimen assessed by interpolation of the equivalence point in both products.

Single-strand Conformation Polymorphism PCR (SSCP-PCR)

In SSCP-PCR, the target sequence is first labelled and amplified simultaneously by the PCR of the genomic DNA and cDNA using labelled, substrates. The PCR product is then denatured and resolved by PAGE and mutations are detected as altered mobility of separated single strands in the autoradiogram. Thus, the overall procedure is rapid and simple. Primers or deoxynucleotides are also labelled efficiently in a batch sufficient for many analyses using g^{32}P-ATP and polynucleotide kinase. The amount of product required in SSCP-PCR for detection is much less than that in some other PCR-based techniques. One obvious use of SSCP-PCR is detection of DNA polymorphisms (Hayashi, 1991).

Vectorate PCR (vPCR)

This method enables the amplification of specific DNA fragments in situations where the sequence of only one primer is known. Thus, it extends the application of PCR to stretches of DNA where the sequence information is only available at one end. Three basic steps are involved in vPCR:

1. digestion of target DNA with a suitable restriction enzyme,
2. ligation of suitable synthetic oligonucleotide on to the digested DNA and
3. PCR assay using a specific primer and a primer directed towards the synthetic oligo.

In the first cycle of vPCR, only the known primer which is directed towards the sequence of interest will prime DNA synthesis. This will produce a complementary strand for the vectorate PCR primer to anneal to in the second cycle of PCR. In the second and subsequent cycles of PCR, both primers can prime the synthesis with the end result being that only the fragment amplified contains the sequence of interest. This technique can be used for (a) sequencing the termini of yeast artificial chromosome clone inserts, (b) to amplify human genome DNA, (c) for genomic walking, (d) mapping of introns in genomic DNA from cDNA clones, (e) sequencing of large clones without sub-cloning and (f) sequencing and mapping of regions containing deletions, insertions and translocations (Arnold and Hodgson, 1991).

Restriction Site-specific PCR (RSS-PCR)

RSS-PCR is a technique that is based on the principle of restriction fragment length polymorphism (RFLP), but which is unique in that it does not require the use of restriction endonucleases. The RSS-PCR method is based on the use of primers that are homologous to specific restriction enzyme recognition sequences that are 10–18 bp long. The primers are designed in such a way that they will amplify genomic DNA segments that lie between the restriction site sequences on which the primers are based. The rationale for this procedure, is that genetically different bacteria exhibit variations in the number and location of different restriction site sequences throughout the genome. The application of this method allows amplification of fragments of various lengths, yielding a unique collection of DNA fragments of "fingerprint" pattern for each different serotype. RSS-PCR is a modification of AP-PCR method which is used for rapid detection and characterization of pathogens. This method has been used for rapid detection and characterization of enterohaemorrhagic *E. coli* 0157: H7 strains in environmental samples (Kimura *et al.*, 2000).

Degenerate PCR

This is a technique in which nucleotide sequence of the primer is based on the sequence of the encoded protein. The genetic code is said to be degenerate because some amino acids are encoded by more than one codon. From the amino acid sequence of a protein, it is possible to design PCR primers based on all the principal codon sequences for each amino acid. Degenerate primers have a number of options at several positions in the sequence so as to allow annealing to and amplification of a variety of related sequences. Degenerate PCR is used to amplify (fish out) conserved sequences of a gene or genes from the genome of an organism and to get the nucleotide sequence after having sequenced some amino acids from a protein of interest. This method is used when there is evidence of highly conserved regions or motifs of amino acids that can be designed into degenerate primers; these regions may be conserved interspecies. Degenerate primers can then be used to fish out these sequences. Sequences amplified this way can then be sequenced to confirm that the sequence is correct. They can then be used as probes to fish out the gene of interest from a genomic library (prokaryotic) or a cDNA library (eukaryotic). This method is also used when there has been a successful isolation of a protein of interest. The terminals of this protein (or some amino acids) were then sequenced. The amino acid sequence can then be used to design degenerate primers. The PCR product can then be sequenced. If, the primers produce a truncated (gene) sequence or are partial (short), they can further be utilized as the probes mentioned above.

Allele-specific PCR

This method differentiates two sequences solely on the basis of their ability to bind to an oligo primer. Under optimal conditions for primer binding, it is possible to detect a single base mismatch because only precise primer-target complementarity will support amplification. This is a selective PCR amplification of one of the alleles to detect single nucleotide polymorphism (SNP). Selective amplification is usually achieved by designing a primer such that the primer will match/mismatch one of the alleles at the 3´-end of the primer.

Rep-PCR

This method is based on the presence of repetitive DNA sequences that are scattered throughout the prokaryotic genome and conserved across a wide range of bacterial species. The intervening stretches of DNA between these repeats vary among different species and strains and therefore allowed for the generation of a genetic fingerprint. Primers specific for the repeated sequence are used to amplify these polymorphic intervening sequences by PCR. Such amplification will occur whenever two of these repeats are sufficiently close to each other.

Rep-PCR genomic fingerprinting makes use of DNA primers complementary to naturally occurring, highly conserved, repetitive DNA sequences, present in multiple copies in the genomes of most gram-negative and several gram-positive bacteria (Lupski and Weinstock, 1992). Three families of repetitive sequences have been identified, including the 35–40-bp repetitive extragenic palindromic (REP) sequence, the 124–127-bp enterobacterial repetitive intergenic consensus (ERIC) sequence, and the 154-bp BOX element (Versalovic *et al*, 1994). These sequences appear to be located in distinct, intergenic positions around the genome. The repetitive elements may be present in both orientations, and oligonucleotide primers have been designed to prime DNA synthesis outward from the inverted repeats in REP and ERIC, and from the box A subunit of BOX, in the polymerase chain reaction (PCR) (Versalovic *et al*, 1994). The use of these primer(s) and PCR leads to the selective amplification of distinct genomic regions located between REP, ERIC or BOX elements. The corresponding protocols are referred to as REP-PCR, ERIC-PCR and

BOX-PCR genomic fingerprinting respectively, and rep-PCR genomic fingerprinting collectively (Versalovic *et al.*, 1991, 1994). The amplified fragments can be resolved in a gel matrix, yielding a profile referred to as a rep-PCR genomic fingerprint (Versalovic *et al.*, 1994).

The rep-PCR genomic fingerprints generated from bacterial isolates permit differentiation up to the species, subspecies and strain level. Rep-PCR genomic fingerprinting protocols have been developed and have been applied successfully in many medical, agricultural, industrial and environmental studies of microbial diversity (Versalovic *et al.*, 1994). In addition to studying diversity, rep-PCR genomic fingerprinting has become a valuable tool for the identification and classification of bacteria, and for molecular epidemiological studies of human and plant pathogens (van Belkum *et al.*, 1994; Louws *et al.*, 1996 and Versalovic *et al.*, 1997).

In situ PCR

This method combines the sensitivity of PCR with the histological localization of *in situ* hybridization. With the use of a specialized thermal cycler block, PCR can be performed directly on slides containing fixed, thin tissue sections. This method reduces the sample contamination, which can lead to false positive results in PCR. This problem does not arise in *in situ* PCR as the positive signal is localized to specific areas within the cell. It also provides enormous amount of information relative to the histological distribution of the amplified PCR product.

The PCR reagents are placed on top of fixed and permeabilized tissue specimens or cells attached to glass slides. The reaction proceeds by incorporating labelled oligonucleotides such as digoxigenin-*n*-dUTP, biotin-11-dATP or fluorescein-labelled nucleotides into newly synthesized DNA during amplification. The fixation process preserves the cell morphology and proteolytic digestion facilitates access of PCR reagents to their target DNA sequences. Subsequently, the labelled amplicon is detected within the cells by means of standard immuno-cytochemical protocol, such as one-step detection with an anti-digoxigenin antibody conjugated with alkaline phosphatases (Uhlmann *et al.*, 1998).

Real-time PCR

All real-time PCR systems rely upon the detection and quantitation of a fluorescent reporter, the signal of which increased in direct proportion to the amount of PCR product in a reaction. The reporter is a fluorescent probe or a double-stranded DNA-specific dye SYBR Green which upon excitation emits light. Thus, as a PCR product accumulates, fluorescence increases. Real-time

PCR assays used for quantitative RT-PCR combine the best attributes of both relative and competitive (end-point) RT-PCR in that they are accurate, precise, capable of high throughput, and relatively easy to perform.

In the basic PCR protocol, at the start of a PCR reaction, reagents are in excess, and template and product are at low-enough concentrations that product renaturation does not compete with primer binding, and amplification proceeds at a constant, exponential rate. Exactly when the reaction rate ceases to be exponential and enters a linear phase of amplification it is extremely variable, even among replicate samples. At some later cycle, the amplification rate drops to near zero (plateaus), and little more product is made. It is necessary to collect quantitative data at a point in which every sample as in the exponential phase of amplification (since it is only in this phase that amplification is extremely reproducible). Analysis of reactions during exponential phase at a given cycle number should theoretically provide several orders of magnitude of dynamic range. Rare targets will probably be below the limit of detection, while abundant targets will be past the exponential phase. In practice, a dynamic range of 2–3 logs can be quantitated during end-point relative RT-PCR. In order to extend this range, replicate reactions may be performed for a greater or lesser number of cycles, so that all of the samples can be analysed in the exponential phase. A typical real-time PCR separation is shown in Figure 12.7.

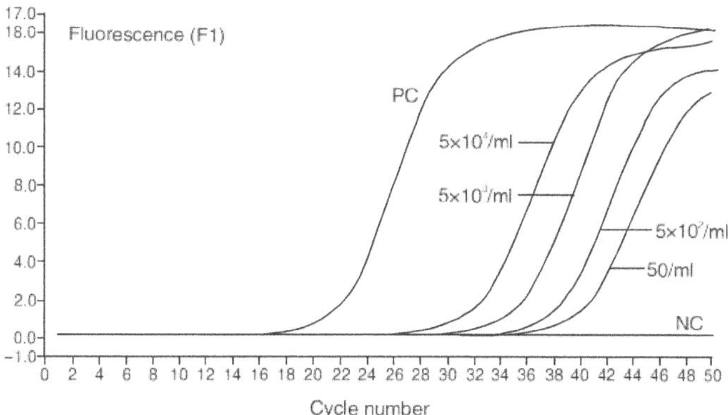

Figure 12.7 Representative results of *Leptospira* amplicon (strain verdun) detection by real-time PCR assay (Merien *et al.*, 2005, *FEMS Microbiol Lett.* 249: 139–47)

Real-time PCR automates this otherwise laborious process by quantitating reaction products for each sample in every cycle. The result is an amazingly broad 107-fold dynamic range, with no user intervention or replicates required. Data analysis, including standard curve generation and copy number calculation, is performed automatically.

There are two general methods for the quantitative detection of amplification fluorescent probes or DNA-binding agents. The TaqMan probes and molecular beacons (and more recently, scorpions) use the fluorogenic 5´ exonuclease activity of *Taq* polymerase to measure the amount of target sequences in cDNA samples. TaqMan probes are 20–30-base long oligonucleotides that contains a reporter fluorescent dye usually on the 5´ base, and a quenching dye (usually TAMRA) on the 3´ base (because the 3´ end is blocked, it cannot act as a primer). The close proximity of the reporter and quencher prevents emission of any fluorescence while the probe is intact. During the reaction when the polymerase replicates a template on which a TaqMan probe is bound, its 5´ exonuclease activity cleaves the probe. This ends the activity of the quencher, and the reporter dye starts to emit fluorescence which increases in each cycle proportional to the rate of probe cleavage. Accumulation of PCR products is detected by monitoring the increase in flurescence of the reporter dye. TaqMan assay uses universal thermal cycling parameters and PCR reaction conditions. Because the cleavage occurs only if the probe hybridizes to the target, the fluorescence detected originates from specific amplification. The process of hybridization and cleavage does not interfere with the exponential accumulation of the product. One specific requirement for fluorogenic probes is that there be no G at the 5´ end. A 'G' adjacent to the reporter dye quenches reporter fluorescence even after cleavage.

Advantages   The main advantage of using this technology is sensitivity and precision. Multicolour detection provides flexibility for multiplex quantitation assays, allelic discrimination assays and plus/minus assays, utilizing an internal positive control. The turn-around time for data acquisition and analysis by real-time PCR is short, the set-up and thermal cycling requiring less than 10 additional minutes. Traditional PCR quantitation using competitive PCR require several days.

Applications   The applications for quantitative real-time PCR are innumerable. Detection of genomic or viral DNA in tissues can be a valuable diagnostic tool. Gene expression can be measured after extraction of total RNA and preparation of cDNA by reverse transcription (RT) step. Set-up and analysis are simple and can more easily be extended to the clinical environment than traditional PCR techniques.

One problem with developing applications is that most research conducted with this technology is done by pharmaceutical companies. Because their work remains proprietary, successful TaqMan primers and probes, as well as optimization conditions, are often not available. As a result, de novo design and optimization still require an investment of time and money.

MAJOR ADVANTAGES OF PCR

Because of its simplicity, PCR is a popular technique with a wide range of applications which depend on essentially three major advantages of the method.

Speed and Ease of Use

DNA cloning by PCR can be performed in a few hours, using relatively unsophisticated equipment. Typically, a PCR reaction consists of 30 cycles containing a denaturation, synthesis and re-annealing step, with an individual cycle typically taking 3–5 minutes in an automated thermal cycler. This compares favourably with the time required for cell-based DNA cloning, which may take weeks. Clearly, some time is also required for designing and synthesizing oligonucleotide primers, but this has been simplified by the availability of computer software for primer design and rapid commercial synthesis of custom oligonucleotides. Once the conditions for a reaction have been tested, the reaction can be repeated simply.

Sensitivity

PCR is capable of amplifying sequences from minute amounts of target DNA, even the DNA from a single cell. Such exquisite sensitivity has afforded new methods of studying molecular pathogenesis and has found numerous applications in forensic science, in diagnosis, in genetic linkage analysis using single-sperm typing and in molecular palaentology studies, where samples may contain minute numbers of cells. However, the extreme sensitivity of the method means that great care has to be taken to avoid contamination of the sample under investigation by external DNA, such as from minute amounts of cells from the operator.

Robustness

PCR can permit amplification of specific sequences from material in which the DNA is badly degraded or embedded in a medium from which conventional DNA isolation is problematic. As a result, it is again very suitable for molecular anthropological and palaentological studies, for example, the analysis of DNA recovered from archaeological remains. It has also been used successfully to amplify DNA from formalin-fixed tissue samples, which has important applications in molecular pathology and, in some cases, genetic linkage studies.

APPLICATIONS OF PCR

Although PCR was first developed only a decade and a half ago, the simplicity and the versatility of the technique have ensured that it is among the most ubiquitous of molecular genetic methodologies, with a wide range of general applications. During the last few years, with the improvement of PCR protocols, and due to the availability of automatic thermal cyclers commercially, the application of PCR have increased manifold. However for application of PCR, we need a pair of primers which could be based on the knowledge of nucleotide sequence of the DNA to be amplified. Therefore, non-availability of this information about the DNA segment or gene to be amplified becomes a limitation in the application of PCR, although in the study of DNA polymorphisms, this difficulty has been overcome through the use of random DNA primers.

Detecting Pathogens

PCR may be employed to detect a wide range of organisms, whether they are present in foodstuff, the environment or biological or histological materials. The detection of pathogens is therefore an essential application of PCR with relevance to the food industry, environmental monitoring, medical, veterinary and botanical sciences. The DNA sequence (primer) upon which a diagnostic test is designed must be unique to the organism. If the sequence is not unique, misdiagnosis is possible through false positive results.

The main advantage of using PCR in the detection of pathogens is that a single cell or viral particle can be detected. Conventional method of characterization of pathogens involving the culture of some organisms may take weeks, whereas a PCR assay can be performed in hours. Furthermore, some organisms cannot be grown *in vitro*, while others are difficult to grow and slow to culture, therefore PCR has opened a way to detect some organisms that were hitherto particularly difficult to detect.

Human retroviruses, e.g. HIV-1 and 2 replicate through RNA intermediate. The RNA intermediate can be detected after performing an RT step to generate a DNA template for PCR. However during latency of infection, transcriptions are dormant. Nevertheless, latent infections can be detected by virtue of the retroviral replicative cycle. This cycle comprises conversion of the single-stranded RNA viral genome to a double-stranded circular proviral DNA molecule after entry into the host cell. The proviral DNA then integrates into the host cell DNA. Therefore, latent infection can be distinguished from proliferating infections allowing disease progression to be monitored. The distinction between latent and proliferative infection is made by the ability to detect viral RNA by the RT-PCR in proliferating infection and viral DNA in both latent and proliferating infections. PCR method has also been used to detect hepatitis B virus and human papilloma virus.

Meyer *et al.* (1991) used PCR for detection of foot-and-mouth disease virus in infected bovine and porcine tissues. The oligonucleotide primers used for PCR yielded a 454-base-pair target amplification product. The technique was specific, as determined by the examination of at least 12 other viruses. Silva (1992) used PCR for differentiation of pathogenic and non-pathogenic serotypes of Marek's disease viruses (MDV). The PCR oligos were chosen to flank the 132-base-pair tandem direct repeats in the serotype 1 Marek's disease virus genome. The PCR was specific for serotype 1 MDVs amplifying fragments corresponding to 1 to 3 copies of the tandem repeats. Use of the PCR technique allowed the detection of two copies of the 132-base pair repeat in the DNA extracted from MDV-induced lymphomas removed from the chicken. No DNA was amplified from the DNA extracted from lymphomas induced by either an avian leucosis virus (RAV-1) or reticulo-endothelial virus (chick syncytial virus). Detection of DNA and RNA viruses using PCR method by various authors are given in Table 12.2 and 12.3 respectively.

Table 12.2 Detection of DNA viruses with polymerase chain reaction

Virus	Gene region	Reference
Bovine herpesvirus	*Eco* RI L	Naeem *et al.*, 1991
Bovine papilloma type 1	*Nt* 3759, 3002	Von Teifke Weiss, 1991
Canine parvovirus	*VP* 2	Mochizuki *et al.*, 1993 Meerarani *et al.*, 1996 Latha and Ramadass, 2002 Decaro *et al.*, 2005
Canine distemper virus		Amude *et al.*, 2006
Canine adenovirus 1 and 2	*E* 3	Hu *et al.*, 2001
Chicken anaemia agent	*Caps*	Todd *et al.*, 1992
Coronavirus	*M*	Hamberger *et al.*, 1991
Duck hepatitis	*Nt* 2594–3000	Qiao *et al.*, 1990
Equine herpesvirus type 1	*gp* 13	Ballagi-Prodany *et al.*, 1990
Infectious laryngotracheitis	*gB*	Poulsen *et al.*, 1991
Marek's disease virus		Silva, 1992
Porcine parvovirus	*VP* 2	Molitor *et al.*, 1991
Pseudorabies virus	*g* II *gp* 50	Belak *et al.*, 1989 Jestin *et al.*, 1990

PCR method has been used for detection of sites of latency of infectious laryngotracheitis (ILT) virus in chicken (Williams *et al.*, 1992). PCR was used to detect a DNA sequence from the ILT thymidine kinase gene. After experimental infection, birds developed mild respiratory infection with clinical signs of ILT. After recovery from disease, when viral excretions were analysed using PCR, from tissues of upper respiratory tract, ocular tissues, trigeminal, proximal and distal ganglia. All tissues were negative for virus except the trigeminal ganglion and thus it was concluded that the trigeminal ganglion was the main site of latency of ILT virus.

PCR was compared with virus isolation (VI) and hemagglutination assay (HA) for the detection of canine parvovirus (CPV) in faecal specimen. PCR was found to be more sensitive than VI and HA. It was found to be a rapid, sensitive and specific technique for CPV detection in contaminated faecal specimens. Another advantage of PCR was that it can be used with even spoiled samples (Mochizuki *et al.*, 1993).

Pratelli *et al.* (2000) reported that the nPCR was more sensitive than regular PCR for detection of canine coronavirus from faecal samples of pups. Of 71 samples examined, 14 were positive by PCR, whereas 30 samples were positive by nPCR assay. CCV was detected by electron microscopic examination in only 4 out of 45 samples and by virus isolation in three out of 30 samples in nPCR-positive samples.

An improved PCR assay was used for the detection of canine parvovirus from faecal samples (Latha and Ramadass, 2002). A method of preparation of DNA from faecal samples, prior to PCR amplification was standardized. Dot ELISA was used to compare the results, which detected 75% of positive cases, while PCR with faecal samples detected 79% of cases, whereas PCR assay using DNA samples purified from faecal samples detected 92% of cases.

Ten decomposed brain samples that were collected between 1998 and 2000 were found negative by direct fluorescent antibody test (FAT) but were found positive by RT-PCR (David *et al.*, 2002). Of the ten samples, only three samples were found positive for rabies virus isolation and by mouse inoculation methods. These results demonstrated the importance of the RT-PCR in the detection of rabies virus in decomposed naturally infected brains, especially in cases when the sample is not suitable for other laboratory assays. A strain-specific RT-PCR and RFLP analysis was used for differentiating dog-related (DRRV) and vampire-bat-related rabies viruses (VRRV) in Brazil (Ito *et al.*, 2003). The PCR products obtained from DRRV were cut at one site by *Blp* I, but not by *Bsu* 36 I. The PCR products obtained from VRRV were cut at one or two sites by *Bsu* 36 I, but not by *Blp* I. SS RT-PCR and RFLP assays have been effective in discriminating the origin of rabies virus isolates in Brazil.

Sato *et al.* (2005) investigated the usefulness of multiplex reverse transcription-polymerase chain reaction (RT-PCR) for determining the origin of 54 rabies virus (RV) isolates from various host species in Brazil. Results were compared with the results of a phylogenetic tree developed from sequences of the RV glycoprotein (G protein) gene. Multiplex RT-PCR products showed five different sizes of products, whereas the phylogenic tree showed six groups. Of these six groups, four corresponded with the four sizes of the multiplex RT-PCR products. The other two groups showed correspondence with another one size of the multiplex RT-PCR products, indicating that multiplex RT-PCR results reflected the lineage of the 54 isolates. This study also showed that this method can detect trace amounts of RNA. In conclusion, this multiplex RT-PCR method allowed the rapid, specific, and simultaneous detection of RVs isolated from various host species in Brazil.

In dogs with neurological disturbances without myoclonus and extraneural signs, the clinical diagnosis of distemper is difficult to perform. Considering the great infectious potential of the disease, the possibility of carrying out an ante-mortem diagnosis of distemper is important, particularly in hospitalized patients with neurological disease. Amude *et al.* (2006) evaluated RT-PCR for ante-mortem CDV detection in hospitalized dogs with neurological disturbances without the typical findings of distemper. They investigated five dogs with canine distemper virus (CDV) encephalomyelitis, in which the clinical diagnosis was not performed owing to the absence of characteristic signs of the disease, such as myoclonus and systemic signs. They observed an apparent high sensitivity of RT-PCR in urine samples for detection of CDV: four out of five urine samples were RT-PCR-positive. The results of the present study suggested that urine was a good biological sample for ante-mortem CDV detection by RT-PCR in dogs with distemper encephalomyelitis in which the clinical diagnosis is likely to be difficult owing to the absence of suggestive distemper signs. The use of two different body fluids (urine and CSF) might increase the RT-PCR sensitivity for ante-mortem diagnosis of distemper in such cases.

Antognoli *et al.* (2001) used one tube nPCR for detection of *Mycobacterium bovis* in milk samples. PCR method was used for the detection and differentiation of canine adenovirus type 1 (CAV1) and type 2 (CAV2) from liver samples (CAV1) and laryngotracheal swab (CAV2) of dogs (Hu *et al.*, 2001). One pair of common primer was designed, which was used to amplify the virus-specific DNA fragment from clinical samples. After electrophoresis, under the same amplification conditions, 508 bp and 1030 bp PCR products were obtained from CAV1 and CAV2, respectively.

Handberg *et al.* (1999) developed RT-PCR technique using common and strain-specific primers for infectious bronchitis virus in allantoic fluid and tracheal tissue preparations. They differentiated IBV types by the use of S gene-specific oligonucleotides. Results indicated that the direct detection of IBV in tracheal tissues by RT-PCR was more sensitive than immunohistochemistry and that the RT-PCR technique was able to distinguish between types of IBV.

Ramadass *et al.* (2003) used RT-PCR method for detection of infectious bursal disease virus from bursal samples collected from suspected flocks for amplification of 474-bp product from variable region of the *VP2* gene. Among 53 bursal samples examined by RT-PCR, 40 showed positive reactions.

A nested RT-PCR was used for detection of infectious bronchitis virus from poultry using primers to amplify a part of *N*-gene of IBV (Suresh Kumar *et al.*, 2007). Of 47 samples screened, ten samples were found positive.

In situ RT-PCR method was used for early detection of IBDV in chicken experimentally infected (Zhang *et al.*, 2002). A typical positive signal was detected in the liver, the kidney and the spleen of the chicken inoculated with the very virulent IBDV H strain at 4 h p.i, but not in the thymus, the caecal tonsil or the thigh muscle until 8 h p.i.

A duplex reverse transcription-polymerase chain reaction (dRT-PCR) assay was developed for the simultaneous, rapid and specific detection/discrimination of avian influenza virus (AIV) and Newcastle disease virus (NDV) (Farkas *et al.*, 2007). Primers targeting the matrix protein gene (M) of AIV and the fusion protein gene (F) of NDV were evaluated experimentally with 13 AIV and 19 NDV strains. PCR products of the expected size of 144 bp and 316 bp were amplified from AIV/NDV samples, respectively, while no cross-reaction was observed with negative controls or with 16 other avian pathogens. The assay was able to detect AIV/NDV with similar sensitivity in spiked stool samples and in specimens from vaccinated birds. The developed dRT-PCR assay is a rapid, cost-effective tool, which provided powerful novel means for the early diagnosis of avian influenza and Newcastle disease.

PCR method has been used for the detection of various bacterial pathogens including leptospires (Gerritsen *et al.*, 1991), *Mycobacterium tuberculosis* (Sjobring *et al.*, 1990) and *Mycoplasma pneumoniae* (Kai *et al.*, 1993). PCR for detection of leptospires were found to be twice as sensitive as culturing and PCR analysis of urine was more successful for early diagnosis of leptospires than PCR analysis of serum (Bat *et al.*, 1994). As little as 5–10 leptospires per ml of urine sample could be detected by PCR method (Gerristsen *et al.*, 1991). Amplification of 274-bp target DNA could be detected in DNA samples purified from 500 µl of blood collected from experimentally infected gerbils 2 days after infection, while antibodies to *L. interrogans* could be detected by MAT 7 days after infection. The specificity and high sensitivity of the test produced valuable tools for the early diagnosis of leptospires. PCR method has been used for specific amplification of 330-bp product for diagnosis of leptospirosis from urine and blood samples (Figure 12.8) (Ramadass *et al.*, 1997a).

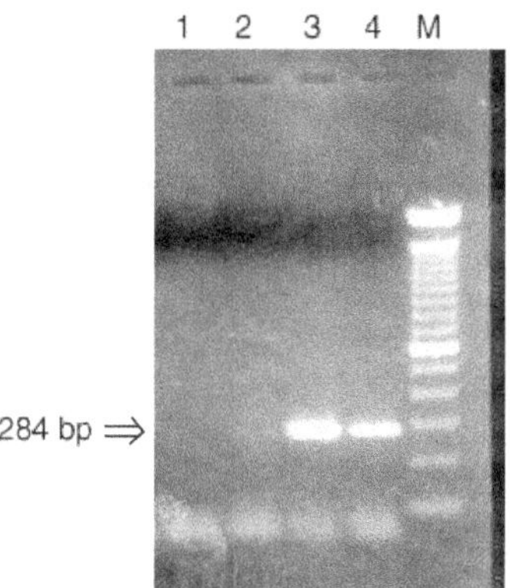

Figure 12.8 PCR for *Leptospira* detection showing 284 bp product in positive cases

PCR was used for detection of leptospires from clinical samples like urine, serum, mastitis milk and cerebrospinal fluids from dogs and cattle (Senthil Kumar *et al.*, 2001). Primers corresponded to nucleotides 348–368 and 38–57 of the primary structure of the *L. interrogans rrs* (16s) gene, which produced an amplicon of 330 bp. Results indicated that the PCR assay of clinical samples could be a potentially useful, quick and specific diagnostic method for confirming active infection with leptospires.

Berg *et al.* (2006) compared microscopy, culture and PCR for the diagnosis of anthrax in blood samples from sheep and cattle. Blood samples were stored at room temperature and at 37°C after receipt, over a period of 15–17 days. Aliquots were plated onto blood agar and blood smears were prepared. Following microscopic examination, DNA was extracted from blood smears and subjected to a multiplex PCR assay targeting the Ba813, *cap* and *lef* markers. PCR provided the most reliable means for the detection of *Bacillus anthracis* in deteriorating blood samples (15–17 days) and was also successful in diagnosing anthrax in blood smears that had been stored for 6 years and a blood sample which had been stored for 18 months at –20°C. While less successful than PCR, culture for *B. anthracis* on 7% sheep blood agar was typically more reliable (2–17 days) than the examination of blood smears (2–6 days) for encapsulated bacilli. This work demonstrated the superiority of PCR for the diagnosis of anthrax from blood smear scrapings, particularly when microscopy is unreliable.

Ryu *et al.* (2003) developed a sensitive and rapid quantitative detection of anthrax spores isolated from soil samples. Using TaqMan real-time PCR, specific primers and probes were designed for the identification of pathogenic *Bacillus anthracis* strains. Serial dilutions of *B. anthracis* DNA and spore were detected up to a level of 0.1 ng/µl and 10 spores/ml, respectively. Spores added to soil samples were detected up to 10^4 spores/g soil within 3 hours by real-time PCR.

The efficacy of bacterial culture and IS900-specific PCR was compared for the detection of *Mycobacterium avium* subsp. *paratuberculosis* (*Map*) from the intestinal and mesenteric lymph node tissues of water buffaloes (*Bubalus bubalis*) showing lesions of paratuberculosis (Sivakumar *et al.*, 2005). Out of 20 animals showing histological lesions suggestive of paratuberculosis, 14 (70%) and 6 (30%) were positive in the PCR and bacterial culture, respectively. The results of this study suggested that PCR was more sensitive than bacterial culture in detection of subclinical paratuberculosis in water buffaloes. The specificity of the PCR was confirmed by the product size and restriction digestion pattern of the amplicons. The sequence analysis of the amplified products (626 bp of IS900 gene) from buffalo strain showed more than 97% homology with the published sequences.

PCR assay was evaluated for the detection of *Brucella canis* in canine semen and the results were compared with that of bacterial isolation, serological tests and PCR assay of blood (Keid *et al.*, 2007). Fifty-two male dogs were examined clinically to detect reproductive abnormalities and their serum was tested by the rapid slide agglutination test, with and without 2-mercaptoethanol (2ME-RSAT and RSAT, respectively). In addition, microbiological culture and PCR assays were performed on blood and semen samples. The findings of the semen PCR were compared (Kappa coefficient and McNemar test) to those of blood PCR, culture of blood and semen, RSAT, and 2ME-RSAT. Nucleic acid extracts from semen collected from dogs not infected with *B. canis* were spiked with decreasing amounts of *B. canis* RM6/66 DNA and the resulting samples subjected to PCR. In addition, semen samples of non-infected dogs were spiked with decreasing amounts of *B. canis cfu* and the resulting suspensions were used for DNA extraction and amplification. Of the 52 dogs that were examined, the following tests were positive: RSAT, 16 (30.7%); 2ME-RSAT, 5 (9.6%); blood culture, 14 (26.9%); semen culture, 11 (21.1%); blood PCR, 18 (34.6%); semen PCR, 18 (34.6%). The PCR assay detected as few as 3.8 fg of *B. canis* DNA experimentally diluted in 444.9 ng of canine DNA (extracted from semen samples of non-infected dogs). In addition, the PCR assay amplified *B. canis* genetic sequences from semen samples containing as little as 10 cfu/mL. Authors concluded that PCR assay of semen was a good candidate as a confirmatory test for the diagnosis of brucellosis in dogs; its diagnostic performance was similar to blood culture or blood PCR. Furthermore, the PCR assay of semen was more sensitive than the 2ME-RSAT or semen culture.

The utility of the *hupB* gene (Rv2986c in M. *tuberculosis*, or Mb3010c in M. *bovis*) to differentiate M. *tuberculosis* and M. *bovis* was evaluated by a PCR-restriction fragment length polymorphism (RFLP) assay with 56 characterized bovine isolates (Mishra *et al.*, 2005). The degree of concordance between the PCR-RFLP assay and the microbiological characterization was 99.0% (P < 0.001). A nested PCR (n-PCR) assay was developed, replacing the PCR-RFLP assay for direct detection of M. *tuberculosis* and M. *bovis* in bovine samples. The n-PCR products of M. *tuberculosis* and M. *bovis* corresponded to 116 and 89 bp, respectively. The detection limit of mycobacterial DNA by n-PCR was 50 fg, equivalent to five tubercle bacilli. M. *tuberculosis* and/or M. *bovis* was detected in 55.5% (105/189) of the samples by n-PCR, compared to 9.4% (18/189) by culture. The sensitivities of n-PCR and culture were 97.3 and 29.7, respectively, and their specificities were 22.2 and 77.7%, respectively. The percentages of animals or samples identified as infected with M. *tuberculosis* or M. *bovis* by n-PCR and culture reflected the clinical categorization of the cattle (P of <0.05 to <0.01). Mixed infection by n-PCR was detected in 22 animals, whereas by culture, mixed infection was detected in 1 animal.

Table 12.3 Detection of RNA viruses with the polymerase chain reaction

Virus	Gene region	Reference
Avian leucosis	*gp* 85	Van Woensel *et al.*, 1992
Bluetongue	VP7 RNA Seg. 3	Gould *et al.*, 1989 Mc Coll & Gould, 1991
Bovine leukemia	*Env*-gp51	Naïf *et al.*, 1990
Bovine rotavirus	gene no. 8	Xu *et al.*, 1990
Bovine viral diarrhoea	*gp* 48 p 80	Belak & Ballagi-Pordany, 1991; Ward & Misra, 1991
Canine corona virus	M protein gene	Pratelli *et al.*, 2000
Foot-and-mouth disease	RNA Pol	Meyer *et al.*, 1991
Hog cholera	*gp* 55	Roche and Woodward, 1991
Infectious bronchitis	M, N N	Anderason *et al.*, 1991 Zwaagstra *et al.*, 1992
	N	Callison *et al.*, 2006 Suresh Kumar *et al.*, 2007

(Contd.)

Table 12.3　(Continued)

Virus	Gene region	Reference
Infectious bursal disease	Nt 1730–1879	Wu *et al.*, 1992
	VP2	Ramadass *et al.*, 2003
		Hairul Aini *et al.*, 2006
Newcastle disease	F	Jstin and Jestin, 1991
	M	Farkas *et al.*, 2007
Canine distemper		Amude *et al.*, 2006
		Elia *et al.*, 2006
Phocid distemper	P	Haas *et al.* 1991
Rabies	N	Shankar *et al.*, 1991
	G protein gene	Sato *et al.*, 2005

Table 12.4　Detection of bacteria with polymerase chain reaction

Bacteria	Gene region	Reference
Leptospira	rrs	Bal *et al.*, 1994; Ramadass *et al.*, 1997a
		Senthil Kumar *et al.*, 2001
Mycoplasma gallisepticum *Mycoplasma synoviae*	rRNA	Ramadass *et al.*, 2006
Mycoplasma pneumoniae	16S rRNA	Kai *et al.*, 1993
Streptococcus pneumoniae	1 yt A	Gillespie *et al.*, 1994
Camplyobacter jejuni	fla A	Itoh *et al.*, 1995
Brucella sp.		Bricker and Halling, 1994
Brucella canis		Keid *et al.*, 2007
Bacillus anthracis	Ba813	Berg *et al.*, 2006
	pla	Skottman *et al.*, 2007
Yersinia enterocolitica	ail gene	Kwago *et al.* 1992
Mycobacterium tuberculosis		Kolk *et al.*, 1992 De Wit *et al.*, 1990
	HupB	Mishra *et al.*, 2005
Mycobacterium avium subsp. *paratuberculosis*	IS 900 gene	Sivakumar *et al.*, 2005

PCR was used for detection of *Mycoplasma pneumoniae* in throat swab samples. A 88-bp amplified product was obtained which was specific for M. *pneumoniae* (Kai *et al.*, 1993). PCR method was used for specific detection of *Mycoplasma gallisepticum* (MG) and *Mycoplasma synoviae* (MS) from both live birds as well as from dead birds using species-specific primers (Ramadass *et al.*, 2006). Primers specific for MG and MS were derived from rRNA of MG and MS amplifying products of 530 bp and 207 bps for MG and MS, respectively (Figure 12.9). Among 1039 samples tested by PCR, 36 samples were positive for MG and 67 were positive for MS infections.

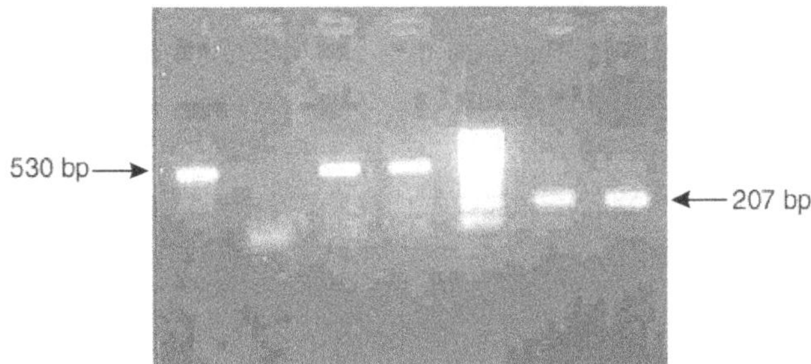

Figure 12.9 PCR for *Mycoplasma gallispeticum* (530 bp) and *Mycoplasma synoviae* (207 bp)

Based on a 10 mer primer, an RAPD method for typing *Campylobacter coli* isolated from pits was developed (Madden *et al.*, 1996). The method proved effective with a high discrimination and good reproducibility. Tcherneva *et al.* (2000) used RAPD method to differentiate different species of *Brucella*. The constructed dendrograms put B. *canis* and B. *suis* bv. 1 in the same cluster and differentiated *Brucella* strains according to their host preferences. They concluded that RAPD can be a useful method to distinguish related bacterial species and under strictly established conditions, the reaction appears to be a simple, quick and sensitive technique for the epidemiological investigation of brucellosis. Butler *et al.* (2001) used AP-PCR method to investigate *Mycoplasma bovis* outbreaks and they characterized different isolates using this method.

PCR method has been used for detection of carrier cattle by detection of *Theileria annulata* from blood samples (D'Oliveira *et al.*, 1995). Iqbal and Rikibrisa (1994) used PCR for detection of *Ehrlichia canis*, in tissues of dogs. A 600-bp product defined by the specific primer was amplified in blood, kidneys, lymph nodes and liver of dogs. Figeroa *et al.* (1992) used PCR method for detection of infected carriers among cattle. This method has also been used for the detection of *Toxoplasma gondii* in cerebrospinal fluids of AIDS patients.

Faggi *et al.* (2001) used PCR fingerprinting for the identification of species and varieties of common dermatophytes, using a single primer $(GACA)_4$. The primer was able to amplify all the strains, producing species-specific profiles for *Microsporum canis, Microsporum gypseum, Trichophyton rubrum, Trichophyton ajelloi* and *Epidermophyton floccosum*.

A multiplex PCR assay was optimized to detect simultaneously four pathogenic species of avian mycoplasmas. Four sets of oligonucleotide primers specific for *Mycoplasma gallisepticum, M. synoviae, M. meleagridis* and *M. iowae* were used in the test. This method was found to be specific, sensitive and cost-effective for simultaneous detection of four avian mycoplasma (Wang *et al.*, 1997). Wang and Khan (1999) developed an mPCR for simultaneous detection of Massachusetts and Arkansas serotypes of infectious bronchitis virus. One common primer and two serotype-specific primers were chosen from the S1 gene sequences of IBV. Two serotype-specific PCR products, 1026 bp for *Mass* and 896 bp for *Ark*, were amplified and detected by agarose gel electrophoresis. A nested mPCR assay was developed for genotyping of bovine viral diarrhoea viruses, BVDV1 and BVDV2 (Gilbert *et al.*, 1999) using cattle blood samples. One common primer and one strain-specific primer were used for amplification.

A nested and real-time PCR was used for detection of *B. anthracis* spores in soil (Cheun *et al.*, 2003). One cell of *B. anthracis* in 1 g soil could be detected by nested and real-time PCR. A real-time quantitative PCR was developed using a 423 bp target on the *lipL32* gene, which is conserved among pathogenic serovars of *Leptospira* (Levbett *et al.*, 2005). Reactions were monitored by SYBR green fluorescence and melting curve analysis. Positive results were obtained with all pathogenic serovars. The analytical sensitivity of this assay was 3 genome equivalents per reaction; approximately 10 genome equivalents were detectable in human urine. The assay successfully detected leptospiral DNA from serum and urine samples of patients with leptospirosis. The assay has the potential to facilitate rapid, sensitive diagnosis of acute leptospirosis.

Real-time PCR and culture methods were compared for detection of *Mycobacterium avium* subsp. *paratuberculosis* in dairy cattle (Bogli-Stuber *et al.*, 2005). Real-time PCR method identified 31 of 310 animals as positive, whereas culture identified 20 positive animals. The specificity of real-time PCR was confirmed by DNA sequencing of the PCR product. Real-time PCR and culture data were in good agreement and the real-time PCR generates data in a short time in contrast to culture. It was concluded that real-time PCR is a suitable alternative method to culture for the detection of MAP in a national surveillance programme. Real-time PCR was developed for fast and sensitive detection of

Mycobacterium avium subsp. *paratuberculosis* (Map) in bovine semen and to make a critical evaluation of the analytical sensitivity (Herthnek *et al.*, 2006). Processed semen was spiked with known amounts of Map. Semen from different bulls as well as semen of different dilutions was tested. The samples were treated with lysing agents and bead beating and the DNA was extracted with phenol and chloroform. Real-time PCR with a fluorescent probe targeting the insertion element IS900 detected as few as 10 organisms per sample of 100 μl semen. PCR-inhibition was monitored by inclusion of an internal control. Real-time PCR was found to be a sensitive method for detection of Map in bovine semen. Lysis by mechanical disruption followed by phenol and chloroform extraction efficiently isolated DNA and removed PCR inhibitors.

Mycoplasma gallisepticum was rapidly detected from chicken tracheal swabs by real-time PCR by LightCycler system (Carli and Evigor, 2003). Out of 96 tracheal swabs, 68 were taken from live chicken and 28 were taken by scraping the mucosal surface of the trachea (SMST) of necropsied chicken. All of the 18 PCR-positive results were from the swabs taken by the SMST method, whereas all of the samples taken from live chickens were negative. Thus, the PCR with the SMST method had a sensitivity and specificity of 64.2% (18 of 28 chickens) and 100%, respectively. Mekkes and Feberwee (2005) used real-time PCR (Q-PCR) for detection of M. *gallisepticum*. No cross-reactivity was observed with DNA from other important avian mycoplasmas, including M. *synoviae* and M. *meleagridis*.

A multiplex real-time PCR assay was developed using minor groove binding probes for simultaneous detection of the *Bacillus anthracis pag* and *cap* genes, the *Francisella tularensis* 23-kDa gene, as well as the *Yersinia pestis pla* gene (Skottman *et al.*, 2007). The sensitivities of these assays were at least 1 fg, except for the assay targeting the *Bacillus anthracis cap* gene, which showed a sensitivity of 10 fg when total DNA was used as a template in a serial dilution. The clinical value of the *Bacillus anthracis-* and *Francisella tularensis*-specific assays was demonstrated by successful amplification of DNA from cases of cow anthrax and hare tularemia, respectively. No cross-reactivity between these species-specific assays or with 39 other bacterial species was noted. These assays may provide a rapid tool for the simultaneous detection and identification of the three categories.

A real-time fluorogenic RT-PCR assay was developed for the specific detection of all seven serotypes of foot-and-mouth disease virus in epithelial suspensions and cell culture virus preparations (Reid *et al.*, 2002). This method specifically detected FMD virus in samples submitted from the UK 2001 FMD outbreak with greater sensitivity than conventional RT-PCR procedure, ELISA and virus isolation in cell culture. The real-time fluorogenic RT-PCR provided relatively fast results, enabled a quantitative assessment to be made of virus amounts and could handle more samples.

Jackwood *et al.* (2003) used real-time RT-PCR to detect IBD virus strains. A mutation probe labelled with fluorescein and an anchor probe labelled with Red-640 dye were prepared for each of the STC, Del E, D78 and Bursine 2 viral sequences. The mutation probes were designed to hybridize to nucleotides that encode the hydrophilic B region of the VP2 for each virus. The anchor probes were designed to a relatively conserved region immediately downstream from the mutation probes. The results indicated that the STC and Variant Vax BD viruses have similar genetic sequences at the hydrophilic B region. Likewise, Bursine 2, Bursine, Bursine+, BioBurs, BioBurs W, BioBurs AB and IBDV Blen have similar nucleotide sequences in this region.

Hairul Aini *et al.* (2006) compared the performances of Sybr Green I real-time PCR, enzyme-linked immunosorbent assay (ELISA) and conventional agarose detection methods in detecting specific IBDV-PCR products. They found the real-time PCR was at least 10 times more sensitive than ELISA detection method with a detection limit of 0.25 pg. The latter was also at least 10 times more sensitive than agarose gel electrophoresis detection method. The developed assay detected both very virulent and vaccine strains of IBDV but not other RNA viruses such as Newcastle disease virus and infectious bronchitis virus. Hence, Sybr Green I-based real-time PCR is considered as a highly sensitive assay for the detection of IBDV.

Callison *et al.* (2006) reported the development and testing of a real-time RT-PCR assay using a TaqMan-labelled probe for early and rapid detection of IBV. The assay amplified a 143-bp product in the 5´-UTR of the IBV genome and has a limit of detection and quantification of 100 template copies per reaction. All 15 strains of IBV tested as well as two Turkey coronavirus strains were amplified, whereas none of the other pathogens examined, tested positive. Evaluation of the assay was completed with 1329 tracheal swab samples. A total of 680 samples collected from IBV antibody negative birds were negative for IBV by the real-time RT-PCR assay. 229 tracheal swabs submitted to two different diagnostic laboratories were tested and it was found that 79.04% of the tracheal swabs were positive for IBV by real-time RT-PCR, whereas only 27.51% of the samples were positive by virus isolation, which is the reference standard test. The real-time RT-PCR test described herein can be used to rapidly distinguish IBV from other respiratory pathogens, which is important for control of this highly infectious virus. The test was extremely sensitive and specific, and can be used to quantitate viral genomic RNA in clinical samples.

A rapid, sensitive and reproducible real-time PCR assay for detecting and quantifying canine parvovirus type 2 (CPV-2) DNA in the faeces of dogs with diarrhoea was described (Decaro *et al.*, 2005). The method was demonstrated to be highly specific and sensitive, allowing a precise CPV-2 DNA quantitation over a range of eight orders of magnitude (from 10^2 to 10^9 copies of standard DNA). Faecal specimens from diarrhoeic dogs were analysed by haemagglutination (HA), conventional PCR and real-time amplification. Comparison between these different techniques revealed that real-time PCR was more sensitive than HA and conventional gel-based PCR, allowing to detect low viral titres of CPV-2 in infected dogs.

A real-time RT-PCR assay was developed for detection and quantitation of canine distemper virus (Elia *et al.*, 2006). The assay exhibited high specificity as all the negative controls (no-template controls and samples from healthy sero-negative dogs) and other canine pathogens were not mis-detected. Up to 1×10^2 copies of RNA were detected by the TaqMan assay, thus revealing a high sensitivity. Quantitative TaqMan was validated on clinical samples, including various tissues and organs collected from dogs naturally infected by canine distemper virus. Urine, tonsils, conjunctival swabs and whole blood were found to contain high virus loads and therefore proved to be suitable targets for detection of canine distemper virus RNA.

PCR-ELISA was used in peripheral whole blood and serum specimens for diagnosis of acute human brucellosis (Vrioni *et al.*, 2004). Diagnosis of brucellosis was established in 179 cases out of 243 tested by isolation of *Brucella* spp. in blood culture, and in 64 cases by clinical and serological investigations. The diagnostic specificity of the PCR-ELISA for both specimen categories was 100%, while the sensitivity was 81.5% for whole blood specimens, 79% for serum specimens and 99.2% for whole blood and serum specimens combined. The results suggested that the detection of *Brucella* DNA in whole blood and serum specimens by PCR-ELISA is a sensitive and specific method that could assist the rapid and accurate diagnosis of acute human brucellosis.

Study of DNA Polymorphism Using PCR

DNA polymorphism can be studied at loci with known DNA sequence (using primers based on known sequences) or at random sites (using primers having unknown to random DNA sequences, called random primers). In either case, DNA may be amplified from one or more genomic DNAs and subjected to digestion with one or more specific restriction endonucleases. The digested DNA may be subjected to electrophoresis and the restriction patterns may be photographed to reveal polymorphisms.

PCR and RAPD Markers

The most important use of PCR for the study of DNA polymorphism, however, involves the use of random primers in random amplified polymorphic DNA (RAPD) fingerprinting. This technique is considered simpler and offers several advantages over the more commonly used technique known as RFLP. RAPD has several advantages over RFLP and can be used as its substitute in several studies. The main advantages include the following.

1. Since same primers with arbitrary sequences can be used for different species, no species-specific probes are needed for different species as required in RFLPs;

2. Collection of data using RAPD proceeds much more quickly than those using RFLP. Since there are fewer steps involved, it has been shown that generation of RAPD data is 5 times quicker than RFLP;

3. RADP technique may also be combined with RFLPs to increase the efficiency and may be labelled and used as a probe for RFLP analysis, thus eliminating the need for recombinant DNA cloning of probes in bacteria. RAPD method has been used for characterization of various bacterial species including leptospirosis (Figure 12.10) (Ramadoss *et al.*, 1997b).

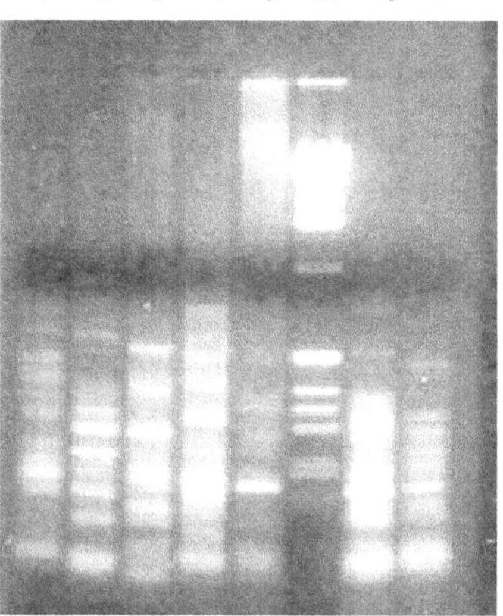

Figure 12.10 The RAPD patterns obtained with various leptospiral serovars. Lanes 1 to 5 and 6 and 7—different leptospiral serovars; Lane M —1 kb DNA ladder

Screening of Uncharacterized Mutations

Because of its rapidity and simplicity, PCR is ideally suited for providing numerous DNA templates for mutation screening. Partial DNA sequences, at the genomic or the cDNA level, from a gene associated with disease, or some other interesting phenotype, immediately enable gene-specific PCR reactions to be designed. Amplification of the appropriate gene segment then enables rapid testing for the presence of associated mutations in large numbers of individuals. Typically, the identification of exon–intron boundaries and sequencing of the ends of introns of a gene of interest offers the possibility of genomic mutation screening.

PCR, VNTR and SSR LOCI

PCR was used to study DNA polymorphism at VNTR (variable number tandem repeats, also known as mini satellites) loci in humans. Primers were developed for the conserved flanking regions of VNTRs loci, so that all available VNTR lock could be amplified. The PCR product differ according to the number of repeat units in the different VNTR loci and thus variation is visualized in the positions occupied by different bands after electrophoresis.

The VNTR loci discussed above consist of repeat units in the range of 11–60 base pairs in length. A much simpler type of repeat units are in the form of dinucleotides, trinucleotides or tetranucleotides, e.g. $(CA)_n$, $(GT)_n$, $(AAT)_n$ or $(AGAT)_n$. These repeat units were described as microsatellites, which is also called as short tandem repeats or simple sequence repeats (SSR). Some of these SSR loci such as $(CA)_n$ or $(GT)_n$ occur in human genome as many as 50,000 times with n varying from 10–60. The DNA sequence flanking SSRs are conserved, allowing the selection of PCR primers that will amplify the intervening SSR in all genotypes of the target electrophoresis. Variation in the length of PCR products has a function of the number of SSR unit.

Restriction site polymorphisms (RSPs) result in alleles possessing or lacking a specific restriction site. Such polymorphisms can be typed using Southern blot hybridization. A DNA probe representing the locus is hybridized against genomic DNA samples that have been digested with the appropriate restriction enzyme and size-fractionated by agarose gel electrophoresis. The resulting RFLPs have two alleles corresponding to the presence or absence of the restriction site. As a convenient alternative to RFLPs, PCR can type RSPs by simply designing primers using sequences which flank the polymorphic restriction site, amplifying from genomic DNA, then cutting the PCR product with the appropriate restriction enzyme and separating the fragments by agarose gel electrophoresis.

Short tandem repeat polymorphisms (STRPs), also called microsatellite markers, consist of a short sequence, typically from one to four nucleotides long that is tandemly repeated several times, and often characterized by many alleles. For example, $(CA)_n/(TG)_n$ repeats are often polymorphic when n exceeds 12, and have been widely used as polymorphic markers in the human genome. Increasingly, however, trinucleotide and tetranucleotide marker polymorphisms are being typed. In each case the STRPs can be typed conveniently by PCR. Primers are designed from sequences known to flank a specific STRP locus, permitting PCR amplification of alleles whose sizes differ by integral repeat units. The PCR products can then be size-fractionated by polyacrylamide gel electrophoresis. The PCR normally includes a radioactive or fluorescent nucleotide precursor which becomes incorporated into the small PCR products and facilitates their detection.

Molecular Mapping Using PCR

Genetic and physical chromosome maps have also been prepared, both in plants and animals, using PCR. Once RAPDs are detected, recombinant inbreds or backcross-segregating populations or double haploids derived from haploids can be used to detect linkage and recombination frequencies.

Gene Tagging Using PCR

PCR has also been utilized for developing molecular markers closely linked to specific genes of economic importance. For instance, in tomato, 144 random primers were used to produce 625 PCR products from a set of near isogenic lines (NILs, which differ only in the presence or absence of specific gene) for *Pseudomonas* resistance gene (*Ptg*). These PCR products could be mapped close to *Ptg*. Such tagging to genes through PCR will be used in future for plant breeding and also for isolation of genes.

CR for Confirming the Presence of Transferred Gene

When a gene is transferred with a vector to cultured cells or organisms, primers can be designed to conduct PCR for amplification of the gene sequence so transformed. This technique allows the confirmation of transfer and maintenance of the gene of interest. In gene therapy experiments, the transfer of a marker gene for neomycin resistance could be detected in the blood of patients. Similarly, *ADA* (adenosine deaminase deficiency) gene was transferred for therapeutic purposes and PCR was used to detect the presence of this gene in 6.5 months. Thus, PCR proves to be of immense help in monitoring a gene in genetic engineering or gene therapy experiments.

Human Genetics Using PCR

PCR has also found extensive use in human genetics. Some of these uses are elaborated below.

Prenatal diagnosis using PCR   Prenatal diagnosis of sickle cell anaemia with enhanced sensitivity was perhaps the first application of PCR. Using PCR, this is done in less than one day in contrast to several weeks needed, when Southern blots are used for hybridization with a probe. The test has also been used for diagnosis of phenylketonuria (PKU), β-thalassemia, haemophilia, etc. the PCR product in all these mutant sequences causing the disease are found. Sometimes, RFLP pattern of PCR products in healthy and defective feti differ thus enabling prenatal diagnosis. In still other cases, PCR products may be sequenced to reveal the differences.

Recombination data using PCR   PCR can be used for DNA amplification, utilizing individual sperms, which are the products of meiosis and recombination. Therefore, by studying PCR products from a large number of individual sperms, one can calculate the proportion of sperms that are recombinant for linked markers. This allows construction of genetic maps at a resolution which is not possible with pedigree analysis.

Sexing of embryo using PCR   Since DNA sequences in single cells can be studied using PCR, sex of human or livestock embryos, fertilized *in vitro*, can be determined before implantation. For this purpose, PCR primers and probes specific for sex chromosomes (e.g. ZFY or ZFX sequences) can be used. Y-specific primers and probes in humans are available for this purpose. This technique can also be used to detect sex-linked disorders in the fertilized embryos.

DNA Fingerprinting Using PCR

Well-characterized sequences of microsatellites are being used for designing primers, so that DNA fingerprints may be achieved through PCR. PCR allows amplification of DNA from individual hairs, stains of blood or seminal fluid having partially degraded DNA, which could not be used earlier for characterization of individuals.

FUTURE DEVELOPMENT IN PCR

Currently, nucleic acid amplification techniques are still labour-intensive and expensive to perform, which limits their use to large commercial laboratories or research-oriented diagnostic laboratories at universities. The ability to automate PCR procedures is therefore considered a key factor in determining how large the role of PCR will be in the future for the average laboratory. For a variety of human pathogens including viruses, commercially available PCR kits have been introduced, e.g. Amplicor microwell plate assay for hepatitis C virus (Roche Diagnostic Systems, Branchburg, NJ, USA). These kits are designed for amplification in conventional thermal cyclers with capture ELISA-like colorimetric detection of the biotinylated PCR products.

The present technology for doing PCR, about the size of a microwave oven and costing several thousand dollars seems destined for further radical improvement. Extraordinary miniaturization of the hardware is also underway, as experimenters squeeze the PCR onto chip-sized devices. Criss-crossed with the tiniest of troughs to hold the reagents and the DNA, the chips are heated electrically and cooled down much faster than the present generation of machines, so amplification is even speedier than today's swift process. Already researchers have reported using a handheld battery-powered gadget to copy pieces of DNA that contained eight different cystic fibrosis mutation sites. While such experimental chip-based devices are not yet ready for prime time, they are hastening the day when scientists can take them on the road, and patients will be able to get on-the-spot readouts of their DNA. Before long it may be quite routine to diagnose an infectious or genetic disorder, or even detect an inherited predisposition to cancer or heart disease, right in the doctor's office.

REVIEW QUESTIONS

1.   What is polymerase chain reaction? What are the compounds of PCR?

2.   What are the different types of enzymes used in PCR array?

3.   What are the different modifications of PCR? Give their advantages and disadvantages.

4.   What are the advantages of PCR array? Enumerate its uses.

REFERENCES

Amude, A.M., Alfieri, A.A. and Alfieri, A.F. (2006). "Antemortem diagnosis of CDV infection by RT-PCR in distemper dogs with neurological deficits without the typical clinical presentation." *Vet. Res. Commun.* 30:679–687.

Anderason, S.R., Jackwood, M.W. and Hilt, D.A. (1991). "PCR amplification of the genome of infectious bronchitis virus." *Avian Dis.* 35: 216–220.

Antognoli, M.S., Salmon, M.D., Triantis, J., Hernandez, J. and Keefe, T. (2002). "A one tube nested PCR for detection of *Mycobacterium bovis* in spiked milk samples: An evaluation of concentration and lytic techniques." *J. Vet. Diagn. Invest.* 13: 111–116.

Aradaib, I.E., Smith, W.L., Osburn, B.I. and Cullor, J.S. (2003). "A multiplex PCR for simultaneous detection and differentiation of North American serotypes of bluetongue and epizootic hemorrhagic disease viruses." *Comp. Immunol. Microbiol. Infect.* Dis. 26: 77–87.

Arnol, C. and Hodgson, I.J. (1991). "Vectorate PCR." *PCR Methods & Applications*. 1: 39–42.

Bal, A.E., Gravekamp, C., Hartskeerl, R.A., De Meza-Brewstor, J., Korver, H. and Terpstra, W.J. (1994). "Detection of leptospires in urine by PCR for early diagnosis of leptospires." *J. Clin. Microbiol* 32: 1894–1898.

Ballagi-Pordany, A., Klingeborn, B. Flensburg, J. and Belak, S. (1990). "Equine herpesvirus type 1: Detection of viral RNA sequences in aborted foetuses with the PCR." *Vet. Microbiol.* 22: 373–381.

Belak, S. and Ballagi-Pordany, A. (1991). "Bovine viral diarrhoea virus infection: Rapid diagnosis by the PCR." *Arch. Virol. Suppl.* 3. 181–190.

Belak, S. and Ballagi-Pordany, A. (1993). "Application of the PCR in veterinary diagnostic virology." *Vet. Res. Comm.* 17: 55–72.

Belak, S., Ballagi-Pordany, A., Flensburg, J. and Virtanen, A. (1989). "Detection of pseudorabies virus DNA sequences by the PCR." *Arch. Virol.* 108: 279–286.

Berg, T., Suddes, H., Morrice, G. and Hornitzky, M. (2006). "Comparison of PCR, culture and microscopy of blood smears for the diagnosis of anthrax in sheep and cattle." *Lett. Appl. Microbiol.* 43:181–186.

Bogli-Stuber, K., Kohler, C., Seitert, G., Glanemann, B., Antognoli, M.C., Salman, M.D., Wittenbrink, M.M., Wittwer, M., Wassenaar, T., Jemmi, T. and Bissig-Choisat, B. (2005). "Detection of *Mycobacterium avium* subsp. *paratuberculosis* in Swiss dairy cattle by real-time PCR and culture: A comparison of the two assays." *J. Appl. Microbiol.* 99: 587–597.

Brakstad, O.G., Aasbakk, K. and Maeland, J.A. (1992). "Detection of *Staphylococcus aureus* by PCR amplification of the *nuc* gene." *J. Clin. Microbiol.* 30: 1654–1660.

Bricker, B.J. and Halling, S.M. (1994). "Differentiation of *Br. abortus* 1, 2 and 4, *Br. mellitensis*, *Br. ovis* and *Br. suis* 1 by PCR." *J. Clin. Microbiol.* 32: 2660–2666.

Butler, J.A., Pinnow, C.C., Thomson, J.V., Levisohn, S. and Rosenbeusch, R.F. (2001). "Use of arbitrarily primed PCR to investigate *Mycoplasma bovis* outbreaks." *Vet. Microbiol.* 26: 175–181.

Callison, S.A., Hilt, D.A., Boynton, T.O., Sample, B.F. Robison, R., Swayne, D.E. and Jackwood, M.W. (2006). "Development and evaluation of a real-time TaqMan RT-PCR assay for the detection of infectious bronchitis virus from infected chickens." *J. Virol. Methods.* 138:60–65.

Carli, K.T. and Evigor, A. (2003). "Real-time PCR for *Mycoplasma gallisepticum* in chicken trachea." *Avian Dis.* 47: 712–717.

Cheun, H.I., Makino, S.I., Watarai, M., Erdenebaatar, J., Kawamoto, K. and Uchida, I. (2003). "Rapid and effective detection of anthrax spores in soil by PCR." *J. Appl. Microbiol.* 95: 728–733.

Cho, H.S., Kang, J.I. and Park, N.Y. (2006). "Detection of canine parvovirus in fecal samples using loop-mediated isothermal amplification." *J. Vet. Diagn. Invest.* 18:81–84.

Cobos-Marin, L., Montes-Vargas, J., Mivera-Gutierrez, S., Licca-Navarro, A., Gonzalex-y-Merchand, J.A. and Estrada-Garcia, I. (2003). "A novel multiple-PCR for the rapid identification of *Mycobacterium bovis* in clinical isolates of both veterinary and human origin." *Epidemiol. Infect.* 30: 485–490.

Danglar, C.A., Deaver, R.E., Kolodziej, C.M. and Rupprecht, J.D. (1992). "Genotypic screening of pseudorabies virus strains for thymidine kinase deletions by use of the PCR." *Amer. J. Vet. Rest.* 53: 904–908.

David, D., Yakobson, B., Rotenberg, D., Dveres, N., Davidson, I. and Stream, Y. (2002). "Rabies virus detection by RT-PCR in decomposed naturally infected brains." *Vet. Microbiol.* 87: 111–118.

Decaro, N., Elia, G., Martella, V., Desario, C., Campolo, M., Trani, L.D., Tarsitano, E., Tempesta, M. and Buonavoglia, C. (2005). "A real-time PCR assay for rapid detection and quantitation of canine parvovirus type 2 in the feces of dogs." *Vet. Microbiol.* 105 : 19–28.

De Wit, D., Steya, L., Shoemaker, S. and Sogin, M. (1990). "Direct detection of *Mycobacterium tuberculosis* in clinical specimen by DNA amplification." *J. Clin. Microbiol.* 28: 2437–2441.

Elia, G., Decaro, N., Martella, V., Cirone, F., Lucente, M.S., Lorusso, E., Di Trani, L. and Buonavoglia, C. (2006). "Detection of canine distemper virus in dogs by real-time RT-PCR." *J. Virol. Methods.* 136:171–176.

Faggi, E., Pini, G., Campisi, E., Bertellini, C., Difonza, E. and Manianti, F. (2001). "Applications of PCR to distinguish common species of dermatophytes." *J. Clin. Microbiol.* 39: 3382–3385.

Fahrimal, Y., Goff, W.L. and Jasmer, D.P. (1992). "Detection of *Babesia bovis* in carrier cattle by using PCR to distinguish common species of dermatophytes." *J. Clin. Microbiol.* 30: 1374–1379.

Farkas, T., Antal, M., Sami, L., German, P., Kecskemeti, S., Kardos, G., Belak, S. and Kiss, I. (2007). "Rapid and simultaneous detection of avian influenza and Newcastle disease viruses by duplex polymerase chain reaction assay." *Zoonoses Public Health.* 54:38–43.

Ferre, F. (1992). "Quantitative or semi-quantitative PCR: reality versus myth." *PCR Methods & Applications.* 2: 1–9.

Figueroa, J.V., Chieves, L.P., Johnson, G.S. and Guening, G.M. (1992). "Detection of *Babesia bigemina*-infected carriers by PCR amplification." *J. Clin. Microbiol.* 30: 2570–2580.

Fratamico, P.M., Bagi, L.K. and Pepe, T. (2000). "Multiplex PCR assay for rapid detection and identification of *E. coli* 0157: H7 in food and bovine faeces." *J. Food Prot.* 63: 1032–1037.

Gerritsen, M.J., Olyhoek, T., Smits, M.A. and Bokhout, B.A. (1991). "Sample preparation method for polymerase chain reaction-based semiquantitative detection of *Leptospira interrogans* serovar *hardjo* subtype Hardjobovis in bovine urine." *J. Clin. Microbiol.* 29: 2805–2808.

Gilbert, S.A., Burton, K.M., Prins, S.E. and Deregt, D. (1999). "Typing of bovine viral diarrhea viruses directly from blood of persistently infected cattle by multiplex PCR." *J. Clin. Microbiol.* 37: 2020–2023.

Gillespie, S.H., Ullman, C., Smith, M.D. and Emery, V. (1994). "Detection of *Streptococcus pneumoniae* in sputum samples by PCR." *J. Clin. Microbiol.* 32: 1308–1311.

Gould, A.R., Hyatt, A.D., Eaton, B.T., White, J.R., Hooper, P.T., Blackwell, S.D. and Smith, P.M.L. (1989). "Current technique in rapid bluetongue diagnosis." *Aust. Vet. J.* 66: 450–454.

Greco, G., Corrente, M., Martella, V., Pratelli, A. and Buonavoglia, D. (2001). "A multiplex PCR for the diagnosis of contagious agalactia of sheep and goats." *Mol. Cell Probes.* 15: 20–25.

Haas, L., Saila, M.S., Harder, T.K. Liess, B. and Barrett, T. (1991). "Detection of phocid distemper virus RNA in seal tissues using slot blot hybridization and the PCR amplification assay: Genetic evidence that the virus is distinct from canine distemper virus." *J. Gen. virol.* 72: 825–832.

Hairul Aini, H., Omar, A.R., Hair-Bejo, M. and Aini, I. (2006). "Comparison of Sybr Green I, ELISA and conventional agarose gel-based PCR in the detection of infectious bursal disease virus." *Microbiol Res.* Sep 11.

Handberg, K.J., Nielsen, O.L., Pedersen, M.W. and Jorgensen, P.H. (1999). "Detection and strain differentiation of infectious bronchitis virus in tracheal tissues from experimentally infected chickens by RT-PCR. Comparison with an immuno-histochemical technique." *Avian Pathol.* 28: 327–335.

Harmon, K.M., Ransom, G.M. and Wesley, I.V. (1997). "Differentiation of *Campylobacter jejuni* and *C. coli* by PCR." *Mol. Cell Probes.* 11: 195–200.

Hayashi, K. (1991). "PCR-single stranded conformation polymorphism." *PCR Methods & Applications.* 1: 34–38.

Herthnek, D., Englund, S., Willemsen, P.T. and Bolske, G. (2006). "Sensitive detection of *Mycobacterium avium* subsp. *paratuberculosis* in bovine semen by real-time PCR." *J. Appl. Microbiol.* 100:1095–1102.

Hombaerger, F.R., Smith, A.L. and Barthold, S.W. (1991). "Detecting rodent coronaviruses in tissues and cell cultures using PCR." *J. Clin. Microbiol.* 29: 2789–2793.

Hill, W.E. (1996). "The polymerase chain reaction: application for the detection of food borne pathogens." *Clin. Rev. Food Sci. Nutr.* 36: 123–173.

Hu, R.L., Huang, G., Qui, W., Zhong, Z.H. Xia, X.Z. and Yin, Z. (2001): "Detection and differentiation of CAV1 and CAV2 by PCR." *Vet. Res. Comm.* 25: 77–84.

Itoh, R., Saitoh, S. and Yatsuyanagi, J. (1995). "Specific detection of *Campylobacter jejuni* by polymerase chain reaction in chicken litter." *J. Vet. Med. Sci.* 57: 125–127.

Ito, M., Itou, T., Shoji, Y., Sakai, T., Ito, F.H., Arai, Y.T., Takasaki, T. and Kurane, I. (2003). "Discrimination between dog-related and vampire bat-related rabies viruses in Brazil by strain-specific RT-PCR and RFLP analysis." *J. Clin. Virol.* 26: 317–330.

Iqbal, Z. and Rikihisa, Y. (1994). "Application of the PCR for the detection of *Ehrlichia canis* in tissues of dogs." *Vet. Microbial.* 42: 281–287.

Jackwood, D.J., Spalding, B.D. and Sommer, S.E. (2003). "Real-time RT-PCR detection and analysis of nucleotide sequences coding for a neutralizing epitope on IBD viruses." *Avian Dis.* 47: 738–744.

Jestin, V. and Jestin, A. (1991). "Detection of Newcastle disease virus RNA in infected allantoic fluid by *in vitro* enzymatic amplification (PCR)." *Arch. Virol.* 118: 151–161.

Jestin, V., Foulon, T., Pertuiset, B., Blachard, P. and Labourdet, M. (1990). "Rapid detection of pseudorabies virus genomic sequences in biological samples from infected pigs using PCR DNA amplification." *Vet. Microbiol.* 23: 317–328.

Kai, M., Kamiyo, S., Yabe, H., Takakura, I., Shiozawa, K. and Ozawa, A. (1993). "Rapid detection of *Mycoplasma pneumoniae* in clinical samples by the PCR." *J. Med. Microbial.* 38: 166–170.

Kee, S.H., Kim, I.S., Choi, M.S. and Chang, W.H. (1994). "Detection of leptospiral DNA by PCR." *J. Clin. Microbiol.* 32: 1035–1039.

Keid, L.B., Soares, R.M., Vasconcellos, S.A., Chiebao, D.P., Megid, J., Salgado, V.R. and Richtzenhain, L.J. (2007). "A polymerase chain reaction for the detection of *Brucella canis* in semen of naturally infected dogs." *Theriogenology.* 67:1203–1210.

Kim, S., Lee, D.S., Suzuki, H. and Watarai, M. (2006). "Detection of *Brucella canis* and *Leptospira interrogans* in canine semen by multiplex nested PCR." *J. Vet. Med. Sci.* 68: 615–618.

Kimura, R., Madrell, R.E., Galland, J.C., Hyatt, D. and Riley, L.W. (2000). "Restriction site specific PCR as a rapid test to detect enterohaemorrhagic *E.coli* 0157: H7 strains in environmental samples." *Appl. Environ. Microbiol.* 66: 2513–1519.

Kolk, A.H.J., Schutema, A.R.J., Kuijper, S., Van Leeuwen, J., Hermans, P.W.M., Van Embden J.D.A. and Hartskeerl, R.A. (1992). "Detection of *Mycobacterium tuberculosis* in clinical samples by using PCR and a non radioactive detection system." *J. Clin. Microbiol.* 30: 2567–2575.

Kwaga, J., Iversen, J.O. and Misra, V. (1992). "Detection of pathogenic *Yersinia enterocolitica* by PCR and digoxigenin-labeled polynucleotide probes." *J. Clin. Microbiol.* 30: 2668–2673.

Latha, D. and Ramadass, P. (2002). "Improved PCR assay for the detection of canine parvovirus from faecal samples." *Ind. Vet. Med. J.* 26: 335–338.

Loa, C.C., Lin, T.L., Wu, C.C., Bryan, T.A., Hooper, T.A. and Schrader, D.L. (2006). "Differential detection of turkey coronavirus, infectious bronchitis virus, and bovine coronavirus by a multiplex polymerase chain reaction." *J. Virol. Methods.* 131:86–91.

Lebech, M., Lebech, A.M., Nelsing, S., Vuust, J., Mathiesen, L. and Petersen, E. (1992). "Detection of *T. gondii* DNA by PCR in CSF from AIDS patients with cerebral toxoplasmosis." *J. Infect. Dis.* 165: 164–165.

Louws, F.J., Schneider, M. and de Bruijn, F.J. (1996) In: *Nucleic Acid Amplification Methods for the Analysis of Environmental Samples.* Toranzos, G. (ed.). Technomic Publishing Co. 63–94.

Levett, P.N., Morey, R.E., Galloway, R.L., Turner, D.E., Steigerwalt, A.G. and Mayer, L.W. (2005). "Detection of pathogenic leptospires by real-time quantitative PCR." *J. Med. Microbiol.* 54: 45–49.

Macedo, C.I., Carnieli, P. Jr. Brandao, P.E., De Rosa, E.S.T., Oliveira, R.N., Castilho, J.G., Medeiros, R., Machado, R.R., De Oliveira, R.C., Carrier, M.L. and Kotait, I. (2006). "Diagnosis of human rabies cases by PCR reaction of neck-skin samples." *Brazilian J. Infect. Dis.* 10: 341–345.

Madden, R.H., Moran, L. and Scates, P. (1996). "Subtyping of animal and human *Campylobacter* sp. using RAPD." *Lett. Appl. Microbiol.* 23: 167–170.

McColl, K.A. and Gould, A.R. (1991). "Detection and characterization of bluetongue virus using the PCR." *Virus Res.* 21: 19–34.

Mekkes, D.R. and Feberwee, A. (2005). "Real-time PCR for the qualitative and quantitative detection of *Mycoplasma gallisepticum*." *Avian Pathol.* 34: 348–354.

Meer, R.R. and Songer, J.G. (1997). "Multiplex PCR assay for genotyping *Clostridium perfringens*." *Am. J. Vet. Res.* 58: 702–705.

Meerarani, S., Ramadass, P., Subhashini, C.R. and Nachimuthu, K. (1996). "Polymerase chain reaction assay for early detection of canine parvovirus." *Ind. Vet. J.* 73: 1013–1016.

Mehrotra, M., Wang, G. and Johnson, W.M. (2000). "Multiplex PCR for detection of genes for *Staph. aureus* enterotoxins, exfoliative toxins, toxic shock syndrome toxin and methicillin resistance." *J. Clin. Microbiol.* 38: 1032–1035.

Meyers, R.F., Brown, C.C., House, C., House, J.A. and Molitor, T.W. (1991). "Rapid and sensitive detection of foot and mouth disease virus in tissues by enzymatic RNA amplification of the polymerase gene." *J. Virol. Meth.* 34: 161–172.

Mishra, A., Singhal, A., Chauhan, D.S., Katoch, V.M., Srivastava, K., Thakral, S.S., Bharadwaj, S.S., Sreenivas, V. and Prasad, H.K. (2005). "Direct detection and identification of *Mycobacterium tuberculosis* and *Mycobacterium bovis* in bovine samples by a novel nested PCR assay: Correlation with conventional techniques." *J. Clin. Microbiol.* 43:5670–5678.

Mochizuki, M., San Gabriel, M.C., Nakatani, H., Yoshida, M. and Harasawa, R. (1993). "Comparison of PCR within virus isolation and hemagglutination assays for the detection of canine provirus in faecal specimens." *Res. Vet. Sci.* 55: 60–63.

Moliter, T.W., Oraverrakul, K., Zhang, Q.Q., Chot, C.S. and Ludemann. (1991). "PCR amplification for the detection of porcine parvovirus." *J. Virol. Meth.* 32: 201–211.

Naeem, K., Murtaugh, M.P. and Goyal, S.M. (1991). "Tissue distribution of bovine herpesvirus 4 in inoculated rabbits and its detection by DNA hybridization and PCR." *Arch. Virol.* 119: 239–255.

Naïf, H.M., Braden, R.B., Daniel, R.C.W. and Lavin, M.F. (1990). "Bovine leukemia proviral DNA detection in cattle using the PCR." *Vet. Microbiol.* 25: 117–129.

Oliveira, D.C., Van der Weide, M., Habela, M.A., Jacquiet, P. and Jongejan, F. (1995). "Detection of *Theileria annulata* in blood samples of carrier cattle by PCR." *J. Clin. Microbiol.* 33: 2665–2669.

Pershina, M.I., Shaginian, I.A., Ananina, I.V. and Prozorovskii, S.V. (1998). "Analysis of genomic polymorphism in leptospira by PCR with random primers." *Mol. Gen. Mikrobiol. Virusol.* 1: 29–32.

Piatak, M.J., Luk, K.C., Williams, B. and Lifson, J.D. (1992). "Quantitative competitive polymerase chain reaction for accurate quantitation of HIV DNA and RNA species." *Biotechniques.* 14: 70–81.

Pratelli, A. Buonavoglia, D., Martella, V., Tempesta, M., Lavazza, A. and Buonavoglia, C. (2000). "Diagnosis of canine coronovirus infection using nested PCR." *J. Virol. Methods.* 84: 91–94.

Poulsen, D.J., Burton, C.R.A., O'Brian, J.J., Rabin, S.J. and Keeler, C.L. Jr. (1991). "Identification of the infectious laryngotracheitis virus glycoprotein *gB* gene by the PCR." *Virus Gene.* 5: 335–347.

Qiao, M., Fowans, E.J., Bailey, S.E., Jilbert, A.R. and Burrele, C.J. (1990). "Serological analysis of duck hepatitis B virus infection." *Virus Res.* 17: 3–14.

Ramadass, P. and Latha, D. (2005). "Multiplex PCR assay for diagnosis of canine parvovirus infection and leptospirosis." *Ind. Vet. J.* 82: 365–368.

Ramadass, P., Meerarani, S., Venkatesha, M.D., Senthil Kumar, A. and Nachimuthu, K. (1997a). "Rapid diagnosis of leptospirosis by polymerase chain reaction." *Ind. Vet. J.* 74: 457–460.

Ramadass, P., Meerarani, S., Senthil Kumar, A., Venkatesha, M.D. and Nachimuthu, K. (1997b). "Characterization of leptospiral serovars by randomly amplified polymorphic DNA fingerprinting." *Int. J. Syst. Bacteriol.* 47: 575–576.

Ramadass, P., Thiagarajan, V., Parthiban, M., Senthil Kumar, T.M., Latha D., Anbalagan, S., Krishnakumar, M. and Nachimuthu, K. (2003). "Sequence analysis of infectious bursal disease virus isolates from India: Phylogenetic relationships." *Acta. Virol.* 47: 131–135.

Ramadass, P., Ananthi, R., Senthilkumar, T.M.A., Venkatesh, G. and Ramaswamy, V. (2006). "Isolation and characterization of *Mycoplasma gallisepticum* and *Mycoplasma synoviae* from poultry." *Ind. J. Ani. Sci.* 76: 796–798.

Ramasoota, P., Chansiripornchai, N., Kallenius, G., Hoffner, S.E. and Svenson, S.B. (2001). "Comparison of *Mycobacterium avium* complex strains from pigs and humans in Sweden by RAPD using standardized reagents." *Vet. Microbiol.* 78: 251–259.

Reid, S.M., Ferris, N.P., Hutchings, G.H., Zhang, Z., Belsham, G.J. and Alexandersen, S. (2002). "Detection of all seven serotypes of FMD virus by real-time, fluorogenic RT-PCR assay." *J. Virol. Methods.* 105: 67–80.

Roche, P.M. and Woodward, M.J. (1991). "PCR amplification of segments of pestivirus genomes." *Arch. Virol. Suppl.* 3: 231–238.

Ryu, C., Lee, K., Yoo, C., Seong, W.K. and Oh, H.B. (2003). "Sensitive and rapid quantitative detection of anthrax spores isolated from soil samples by real-time PCR." *Microbiol. Immunol.* 47: 693–699.

Saiki, R.F., Scharf, S., Faloona, F.A., Mullis, K.B., Horn, GT., Erlich, H.A. and Arnheim, N. (1985). "Enzymatic amplification of β-globin genome sequences and restriction site analysis for diagnosis of sickle cell anaemia." *Science.* 230: 1850–1854.

Saito, T.B., Alfieri, A.A., Wosiacki, S.R., Negrao, F.J., Morais, H.S. and Alfieri, A.F. (2005). "Detection of canine distemper virus by RT-PCR in the urine of dogs with clinical signs of distemper encephalitis." *Res. Vet. Sci.* 80: 116–119.

Saravanan, M., Parthiban, M. and Ramadass, P. (2006). "Genotyping of rotavirus of neonatal calves by nested-multiplex PCR in India." *Veterinarski. Archiv.* 76: 497–505.

Sato, G, Tanabe, H., Shoji, Y., Itou, T., Ito, F.H., Sato, T. and Sakai, T. (2005). "Rapid discrimination of rabies viruses isolated from various host species in Brazil by multiplex reverse transcription-polymerase chain reaction." *J. Clin. Virol.* 33:267–273.

Saunders, V.F., Reddacliff, L.A., Berg, T. and Hornitzky, M. (2007). "Multiplex PCR for the detection of *Brucella ovis*, *Actinobacillus seminis* and *Histophilus somni* in ram semen." *Aust. Vet. J.* 85:72–77.

Senthil Kumar, A., Ramadass, P. and Nachimuthu, K. (2001). "Use of PCR for the detection of leptospires in clinical samples." *Ind. Vet. J.* 78: 1087–1090.

Shankar, V., Dietzschold, B. and Koprowski, H. (1991). "Direct entry of rabies virus into the central nervous system without prior local multiplication." *J. Virol.* 65: 2736–2738.

Shin, Y.J., Cho, K.O., Cho, H.S., Kang, S.K., Kim, H.J., Kim, Y.H., Park, H.S. and Park, N.Y. (2004). "Comparison of one-step RT-PCR and a nested PCR for the detection of canine distemper virus in clinical samples." *Aust. Vet. J.* 82: 83–86.

Silva, R.F. (1992). "Differentiation of pathogenic and non-pathogenic serotype 1 Marek's disease viruses (MDVs) by the PCR amplification of the tandem direct repeats within the MDV genome." *Avian Dis.* 36: 521–528.

Sivakumar, P., Tripathi, B.N. and Singh, N. (2005). "Detection of *Mycobacterium avium* subsp. *paratuberculosis* in intestinal and lymph node tissues of water buffaloes (*Bubalus bubalis*) by PCR and bacterial culture." *Vet. Microbiol.* 108: 263–270.

Skottman, T., Piiparinen, H., Hyytiainen, H., Myllys, V., Skurnik, M. and Nikkari, S. (2007). "Simultaneous real-time PCR detection of *Bacillus anthracis*, *Francisella tularensis* and *Yersinia pestis*." *Eur. J. Clin. Microbiol. Infect. Dis.* 26:207–211.

Sreevatsan, S., Bookout, J.B., Ringspis, F., Penimasla, V.S., Ficht, T.A., Adams, L.G., Hagins, S.D., Elzer, P.H., Bricker, B.J., Kumar, G.K., Rajasekar, M., Isloor, S. and Barathur, R.R. (2000). "A multiplex approach to molecular detection of *Br. abortus* and/or *Mycobacterium bovis* infection in cattle." *J. Clin. Microbiol.* 38: 2602–2610.

Suresh Kumar, K., Dhinakar Raj, G., Raja, A. and Ramadass, P. (2007). "Genotypic characterization of infectious bronchitis viruses from India." *Ind. J. Biotechnol.* 6: 41–44.

Tcherneva, E., Rijpens, N., Jersek, B. and Herman, L.M. (2000). "Differentiation of *Brucella* species by RAPD analysis." *J. Appl. Microbiol.* 88: 69–80.

Todd, D., Mawhinney, K.A. and McNulty, M.S. (1992). "Detection and differentiation of chicken anaemia virus isolates by using the PCR." *J. Clin. Microbiol.* 30: 1661–1666.

Uhlmann, V., Silva, I., Luttich, K., Picton, S. and O' Leavy, J.J. (1998). "In cell amplification." *J. Clin. Pathol.* 51: 1190–130.

van Belkum, A. (1994). "DNA fingerprinting of medically important microorganisms by use of PCR." *J. Clin. Microbiol. Rev.* 7:174–184.

van Woensel, P.A.M., van Bleaderen, A., Moormann, R.J.M. and DdeBoer, G.F. (1992). "Detection of proviral DNA and viral RNA in various tissues early after avian leucosis infection." *Leukemia.* 6: 1355–1375.

Versalovic, J., Koeuth, T., and Lupski, J.R. (1991) "Distribution of repetitive DNA sequences in eubacteria and application to fingerprinting of bacterial genomes." *Nucl. Acids Res.* 19: 6823–6831.

Versalovic J., Schneider, M., de Bruijn, F.J. and Lupski, J.R. (1994) "Genomic fingerprinting of bacteria using repetitive sequence-based polymerase chain reaction." *Meth. Cell. Mol. Biol.* 5: 25–40.

Versalovic, J., de Bruijn, F.J. and Lupski, J.R. (1997) In: *Bacterial Genomes: Physical Structure and Analysis.* de Bruijn, F.J., Lupski, J.R. and Weinstock, G.M. (eds.). Chapman and Hall, New York.

Von Teifke, J.P. and Weiss, E. (1991). "Nachweis bonner Papilloma virus DNA in Sarkoiden des Pferdes Mitteles der Polymerase Kettenreaktion (PCR). *"Berliner and Munchener Tierarztliche Wochenschrift*. 104: 185–187. Cited by Belak and Ballagi Pordany (1993).

Vrioni, G., Gartzonika, C., Kostoula, A., Boboyianni, C., Papadopoulou, C. and Levidiotou, S. (2004). "Application of a PCR-enzyme immunoassay in peripheral whole blood and serum specimens for diagnosis of acute human brucellosis." *Eur. J. Clin. Microbiol. Infect. Dis.* 23: 194–199.

Wang, X. and Khan, M.I. (1999). "A multiplex PCR for Massachusetts and Arkansas serotypes of infectious bronchitis virus." *Mol. Cell Probes.* 13: 1–7.

Wang, H., Fadl, A.A. and Khan, M.I. (1997). "Multiplex PCR for avian pathogenic mycoplasmas." *Mol. Cell Probes.* 11: 211–216.

Ward, P. and Misra, V. (1991). "Detection of bovine diarrhoea virus, using degenerate oilgonucleotide primers and the PCR." *Amer. J. Vet. Res.* 52: 1231–1236.

Welsh J., and McClelland, M. (1990). "Fingerprinting genomes using PCR with arbitrary primers." *NAR.* 18:7213–7218.

Williams, R.A., Bennel, M., Bradbury, J.M., Gaskell, R.M., Jones R.C. and Jordan, F.T.W. "Demonstration of sites of latency of infectious laryngotracheitis virus using the polymerase chain reaction." *J. Gen. Virol.* 73: 2415–2420.

Wu, C.C., Lin, T.L., Zhang, H.G., Davis, V.S. and Boyle, J.A. (1992). "Molecular detection of infectious bursal disease virus by PCR." *Avian Dis.* 36: 221–226.

Xu, L., Harbour, D. and McCrae, M.A. (1990). "The application of PCR to the detection of rotaviruses in faeces." *J. Virol. Meth.* 27: 29–38.

Xwaagstra, K.A., van der Zeijst, B.A.M. and Kusters, J.G. (1992). "Rapid detection and identification of avian infectious bronchitis virus." *J. Clin. Microbiol.* 30: 79–84.

Zhang, M.F., Huang, G.M. and Qiao, S. (2002). "Early stages of IBDV infection in chickens detected by *in situ* RT-PCR." *Avian Pathol.* 31: 593–597.

Zucker, B., Krugera, M. and Horsch, F. (1996). "Differentiation of *Pasteurella multocida* subspecies multocida isolates from respiratory system of pigs by using PCR fingerprinting technique." *Zentralbl. Veterinarmed* [B]. 43: 585–591.

13

NUCLEIC ACID SEQUENCING

INTRODUCTION

The ultimate description of a gene is the determination of the sequence of nucleotide bases of which it is comprised and which, in turn, determines the nature of the encoded protein product. From the first development of an accessible and routine technique to determine the nucleotide sequences of only small stretches of DNA, we have now advanced to the stage where nucleotide sequences are determined at a prodigious rate. This has resulted in the sequencing of the complete genomes of a number of bacteria and the eukaryote yeast *Saccharomyces cerevisiae*. Such is the pace of progress in the application of nucleotide sequencing technology that the sequence of the complete human genome is available now. This remarkable progress has been made possibly by the application of relatively simple technologies. Such was the importance of these techniques that their proponents, Walter Gilbert and Fred Sanger, were jointly awarded the Nobel Prize for chemistry in 1980.

Methods of DNA sequencing were developed in the late 1970s and have revolutionized the science of molecular genetics. The DNA sequences of many different genes from diverse sources have been determined. Projects are already underway to map the sequence of the entire genome of organisms such as *Escherichia coli*, *Saccharomyces cerevisiae* and *Homo sapiens*. Large-scale sequencing projects such as the Human Genome Project, produce the DNA sequence of many unknown genes. DNA sequencing can act as a catalyst to stimulate future research into many diverse areas of science.

The two original methods of DNA sequencing described in 1977 differ considerably in principle. The enzymatic (or dideoxy chain termination) method of Sanger (Sanger *et al.*, 1977) involves the synthesis of a DNA strand from a single-stranded template by a DNA polymerase. The Maxam and Gilbert (or chemical degradation) method (Maxam and Gilbert, 1977) involves chemical degradation of the original DNA. Both methods produce populations of radioactively labelled polynucleotides that begin from a fixed point and terminate at points dependent on the location of a particular base in the original DNA strand. The polynucleotides are separated by polyacrylamide gel electrophoresis, and the order of nucleotides in the original DNA can be read directly from an autoradiography of the gel.

CHAIN TERMINATION SEQUENCING

There are three stages in a chain termination sequencing experiment.

Stage 1   Preparing the single-stranded DNA that will be the template for the strand synthesis reaction.

Stage 2   Carrying out the strand synthesis reaction.

Stage 3   Separating the chain-terminated molecules by polyacrylamide gel electrophoresis and reading the gel.

Preparation of Single-stranded DNA Template

M13 phages are the commonly used cloning vectors for making single-stranded DNA. M13 is a filamentous phage with a genome size of 6407 nucleotides in length. Within the phage particle, this genome exists as a single-stranded DNA (Marvin and Wachtal, 1975). M13 phage particles infect *Escherichia coli* bacteria by injecting their DNA into the cell. Once inside the cell, the single-stranded M13 DNA is converted into a double-stranded molecule which replicates until over 100 copies are present. This double-stranded version of the phage genome is called the replicative form or RF. When the bacterium divides, each daughter receives copies of the phage RF, which continues to replicate, maintaining its overall numbers per cell.

The vectors constructed from M13 for cloning are called the M13mp series. Each of these vectors (M13mp 7&8, 9&10 and 18&19) have a different set of unique restriction sites that can be used for the insertion of new DNA during a cloning experiment. The restriction sites are contained within a copy of the *lacz* gene that has been transferred to the M13 genome. The *lacZ* gene codes for the first 146 amino acids of the *E. coli* β-galactosidase enzyme. This is the enzyme responsible for converting lactate into glucose plus galactose. After transformation with a *lacZ* vector, the bacteria are plated on an agar containing the lactose analogue X gal (5-bromo-4-chloro-3-indolyl-β-D-galactopyroside) (Horwitz *et al.*, 1964). Then the colonies containing the recombinant plasmids are selected as follows:

i.   Cells that contain a cloning vector with an intact *lacZ* gene (i.e., no inserted DNA) are able to synthesize β-galactosidase, which coverts the X-gal into a compound bromochloroindole, which is blue in colour. These cells therefore give rise to blue colonies (with a plasmid vector) or blue plaques (with a phage vector).

ii. Conversely, recombinant cells, in which the *lacZ* gene is disrupted by the inserted DNA, and cannot synthesize β-galactosidase and so gives rise to white colonies or clear plaques.

The *lacZ* gene transfer, provides an easy means of identifying recombinant bacteria.

There are a number of *E. coli* strains that could be used as host bacteria for the M13mp vectors. These strains must be male, so they can be infected by M13 phage and be able to participate in *lac* selection. These include *E. coli* JM 109, 110, TG2 and XL1-Blue.

In a typical cloning procedure, there are 3 steps:

1.   restriction enzyme digestion and ligation of the insert DNA into a double-stranded replicative form of cloning vector.

2.   introduction of the recombinant cloning vector into the host bacteria by transfection and

3.   plating of the transformants on to agar-containing X-gal and IPTG.

After incubating the plates overnight at 37°C, a collection of blue and clear plaques are obtained. Single-stranded DNA from the clear plaque is prepared by transforming toothpick of single clear plaques into a liquid medium and incubating at 37°C with shaking for about 5–8 hours. The idea is to grow the culture to the highest possible cell density without allowing the cells to enter the stationary phase. The cells are pelleted, supernatant collected and phage particles precipitated with polyethylene glycol and salt. The phage pellet is suspended in buffer and phenol is added to remove the phage protein coat. After centrifugation, the upper aqueous phase containing phage DNA is removed. The DNA is precipitated with ethanol and dissolved in a small amount of buffer. Usually, in a pure preparation of single-stranded DNA, the pellet will be invisible.

Strand Synthesis Reaction for Chain Termination Sequencing

The first step in the strand synthesis reaction is to anneal a short oligonucleotide primer to each single-stranded template molecule. The primer acts as a starting point for synthesis of a new polynucleotide chain, complementary to the template strand. The strand synthesis is carried out by a DNA polymerase enzyme and requires deoxynucleoside triphosphates as substrates. One of these dNTPs is labelled with a radioactive marker, so the results of the gel electrophoresis could be visualized by autoradiography. Strand synthesis is not allowed to continue to completion as a chain-terminating nucleotide is included in the reaction mixture. Four families of chain-terminated molecules are produced, one family containing the polynucleotides that terminate in A, a second family of C-terminated polynucleotides and so on. The four families are then fractionated by polyacrylamide gel electrophoresis to obtain a banding pattern from which the sequence is read.

Primer   The primer is needed because template-dependent DNA polymerase is unable to initiate DNA synthesis on an entirely single-stranded molecule. A short double-stranded region is needed to provide a 3´ end onto which the polymerase can add new nucleotides. The primer provides this 3´ end, without it, strand synthesis cannot take place. Primer anneals to the template immediately adjacent to one of the junctions between the vector and inserted DNA. This is the ideal position at which to begin strand synthesis as the DNA polymerase immediately starts to read the inserted DNA.

The priming site is within the *lacZ* region of the vector molecule (M13). This region has the same sequence in all vectors that carry the *lacZ* gene and hence a single universal primer, whose sequence is complementary to the relevant segment of the Z gene is suitable for all experiments regardless of the origin of the cloned DNA.

DNA polymerase   Most DNA polymerases are template-dependent, meaning that the new DNA strand is synthesized on existing single-stranded template molecule, the sequence of the new strand being complementary to the sequence of the template. Some enzymes use cDNA template and so are DNA-dependent DNA polymerases, others use RNA-dependent DNA polymerase. All DNA polymerases synthesize DNA in the 5´ to 3´ direction, meaning that they add nucleotides on the 3´ end of the strand that is being synthesized.

For use in DNA sequencing, a DNA polymerase must satisfy a number of criteria. The most important of these are they should possess high processivity and low exonuclease activity. Processivity refers to the length of the new strand that a DNA polymerase is able to synthesize before the reaction terminates through natural causes. Many DNA polymerases are dual-function enzymes, being able to degrade DNA as well as synthesize it. DNA polymerase I of *E. coli*, for example, is a DNA repair enzyme, which means that it is able to remove nucleotides from a DNA strand in order to correct mistakes made during DNA replication or arising from mutation. To do this, enzyme needs a 5´ to 3´ exonuclease activity. In addition, DNA polymerase I like most DNA polymerases, has a proof-reading function that helps it to correct any mistake that it makes during DNA synthesis. This requires a 3´ to 5´ exonuclease activity, as the enzyme has to go in reverse order to remove incorrect nucleotides from the DNA strand it is synthesizing. Few of the other properties of DNA polymerases used in sequencing include activity at 55°C and above and ability to use modified nucleotides as substrates.

The DNA polymerases which are used in sequencing include Sequenase, a modified version of the DNA polymerase encoded by bacteriophage T7; Klenow DNA polymerase I of *E. coli* and *Taq* DNA polymerase.

Deoxynucleoside triphosphates   DNA is synthesized by polymerization of deoxynucleoside triphosphates (dNTPs). In the chain termination sequencing experiment, each of the four-stranded dNTPs are included in the reaction mixture and one of the dNTPs is labelled with either ^{32}P at the α position or with ^{35}S at the α phosphorus. The β particles emitted by ^{32}P are approximately 10 times more energetic than those from ^{35}S. ^{32}P-labelling results in fuzzier bands on the autoradiography, due to scattering within the X-ray film.

Most sequencing is therefore carried out with ^{35}S-dNTP. The autoradiography must be exposed for longer period, but on the whole the results are better.

Chain terminating nucleotide   A ddNTP (dideoxynucleoside triphosphate) lacks the hydroxyl group attached to the 3´ carbon of the sugar component of a normal deoxynucleotide. The 3´ carbon participates in the phosphodiester bond formation but without the hydroxyl group, the reaction cannot take place. Incorporation of the ddNTP therefore blocks further strand elongation. In addition to the normal nucleotide precursors, DNA synthesis is carried out in the presence of base-specific dideoxynucleotides (ddNTPs). The latter are analogs of the normal dNTPs but differ in that they lack a hydroxyl group at the 3´ carbon position as well as the 2´ carbon. A dideoxynucleotide can be incorporated into the growing DNA chain by forming a phosphodiester bond between its 5´ carbon atom and the 3´ carbon of the previously incorporated nucleotide. However, since ddNTPs lack a 3´ hydroxyl group, any ddNTP that is incorporated into a growing DNA chain cannot participate in phosphodiester bonding at its 3´carbon atom, thereby causing abrupt termination of chain synthesis.

Sequencing reactions   Four parallel base-specific reactions are conducted using a mix of all four dNTPs and also a small proportion of one of the four ddNTPs. By setting the concentration of the ddNTP to be very much lower than that of its normal dNTP analog, chain termination will occur randomly at one of the many positions containing the base in question. Each reaction is therefore a **partial reaction**: chain termination occurs randomly at one of the possible bases in any one DNA strand. However, the DNA to be sequenced in a DNA sequencing reaction is a **population** of (usually) identical molecules. As a result, each one of the four base-specific reactions will generate a collection of labelled DNA fragments of different sizes, with a common 5´ end but variable 3´ ends (the common 5´ end is defined by the sequencing primer and the 3´ ends which terminate with the chosen ddNTP are variable because the insertion of the dideoxynucleotide occurs randomly at one of the many different positions that will accept that specific base.

There are two main stages of this procedure:

1. annealing the primer to the template and

2. extending the annealed primer by strand synthesis to produce the chain-terminated molecules.

Primer annealing   The first stage of any chain termination sequencing procedure is to anneal the oligonucleotide primer onto the template DNA molecule. The specificity of the primer is important because it must anneal only to the desired position on the template, immediately upstream of the region that is to be sequenced. Primer annealing is usually carried out at 55–60°C. The different steps involved in dideoxy method of sequencing is shown in Figure 13.1.

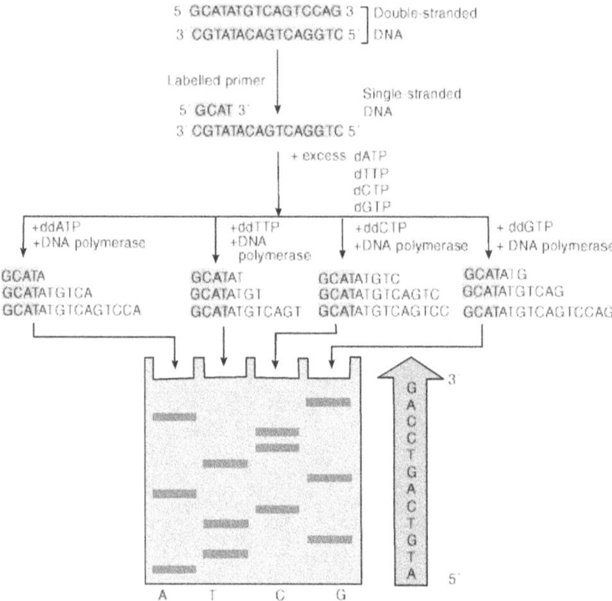

Figure 13.1   Steps in dideoxy method of nucleic acid sequencing

Strand synthesis The strand synthesis is carried out in two steps. First, limited quantities of the four dNTPs, one of which is labelled are added to the DNA polymerase. The enzyme synthesizes new strands, but cannot extend them very far before running out of the four nucleotides.

To set up the labelling step, the annealed primer-template, labelled and unlabelled dNTPs and DNA polymerase in a tris-NaCl-MgCl$_2$ buffer are mixed and incubated at room temperature for about 10 minutes. At the end of the first incubation, the termination reactions are initiated. This is done by adding an additional unlabelled dNTP, plus the appropriate dNTPs, into each reaction tube and incubated for further 5 minutes. During this second incubation period the polymerase extends the strands until chain termination occurs. To complete the experiment, few ml of formamide dye mix which contains formamide plus two dyes, xylene cyanol and bromophenol blue are added.

Running the Gel and Reading the Sequence

The strand synthesis reactions provide with four families of chain terminated molecules. In one family, all the molecules terminate with A, in another they all end in C, in the third with G and in the fourth with T. These molecules are each separated in polyacrylamide gel in order to obtain the banding pattern from which the sequence of the template DNA can be determined.

Gel electrophoresis A 6% gel is usually used for sequencing of molecules in the range of 25–500 nucleotides. Gel is poured in glass plates of size 40 × 20 cm, avoiding air bubbles. When the gel is fully polymerized, it is attached to the electrophoresis apparatus. Polyacrylamide gel is run in a vertical position, with the wells at the top, so that DNA molecules migrate down through the gel.

Before loading the samples, they are denatured at 95°C for 2 minutes to detach the chain terminated molecules from the template DNA. Then the samples are held on ice and 2–4 µl of the samples are loaded. The samples are loaded in the order A-C-G-T in a 6% gel, the bromophenol blue dye moves at the same rate as a single-stranded DNA molecule, 25 nucleotides in length. Xylene cyanol in this gel moves at the rate of a 100-nucleotide molecule. For sequencing more bases, when the bromophenol blue dye reached near the bottom of the gel, electrophoresis can be stopped and second aliquot of samples can be loaded in the neighbouring wells and electrophoresis is continued.

Reading the sequence First the band that has moved the furthest, the one that is closest to the bottom of the autoradiograph is located. This band represents the shortest chain-terminated molecule. The next more mobile band corresponds to a chain-terminated molecule that is one nucleotide longer (Figure 13.2).

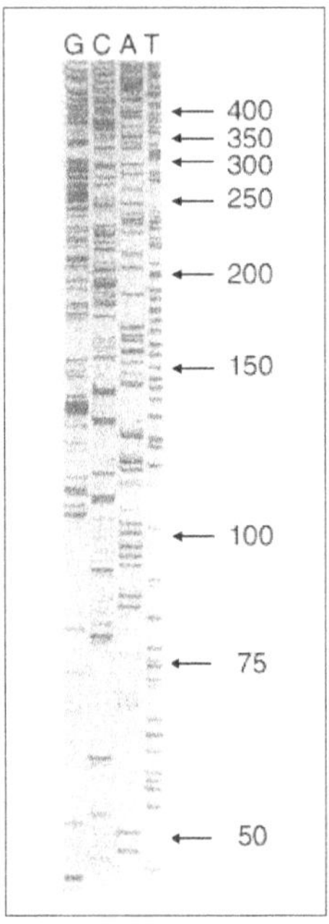

Figure 13.2   A typical sequencing pattern of nucleic acids

Direct Cycle Sequencing

Recently, an automated sequencing method has been used for sequencing of nucleic acids. This method is based on the chain termination method of Sanger in which instead of radiolabelled fragments, fluorescence-labelled fragments are obtained by use of fluorescent-labelled dNTPs and amplified by PCR. Cycle sequencing, like the standard PCR reaction, uses a thermostable DNA polymerase and a temperature-cycling format of denaturation, annealing and DNA synthesis. The difference is that cycle sequencing employs only one primer and includes a ddNTP chain terminator in the reaction. The use of only a single primer means that unlike the exponential increase in product during standard PCR reactions, the product accumulates linearly. Because the product accumulates during the reaction, and because of the high temperature at which the sequencing reactions are carried out, and the multiple heat denaturation steps, small amounts of double-stranded plasmids, cosmids, λDNA and PCR products may be sequenced reliably without a separate heat denaturation step. The amplified fragments are directly loaded on a 4% polyacrylamide gel for fluorescent DNA sequencing. The bands are read when they pass through the laser beam. Final reading of the sequence involves computation.

CHEMICAL DEGRADATION SEQUENCING

The basic idea behind the two DNA sequencing methods is the same. In both cases, the objective is to generate families of DNA molecules, which are separated by PAGE, and produce the banding pattern from which the sequence of the starting molecule can be read. The difference lies in the way in which the A, C, G and T families of molecules are generated. In the chemical degradation experiment, these families are produced not by synthesizing new strands, but by breaking down existing molecules with chemicals that cleave polynucleotides specifically at A, C, G or T nucleotide positions.

For generating G family of molecules in a chemical degradation method, the starting polynucleotides that are to be sequenced must have a radioactive marker at one end. The molecule is treated with a reagent-dimethyl sulphate, which reacts specifically with G nucleotide, resulting in a chemical modification of the purine ring. The amount of dimethyl sulphate that is added is carefully controlled so that on an average, just one G nucleotide in each copy of the molecule to be sequenced is modified. At this stage, the DNA strands are still intact. Strand cleavage occurs when a second chemical, piperidine, is added to the reaction. Piperidine removes the modified G nucleotide and cuts the DNA molecule at the phosphodiester bond immediately upstream of the 'base-less' site that is created.

Each cleavage produces two fragments, one from either side of the cut point. But only one of the fragments is labelled, so the second is in effect 'invisible' as it does not show up after autoradiography of the gel. Each cleavage therefore, gives rise to a single band on the autoradiograph, the position of this band being determined by the length of the fragment, which in turn is determined by the position within the original molecule of the G nucleotide that was modified during the chemical treatment. There are many copies of starting molecules each of which is now cleaved at the phosphodiester bond adjacent to a G nucleotide. Similarly, A, C and T families are produced using different chemicals. Gel electrophoresis and reading of sequence is similar to chain termination method.

There are three steps in a chemical degradation method.

1. Preparation of the end-labelled DNA

2. The chemical degradation reaction

3. Gel electrophoresis, autoradiography and reading of the sequence

Preparation of End-labelled DNA

There are two possible ways of end-labelling of DNA molecule, viz.,

1. by attachment of a labelled phosphate or nucleotide to the 5´ or 3´ end of a polynucleotide and

2. by filling in the 5´ overhangs created at the ends of a double-stranded molecule by some restriction enzymes (Muridy *et al.*, 1991).

End labelling is most frequently carried out with the enzyme called T4 polynucleotide kinase (PNK). PNK transfers a phosphate group from the γ position of an ATP molecule on to the 5´ end of a dephosphorylated acceptor molecule. DNA molecules usually have phosphorylated 5´ ends but the exchange reaction can be set up so that PNK removes these 5´ phosphates prior to carrying out the transfer. Alternatively, the phosphate groups can be removed by pretreating the DNA with alkaline phosphatase.

Many restriction enzymes leave a 5´-overhangs when they cut a double-stranded DNA molecule. DNA polymerase enzymes are able to extend the recessed 3´end, using the overhangs as a template in a "filling-in" reaction. If a labelled nucleotide is included in the reaction, then the polynucleotide becomes 3´ end-labelled, presuming that the labelled nucleotide is complementary to one of the nucleotides in the overhang. Filling-in reaction is usually carried out with Klenow polymerase, as an enzyme that lacks the 5´to 3´exonuclease activity.

Label   For chain-termination sequencing ^{35}S is a better choice than ^{32}P. ^{35}S gives sharper bands on the autoradiography and is less hazardous. With chemical degradation sequencing, ^{35}S makes a less suitable label than ^{32}P. After chemical degradation experiment, the polynucleotides that are loaded on to the sequencing gel each carry just a single label. In contrast, the polynucleotides from a chain termination experiment are multiple-labelled. Hence, the autoradiograph from a chemical degradation experiment must be exposed for longer period than from a chain-termination gel. The weaker energy of ^{35}S now becomes a distinct disadvantage as it means that exposure for a week or more is needed before a clear autoradiography is obtained. ^{32}P is therefore the more popular radioactive label for the chemical degradation method.

Conversion of the end labelled molecule into a form suitable for sequencing   The DNA molecule for chemical degradation sequencing must be in one of the two forms:

1. single-stranded, end-labelled or

2. double-stranded comprising of one end-labelled polynucleotide and an unlabelled polynucleotide.

Obtaining single-stranded DNA is more difficult. After denaturing a double-stranded DNA molecule, single-stranded form is purified by running on a denaturing polyacrylamide gel. As the two polynucleotides may have different purine–pyrimidine contents, they migrate differently. From the gel, single-stranded polynucleotide is purified.

If the double-stranded molecule that has been end-labelled contains an internal restriction site, then it can be cut with an appropriate restriction enzyme and the two fragments separated in an agarose or polyacrylamide gel. The fragment required for sequencing, which is now end-labelled, can then be purified from the gel.

Chemical Degradation Reactions

The chemical degradation reactions are carried out in two steps. First, the DNA is treated with a chemical that modifies a nucleotide base. Then piperidine is added to cut the DNA at the modified position.

Nucleotide modification reactions   The DNA sample to be sequenced is divided into four aliquots and treated as follows:

Tube 1   add dimethyl sulphate ('G' reaction)

Tube 2   add formic acid ('A + G' reaction)

Tube 3   add hydrazine ('C + T' reaction)

Tube 4   add NaCl + hydrazine ('C' reaction)

Two of the reactions are not entirely base-specific. It has been proved impossible to devise modification reactions that recognize each base individually and in this standard procedure, two of the reactions give mixed families of molecules, a "A + G" family and a "C + T" family. A fifth reaction (NaOH + EDTA) that cleaves at A and C nucleotides with preference for A's could also be carried out.

Each modification is complete after few minutes at 25°C. Reactions are stopped by adding acetate buffer, and DNA is recovered by ethanol precipitation.

Strand cleavage   For strand cleavage, the DNA pellets are resuspended in 1 M piperidine and incubated for 30 minutes at 90°C. During this incubation a modified purine and pyrimidine bases are removed from the polynucleotides and each strand is cut at the phosphodiester bond immediately upstream of the base-less position. It is important to make sure that the cleavage reaction goes to completion. If not, the bands on the autoradiograph will be weak and indistinct. After incubation, the piperidine is removed in a rotary evaporator and the DNA sample resuspended in loading buffer, ready for electrophoresis in the sequencing gel.

Running the Gel and Reading the Sequence

The sequencing gel is prepared, electrophoresed and autoradiographed exactly as described for the chain termination method. There is a difference only in reading the autoradiograph. It should be remembered that two of the modification reactions are specific not for one, but for two different nucleotides. As a result, there will be some bands on the autoradiograph that span two tracks rather than just one. The band which has moved furthest is read first and then the next band above.

AUTOMATED NUCLEIC ACID SEQUENCING

Recent developments have focused on making the sequencing reactions and characterization of products easier. One major improvement in recent years has been the development of automated procedures for fluorescent DNA sequencing. These procedures generally use primers or dideoxynucleotides to which are attached fluorophores (chemical groups capable of fluorescing). During electrophoresis, a monitor detects and records the fluorescence signal as the DNA passes through a fixed point in the gel. The use of different fluorophores in the four base-specific reactions means that, unlike conventional DNA sequencing, all four reactions can be loaded into a single lane. The output is in the form of intensity profiles for each of the differently coloured fluorophores, but the information is simultaneously stored electronically. This precludes transcription errors when an interpreted sequence is typed by hand into a computer file. Recent advances in technology mean that the accuracy of DNA sequencing using automated methods is acceptably high. There are new commercially available, and widely used, automated sequencing machines that combine the initial running of samples and the reading of tracks to generate the nucleotide sequence. These machines are designed for high throughput of DNA sequencing and require the use of a unique set of nucleoside analogues in conjunction with the chain-termination reaction. The principle of the reaction is as described above. Attached to each dideoxy nucleoside triphosphate is a fluorescent dye that, when excited by illumination with light of an appropriate wavelength, emits light at a different wavelength. The automated sequence reaction takes place in a single tube because each nucleotide analogue is attached to a different fluorescent dye. The reaction uses thermophilic DNA polymerases in the cycle sequencing reaction. When products of the reaction are separated by electrophoresis, each DNA molecule migrates, according to its size, towards the bottom of the gel. Light from a laser is used to excite the fluorescent dye as the DNA fragment bearing it migrates past. The fluorescent light emitted from each dye is detected by a photomultiplier and converted into an electrical signal, which is then interpreted by computer software at the appropriate nucleotide base. The strength of the fluorescent light determines the strength of the electrical signal. The chemistry of this process is slightly different because only DNA molecules terminated with a fluorescent dye analogue are detected. The length of the sequence determined can be as much as 700–1000 bases in each run. The main disadvantage of this system is the very high cost of equipment required for the running, detection and analysis of DNA fragments. For this reason, automated sequencing is normally available as a core facility within a research institute, university or company. A typical nucleic acid separation pattern in an automated sequences is shown in Figure 13.3.

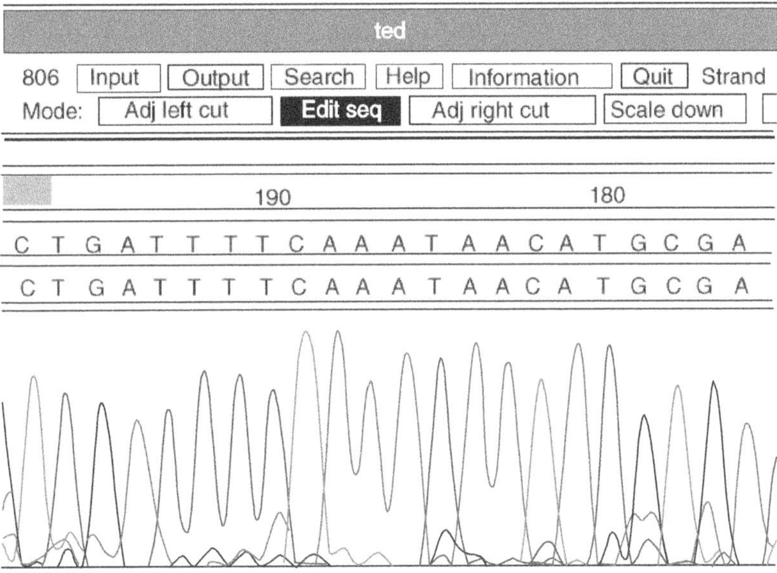

Figure 13.3 Automated nucleic acid sequence separation

APPLICATIONS OF GENE SEQUENCING

Remond *et al.* (1992) cloned and sequenced an *Eco* RI–*Pst* I fragment derived from the replicative form of a canine parvovirus (CPV) vaccine strain. M13 dideoxynucleotide sequencing was carried out using Sequenase Kit (USB). The variability of the 5′ end of NSI protein gene in the genome is confirmed by comparison with previously determined DNA sequences. Comparison of this sequence with that of three other CPV strains and one FPLV strain revealed point mutations between different isolates. A 15-nucleotide deletion was also observed in this vaccine strain.

Yehuda *et al.* (1999) cloned and sequenced VP1 (encoded by segment B) of a very virulent Israeli virus IL3, its attenuated strain, IL4 and the attenuated Winterfield vaccine 2512 of IBDV. A comparison was made among them and with six other published sequences of segment B. Six nucleic acids were distinguished between IL3 and IL4, three of which were predicted to be expressed as amino acids. A striking similarity between the VP1 sequences of 2512 and P2 (an attenuated German strain) was discovered.

Bordignon *et al.* (2005) described an RT-PCR assay and sequencing for rabies diagnosis and characterization. A total of 75 samples obtained from a variety of animal species in the state of Santa Catarina (SC), Southern Brazil, were comparatively studied by fluorescence antibody test (FAT), mouse inoculation test (MIT), cell infection assay and RT-PCR, which revealed itself to be as sensitive as FAT and MIT and less time-consuming than MIT. Direct sequencing of the 5´ end of the N gene allowed the clustering of the SC samples with samples from the vampire bat-related or sylvatic cycle through comparative sequence analysis.

Variable cDNA regions in the *VP2* gene of five highly virulent infectious bursal disease viruses (IBDVs) isolated in Japan were amplified by PCR and sequenced (Lin *et al.*, 1993). To prepare single-stranded cDNA for the sequencing, asymmetric PCR was carried out and the amplified product was sequenced by dideoxynucleotide chain termination method of sequencing. The nucleotide sequences of five highly virulent IBDVs were identical. Comparison of the nucleotide and the deduced amino acid sequences with those of other strains of IBDV indicated that Japanese highly virulent IBDV is different from all other strains of IBDV that were compared. The number of amino acid substitutions were scored between isolate 90–11 and the reference IBDV strains. The score between isolate 90–11 and the virulent European strain 52/70 was shown to be low at 4, indicating that the recent highly virulent IBDV in Japan is closely related to strain 52/70. These results strongly suggest that a single strain of highly virulent IBDV that might have originated from a European strain is prevalent in Japan.

The VP2 hypervariable region of P97/302 local infectious bursal disease virus (IBDV) isolate was amplified by the reverse transcriptase (RT) nested polymerase chain reaction (PCR) and cloned (Phong *et al.*, 2003). This region of P97/302 local isolate was sequenced and compared with eight other reported IBDV sequences. The result showed that P97/302 IBDV was most identical to the reported very virulent IBDV strains because it has amino acid substitutions at positions 222, 256, 294, and 299, which encode alanine, isoleucine, and serine, respectively. A very virulent infectious bursal disease virus (vvIBDV) field strain, named SH95, was identified and characterized from flocks with

vaccination failure in Shanghai (Sun *et al.*, 2003). The use of random primer and a reverse transcriptase lacking RNase-H activity produced full-length cDNA copies of the viral genomic A and B segments of SH95. The 3259 base pairs (bp) of segment A and 2827 bp of segment B were amplified by long and accurate PCR in a single step, then successfully cloned and sequenced. There were 5–27 amino acid substitutions compared with other IBDV strains within the segment A polyprotein (of these, three are unique) and about nine to 38 amino acid substitutions within VP1 (of which, four are unique). The comparison of sequences encoding the polyprotein showed that vvIBDV SH95 was most closely related to Asiatic vvIBDVs, which formed a closely related group clearly distinguishable from other classical strains of IBDV. Phylogenetic analysis suggested that vvIBDV SH95 and some other Asiatic vvIBDVs were derived from similar origin.

Among 53 bursal samples collected from different flocks in different parts of Tamil Nadu and examined by RT-PCR, 40 showed a positive reaction (Ramadass *et al.*, 2003). The amplified products were subjected to nucleotide sequencing and the obtained sequences were compared with those of IBD virus (IBDV) vaccine strain Georgia, the classical virulent strain 52/70 and the very virulent Japanese OKYM strain. Nucleotide homology data indicated that all the Tamil Nadu isolates showed homology ranging from 91 to 99.6% among themselves. When compared with the very virulent Japanese OKYM strain, four isolates grouped with that strain. Majority of the isolates clustered with the very virulent OKYM strain as evident from phylogenetic analysis performed using the MEGA program. Comparison of the deduced amino acid sequences of IBDV isolates with those of the vaccine strain Georgia, the classical virulent strain 52/70 and the very virulent strain OKYM also revealed the presence of conserved serine-rich heptapeptide sequence in most of the isolates. Results of this study indicated that majority of the IBDV isolates are very virulent, which is evident from heavy mortality that has been reported in few flocks of poultry in spite of regular vaccination.

Mittal *et al.* (2006) characterized field isolates of infectious bursal disease virus (IBDV) by reverse transcription-polymerase chain reaction (RT-PCR) and partial sequencing of *VP2* gene. The virus could be detected in 17 of 20 field samples from broiler chickens in Haryana state, India as well as in all the four vaccine strains. Nucleotide sequences of four field isolates and one vaccine strain were compared with 10 reported IBDV strains from different parts of the world. Nucleotide substitutions at 795G, 827T, 833C, 857C, 897A, 905T, 908T, 1011A and 1094G specific for very virulent (vv) strains, were maintained in all the four field isolates. However, unique nucleotide substitutions at 806A-G, 851 C-T, 1010 T-C, 1019T-C and 1082T-C showed further divergence of these isolates from already reported vvIBDVs. Deduced amino acid substitutions at 222P-A, 256V-I, 279N-D, 294L-I and 299N-S specific for vvIBDV strains were also present in all the four isolates. The vaccine strain showed amino acid change 279D-N, a characteristic of attenuated vaccine strains. Phylogenetic analysis showed that all the field isolates in the present study were closely related to reported UK (UK661) and Japan (OKYM) field isolates. All the four field IBDV strains of the present study were closely related to each other but distinct from already reported vvIBDVs of India. On the basis of nucleotide sequencing and phylogenetic analysis, it was observed that IBD-causing strains in this part of India were of very virulent character and were still undergoing changes at genetic level.

The nucleotide sequences of the genome segments A and B encoding the precursor polyprotein and VP1 were determined for a highly virulent strain of infectious bursal disease virus (IBDV) (Yamaguchi *et al.*, 1997). The precursor polyprotein and VP1 coding regions of highly virulent OKYM strain consisted of 3039 nucleotides and 2640 nucleotides, respectively. Comparison of the deduced amino acid sequences of the highly virulent IBDV (HV0IBDV) with other serotype 1 and 2 sequences revealed 17 amino acid residues which were conserved only in the HV-IBDV. Among the 17 unique amino acid differences, 8 were in VP1, 4 in VP2, 3 in VP3 and 2 were in VP4.

Kwon and Jackwood (1995) cloned and sequenced *S1* glycoprotein genes of the Gray and JMK strains of avian infectious bronchitis virus (IBV) and compared with published sequences for IBV. The IBV Gray and JMK strains had 99% nucleotide sequence similarity. The overall nucleotide sequence similarity of the Gray and JMK strains compared with other IBV strains were between 82.0 and 87.4% respectively. The similarity of the predicted amino acid sequence for the S1 glycoprotein of the Gray and JMK strains was 98.8%. Six of the 10 differences in the amino acid sequence were found between residues 99 and 127, which suggested a possible role for that region in the tissue tropisms of the viruses. The Gray and JMK strains of IBV are the same serotype, indicating that they are very similar antigenically. However, the pathogenicity of these viruses is different because the Gray strain can produce nephritis. It follows that the amino acids located between residues 99 and 127 may play a role in the difference observed in the pathogenesis of these viruses.

An approximately 450-bp region containing HVR 1 and HVR 2 in the S1 gene of the infectious bronchitis virus (IBV) of 7 untyped field isolates obtained in 1999 and 2000 was amplified (Lee *et al.*, 2003). Direct sequencing followed by phylogenetic analysis on that region allowed those field isolates that were not typable to be typed by reverse transcriptase-polymerase chain reaction (RT-PCR) and restriction fragment length polymorphism (RFLP). Furthermore, it was found that typing by phylogenetic analysis of that region correlates with virus neutralization results. Together with RT-PCR and RFLP, this method will serve as a fast typing method for IBV diagnosis. Infectious bronchitis virus was isolated from poultry samples obtained from different parts of India and amplified *S1* gene from different isolates were sequenced for further characterization (Suresh Kumar *et al.*, 2007). All the seven isolates had sequences of 94.8 to 98.8% homology with the vaccine strain H120. All the isolates sequenced had 1–4 nucleotide differences from that of the H120 vaccine virus. Thus, it appeared that the vaccine virus is acquiring point mutations and insertions during its spread in the field. The sequencing results demonstrated the co-circulation of vaccine and wild type infectious bronchitis viruses belonging to Mass 41 serogroup.

A 695-fragment of Newcastle disease virus (NDV) was amplified by polymerase chain reaction between matrix protein gene and fusion protein gene of 30 Korean NDV isolates, which were isolated from field outbreaks of Newcastle disease between 1949 and 2002 (Lee *et al.*, 2004). All isolates showed the amino acid sequence 112R-R-Q/R-K-R116 at the C-terminus of the F2 protein and phenylalanine (F) at the N-terminus of the F1 protein residue 117. These amino acid sequences were identical to a known virulent motif. The region of the *F* gene between nucleotides 47 and 435 was compared by phylogenetic analysis. The five Korean Newcastle disease epizootics were closely related with the Newcastle disease panzootics or Newcastle disease epizootics in other countries. But the Korean genotype V isolated before 1984 was related with European Newcastle disease epizootics in the 1970s, whereas the Korean genotypes VI and VII isolated after 1988 were more closely related with Far East Newcastle disease epizootics, especially Newcastle disease epizootics in Japan, Taiwan and China.

Five Newcastle disease virus (NDV) isolates from pigeons were characterized by biological and molecular methods (Kumanan *et al.*, 2005). Four of the five isolates were found to be velogenic with high intracerebral pathogenicity indices (ICPI). The fusion protein cleavage site (FPCS) sequences of these isolates had multiple basic amino acids RRQKRF at positions 112–116 and a phenylalanine at position 117 characteristic of velogenic isolates. Three of these velogenic isolates were phylogenetically related to mesogenic vaccine virus strain and the fourth one to a few exotic velogenic isolates. The lentogenic isolate obtained in this study was identical with the LaSota strain.

The amino acid composition of the two surface proteins of peste-des-petits ruminants vaccine virus belonging to lineage four from India were deduced from the nucleotide sequence (Dhar *et al.*, 2006). The fusion (F) protein gene of PPRV Sungri/96 is 2405 nucleotides long and in relation to the length, it is

80 nucleotides longer than that of PPRV Nigeria/75/1 which are found to be present at the 5′ UTR of this virus. The complete *F* gene alignment with other morbillivirus reveals a homology of 89% with PPRV/Nigeria/75/1 and 48–51% with other morbilliviruses. The F protein of PPRV Sungri/96 exhibited characteristic similarity to those of other morbillivirus F proteins. The overall amino acid similarity with its counterpart PPRV Nigeria/75/1 was 96%; with other morbilliviruses, it is 65–74%. The PPRV Sungri/96 haemagglutinin (H) protein gene is 1954 nucleotides long and showed a sequence homology of 90.7% with PPRV/Nigeria/75/1 and with other morbilliviruses it ranged from 33% to 45%. At amino acid level, PPRV Sungri/96 showed a homology of 92.3% with PPRV/Nigeria/75/1 and 34–49% with other morbilliviruses.

Partial nucleotide sequences of 1D gene of 38 isolates of foot-and-mouth disease virus (FMDV) of serotypes O, A and Asia 1 originating from various parts of India were determined (Muthuchelvan *et al.*, 2001). Field materials were subjected straight to RNA extraction, reverse transcription PCR (RT-PCR) and sequencing. Also 3 FMDV vaccine strains, IND R2/75 (serotype O), IND 63/72 (serotype Asia 1) and IND 17/77 (serotype A) were included in the analysis. The sequences were compared mutually as well as with available corresponding sequences of other FMDV isolates, and their phylogenetic relationships were calculated. The deduced amino acid sequences showed that the serotype O isolates were relatively conserved as compared to serotype Asia 1 or A isolates from India. In phylogenetic analysis, the serotype O viruses clustered in two genotypes, one including the European vaccine strain (O1/K) and the other represented by the isolates from Bangladesh, India, Nepal and Turkey. The serotype Asia 1 viruses clustered in two groups of single genotype where the prototype strain from Pakistan (PAK 1/54) formed one group and the other was formed by the isolates from Bangladesh, Bhutan, India, Israel and Nepal. In serotype A viruses three well-differentiated genotypes were observed. The isolates from Azerbaijan, Bangladesh, Malaysia and India formed the first genotype. The second genotype was formed by isolates from Iran, Saudi Arabia and Turkey, while two recent Iranian isolates represented the third genotype. This evolutionary clustering of isolates from the neighbour countries is not surprising, since these countries share border with India. The genetic relatedness between sequences of isolates from India and those from distant places is indicative of spread of the virus between the countries. Of importance is the fact that clinical materials proved useful for rapid generation of sequences and subsequent studying of molecular epidemiology of the disease.

Saiz *et al.* (2003) detected foot-and-mouth disease virus from culture and clinical samples by reverse transcription-PCR coupled to restriction enzyme and sequence analysis. A primer pair flanking a region of the viral polymerase gene (3D) corresponding to the C-terminus of the protein was designed and a single-step RT-PCR reaction was developed. The assay allowed the detection of viral RNA from a variety of animal samples and from a wide range of FMDV isolates of different origins and serotypes. The presence of an *Ahd* I restriction site within the amplicon in 96% of the isolates analysed allowed an additional confirmation step of the positive reactions by a simple digestion yielding characteristic fragment sizes. The set of primers described here was suitable for direct sequencing of the PCR product (290 bp), and the nucleotide sequences corresponding to the SAT 1 and SAT 3 strains were determined. The segment amplified, when used in phylogenetic studies, allowed the clustering of SAT isolates and the rest of FMDV strains as two separate lineages.

The genetic diversity among the Indian serotype A foot-and-mouth disease virus (FMDV) isolates sampled over a period of 24 years (1977–2000) was studied by sequencing the *VP1* gene (Tosh *et al.*, 2002). In the phylogenetic tree constructed from 83 Indian and 37 other available sequences, the FMDV type A isolates were distributed into 10 major genotypes (designated as I-X). The Indian isolates were distributed in 4 genotypes (I, IV, VI and VII), and co-circulation of at least 2 genotypes (VI and VII) in different states of the country in recent years is evident from the result. The study also revealed differential geographic distribution of genotypes, for example, some (genotypes I and VII) were recovered from large geographical areas, sometimes even across the continents, suggesting the spread of the viruses beyond continental barriers.

A conventional and real-time PCR amplification and sequencing of DNA gyrase subunit B (*gyrB*) gene was developed for identification of pathogenic *Leptospira* species (Slack *et al.*, 2006). Phylogenetic comparisons were undertaken between pathogenic *Leptospira* 16s rRNA and *gyrB* genes using clustering and minimum evolution analysis. In addition, 50 unidentified *Leptospira* isolates were characterized by *gyrB* sequencing and compared with conventional 16S rRNA sequencing. The *gyrB* gene showed greater nucleotide divergence (3.5% to 16.1%) than the 16S rRNA gene (0.1% to 1.4%). Minimum evolution analysis revealed that the *gyrB* had a different evolution topology for *L. kirschneri* and *L. interrogans*. When the two genes were compared for the identification of the 50 unknown isolates, there was 100% agreement in the results. The *gyrB* encoding gene showed higher nucleotide/evolutionary divergence allowing for superior identification and also the potential for the development of DNA-probe-based identification.

16S rRNA gene sequences are potentially useful for species identification of *Leptospira*, but there are a large number of sequences of various lengths and quality in the public databases. 16S rRNA gene sequences of near full length and bidirectional high redundancy were determined for all type strains of the species of Leptospiraceae (Morey *et al.*, 2006). Three clades were identified within the genus *Leptospira*, composed of pathogenic species, non-pathogenic species, and another clade of undetermined pathogenicity with intermediate 16S rRNA gene sequence relatedness. All type strains could be identified by 16S rRNA gene sequences, but within both pathogenic and non-pathogenic clades as few as two or three base pairs separated some species. Sequences within the non-pathogenic clade were more similar, and in most cases < or =10 bp distinguished these species. These sequences provide a reference standard for identification of *Leptospira* species and confirm previously established relationships within the genus. 16S rRNA gene sequencing is a powerful method for identification in the clinical laboratory and offers a simplified approach to the identification of *Leptospira* species.

Members of the genus *Brucella* are categorized as biothreat agents and pose a hazard for both humans and animals. Current identification methods rely on biochemical tests that may require up to 7 days for results. Gee *et al.* (2004) sequenced the 16S rRNA genes of 65 *Brucella* strains along with 17 related strains likely to present a differential diagnostic challenge. All *Brucella* 16S rRNA gene sequences were determined to be identical and were clearly different from the 17 related strains, suggesting that 16S rRNA gene sequencing is a reliable tool for rapid genus-level identification of *Brucella* spp. and their differentiation from closely related organisms. The *rpoB* gene encoding the beta subunit of the DNA-dependent RNA polymerase was molecularly characterized by PCR amplification and DNA sequencing in 26 *Brucella* reference strains by using primers selected according to the *B. melitensis* 16 M *rpoB* published sequence (Marianelli *et al.*, 2006). Comparison of the *rpoB* nucleotide sequence of all *Brucella* strains analysed revealed specific nucleotide variations associated with different *Brucella* species and biovars. 17 *rpoB* alleles were recognized and new *Brucella* typing is proposed. Our results probably, the *rpoB* gene polymorphism can be used to identify all *Brucella* species and most of the biovars, offering an improvement over conventional typing methods.

Gorkiewicz *et al.* (2003) evaluated the reliability of 16S rDNA sequencing for the species-specific identification of campylobacters. Sequence analyses were performed by using almost 94% of the complete 16S rRNA genes of 135 phenotypically characterized *Campylobacter* strains, including all known taxa of this genus. It was shown that 16S rDNA analysis enabled specific identification of most *Campylobacter* species. The exception was a lack of discrimination among the taxa *Campylobacter jejuni* and *C. coli* and atypical *C. lari* strains, which shared identical or nearly identical 16S rDNA sequences. Subsequently, it was investigated whether partial 16S rDNA sequences are sufficient to determine species identity. Sequence alignments led to the identification of four 16S rDNA regions with high degrees of interspecific variation but with highly conserved sequence patterns within the respective species. A simple protocol based on the analysis of these sequence patterns was developed, which enabled the unambiguous identification of the majority of *Campylobacter* species. Thus 16S rDNA sequence analysis is an effective, rapid procedure for the specific identification of campylobacters.

Price *et al.* (2006) described the development of a generally applicable, bioinformatics-driven, single-nucleotide polymorphism (SNP) genotyping assay for the common bacterial gastrointestinal pathogen *Campylobacter jejuni*. SNPs were identified *in silico* using the program "Minimum SNPs", which selected for polymorphisms providing the greatest resolution of bacterial populations based on Simpson's index of diversity (D). The high-D SNPs identified in this study were derived from the combined *C. jejuni/Campylobacter coli* multilocus sequence typing (MLST) database. Seven SNPs were found that provided a D of 0.98 compared with full MLST characterization, based on 959 sequence types (STs). The total-turnaround time of the SNP typing assay was approximately 2 hours. Concurrently, 69 *C. jejuni* isolates were subjected to MLST and flagellin A short variable region (flaA SVR) sequencing and combined with a population of 84 *C. jejuni* and *C. coli* isolates previously characterized by these methods. Within this collection of 153 isolates, 19 flaA SVR types (D=0.857) were identified, compared with 40 different STs (D=0.939). When MLST and flaA SVR sequencing were used in combination, the discriminatory power was increased to 0.959. This investigation had showed that a seven-member *C. jejuni* SNP typing assay, used in combination with sequencing of the flaA SVR, efficiently discriminated *C. jejuni* isolates.

Ramadass *et al.* (2006) isolated *Mycoplasma gallisepticum* (MG) and *Mycoplasma synoviae* (MS) from poultry samples in Tamil Nadu, used PCR for confirmation of the isolates and the amplicons were sequenced. Nucleotide homology data indicated that all the MS isolates showed 100% homology with that of reference strains from GenBank. MG isolates showed 96 to 99% homology with that of reference strains.

REVIEW QUESTIONS

1. Write in detail, the two methods of nucleic acid sequencing with their merits and demerits.

2. What are the different applications of nucleic acid sequencing?

REFERENCES

Bordignon, J., Brasil-Dos-Anjos, G., Bueno, C.R., Salvatiera-Oporto, J., Dávila, A.M., Grisard, E.C. and Zanetti, C.R. (2005). "Detection and characterization of rabies virus in Southern Brazil by PCR amplification and sequencing of the nucleoprotein gene." *Arch. Virol.* 150:695–708.

Dhar, P., Muthuchelvan, D., Sanyal, A., Kaul, R., Singh, R.P., Singh, R.K. and Bandyopadhyay, S.K. (2006). "Sequence analysis of the haemagglutinin and fusion protein genes of peste-des-petits ruminants vaccine virus of Indian origin." *Virus Genes.* 32:71–78.

Gee, J.E., De, B.K., Levett, P.N., Whitney, A.M., Novak, R.T. and Popovic, T. (2004). "Use of 16S rRNA gene sequencing for rapid confirmatory identification of *Brucella* isolates." *J. Clin. Microbiol.* 42:3649–3654.

Gorkiewicz, G., Feierl, G., Schober, C., Dieber, F., Kofer, J., Zechner, R. and Zechner, E.L. (2003). "Species-specific identification of campylobacters by partial 16S rRNA gene sequencing." *J. Clin. Microbiol.* 41:2537–2546.

Horwitz, J.P. *et al.* (1964). "Substrate for cytochemical demonstration of enzyme activity. I: Some substituted 3-indolyl, β–D–glycopyranosides." *J. Med. Chem.* 7: 574.

Kumanan, K., Mathivanan, B., Vijayarani, K., Gandhi, A.A., Ramadass, P. and Nachimuthu, K. (2005). "Biological and molecular characterization of Indian isolates of Newcastle disease virus from pigeons." *Acta Virol.* 49:105–109.

Kwon, H.M. and Jackwood, M.W. (1995). "Molecular cloning and sequencing comparison of the S1 glycoprotein of the Gray and JMK strains of avian infectious bronchitis virus." *Virus Genes.* 9: 219–229.

Lee, C.W., Hilt, D.A. and Jackwood, M.W. (2003a). "Typing of field isolates of infectious bronchitis virus based on the sequence of the hypervariable region in the S1 gene." *J. Vet. Diagn. Invest.* 15:344–348.

Lee, Y.J., Sung, H.W., Choi, J.G., Kim, J.H. and Song, C.S. (2003b). "Molecular epidemiology of Newcastle disease viruses isolated in South Korea using sequencing of the fusion protein cleavage site region and phylogenetic relationships." *Avian Pathol.* 33:482–491.

Lin, Z., Kato, A., Otaki, Y., Nakamura, T., Sasmaz, E. and Ueda, S. (1993). "Sequence comparisons of a highly virulent infectious bursal disease virus prevalent in Japan." *Avian Dis.* 37: 315–323.

Marianelli, C., Ciuchini, F., Tarantino, M., Pasquali, P. and Adone, R. (2006). "Molecular characterization of the *rpoB* gene in *Brucella* species: new potential molecular markers for genotyping." *Microbes Infect.* 8:860–865.

Marvis, D.A. and Wachtel, E.J. (1975). "Structure and assembly of filamentous bacterial viruses." *Nature.* 253: 19.

Maxam, A.M. and Gilbert, W. (1977). "Structure and assembly of filamentous bacterial viruses." *Nature*. 253: 19.

Maxam, A.M. and Gilbert, W. (1977). "A new method for sequencing DNA." *Proc. Natl. Acad. Sci. USA*. 74: 560–564.

Mittal, D., Jindal, N., Gupta, S.L., Kataria, R.S., Singh, K. and Tiwari, A.K. (2006). "Molecular characterization of Indian isolates of infectious bursal disease virus from broiler chickens." *DNA Seq*. 17:431–439.

Morey, R.E., Galloway, R.L., Bragg, S.L., Steigerwalt, A.G., Mayer, L.W. and Levett, P.N. (2006). "Species-specific identification of Leptospiraceae by 16S rRNA gene sequencing." *J. Clin. Microbiol*. 44:3510–3516.

Muthuchelvan, D., Venkataramanan, R., Hemadri, D., Sanyal, A. and Tosh, C. (2001). "Sequence analysis of recent Indian isolates of foot-and-mouth disease virus serotypes O, A and Asia 1 from clinical materials." *Acta Virol*. 45:159–167.

Mundry, C.R., Cunningham, M.W. and Read, C.A. (1991). "Nucleic acid labeling and detection." In: *Essential Molecular Biology: A Practical Approach* vol. II. (ed.). Brown, T.A. IRL Press Oxford University Press. pp. 57–109.

Phong, S.F., Hair-Bejo, M., Omar, A.R. and Aini, I. (2003). "Sequence analysis of Malaysian infectious bursal disease virus isolate and the use of reverse transcriptase nested polymerase chain reaction enzyme-linked immunosorbent assay for the detection of VP2 hypervariable region." *Avian Dis*. 47:154–62.

Price, E.P., Thiruvenkataswamy, V., Mickan, L., Unicomb, L., Rios, R.E., Huygens, F. and Giffard, P.M. (2006). "Genotyping of *Campylobacter jejuni* using seven single-nucleotide polymorphisms in combination with flaA short variable region sequencing." *J. Med. Microbiol*. 55(Pt 8):1061–1070.

Ramadass, P., Thiagarajan, V., Parthiban, M., Senthil Kumar, T.M., Latha, D., Anbalagan, S., Krishnakumar, M. and Nachimuthu, K. (2003). "Sequence analysis of infectious bursal disease virus isolates from India: phylogenetic relationships." *Acta Virol*. 47:131–135.

Ramadass, P., Ananthi, R., Senthilkumar, T.M.A., Venkatesh, G. and Ramaswamy, V. (2006). "Isolation and characterization of *Mycoplasma gallisepticum* and *Mycoplasma synoviae* from poultry." *Ind. J. Ani. Sci*. 76: 796–798.

Remond, M., Boireau, P. and Lebreton, F. (1992). "Partial DNA cloning and sequencing of a canine parvovirus vaccine strain: application of nucleic acid hybridization to the diagnosis of canine parvovirus disease." *Arch. Virol*. 127: 257–269.

Sáiz, M., De La Morena, D.B., Blanco, E. Núñez, J.I., Fernández, R. and Sánchez-Vizcaíno, J.M. (2003). "Detection of foot-and-mouth disease virus from culture and clinical samples by reverse transcription-PCR coupled to restriction enzyme and sequence analysis." *Vet. Res*. 34:105–117.

Sanger, F., Nicklen, S. and Coulson, A.R. (1971). "DNA sequencing with chain termination inhibitors." *Proc. Natl. Acad. Sci. USA*. 74: 5463–5467.

Slack, A.T., Symonds, M.L., Dohnt, M.F. and Smythe, L.D. (2006). "Identification of pathogenic Leptospira species by conventional or real-time PCR and sequencing of the DNA gyrase subunit B encoding gene." *BMC. Microbiol.* 6:95.

Sun, J.H., Lu, P., Yan, Y.X., Hua, X.G., Jiang, J. and Zhao, Y. (2003). "Sequence and analysis of genomic segment A and B of very virulent infectious bursal disease virus isolated from China." *J. Vet. Med. B Infect. Dis. Vet. Public Health.* 50:148–154.

Suresh Kumar, K., Dhinakar Raj, G., Raja, A. and Ramadass, P. (2007). "Genotype characterization of infectious bronchitis viruses from India." *Ind. J. Biotechnol.* 6: 41–44.

Tosh, C., Sanyal, A., Hemadri, D. and Venkataramanan, R. (2002). "Phylogenetic analysis of serotype A foot-and-mouth disease virus isolated in India between 1977 and 2000." *Arch Virol.* 147:493–513.

Yamaguchi, T., Ogawa, M., Miyoshi, M., Inoshima, Y., Fukushi, H. and Hirai, K. (1997). "Sequence and phylogenetic analyses of highly virulent infectious bursal disease virus." *Arch. Virol.* 142: 1441–1458.

Yehuda, H., Pitcovski, J., Michael, A., Gutter, B. and Goldway, M. (1999). "Viral protein 1 sequence analysis of three infectious bursal disease virus strains: A very virulent virus, its attenuated form and an attenuated vaccine." *Avian Dis.* 43: 55–64.

14

ENZYME TECHNOLOGY

THE NATURE OF ENZYMES

Enzymes are catalysts which increase the rate of otherwise slow or imperceptible reactions without undergoing any net change in their structure. These enzymes are proteins and mediate all synthetic and degradative reactions carried out by living organisms. They are very efficient catalysts, often far superior to conventional chemical catalysts, for which reason they are being employed increasingly in today's high-technological society, as a highly significant part of biotechnological expansion. Enzymes have a number of distinct advantages over conventional chemical catalysts. Foremost among these are their specificity and selectivity not only for particular reactions but also in their discrimination between similar parts of molecules (**regiospecificity**) or optical isomers (**stereospecificity**). They catalyse only the reactions of very narrow ranges of reactants (**substrates**), which may consist of a small number of closely related classes of compounds (e.g. trypsin catalyses the hydrolysis of some peptides and esters in addition to most proteins), a single class of compounds (e.g. hexokinase catalyses the transfer of a phosphate group from ATP to several hexoses), or a single compound (e.g. glucose oxidase oxidizes only glucose amongst the naturally occurring sugars).

Enzymes are complex protein molecules present in living cells, where they act as catalysts in bringing about chemical changes in substances. These are biological catalysts used by living cells to achieve a variety of chemical conversions recognized as the "chemistry of life". Every cell contains a large number of enzymes, each with an ability limited to the conversion of a particular chemical molecule or portion of molecule to a modified or cleaved version of that molecule. More than 3000 enzymes catalysing a wide array of reactions are known to exist. The disintegration of foodstuff to amino acids, sugars, lipids is normally accomplished within 3 to 6 hours, depending on the amount and type of food. In the absence of enzymes, hydrolysis by digestive enzymes would take more than 30 years. With the development in the science of biochemistry, a fuller understanding of the wide range of enzymes present in living cells and of their mode of action has come to light. Without enzymes there can be no life. Although enzymes are formed only in living cells, many can be separated from the cells and can continue to function *in vitro*. This unique ability of enzymes to perform their specific chemical transformations in isolation has led to an ever-increasing use of enzymes in industrial processes, collectively termed enzyme technology.

All enzymes contain a protein backbone. In some enzymes this is the only component in the structure. However there are additional non-protein moieties usually present which may or may not participate in the catalytic activity of the enzyme. Covalently attached carbohydrate groups are commonly encountered structural features which often have no direct bearing on the catalytic activity, although they may well affect an enzyme's stability and solubility. Other factors often found are metal ions (**cofactors**) and low molecular weight organic molecules (**coenzymes**). These may be loosely or tightly bound by noncovalent or covalent forces. Enzymes are classified according to the report of a Nomenclature Committee appointed by the International Union of Biochemistry (1984). The enzyme commission (**EC**) numbers divide enzymes into six main groups according to the type of reaction catalysed:

1. **Oxidoreductases** which are involved in redox reactions in which hydrogen or oxygen atoms or electrons are transferred between molecules. This extensive class includes the dehydrogenases (hydride transfer), oxidases (electron transfer to molecular oxygen), oxygenases (oxygen transfer from molecular oxygen) and peroxidases (electron transfer to peroxide). For example: glucose oxidase (EC 1.1.3.4, systematic name, β-D-glucose:oxygen 1-oxidoreductase).

2. **Transferases** which catalyse the transfer of an atom or group of atoms (e.g. acyl-, alkyl- and glycosyl-), between two molecules, but excluding such transfers as are classified in the other groups (e.g. oxidoreductases and hydrolases). For example: aspartate aminotransferase (EC 2.6.1.1, systematic name, L-aspartate:2-oxoglutarate aminotransferase; also called glutamic-oxaloacetic transaminase or simply GOT).

3. **Hydrolases** which involve hydrolytic reactions and their reversal. This is presently the most commonly encountered class of enzymes within the field of enzyme technology and includes the esterases, glycosidases, lipases and proteases. For example: chymosin (EC 3.4.23.4, no systematic name declared; also called rennin).

4. **Lyases** which involve elimination reactions in which a group of atoms is removed from the substrate. This includes the aldolases,

decarboxylases, dehydratases and some pectinases but does not include hydrolases. For example: histidine ammonia-lyase (EC 4.3.1.3, systematic name, L-histidine ammonia-lyase; also called histidase).

5. **Isomerases** which catalyse molecular isomerisations and include the epimerases, racemases and intramolecular transferases. For example: xylose isomerase (EC 5.3.1.5, systematic name, D-xylose ketol-isomerase; commonly called glucose isomerase).

6. **Ligases**, also known as synthetases, form a relatively small group of enzymes which involve the formation of a covalent bond joining two molecules together, coupled with the hydrolysis of a nucleoside triphosphate. For example: glutathione synthase (EC 6.3.2.3, systematic name, γ-L-glutamyl-L-cysteine:glycine ligase (ADP-forming); also called glutathione synthetase).

Enzymes carry out a rapid conversion of their substrate at moderate temperatures and near-neutral pH, usually with high specificity both in selection of substrate from a complex mixture, and in the chemical conversion actually affected. Foodstuff are often the substrates, as for the well-known enzyme produced in the stomach or intestine of animals. The activity of an enzyme is due to its catalytic nature. An enzyme carries out its activity without being consumed in the reaction, while the reaction occurs at a much higher rate when the enzyme is present. Enzymes are highly specific and function only on certain types of compound, the substrates. Some enzymes require additional factors, termed cofactors that can be metal ions, nucleotides, etc.

Enzyme technology is best described as the technology associated with the application of enzymes as the tools of industry, agriculture and medicine. Although the earliest reports concerning the exploitation of enzymes were documented in the late 1800s, true industrial application of enzymes only began in earnest in the 1960s. The majority of enzymes used in industrial/ biotechnological applications are derived from particular fungi (*Aspergillus*) and bacteria (*Bacillus*). Safe organisms must be used for consumer related applications.

Enzyme technology embraces production, isolation, purification, use in soluble form and finally the immobilization and use of enzymes in a wide range of bioreactor systems.

NOVEL APPLICATIONS AND FUTURE USES

1. The exploitation of enzymes as electrocatalysts (specific biosensors);
2. Enzymes as analytical tools to measure specific compounds, for the regeneration of specific metabolites;

3. Enzyme utilization in the synthesis of bulk organic materials and the production of fragrances and cosmetics;

4. Enzyme utilization in the formation of food flavours and aroma compounds;

5. The use of enzymes as tools for the detoxification of pesticide residues;

6. Enzymes as monitors of toxic chemical levels in food and water.

BIOMEDICAL APPLICATIONS OF ENZYME TECHNOLOGY

1. The synthesis of new anti-microbial compounds

2. Enzyme replacement therapy

3. Enzymes in the treatment of cancer

4. Enzyme graft and dermatological applications

5. Enzymes as activators of precursor biomolecules

6. Enzyme technology in the prevention of dental cavities

ENZYME SOURCES

Biologically active enzymes may be extracted from any living organism. A very wide range of sources are used for commercial enzyme production from *Actinoplanes* to *Zymomonas*, from spinach to snake venom. Of the hundred or so enzymes being used industrially, over a half is from fungi and yeast and over a third are from bacteria with the remainder divided between animal (8%) and plant (4%) sources (Table 14.1). A larger number of enzymes find use in chemical analysis and clinical diagnosis. Non-microbial sources provide a larger proportion of these, at the present time. Microbes are preferred to plants and animals as sources of enzymes because:

1. They are generally cheaper to produce.

2. Their enzyme contents are more predictable and controllable.

3. Reliable supplies of raw material of constant composition are more easily arranged, and

4. Plant and animal tissues contain more potentially harmful materials than microbes, including phenolic compounds (from plants), endogenous enzyme inhibitors and proteases.

Attempts are being made to overcome some of these difficulties by the use of animal and plant cell culture.

Enzymes are present in all living things and, if sufficient care is taken to protect them, they can be isolated and purified from any organism. The great bulk of enzymes used in industry are microbial in origin, but there are exceptions, such as the various plant proteases (including papain, bromelain and ficin) and the animal proteases such as rennin and pepsin. In general, higher plant tissues are not satisfactory materials from which to attempt to isolate enzymes. The primary hazard to animal cell enzymes is hydrolysis by proteases resulting in the disruption of the cells. Plants have no means of excreting waste materials and accumulate these in vacuoles. On disruption of cells, the vacuole contents are released and come into contact with enzymes with undesirable effects. Their effects can be reduced by inclusion of additional proteins, such as albumins, in extraction media, by keeping the tissue cool during extraction, and by isolating the enzyme required as rapidly as possible. While isolating enzymes from microbial cells, isolation should be done rapidly and at low temperatures.

Table 14.1   Some important industrial enzymes and their sources

Enzyme[1]	EC number[2]	Source	Intra/extra -cellular[3]	Scale of production[4]	Industrial use
Animal enzymes					
Catalase	1.11.1.6	Liver	I	–	Food
Chymotrypsin	3.4.21.1	Pancreas	E	–	Leather
Lipase[5]	3.1.1.3	Pancreas	E	–	Food
Rennet[6]	3.4.23.4	Abomasum	E	+	Cheese
Trypsin	3.4.21.4	Pancreas	E	–	Leather
Plant enzymes					
Actinidin	3.4.22.14	Kiwi fruit	E	–	Food
α-Amylase	3.2.1.1	Malted barley	E	+++	Brewing
β-Amylase	3.2.1.2	Malted barley	E	+++	Brewing
Bromelain	3.4.22.4	Pineapple latex	E	–	Brewing
β-Glucanase[7]	3.2.1.6	Malted barley	E	++	Brewing
Ficin	3.4.22.3	Fig latex	E	–	Food
Lipoxygenase	1.13.11.12	Soybeans	I	–	Food
Papain	3.4.22.2	Pawpaw latex	E	++	Meat

(Contd.)

Table 14.1 (Continued)

Enzyme[1]	EC number[2]	Source	Intra/extra -cellular[3]	Scale of production[4]	Industrial use
Bacterial enzymes					
α-Amylase	3.2.1.1	*Bacillus*	E	+++	Starch
β-Amylase	3.2.1.2	*Bacillus*	E	+	Starch
Asparaginase	3.5.1.1	*Escherichia coli*	I	–	Health
Glucose isomerase[8]	5.3.1.5	*Bacillus*	I	++	Fructose syrup
Penicillin amidase	3.5.1.11	*Bacillus*	I	–	Pharmaceutical
Protease[9]	3.4.21.14	*Bacillus*	E	+++	Detergent
Pullulanase[10]	3.2.1.41	*Klebsiella*	E	–	Starch
Fungal enzymes					
α-Amylase	3.2.1.1	*Aspergillus*	E	++	Baking
Aminoacylase	3.5.1.14	*Aspergillus*	I	–	Pharmaceutical
Glucoamylase[11]	3.2.1.3	*Aspergillus*	E	+++	Starch
Catalase	1.11.1.6	*Aspergillus*	I	–	Food
Cellulase	3.2.1.4	*Trichoderma*	E	–	Waste
Dextranase	3.2.1.11	*Penicillium*	E	–	Food
Glucose oxidase	1.1.3.4	*Aspergillus*	I	–	Food
Lactase[12]	3.2.1.23	*Aspergillus*	E	–	Dairy
Lipase[5]	3.1.1.3	*Rhizopus*	E	–	Food
Rennet[13]	3.4.23.6	*Mucor miehei*	E	++	Cheese
Pectinase[14]	3.2.1.15	*Aspergillus*	E	++	Drinks
Pectin lyase	4.2.2.10	*Aspergillus*	E	–	Drinks
Protease[13]	3.4.23.6	*Aspergillus*	E	+	Baking
Raffinase[15]	3.2.1.22	*Mortierella*	I	–	Food
Yeast enzymes					
Invertase[16]	3.2.1.26	*Saccharomyces*	I/E	–	Confectionery
Lactase[12]	3.2.1.23	*Kluyveromyces*	I/E	–	Dairy
Lipase[5]	3.1.1.3	*Candida*	E	–	Food
Raffinase[15]	3.2.1.22	*Saccharomyces*	I	–	Food

[1] The names in common usage are given. As most industrial enzymes consist of mixtures of enzymes, these names may vary from the recommended names of their principal component. Where appropriate, the recommended names of this principal component are given.

[2] EC number of the principal component. [3] I—intracellular enzyme; E—extracellular enzyme. [4] +++ > 100 ton year^{-1}; ++ > 10 ton year^{-1}; + > 1 ton year^{-1}; - < 1 ton year^{-1}. [5] triacylglycerol lipase; [6] chymosin; [7] endo-1,3(4)-β-glucanase; [8] xylose isomerase; [9] subtilisin; [10] α-dextrin endo-1,6- α-glucosidase; [11] glucan 1,4- α-glucosidase; [12] β-galactosidase; [13] microbial aspartic proteinase; [14] polygalacturonase; [15] α-galactosidase; [16] β-fructofuranosidase.

In practice, the great majority of microbial enzymes come from a very limited number of genera, of which *Aspergillus* species, *Bacillus* species and *Kluyveromyces* (also called *Saccharomyces*) species predominate. Most of the strains used have either been employed by the food industry for many years or have been derived from such strains by mutation and selection. There are very few examples of the industrial use of enzymes having been developed for one task. Shining examples of such developments are the production of high fructose syrup using glucose isomerase and the use of pullulanase in starch hydrolysis.

CLARIFICATION OF THE SOLUBLE ENZYMES

The solubilized enzymes, or any enzyme produced extracellularly, exists in the presence of cells (or their debris) which produced it, together with many other enzymes and proteins, nucleic acids, salts, other low molecular weight metabolites and medium components.

Centrifugation   On a laboratory scale, centrifugation is the primary clarification method most often selected because effective equipment is readily available.

Flocculation and coagulation   Centrifugation and filtration are aided by aggregation of the particles. Flocculation occurs when an agent, often in very dilute solution, bridges particles to produce a loose aggregate. Many natural and synthetic polymers such as gelatin or polyacrylamides contain various proportions of acrylic acid and inorganic salts such as alum, ferric and calcium salts and may be used to flocculate microbial cells. Flocculation or coagulation is a potentially useful technique for removing cell wall debris.

Filtration   The rate of passage of a liquid through a filter of unit area is dependent on the pressure difference applied, the resistance of the filter material, viscosity of the liquid and the resistance produced by cake already present. Filter aids like diatomaceous earth, retain finer particles and are valuable in enzyme isolation, but they tend to occlude liquor containing the enzyme and will damage downstream equipment if allowed to pass into the filtrate. The commonest forms of industrial filter are the plate and frame press and the rotary drum filter. The former consists of filter clothes trapped between corrugated plates; fluid passes in at one side of the cloth and out, via the corrugations, to a pipe serving all the units in the battery. To remove the solids the plates must be parted manually or semi-automatically and the clothes thoroughly cleaned. In the rotary drum filter, vacuum is applied to the inside of a hollow drum rotating in a trough containing the material to be filtered. Sediment accumulates on a filter cloth from which it may be removed by a multiplicity of methods.

ENZYME CONCENTRATION

Nucleic acids may be removed from cell-free extracts by a variety of means. An obvious and inexpensive technique is enzyme hydrolysis using nucleases. The nucleic acid may also be precipitated using high molecular weight cations such as polyethylenamine, streptomycin sulphate, cetyltrimethylammonium bromide or protamine sulphate.

Precipitation

Ammonium sulphate   Enzymes may be precipitated and fractionated by 'salting out', usually by ammonium sulphate which is cheap, very soluble, self-cooling on dissolving in water and harmless to most enzymes.

Organic solvents   Organic solvents reduce the dielectric constants of aqueous media, and thus may lower the solubility of proteins by allowing the protein molecules to interact more readily with each other than with the water. Proteins can be precipitated very satisfactorily using organic solvents provided that the temperature is below 4°C. Above this, denaturation may occur.

High molecular weight polymers   Polyethylene glycol may be used very satisfactorily to precipitate proteins. Unlike solvents, it has a protein-stabilizing effect and thus may be used at ambient temperatures. It is effective at relatively low concentrations, most proteins being precipitated at polyethylene glycol concentrations of 6–12%.

Ultrafiltration and Reverse Osmosis

In ultrafiltration, molecules are forced hydraulically through a membrane of very small pore size. Reverse osmosis is simply ultrafiltration using a membrane with pores small enough to allow the passage of solvent molecules only, as in the desalination of sea water. Reverse osmosis may thus be used to concentrate enzyme solutions. Slightly less fine-pored membranes can be used for low molecular weight solutes, as in the removal of ammonium sulphate after enzyme precipitation. Ultrafiltration membranes are available with molecular weight cut-offs between 500 and 300,000. There are two types of ultrafiltration membranes, microporous and diffusive. The former is a rigid membrane with small pores running through it of average diameter 500–5000 Å. Very small molecules will pass through the membrane and large ones will be retained at the filter surface. Intermediate-sized molecules will be retained with the structure of the membrane and will eventually block the pores.

Other Methods

Freeze-drying Freeze-drying (lyophilization) relies on the ability of ice to sublime. At reduced pressures with a suitably cold sink, the process occurs rapidly and solutes are dried very effectively, remaining as fluffy, readily soluble powders.

Evaporation Simple evaporation is a fairly useful technique for those enzymes which are not harmed by increased concentrations of low molecular weight solutes, and also stable to the somewhat elevated temperatures which are necessary for the process to occur at a practically useful rate.

Freezing It is possible to concentrate any water-soluble substance by freezing the water and separating the ice crystals from the remaining solution. The conditions used must allow the formation of ice crystals which do not contain the enzyme in inclusions: this is best done by stirring the solution and allowing ice crystals to grow reasonably slowly. Once this is done, the ice crystals may be removed from the liquid phase by centrifugation or filtration. The technique has proved very successful in concentrating culture filtrates of various organisms with very little loss of enzyme activity.

ENZYME PURIFICATION

Purification of enzymes is best achieved by chromatography and related techniques. The basis of purification by chromatography is the retardation of solute molecules during the passage of a solution through a column containing particles of solid material.

Gel Chromatography

This technique serves to purify proteins. The materials most frequently used are the various grades of Sephadex (cross-linked dextrans), Sepharose (agarose) and Bio-Rad gels (polyacrylamides), among many others. Sephadex gels consist of dextran molecules cross-linked with epichlorhydrin. The amount of epichlorhydrin used determines the extent of cross-linking which in turn determines the degree of water regain which is possible. The greater the water gain the greater the porosity, i.e., the larger the molecular species which can be fractionated. Bio-Gels consist of beads of polyacrylamide which again can be manufactured with a range of porosities determined by the initial concentration of monomers and by the amount of cross-linking. Agarose gels have larger pore sizes than dextrans or polyacrylamide gels, and therefore are capable of fractionating larger proteins.

Ion-exchange Chromatography

Ion-exchange resins have high capacities compared with ion-exchange celluloses. Their high capacity means that they are potentially very useful for protein concentration and purification. Protein could also be purified mainly by electrostatic forces. The magnitude of these factors may be changed by altering the pH of the environment, which will change the net charge by altering the pH of the environment, which will change the net charge on the proteins and the extent of association of weakly acidic or basic ion exchangers, or by changing the ionic strength of the environment. Thus, proteins may be adsorbed tightly to ion exchangers and desorbed by changing the pH or ionic strength or both. Proteins may be separated very effectively by column chromatography using gradients of pH or ionic strength.

Affinity Purification

This is an extremely powerful technique which can be highly specific for individual enzymes. In the most specific form of affinity purification, an analogue of the substrate or cofactor of an enzyme is attached to a support material and used as a ligand. On contacting with this, an enzyme–ligand complex forms. This is highly specific, and no other proteins in the mixture will be bound. The enzyme may be released from the complex by treatment with a solution of the normal substrate of the enzyme or by changing the pH sufficiently to change the conformation of the enzyme and breaking the complex. Specific antibodies are equally effective as ligands for enzyme purification.

APPLICATIONS OF ENZYMES IN BIOTECHNOLOGY

For thousands of years processes such as brewing, bread-making and the production of cheese have involved the unrecognized use of enzymes. In the West the industrial understanding of enzymes revolved around yeast and malt, where traditional baking and brewing industries were rapidly expanding. Much of the early development of biochemistry was centred on yeast fermentations and processes for conversion of starch to sugar.

Several enzymes, especially those used in starch processing, high-fructose syrup manufacture, textile desizing and detergent formulation, are now traded as commodity products in the world market. Relatively few enzymes, notably those in detergents, meat tenderizers and garden composting agents, are sold directly to the public. Most are used by industry to produce improved or novel products, to bypass long and involved chemical synthetic pathways or for use in the separation and purification of isomeric mixtures. Many of the most useful, but least-understood, uses of free enzymes are in the food industry.

There is now a rapid proliferation of uses and potential uses for more highly purified enzyme preparations in industrial processing, clinical medicine and laboratory practice. The range of pure enzymes now available commercially is rapidly increasing. Most of the enzymes used on an industrial scale are extracellular enzymes, i.e., enzymes that are normally excreted by the microorganisms to act upon their substrate in an external environment, and are analogous to the digestive enzymes of human beings and animals. Thus, when microorganisms produce enzymes to split large external molecules into an assimilable form, the enzymes are usually excreted into the fermentation media. In this way the fermentation broth from the cultivation of certain microorganisms, e.g. bacteria, yeasts or filamentous fungi, then becomes a major source of proteases, amylases and (to a lesser extent) cellulases, lipases, etc.

Some intracellular enzymes are now being produced industrially and include glucose oxidase for food preservation, asparaginase for cancer therapy and penicillin acylase for antibiotic conversion. Since most cellular enzymes are by nature intracellular, more advances can be expected in this area.

Enzymes in soluble form have been used in the food industry for many years. This is especially so in the baking and brewing industries, the latter being the best example of traditional biotechnology.

Uses of Enzymes in Starch Hydrolysis

The ability of α-amylases to cause mid-chain random degradation of starch has been of vital importance to several industries, including baking, brewing, distilling, textile and paper manufacturing. In the baking industry the production of carbon dioxide by yeast requires the fermentative breakdown of glucose (the other product is ethanol, most of which evaporates off, but which provides the flavour of newly-baked bread). Extra quantities of small molecule sugars, including glucose, are made available to the yeast by the addition of α-amylase produced from fungi (*Aspergillus* species) to the wheat flour.

The α-amylases (1,4-α-D-glucan glucanohydrolases) are endohydrolases which cleave 1,4-α-D-glucosidic bonds and can bypass but cannot hydrolyse 1,6-α-D-glucosidic branchpoints. Commercial enzymes used for the industrial hydrolysis of starch are produced by *Bacillus amyloliquefaciens* (supplied by various manufacturers) and by *B. licheniformis* (supplied by Novo Industri A/S as Termamyl). Fungal α-amylase also finds use in the baking industry. It often needs to be added to bread-making flours to promote adequate gas production and starch modification during fermentation.

Bacillus licheniformis alpha-amylase (BLA) is widely used in various procedures of starch degradation in the food industry, and a BLA species with improved activity at higher temperature and under acidic conditions is desirable. Two BLA species, designated as PA and MA, had been isolated from the wild-type *B. licheniformis* strain and a mutant strain respectively, and their starch-hydrolysis activity and thermal stability were examined (Lee *et al.*, 2006). MA showed higher activity than PA, especially at acidic pH (pH 5.0–5.5), and even after 1 hour of treatment at 90°C, MA was active in the range of pH 4.0–8.0, which is much wider than that (pH 4.5–7.5) of PA. The activation energy and thermodynamic parameters for their thermal inactivation indicated that MA was more thermally stable and catalytically active than PA, suggesting that MA could be useful for glucose-production process coupled with reactions catalysed by beta-amylase.

Barley alpha-amylase 1 mutant (AMY) and *Lentinula edodes* glucoamylase (GLA) were cloned and expressed in *Saccharomyces cerevisiae* (Wong *et al.*, 2007). The purified recombinant AMY hydrolysed corn and wheat starch granules, respectively, at rates 1.7 and 2.5 times that of GLA under the same reaction conditions. AMY and GLA synergistically enhanced the rate of hydrolysis by approximately $3\times$ for corn and wheat starch granules, compared to the sum of the individual activities. The exo–endo synergism did not change by varying the ratio of the two enzymes when the total concentration was kept constant. A yield of 4% conversion was obtained after 25 min. 37°C incubation (1 unit total enzyme, 15 mg raw starch granules, pH 5.3). The temperature stability of the enzyme mixtures was $<\!/\!=50$°C, but the initial rate of hydrolysis continued to increase with higher temperatures.

The gene for a novel glucanotransferase, isocyclomaltooligosaccharide glucanotransferase (*IgtY*), involved in the synthesis of a cyclomaltopentaose cyclized by an alpha-1,6-linkage [ICG5; cyclo-{ →6)-alpha-D-Glcp-(1→4)-alpha-D-Glcp-(1→4)-alpha-D-Glcp-(1→4)-alpha-D-Glcp-(1→4)-alpha-D-Glcp-(1→ }] from starch, was cloned from the genome of *B. circulans* AM7 (Watanabe *et al.*, 2006). The *IgtY* gene, designated *igtY*, consisted of 2,985 bp encoding a signal peptide of 35 amino acids and a mature protein of 960 amino acids with a calculated molecular mass of 102,071 Da. The DNA sequence of 8,325bp analysed in this study contained two open reading frames (ORFs) downstream of *igtY*. The first ORF, designated *igtZ*, formed a gene cluster, *igtYZ*. The amino-acid sequence deduced from *igtZ* exhibited no similarity to any protein with known or unknown functions. *IgtZ* was expressed in *Escherichia coli*, and the enzyme was purified. The enzyme acted on maltooligosaccharides that have a degree of polymerization (DP) of 4 or more, amylose, and soluble starch to produce glucose and malto-oligosaccharides up to DP5 by a hydrolysis reaction. The enzyme (*IgtZ*), which has a novel amino-acid sequence, should be assigned to alpha-amylase. It is notable that both IgtY and IgtZ have a tandem sequence similar to a carbohydrate-binding module belonging to a family 25. These two enzymes jointly acted on raw starch, and efficiently generated ICG5.

A gene corresponding to a maltogenic amylase (MAase) in *Lactobacillus gasseri* ATCC 33323 (*lgma*) was cloned and expressed in *Escherichia coli* (Oh *et al.*, 2005). The recombinant LGMA was efficiently purified 24.3-fold by one-step Ni-NTA affinity chromatography. The purified enzyme exhibited optimal activity for beta-CD hydrolysis at 55°C and pH 5. LGMA possessed some unusual properties distinguishable from typical MAases, such as being in a tetrameric form, having hydrolysing activity towards the alpha-(1,6)-glycosidic linkage and being inhibited by acarbose.

Fungi were screened for their ability to produce alpha-amylase by a plate culture method (Balkan and Ertan, 2005). *Penicillium chrysogenum* showed high enzymatic activity. Alpha-amylase production by *P. chrysogenum* cultivated in liquid media containing maltose (2%) reached its maximum at 6–8 days, at 30°C, with a level of 155 U ml^{-1}. Some general properties of the enzyme were investigated. The optimum pH and temperature for the reaction were 5.0 and 30–40°C, respectively. The enzyme was stable at a pH range from 5.0–6.0 and at 30°C for 20 minutes and the enzyme's 92.1% activity was retained at 40°C for 20 min. without substrate. Hydrolysis products of the enzyme were maltose, undefined oligosaccharides, and a trace amount of glucose. Alpha-amylase of *P. chrysogenum* hydrolysed starches from different sources.

Ao *et al.* (2007) used recombinant N-terminal subunit enzyme of human small intestinal maltase-glucoamylase (rhMGAM-N) to explore digestion of native starches from different botanical sources. The susceptibility to enzyme hydrolysis varied among the starches. The rate and extent of hydrolysis of amylomaize-5 and amylomaize-7 into glucose were greater than for other starches. Such was not observed with fungal amyloglucosidase or pancreatic alpha-amylase. The degradation of native starch granules showed a surface-furrowed pattern in random, radial, or tree-like arrangements that differed substantially from the erosion patterns of amyloglucosidase or alpha-amylase. The evidence of raw starch granule degradation with rhMGAM-N indicated that pancreatic alpha-amylase hydrolysis was not a requirement for native starch digestion in the human small intestine.

Bacterial α-amylases are used extensively in the other applications of amylases, however. Starch is much more readily dissolved in hot water following a prior gelatinization step using thermostable bacterial α-amylase at 100°C (or higher in some cases).

Table 14.2 Enzymes used in starch hydrolysis

Enzyme	EC number	Source	Action
α-Amylase	3.2.1.1	*Bacillus amyloliquefaciens*	Only α-1,4-oligosaccharide links are cleaved to give α-dextrins and predominantly maltose (G2), G3, G6 and G7 oligosaccharides
		B. licheniformis	Only α-1,4-oligosaccharide links are cleaved to give α-dextrins and predominantly maltose, G3, G4 and G5 oligosaccharides
		Aspergillus oryzae, A. niger	Only α-1,4 oligosaccharide links are cleaved to give α-dextrins and predominantly maltose and G3 oligosaccharides
Saccharifying α-amylase	3.2.1.1	*B. subtilis* (*amylosacchariticus*)	Only α-1,4-oligosaccharide links are cleaved to give α-dextrins with maltose, G3, G4 and up to 50% (w/w) glucose
β-Amylase	3.2.1.2	Malted barley	Only α-1,4-links are cleaved, from non-reducing ends, to give limit dextrins and β-maltose
Glucoamylase	3.2.1.3	*A. niger*	α-1,4 and α-1,6-links are cleaved, from the non-reducing ends, to give β-glucose
Pullulanase	3.2.1.41	*B. acidopullulyticus*	Only α-1,6-links are cleaved to give straight-chain maltodextrins

Uses of Enzymes in Detergents

The use of enzymes in detergent formulations is now common in developed countries, with over half of all detergents presently available containing enzymes. Dirt comes in many forms and includes proteins, starches and lipids. In addition, clothes that have been starched must be freed of the starch. Using detergents in water at high temperatures and with vigorous mixing, it is possible to remove most types of dirt but the cost of heating the water is high and lengthy mixing or beating will shorten the life of clothing and other materials. The use of enzymes allows lower temperatures to be employed and shorter periods of agitation are needed, often after a preliminary period of soaking. In general, enzyme detergents remove protein from clothes soiled with blood, milk, sweat, grass, etc. far more effectively than non-enzyme detergents. Detergent enzymes must be cost-effective and safe to use. Early attempts to use proteases foundered because of producers and users developing hypersensitivity. This was combated by developing dust-free granulates (about 0.5 mm in diameter) in which the enzyme is incorporated into an inner core, containing inorganic salts (e.g. NaCl) and sugars as preservative, bound with reinforcing fibres of carboxymethyl cellulose or similar protective colloid. The enzymes used are all produced using species of *Bacillus*, mainly by just two companies. Novo Industri A/S produce and supply three proteases, Alcalase, from *B. licheniformis*, Esperase, from an alkalophilic strain of *B. licheniformis* and Savinase, from an alkalophilic strain of *B. amyloliquefaciens* (often mistakenly attributed to *B. subtilis*). GistBrocades produce and supply Maxatase, from *B. licheniformis*. Alcalase and Maxatase (both mainly subtilisin) are recommended for use at 10–65°C and pH 7–10.5. Savinase and Esperase may be used at up to pH 11 and 12, respectively. The α-amylase supplied for detergent use is Termamyl, the enzyme from *B. licheniformis* which is also used in the production of glucose syrups. α-Amylase is particularly useful in dish-washing and de-starching detergents. Home detergents will probably include both an amylase and a protease and a lengthy warm-water soaking time will be recommended.

Alkaline xylanases from alkaliphilic *Bacillus* strains NCL (87-6-10) and *Sam* III were compared with the commercial xylanases Pulpzyme HC and Biopulp for their compatibility with detergents and proteases for laundry applications (Kamal Kumar *et al.*, 2004). Among the four xylanases evaluated, the enzyme from the alkaliphilic *Bacillus* strain NCL (87-6-10) was the most compatible. The enzyme retained its full activity (40° for 1 h) in the presence of detergents, whereas Pulpzyme HC and *Sam* III showed only 30% and 50% of their initial activity, respectively. Biopulp, though stable to detergents, had only marginal activity (5%) at pH 10. However, all four enzymes retained significant activity (80%) for 60 minutes in the presence of the proteases, alcalase and *Conidiobolus* protease. Supplementation of the enzyme enhanced the cleaning ability of the detergents.

The stability of immobilized and native Esperase, a commercial serine protease, was studied by incubating the enzymes in four formulations containing the same amount of anionic and non-ionic surfactants (Vasconcelos *et al.*, 2006). The results showed that the activity of the immobilized enzyme was not affected by the presence of detergents while the native enzyme lost 50% of activity after 20 minutes of incubation in these four formulations. The washing performance of the detergents prepared with the immobilized Esperase was studied on cotton and wool fabric samples stained with human blood and egg yolk, using as control the detergent containing native Esperase. The best stain removal for cotton samples stained with human blood was achieved using the detergent with immobilized Esperase. Several physical tests confirmed that wool keratin was not degraded by the immobilized Esperase, validating the ability to use formulated detergents containing this immobilized enzyme for safe wool domestic washing.

An enormous variety of applications exist for these protein-degrading enzymes. Washing powders contain a small amount of bacterial protease (from *Bacillus subtilis*) to provide the biological catalytic action needed to dissolve dried blood stains, for example, which are very difficult to remove by any other acceptable means. The same enzyme (subtilisin) is used in the leather industry to loosen hair from the hide, or in the textile industry to recover wool from sheep skin.

Uses of Proteases in the Food Industry

Certain proteases have been used in food processing for centuries and any record of the discovery of their activity has been lost in the midst of time. Rennet (mainly chymosin), obtained from the fourth stomach (abomasum) of unweaned calves has been used traditionally in the production of cheese. Similarly, papain from the leaves and unripe fruit of the pawpaw (*Carica papaya*) has been used to tenderize meats. These ancient discoveries have led to the development of various food applications for a wide range of available proteases from many sources, usually microbial. Proteases may be used at various pH values, and they may be highly specific in their choice of cleavable peptide links or quite non-specific. Proteolysis generally increases the solubility of proteins at their isoelectric points.

The action of rennet in cheese making is an example of the hydrolysis of a specific peptide linkage, between phenylalanine and methionine residues ($-Phe_{105}-Met_{106}-$) in the κ-casein protein present in milk. Calf rennet, consisting of mainly chymosin with a small but variable proportion of pepsin, is a relatively expensive enzyme and various attempts have been made to find cheaper alternatives from microbial sources. These have ultimately proved to be successful and microbial rennets are used for about 70% of US cheese and 33% of cheese production worldwide. The development of unwanted bitterness in ripening cheese is an example of the role of proteases in flavour production in foodstuff. The action of endogenous proteases in meat after slaughter is complex but "hanging" meat allows flavour to develop, in addition to tenderizing it. It has been found that peptides with terminal acidic amino acid residues give meaty, appetizing flavours akin to that of monosodium glutamate. The presence of proteases during the ripening of cheese is not totally undesirable and a protease from *Bacillus amyloliquefaciens* may be used to promote flavour production in

Cheddar cheese. Lipases from *Mucor miehei* or *Aspergillus niger* are sometimes used to give stronger flavours in Italian cheese by a modest lipolysis, increasing the amount of free butyric acid. Meat tenderization by the endogenous proteases in the muscle after slaughter is a complex process which varies with the nutritional, physiological and even psychological (i.e., frightened or not) state of the animal at the time of slaughter. Meat of older animals remains tough but can be tenderized by injecting inactive papain into the jugular vein of the live animals shortly before slaughter. Proteases are also used in the baking industry. Where appropriate, dough may be prepared more quickly if its gluten is partially hydrolysed. A heat-labile fungal protease is used so that it is inactivated early in the subsequent baking. Weak-gluten flour is required for biscuits in order that the dough can be spread thinly and retains decorative impressions.

EPg222 protease is a novel extracellular enzyme produced by *Penicillium chrysogenum* (Pg222) isolated from dry-cured hams that has the potential for use over a broad range of applications in industries that produce dry-cured meat products. The gene encoding EPg222 protease has been identified, amplified and cloned and expressed in *Pichia pastoris* (Benito *et al.*, 2006). The recombinant enzyme exhibited similar activities to the native enzyme against a wide range of protein substrates including muscle myofibrillar protein.

Papain is a widely used protease prepared from the latex of the fruit of the tropical fruit tree *Carica papaya*. It can be used to tenderize meat by preferential degradation of tough connective tissue rather than the actin and myosin components of meat. Canned meat or steak may be tenderized in this way, and the action of the added papain continues for a significant time while the meat is cooking, before this relatively thermostable enzyme is destroyed. Another application of papain is in the chill proofing of beer. Precipitation of a protein–tannin complex is likely to occur if beer is stored cold. Prior degradation of this protein component, which may be derived for example from the barley malt used in brewing, prevents the 'chill haze' problem which the consumer finds unattractive.

Several useful proteases have been obtained from animal sources. These include pepsin, trypsin and chymotrypsin, used as digestive aids and for predigestion of baby foods. Rennin, derived from the fourth stomach of the unweaned calf, breaks a single peptide bond in the casein component of milk protein, causing it to co-precipitate with calcium ions.

Very large amounts of glucose are produced by the degradation of starch from a variety of sources on a worldwide basis. Fungal amylases are used that have an amyloglucosidase content adequate to degrade the α-1,6 cross links in the starch molecule. α-Amylases break only the α-1, 4 main chain links in the starch, so that it is not completely degraded to glucose. The product also contains dextrins, whose amyloglucosidase activity is able to degrade. Complete conversion of starch to glucose is an important industry, but as glucose is not very sweet, the worldwide demand for sweetness has led to another process, the conversion of starch to glucose which is an important industry. But as glucose is not very sweet, the worldwide demand for sweetness has led to another process, the conversion of glucose to fructose (very sweet) using glucose isomerase. Normally this enzyme proceeds to the position of equilibrium of the isomerization, so that a roughly equal mixture of glucose and fructose is obtained. This mixture, known as invert sugar, has been produced for many years by the acid hydrolysis of sucrose and could also be achieved by the enzymatic splitting of sucrose using yeast invertase.

Another enzyme which attacks the disaccharide lactose, is β-galactosidase (lactase, produced by some yeast). Lactose is hydrolysed to its constituent monosaccharides, glucose and galactose, by lactase. Glucose itself can be removed from some foodstuff, for example, prior to drying where glucose would cause discoloration, by fungal glucose oxidase. Some enzymes used in industry are listed in Table 14.3.

Table 14.3 Enzyme used in industry

Enzyme	Uses
Bacterial glucose isomerase	Glucose $\longrightarrow$ Invert sugar (i.e., fructose formation)
Bacterial α-Amylase Fungal amyloglucosidase	Starch $\longrightarrow$ Glucose
Fungal α-Amylase	Partial degradation of starch in supplementation of amylase-deficient flour for bread making
Microbial rennets	κ-casein $\longrightarrow$ para-casein (in milk curdling for cheese manufacture)
Bacterial protease	Removal of protein-based stains and laundering (in biological washing powders)
Papain (from papaya melon)	Several protease applications including meat tenderization and dehazing of beer
Cellulose	Cellulose $\longrightarrow$ Glucose
Fungal pectinase	Pectin degradation (in fruit and vegetable processing)
Aminoacylase (immobilized)	Resolution of DL-amino acids to produce L-amino acids for food supplementation
Glucose isomerase (immobilized)	Production of invert sugar and of high fructose syrups from glucose
Penicillin acylase (immobilized)	Hydrolysis of penicillin-G to make 6-aminopenicillanic acid for production of new penicillins

Uses of Proteases in the Leather and Wool Industries

The leather industry consumes a significant proportion of the world's enzyme production. Alkaline proteases are used to remove hair from hides. This process is far safer and more pleasant than the traditional methods involving sodium sulphide. Relatively large amounts of enzyme are required (0.1–1.0 % (w/w)) and the process must be closely controlled to avoid reducing the quality of the leather. After dehairing, hides which are to be used for producing soft leather clothing and goods are bated, a process, often involving pancreatic enzymes, that increases their suppleness and improves the softness of their appearance.

Proteases have been used, in the past, to 'shrinkproof' wool. Wool fibres are covered in overlapping scales pointing towards the fibre tip. A successful method involves the partial hydrolysis of the scale tips with the protease papain. This method also gave the wool a silky lustre and added to its value.

Microbial keratinases have become biotechnologically important since they target the hydrolysis of highly rigid, strongly cross-linked structural polypeptide "keratin" recalcitrant to the commonly known proteolytic enzymes trypsin, pepsin and papain. These enzymes are largely produced in the presence of keratinous substrates in the form of hair, feather, wool, nail, horn, etc. during their degradation. The complex mechanism of keratinolysis involves cooperative action of sulphitolytic and proteolytic systems. Keratinases are robust enzymes with a wide temperature and pH activity range and are largely serine or metalloproteases. Sequence homologies of keratinases indicate their relatedness to subtilisin family of serine proteases (Gupta and Ramnani, 2006). Their application could also be extended to detergent and leather industries where they serve as speciality enzymes. Besides, they also find application in wool and silk cleaning; in the leather industry, better dehairing potential of these enzymes has led to the development of greener hair-saving dehairing technology and personal care products.

Uses of Lactases in the Dairy Industry

Lactose is present at concentrations of about 4.7% (w/v) in milk and the whey (supernatant) left after the coagulation stage of cheese-making. Its presence in milk makes it unsuitable for the majority of the world's adult population, particularly in those areas which have traditionally not had a dairy industry. Real lactose tolerance is confined mainly to people whose origins lie in Northern Europe or the Indian subcontinent and is due to 'lactase persistence'; the young of all mammals clearly are able to digest milk but in most cases this ability reduces after weaning. Of the Thai, Chinese and Black American populations, 97%, 90% and 73% respectively, are reported to be lactose-intolerant, whereas

84% and 96% of the US White and Swedish populations, respectively, are tolerant. Additionally, and only very rarely, some individuals suffer from inborn metabolic lactose intolerance or lactase deficiency, both of which may be noticed at birth. The need for low-lactose milk is particularly important in food-aid programmes as severe tissue dehydration, diarrhoea and even death may result from feeding lactose-containing milk to lactose–intolerant children and adults suffering from protein-calorie malnutrition. In all these cases, hydrolysis of the lactose-to glucose and galactose would prevent the (severe) digestive problems.

Lactose may be hydrolysed by lactase, a β-galactosidase. Commercially, it may be prepared from the dairy yeast *Kluyveromyces fragilis* (*K. marxianus* var. *marxianus*), with a pH optimum (pH 6.5–7.0) suitable for the treatment of milk, or from the fungi *Aspergillus oryzae* or *A. niger*, with pH optima (pH 4.5–6.0 and 3.0–4.0, respectively) more suited to whey hydrolysis. Lactases are now used in the production of ice cream and sweetened flavoured and condensed milks. When added to milk or liquid whey (2000 U kg^{-1}) and left for about a day at 5°C about 50% of the lactose is hydrolysed, giving a sweeter product which will not crystallize if condensed or frozen.

Uses of Enzymes in the Fruit Juice, Wine, Brewing and Distilling Industries

One of the major problems in the preparation of fruit juices and wine is cloudiness due primarily to the presence of pectins. These consist primarily of α-1,4-anhydrogalacturonic acid polymers, with varying degrees of methyl esterification. They are associated with other plant polymers and, after homogenization, with the cell debris. The cloudiness that they cause is difficult to remove except by enzymic hydrolysis. Such treatment also has the additional benefits of reducing the solution viscosity, increasing the volume of juice produced (e.g. the yield of juice from white grapes can be raised by 15%), subtle but generally beneficial changes in the flavour and, in the case of wine-making, shorter fermentation times. The enzymes used in brewing are needed for saccharification of starch (bacterial and fungal α-amylases), breakdown of barley α-1,4- and α-1,3-linked glucan (β-glucanase) and hydrolysis of protein (neutral protease) to increase the (later) fermentation rate, particularly in the production of high-gravity beer, where extra protein is added. Cellulases are also occasionally used, particularly where wheat is used as adjunct to help break down the barley β-glucans. Due to the extreme heat stability of the *B. amyloliquefaciens* α-amylase where this is used, the wort must be boiled for a much longer period (e.g. 30 minutes) to inactivate it prior to fermentation. Papain is used in the later post-fermentation stages of beer-making to prevent the occurrence of protein- and tannin-containing 'chill-haze' otherwise formed on cooling the beer. Recently, 'light' beers, of lower calorific content, have become more popular. These require a higher degree of saccharification at lower starch concentrations to reduce the alcohol and total solid content of the beer. This may be achieved by the use of glucoamylase and/or fungal α-amylase during the fermentation.

A new enzyme preparation of fungal pectin lyase (EC 4.2.2.10) was shown to be useful for the production of cranberry juice and clarification of apple juice in the food industry (Semenova *et al.*, 2006). A comparative study showed that the preparation of pectin lyase is competitive with commercial pectinase products. The molecular weight of homogeneous pectin lyase was 38 kDa. Properties of the homogeneous enzyme were studied. This enzyme was most efficient in removing highly esterified pectin.

The zygomycete microfungus *R. microsporus* var. *microsporus* produced a 1,3-1,4-beta-D-glucan 4-glucanhydrolase (EC 3.2.1.73) which was able to hydrolyse beta-D-glucan that contains both the 1,3- and 1,4-bonds (barley beta-glucans) (Celestino *et al.*, 2006). Its molecular mass was 33.7 kDa. Maximum activity was detected at pH values in the range of 4–5, and temperatures in the range of 50–60°C. The enzyme was able to reduce both the viscosity of the brewer mash and the filtration time, indicating its potential value for the brewing industry.

Landbo *et al.* (2006) examined the clarification and haze-diminishing effects of alternative clarification strategies on black currant juice including centrifugation and addition of acidic protease and pectinolytic enzyme preparation and gallic acid. The extent of clarification and haze diminishment varied after individual treatment with five different acidic proteases, but one of the protease preparations, Enzeco, derived from *Aspergillus niger*, consistently tended to perform best. The individual and interactive effects on juice turbidity, total phenols, and total anthocyanin content of clarification treatment involving the use of two selected acid proteases (Enzeco and Novozyme 89L), a pectinase (Pectinex BE 3-L), and gallic acid were evaluated in a full factorial 2(4) experimental design. Haze development during cold storage decreased when gallic acid or any of the enzyme preparation was employed individually, but negative interaction effects resulted when the pectinase was employed in combination with any of the proteases. After 28 storage days at 2°C, the lowest levels of haze formation were achieved when the Enzeco protease preparation, added at 0.025 g/L, was added with 0.050 g/L of gallic acid and allowed to react in the juice for 90 min. at 50°C. The corresponding anthocyanin reduction was approximately 12% (compared to approximately 30% with gelatin silica sol treatment). The data support the hypothesis that phenol–protein interactions are involved in juice turbidity development during cold storage of berry juices and demonstrate that pre-centrifugation and protease-assisted clarification show promise as an alternative, phenolic-retaining clarification strategy in black currant juice processing.

A monosodium glutamate (MSG) biosensor with immobilized L-glutamate oxidase (L-GLOD) has been developed and studied for analysis of MSG in sauces, soups, etc. (Basu *et al.*, 2006). The immobilized enzymatic membrane was attached with oxygen electrode with a push-cap system. The detection limit of the sensor was 1 mg/dl and the standard curve was found to be linear up to 20 mg/dl. Response time of the sensor was 2 minutes. Cross-linking with glutaraldehyde in the presence of bovine serum albumin (BSA) as a spacer molecule has been used for immobilization. Optimization of the sensor was done with an increase in L-GLOD concentration (6.3–31.5 IU) and also with increase in loading volume of enzyme solution (5–20 microL). Optimization of pH and temperature was also studied. The permeability of O_2 through different membrane was studied with and without immobilized L-GLOD. The enzymatic membrane was used for over 20 measurements and stability of the membrane was observed.

Uses of Glucose Oxidase and Catalase in the Food Industry

Glucose oxidase is a highly specific enzyme (for D-glucose) from the fungi *Aspergillus niger* and *Penicillium*, which catalyses the oxidation of β-glucose to glucono-1,5-lactone (which spontaneously hydrolyses non-enzymically to gluconic acid) using molecular oxygen and releasing hydrogen peroxide. It finds use in the removal of either glucose or oxygen from foodstuff in order to improve their storage capability. A major application of the glucose oxidase/catalase system is in the removal of glucose from egg-white before drying for use in the baking industry. Other uses are in the removal of oxygen from the head-space above bottled and canned drinks and reducing non-enzymic browning in wines and mayonnaises.

Medical Applications of Enzymes

Development of medical applications for enzymes have been at least as extensive as those for industrial applications, reflecting the magnitude of the potential rewards: for example, pancreatic enzymes have been in use since the nineteenth century for the treatment of digestive disorders. The variety of enzymes and their potential therapeutic applications are considerable. A selection of those enzymes which have realized this potential to become important therapeutic agents is shown in Table 14.4.

Table 14.4 Some important therapeutic enzymes

Enzyme	EC number	Reaction	Use
Asparaginase	3.5.1.1	L–Asparagine + H_2O $\rightarrow$ L-Aspartate + NH_3	Leukaemia
Collagenase	3.4.24.3	Collagen hydrolysis	Skin ulcers
Glutaminase	3.5.1.2	L-Glutamine + H_2O $\rightarrow$ L-Glutamate + NH_3	Leukaemia
Hyaluronidase[1]	3.2.1.35	Hyaluronate hydrolysis	Heart attack
Lysozyme	3.2.1.17	Bacterial cell wall hydrolysis	Antibiotic
Rhodanase[2]	2.8.1.1	$S_2O_3^{2-} + CN^- \rightarrow SO_3^{2-} + SCN^-$	Cyanide poisoning
Ribonuclease	3.1.26.4	RNA hydrolysis	Antiviral
β-Lactamase	3.5.2.6	Penicillin $\rightarrow$ Penicilloate	Penicillin allergy
Streptokinase[3]	3.4.22.10	Plasminogen $\rightarrow$ Plasmin	Blood clots
Trypsin	3.4.21.4	Protein hydrolysis	Inflammation
Uricase[4]	1.7.3.3	Urate + O_2 $\rightarrow$ Allantoin	Gout
Urokinase[5]	3.4.21.31	Plasminogen $\rightarrow$ Plasmin	Blood clots

[1] hyaluronoglucosaminidase   [2] thiosulphate sulphur transferase   [3] streptococcal cysteine proteinase
[4] urateoxidase   [5] plasminogen activator

In contrast to the industrial use of enzymes, therapeutically useful enzymes are required in relatively tiny amounts but at a very high degree of purity and (generally) specificity. Thus the sources of such enzymes are chosen with care to avoid any possibility of unwanted contamination by incompatible material and to enable ready purification. Therapeutic enzyme preparations are generally offered for sale as lyophilized pure preparations with only biocompatible buffering salts and mannitol diluent added. The cost of such enzymes may be quite high but still comparable to those of competing therapeutic agents or treatments. As an example, urokinase (a serine protease) is prepared from human urine (some genetically engineered preparations are being developed) and used to dissolve blood clots. The cost of the enzyme is about £100 mg^{-1}, with the cost of treatment in a case of lung embolism being about £10000 for the enzyme alone.

A major potential therapeutic application of enzymes is in the treatment of cancer. Asparaginase has proved to be particularly promising for the treatment of acute lymphocytic leukaemia. Its action depends upon the fact that tumour cells are deficient in aspartate-ammonia ligase activity, which restricts their ability to synthesize the normally non-essential amino acid L-asparagine. Therefore, they are forced to extract it from body fluids.

Reiff *et al.* (2001) evaluated the safety and efficacy of L-asparaginase as an immunosuppressive agent in a mouse model of rheumatoid arthritis. Male DBA/1 mice with collagen-induced arthritis (CIA) were treated at different intervals with various doses of native and peglyated L-asparaginase from *E. coli.* When native L-asparaginase was administered before the onset of arthritis (days 14-post immunization) the number of mice developing arthritis as well as the number of arthritic paws and the severity of arthritis in the treatment group were significantly decreased (p < 0.0001). Results indicated that L-asparaginase was a potent anti-arthritic agent and might represent an effective second line agent for future treatment studies in juvenile and adult rheumatoid arthritis. Albertsen *et al.* (2005) studied the effect of asparaginase in lidocaine to relieve pain of an intramuscular injection in children. The study was designed as a double-blinded study, randomizing 12 children with acute lymphoblastic leukemia (ALL) to four different combinations of injections, including two injections where asparaginase was dissolved in a lidocaine solution and two in sterile water. Seventeen treatment courses of asparaginase, each consisting of four injections, were evaluated. Pain intensity (Pain Visual Analog Scale, VAS-score) and pharmacokinetics of the drug was evaluated. Asparaginase with addition of lidocaine significantly decreases the pain as measured by the visual analog scale without changing the bioavailability or the absorption rate of the enzyme.

A large retrospective analysis was performed to assess the outcomes of burns and chronic ulcers treated with collagenase in an outpatient setting (Marazzi *et al.*, 2006). The study included 647 patients with burns and 332 had chronic ulcers of various aetiologies. All were treated with collagenase-based ointments once daily (Noruxol or Iruxol, Smith and Nephew). In burns patients the overall average healing time was 17.9 days in the paediatric population and 23.6 days in adults. Burn depth and presence of eschar were the main factors affecting healing probability. Average healing time for ulcers was 15.4 weeks, with ulcers of mixed aetiology showing the shortest average healing time (9.2 weeks). This large retrospective analysis showed that collagenase treatments in outpatient clinics were effective and well accepted in patients with burns affecting < or = 15% BSA or with chronic ulcers of various aetiologies. Implementation of collagenase treatments in outpatient clinics has the potential to improve wound healing and may also decrease the cost of wound care.

Kozak *et al.* (2006) studied the effect of recombinant human hyaluronidase (rhuPH20) on dexamethasone (DM) penetration into the posterior segment of the eye after sub-Tenon's injection. Only the right eye of each rabbit was injected. The first group (n = 16) received an injection of DM and rhuPH20, whereas the second group (n = 16) received DM only. The eyes were enucleated

1, 2, 3, and 6 hours after the injection, and the choroid, retina, vitreous, aqueous, and serum were harvested. DM concentration was assessed by mass spectrometry. This enzyme significantly increased DM level in the choroid and the retina 3 hours after administration.

Bookbinder *et al.* (2006) found that depolymerization of the viscoelastic component of the interstitial matrix in animal models with a highly purified recombinant human hyaluronidase enzyme (rHuPH20) increased the dispersion of locally injected drugs, across a broad range of molecular weights without tissue distortion. rHuPH20 increased infusion rates and the pattern and extent of appearance of locally injected drugs in systemic blood. rHuPH20 may function as an interstitial delivery enhancing agent capable of increasing the dispersion and bioavailability of co-injected drugs that may enable subcutaneous administration of therapeutics and replace intravenous delivery.

ADVANTAGES OF USING ENZYMES IN MANUFACTURE OF PRODUCTS

Process improvement and speed-up   Any aspects of biotechnology-based process can be greatly improved, using the appropriate enzyme. Simple hydrolases, the enzymes that add water to the linking bonds to polymers, will break down these macromolecules, eventually to their constituent monomers in some cases. Thus, amylases will degrade starch, and proteases will degrade proteins. Such degradation, involving a lowering of chain length (and molecular weight) of the polymeric chain, will change many of its features. Increase in solubility in water and decrease in viscosity is the change most likely to be noticed in a process.

Product improvement   Requirements for a saleable product depend upon many factors including consumer acceptability and government safety requirements for avoiding toxicity. An enzymatic step may be useful in changing the taste or texture of a foodstuff, or its colour or aroma. It is the composition of the foodstuff that is being altered for these purposes, and other useful changes may be achieved by the manufacturer for a variety of reasons connected with nutritional and digestibility aspects of the product, or, for its shelf-life.

GENETIC ENGINEERING AND PROTEIN ENGINEERING OF ENZYMES

Recombinant DNA technology has allowed the transfer of useful enzyme genes from one organism to another. Thus, when an enzyme has been identified as a good candidate enzyme for industrial use, the relevant gene can be cloned into a more suitable production host microorganism and an industrial fermentation carried out. In this way, it becomes possible to produce industrial enzymes of very high quality and purity.

A recent example of this technology is the detergent enzyme Lipolase produced by Novo Nordisk A/S, which has improved removal of fat stains in fabrics. The enzyme was first identified in the fungus *Humicola languinosa* at levels inappropriate for commercial production. The gene DNA fragment for the enzyme was cloned into the fungus *Aspergillus oryzae* and commercial levels of enzyme achieved. The enzyme has proved to be efficient under many wash conditions. The enzyme is also very stable at a variety of temperature and pH conditions relevant to washing.

Genetic engineering and protein engineering will have dramatic impacts on the enzyme industry in its many forms. Genetic engineering will ensure better product economy, production of enzymes from rare microorganisms, faster development programmes, etc. Also, extensive tests of the enzymes now used have shown no harmful effects on the environment.

THE TECHNOLOGY OF ENZYME PRODUCTION

Although many useful enzymes have been derived from plant and animal sources, it is clear that most future developments in enzyme technology will rely on enzymes of microbial origin. Even in the malting process of brewing, where the amylases of germinated barely that hydrolyse the starch are relatively inexpensive and around which existing brewing technology has developed, there are now some competitive processes involving microbial enzymes.

Novel enzymes from unusual sources can now be produced by cloning the relevant gene into a well-characterized and easily grown microorganism such as *Aspergillus oryzae*.

Depending on source material, enzymes differ greatly in their stability to temperature and to extremes of pH. Thus, *Bacillus subtilis* proteases are relatively heat-stable and active under alkaline conditions and have been most suitable as soap-powder additives. In contrast, fungal amylases, because of their greater sensitivity to heat, have been more useful in the baking industry.

Industrial enzyme production from microorganisms relies predominantly on either submerged liquid conditions or solid substrate fermentation. Solid-substrate methods of producing fungal enzymes have long-standing historical applications, particularly in Japan and other Far-East countries. In practice, this method uses most wheat or rice bran with added nutrient salts as substrates. The growing environment usually comprises rectangular or circular trays held in constant-temperature rooms. Commercial enzymes of importance produced in this way include fungal amylases, proteases, pectinases and cellulases. In most cases, enzymes are produced in batch fermentations lasting from 30–150 hours; continuous cultivation processes have found little application in industrial enzyme production.

ENZYME IMMOBILIZATION

Enzymes have many advantages over their chemical counterparts in that they are more specific, and generally possess high catalytic properties. Enzymes can be immobilized, i.e., an enzyme can be linked to an inert support material without loss of activity which facilitates reuse and recycling of the enzyme. Enzymes can also be encapsulated or entrapped.

Soluble enzymes have acquired many important commercial uses, such as the microbial proteases used in washing powders. The industrial application of soluble enzymes include the use of alkaline proteases in washing powders and α–amylase, glucoamylase and pullulanase in the production of glucose and fructose syrups from starch. Likewise, maltose syrups are produced from starch using malt (α-amylase) or fungal β-amylase. Microbial and animal chymosin is used for milk coagulation, lipases and proteases for accelerated ripening and flavour development in cheese, and proteases for protein processing, recovery from scrap meat and fish, and also meat tenderization. Proteases are also used for dehairing and rehydrating hides and skins during leather manufacture.

In many cases, the use of biological catalysts, in the form of enzymes of cells, is greatly facilitated by their immobilization and subsequent use in enzyme reactors. The best example of an immobilized enzyme used in industry is fructose isomerase, which, at present, produces several million tonnes of high fructose syrup per year, mainly in the USA.

A new and valuable area of enzyme technology is that concerned with the immobilization of enzymes on insoluble polymers, such as membranes and particles acting as supports or carriers for the enzyme activity. Some enzymes that are rapidly inactivated by heat when in cell-free form become heat-stable by attachment to inert polymeric supports. Whole microbial cells can also be immobilized inside polyacrylamide beads and used for a wide range of catalytic functions.

Immobilization can be defined as the process whereby the movement of enzymes, cells, organelles, etc. in space is completely or severely restricted usually resulting in a water-insoluble form of the enzyme. Immobilized enzymes are also sometimes referred to as bound, insolubilized, supported or matrix-linked enzymes.

Immobilized enzymes are normally more stable than their soluble counterparts and are able to be reused in the purified, semi-purified, or whole-cell form. Catalytic properties of immobilized enzymes can often be altered favourably to allow operation under broader or more rigorous reaction conditions, e.g. immobilized glucose isomerase can be used continuously for over 1000 hours at temperature of between 60–65°C.

How are enzymes immobilized? In practice both physical and chemical methods are routinely used for enzyme immobilization. Physically, enzymes may be adsorbed onto an insoluble matrix, entrapped within a gel or encapsulated within a microcapsule or behind a semi-permeable membrane. Chemically enzymes may be covalently attached to solid supports or cross-linked.

A large number of chemical reactions have been used for the covalent binding of enzymes by way of their non-essential functional groups to inorganic carriers such as ceramics, glass, iron, zirconium and titanium, to natural polymers such as Sepharose and cellulose and to synthetic polymers such as nylon, polyacrylamide and other vinyl polymers and copolymers possessing reactive chemical groups.

As a consequence of successful immobilization techniques in the form of enzyme capsules, enzyme beads, enzyme columns and enzyme membranes, many types of bioreactors have been developed at a laboratory scale and to a lesser extent at industrial scale. These include batch-stirred tank bioreactors, continuous packed-bed bioreactors and continuous fluidized-bed bioreactors.

Methods of Enzyme Immobilization

There are six main methods used for the immobilization of biocatalysts. These are: entrapment in polymer matrixes, adsorption of the charged biocatalyst onto oppositely charged support materials, covalent attachment to chemically activated supports, encapsulation inside semi-permeable membranes, aggregation of the biocatalyst particles into flocs or biospecific attachment to supports by means of lectins, etc. Combinations of these techniques can also be used, such as adsorption of the biocatalyst to a charged support followed by cross-linking into place.

Carrier matrices for enzyme immobilization by adsorption and covalent binding must be chosen with care. Of particular relevance to their use in industrial processes is their cost relative to the overall process costs; ideally they should be cheap enough to discard. The nature of support will also have a considerable effect on an enzyme's expressed activity and apparent kinetics. The form, shape, density, porosity, pore-size distribution, operational stability and particle-size distribution of the supporting matrix will influence the reactor configuration in which the immobilized biocatalyst may be used. The ideal support is cheap, inert, physically strong and stable. It will increase the enzyme specificity whilst reducing product inhibition, shift the pH optimum to the desired value for the process, and discourage microbial growth and non-specific adsorption. Clearly, most supports possess only some of these features, but a thorough understanding of the properties of immobilized enzymes does allow suitable engineering of the system to approach these optimal qualities.

Covalent binding is the most widely used method for immobilizing enzymes. The enzymes are very firmly bound, but are chemically modified and so many are denatured during immobilization. The most common technique is to activate a cellulose-based support with cyanogen bromide, which is then mixed with the enzyme. Immobilization of enzymes by their covalent coupling to insoluble matrices is an extensively researched technique. Only small amounts of enzymes may be immobilized by this method (about 0.02 gram per gram of matrix) although in exceptional cases, as much as 0.3 gram per gram of matrix has been reported. The strength of binding is very strong, however, and very little leakage of enzyme from the support occurs. The most commonly used method for immobilizing enzymes on the research scale (i.e., using less than a gram of enzyme) involves Sepharose, activated by cyanogen bromide. This is a simple, mild and often successful method of wide applicability. Sepharose is a commercially available beaded polymer which is highly hydrophilic and generally inert to microbial attack. Chemically, it is an agarose gel, the hydroxyl groups of this polysaccharide combine with cyanogen bromide to give the reactive cyclic imido-carbonate. This reacts with primary amino groups (i.e., mainly lysine residues) on the enzyme under mildly basic conditions (pH 9–11.5). Glutaraldehyde is another bifunctional reagent which may be used to cross-link enzymes or link them to supports. It is particularly useful for producing immobilized enzyme membranes for use in biosensors, by cross-linking the enzyme plus a non-catalytic diluent protein within a porous sheet (e.g. lens tissue paper or nylon net fabric).

Immobilization of enzymes by **adsorption** is probably the mildest method available, being mediated by ionic, hydrophobic or hydrogen bonds. **Adsorption** of enzymes onto insoluble supports is a very simple method of wide applicability and capable of high enzyme loading (about one gram per gram of matrix). Simple mixing of the enzyme with a suitable adsorbent, under appropriate conditions of pH and ionic strength, followed after a sufficient incubation period, by washing off loosely bound and unbound enzyme will produce the immobilized enzyme in a directly usable form. The driving force causing this binding is usually due to the combination of hydrophobic effects and the formation of several salt-links per enzyme molecule. The particular choice of adsorbent depends principally upon minimizing leakage of the enzyme during use. Although the physical links between the enzyme molecules and the support are often very strong, they may be reduced by many factors including the introduction of the substrate. Examples of suitable adsorbents are ion-exchange matrices, porous carbon, clay, hydrous metal oxides, glasses and polymeric aromatic resins. Manufacture of vinegar by naturally immobilized A. *aceti* cells on birchwood twigs is an established method. Adsorption is also easy to perform simply by stirring the biocatalysts with an ion-exchange resin.

In **entrapment**, the enzymes or cells are not directly attached to the support surface, but simply trapped inside the polymer matrix. Thus, loss of enzyme activity upon immobilization is minimized. Entrapment is carried out by mixing the biocatalyst into a monomer solution, followed by polymerization initiated by a change in temperature or by a chemical reaction. The polymer is formed either in particulate form, or as a block which can be disrupted to form discrete particles. The most common methods of entrapment use polyacrylamide, collagen, cellulose acetate, calcium alginate or carrageenan as the matrices. Entrapment is chiefly used for the immobilization of cells, rather than enzymes, because their larger size means that they are more easily retained in the support. Entrapment of enzymes within gels or fibres is a convenient method for use in processes involving low molecular weight substrate and products. Amounts in excess of 1 g of enzyme per gram of gel or fibre may be entrapped. However, the difficulty which large molecules have in approaching the catalytic sites of entrapped enzymes precludes the use of entrapped enzymes with high molecular weight substrates. The entrapment process may be a purely physical caging or involve covalent binding. As an example of this latter method, the enzymes' surface lysine residues may be derivatized by reaction with acryloyl chloride to give the acryloyl amides. This product may then be co-polymerized and cross-linked with acrylamide and bisacrylamide to form a gel. Enzymes may be entrapped in cellulose acetate fibres by, for example, making up an emulsion of the enzyme plus cellulose acetate in methylene chloride, followed by extrusion through a spinneret into a solution of an aqueous precipitant. Entrapment is the method of choice for the immobilization of microbial, animal and plant cells, where calcium alginate is widely used.

Membrane confinement of enzyme may be achieved by a number of different methods, all of which depend for their utility on the semipermeable nature of the membrane. This must confine the enzyme whilst allowing free passage for the reaction products and, in most configurations, the substrates. The simplest of these methods is achieved by placing the enzyme on one side of the semi-permeable membrane whilst the reactant and product stream is present on the other side.

In Europe, immobilized penicillin acylase is used to prepare 6-amino penicillanic acid (6-APA) from naturally produced penicillin G or V. This compound is an important intermediate in the synthesis of semi-synthetic penicillins so essential in our fight against bacterial diseases. Two types of penicillins are produced by industrial fermentation: penicillin G (phenylacetyl 6-APA) and penicillin V (phenoxyacteyl-6-APA), each containing a nucleus of 6-APA and a side chain. The antibiotic activity of the penicillin molecule is governed by the side-chain and, when removed and replaced with another, can profoundly alter the antibiotic spectrum and other properties.

Immobilized glucose isomerase is used in the USA, Japan and Europe for the industrial production of high-fructose syrups by partial isomerization of glucose derived from starch. Thousands of tonnes of high-fructose syrup are produced annually by this enzyme process, which is undoubtedly the most widely used of all the immobilized enzyme systems.

Another important use of immobilized enzymes is aminoacylase production of amino acids. Aminoacylase columns are used in Japan to produce thousands of kilograms of L-methionine, L-phenylalanine, L-tryptophan and L-valine.

ENZYMATIC ELECTROCATALYSIS

The term "enzymatic electrocatalysis" has been used to define the synergy between the electrochemical and enzymatic oxido-reduction reactions in a configuration where the electrochemical and enzymatic sites are close at the molecular level. The evolution of this concept has resulted in the deposition of chemically modified enzymes onto electrodes, which have become known as **biosensors** or **enzyme electrodes**. The product of the transfer can occur directly between the active site of the enzyme and the electrode or between a conduction polymer and the active site of the enzyme entrapped into it. Examples of enzymatic electrocatalysis systems using a carbon electrode is the enzyme/ substrate/co-substrate combinations of: glucose oxidase/glucose/oxygen, lactic dehydrogenase or lactase/NAF hydrogenase/hydrogen.

BASIC PRINCIPLE OF BIOSENSORS

A biosensor is a sensor that is based on the use of biological material for its sensing function. The biocomponent specifically reacts or interacts with the analyte of interest resulting in a detectable chemical or physical change. A biosensor is an analytical device which converts a biological response into an electrical signal. The term 'biosensor' is often used to cover sensor devices used in order to determine the concentration of substances and other parameters of biological interest even where they do not utilize a biological system directly.

A new biosensing flow injection method for the determination of alpha-amylase activity was developed (Zajoncova *et al.*, 2004). This method was based on the analysis of maltose produced during the hydrolysis of starch in the presence of alpha-amylase. Maltose determination in the flow system was allowed by the application of peroxide electrode equipped with an enzyme membrane. The membrane was obtained by immobilization of glucose oxidase, alpha-glucosidase and optionally mutarotase on a cellophane, co-cross-linked by gelatin-glutaraldehyde together with bovine serum albumin. Alpha-glucosidase hydrolyses maltose to alpha-D-glucose, which is converted to beta-D-glucose by mutarotase. Beta-D-glucose is then determined via glucose oxidase. The new biosensor has the limit of detection of 50 nmol l^{-1} maltose, which means 2 nkat ml^{-1} in alpha-amylase activity units, when the reaction time of amylase was 5 minutes (determined with respect to a signal-to-noise ratio 3 : 1). The biosensor was stable at least two months and retained 70% of its original activity (with mutarotase the stability is decreased to 3 weeks). When the enzyme membrane was stored in a dry state at 4°C in a refrigerator, the lifetime was approximately 6 months (with mutarotase only 3 months).

In a biosensor, the enzymes are simply entrapped with a dialysis membrane on the sensitive top of the detector or immobilized on it by covalent linking or cross-linking. At a given concentration of the substrate, the enzymatic reaction gives a corresponding reaction rate which alters the local physico-chemical environment of the detector. This alteration results in a response of the biosensor. Three main characteristics are important in determining the performance of the biosensors.

Selectivity This characteristic is brought about both by the specificity of the enzymatic reaction towards the substrate and by the selectivity of the detector towards the physico-chemical parameter. When the detector measures directly the concentration of the product of the enzymatic reaction, the improvement in selectivity is realized by the addition of a selective membrane between enzymatic reaction and the detector.

Sensitivity Because the response of the biosensor is not linear in all the ranges of substrate concentration, it is necessary to define the range in which the concentration is determined with good precision. The range depends both on the detector type, particularly to its detection limit, and on the nature of the diffusion and reaction of the enzymatic system.

Stability In addition to the need for general stability of the detector, the stability of the enzyme during measurement or even in storage is crucial. For this, immobilization of the enzyme is not only a way to maintain the enzyme on the detector but also to stabilize its enzymatic activity.

Tsifoulis *et al.* (2002) developed amperometric biosensors for the determination of glycolic acid in real samples. The first enzyme-based biosensors capable of determining glycolic acid in various complex matrixes, such as cosmetics, instant coffee and urine were reported. Recombinant microorganisms were used as environmental biosensors—*E. coli* was used for detection of pollution (Bechor *et al.*, 2002). A set of genetically engineered *E. coli* strains were constructed, in which the promoter of the *fabA* gene is fused to *Vibrio fischeri* luxCDABE either in a multi-copy plasmid or as a single copy chromosomal integration. The *fabA* gene codes for β-hydroxydecanoyl-ACP dehydrase, a key enzyme in the synthesis of unsaturated fatty acids and is induced when fatty acid biosynthesis pathways are interrupted. A dose-dependent and highly sensitive bioluminescent response to a variety of chemicals was controlled by the *fadR* gene.

There are two types of electrodes employed in biosensors, namely the amperometric and the potentiometric electrodes. In the amperometric electrode, a potential is imposed between the sensitive part of the electrode and a reference electrode. At this imposed potential, a molecule intervening in the enzymatic reaction (substrate or product) can transfer an electron with the electrode surface to be oxidized or reduced. The current through the electrode reflects the concentration of this molecule, but because this molecule is consumed at the electrode surface, the current is also a function of the mass-transfer to the surface.

TYPES OF BIOSENSORS

There are different types of biosensors, which have different applications. These are listed below.

Calorimetric Biosensors

Many enzyme-catalysed reactions are exothermic, generating heat (Table 14.5) which may be used as a basis for measuring the rate of reaction and, hence, the analyte concentration. This represents the most generally applicable type of biosensor. The temperature changes are usually determined by means of thermistors at the entrance and exit of small packed-bed columns containing immobilized enzymes within a constant temperature environment. Under such closely controlled conditions, up to 80% of the heat generated in the reaction may be registered as a temperature change in the sample stream. This may be simply calculated from the enthalpy change and the amount reacted. If a 1 mM reactant is completely converted to product in a reaction generating 100 kJ mole^{-1} then each ml of solution generates 0.1 J of heat. At 80% efficiency, this will cause a change in temperature of the solution amounting to approximately 0.02°C. This is about the temperature change commonly encountered and necessitates a temperature resolution of 0.0001°C for the biosensor to be generally useful.

Table 14.5 Heat output (molar enthalpies) of enzyme-catalysed reactions

Reactant	Enzyme	Heat output $-\Delta H$ (kJ mole^{-1})
Cholesterol	Cholesterol oxidase	53
Esters	Chymotrypsin	4–16
Glucose	Glucose oxidase	80
Hydrogen peroxide	Catalase	100
Penicillin G	Penicillinase	67
Peptides	Trypsin	10–30
Starch	Amylase	8
Sucrose	Invertase	20
Urea	Urease	61
Uric acid	Uricase	49

Potentiometric Biosensors

Potentiometric biosensors make use of ion-selective electrodes in order to transduce the biological reaction into an electrical signal. In the simplest terms this consists of an immobilized enzyme membrane surrounding the probe from a pH-meter, where the catalysed reaction generates or absorbs hydrogen ions. The reaction occurring next to the thin sensing glass membrane causes a change in pH which may be read directly from the pH-meter's display. Typical of the use of such electrodes is that the electrical potential is determined at very high impedance allowing effectively zero current flow and causing no interference with the reaction.

Amperometric Biosensors

Amperometric biosensors function by the production of a current when a potential is applied between two electrodes. They generally have response times, dynamic ranges and sensitivities similar to the potentiometric biosensors. The simplest amperometric biosensors in common usage involve the Clark oxygen electrode. This consists of a platinum cathode at which oxygen is reduced and a silver/silver chloride reference electrode. When a potential of –0.6 V, relative to the Ag/AgCl electrode is applied to the platinum cathode, a current proportional to the oxygen concentration is produced. Normally both electrodes are bathed in a solution of saturated potassium chloride and separated from the bulk solution by an oxygen-permeable plastic membrane (e.g. teflon, polytetrafluoroethylene). The following reactions occur:

$$\text{Ag anode} \qquad 4Ag^+ + 4Cl^- \rightarrow 4AgCl + 4e^-$$
$$\text{Pt cathode} \qquad O_2 + 4H^+ + 4e^- \rightarrow 2H_2O$$

The efficient reduction of oxygen at the surface of the cathode causes the oxygen concentration there to be effectively zero. The rate of this electrochemical reduction therefore depends on the rate of diffusion of the oxygen from the bulk solution, which is dependent on the concentration gradient and hence the bulk oxygen concentration. It is clear that a small but significant proportion of the oxygen present in the bulk is consumed by this process; the oxygen electrode measuring the rate of a process which is far from equilibrium, whereas ion-selective electrodes are used close to equilibrium conditions. This causes the oxygen electrode to be much more sensitive to changes in the temperature than potentiometric sensors. A typical application for this simple type of biosensor is the determination of glucose concentrations by the use of an immobilized glucose oxidase membrane. The reaction results in a reduction of the oxygen concentration as it diffuses through the biocatalytic membrane to the cathode, this being detected by a reduction in the current between the electrodes. Other oxidases may be used in a similar manner for the analysis of their substrates (e.g. alcohol oxidase, D- and L-amino acid oxidases, cholesterol oxidase, galactose oxidase, and urate oxidase).

The amperometric biosensor based on lactate oxidase for determination of lactate has been developed, and two methods of immobilization of lactate oxidase on the surface of industrial screen-printed platinum electrodes SensLab were compared (Shkotova *et al.*, 2005). A sensor immobilized in the Resydrol polymer lactate oxidase by the method of physical adsorption was characterized by narrow dynamic range and greater response value in comparison with a biosensor immobilized in poly(3,4-ethylenedioxythiophene) lactate oxidase by the method of electrochemical polymerization. Operational stability of the biosensor developed was studied and it was shown, that the immobilization method did not influence their stability. The developed biosensor could be applied in the food industry for the control and optimization of the wine fermentation process, and quality control of wine.

Optical Biosensors

There are two main areas of development in optical biosensors. These involve determining changes in light absorption between the reactants and products of a reaction, or measuring the light output by a luminescent process. The former usually involves the widely established, if rather low technology, use of colorimetric test strips. These are disposable single-use cellulose pads impregnated with enzyme and reagents. The most common use of this technology is for whole-blood monitoring in diabetes control. In this case, the strips include glucose oxidase, horseradish peroxidase (EC 1.11.1.7) and a chromogen (e.g. *o*-toluidine or 3,3´,5,5´-tetramethylbenzidine). The hydrogen peroxide, produced by the aerobic oxidation of glucose, oxidizes the weakly coloured chromogen to a highly coloured dye.

$$\text{Peroxidase chromogen (2H)} + H_2O_2 \rightarrow \text{Dye} + 2H_2O$$

The evaluation of the dyed strips is best achieved by the use of portable reflectance meters, although direct visual comparison with a coloured chart is often used. A most promising biosensor involving luminescence uses firefly luciferase (*Photinus*-luciferin 4-monooxygenase (ATP-hydrolysing), EC 1.13.12.7) to detect the presence of bacteria in food or clinical samples. Bacteria are specifically lysed and the ATP released (roughly proportional to the number of bacteria present) reacted with D-luciferin and oxygen in a reaction which produces yellow light in high quantum yield.

$$\text{Luciferase ATP} + \text{D-Luciferin} + O_2 \rightarrow \text{Oxyluciferin} + \text{AMP} + \text{Pyrophosphate} + CO_2 + \text{Light (562 nm)}$$

The light produced may be detected photometrically by use of high-voltage, and expensive, photomultiplier tubes or low-voltage cheap photodiode systems. Firefly luciferase is a very expensive enzyme, only obtainable from the tails of wild fireflies. Use of immobilized luciferase greatly reduces the cost of these analyses.

Immunosensors

Biosensors may be used in conjunction with enzyme-linked immunosorbent assays (ELISA). ELISA is used to detect and amplify an antigen–antibody reaction, the amount of enzyme-linked antigen bound to the immobilized antibody being determined by the relative concentration of the free and conjugated antigen and quantified by the rate of enzymic reaction. Enzymes with high turnover numbers are used in order to achieve rapid response. The sensitivity of such assays may be further enhanced by utilizing enzyme-catalysed reactions which give intrinsically greater response; for instance, those giving rise to highly coloured, fluorescent or bioluminescent products. Assay kits using this technique are now available for a vast range of analyses. Recently, ELISA techniques have been combined with biosensors, to form **immunosensors**, in order to increase their range, speed and sensitivity. In the immunosensor the biosensor merely replaces the traditional colorimetric detection system. However more advanced immunosensors are being developed which rely on the direct detection of antigen bound to the antibody-coated surface of the biosensor.

It is not possible to produce a batch of antibody (called a monoclonal antibody) which reacts specifically to one individual antigen. If the substance to be measured is itself antigenic, then a specific antibody against it can be produced and can be incorporated into the biosensor system. The biosensor can in this way be rendered highly specific in its action.

Immunochemical biosensor use immunological specificity with spectrophotometric detection. Two main procedures can be used, **the sandwich method** and the **competitive method**. In the first method, antibody (which reacts specifically with the substance to be measured, the analyte) is immobilized at the surface of a microplate incubated with the analyte. After rinsing, a solution containing the antibody linked to an enzyme is put in the microplate to read with the immobilized antibody–analyte complex from the first step. The spectrophotometric detection involves measuring a coloured product of the enzymatic reaction which reflects the number of analyte molecules and of the antigen coupled to the enzyme when added to the immobilized antibody. After rinsing, the enzymatic reaction is used to measure the number of antibodies which did not have the analyte.

van Regenmortel *et al.* (1998) described the measurement of antigen–antibody interactions with biosensors. In this technique, one of the interacting partners is immobilizing on a sensor chip and the binding of the other is followed by the increase in refractive index caused by the mass of bound species. These biosensors could be used for (a) functional mapping of epitopes and paratopes of mutagenesis; (b) analysis of the thermodynamic parameters of the interaction; (c) measurement of the concentration of biologically active molecules and (d) selection of diagnostic probes.

An optical sensor system was developed, which was particularly well-suited for medical point-of-care diagnostics (Schult *et al.*, 1999). The system allowed for all kinds of immunochemical assay formats and consisted of disposable sensor chip and an optical read-out device. This system allowed the determination of pregnancy hormone chorionic gonadotropin (hCG), in human serum with a detection limit of 1 ng/ml. This system was also found to be useful for assay of serum theophylline using a competitive type immunoassay.

AC voltammetric carbon paste-based enzyme immunosensors was developed by Fernandez-Sanchez *et al.* (2000) for determination of human IgG. Passive adsorption of the appropriate immunochemical reagent was performed onto the electrode surface. Alkaline phosphatase-labelled immunoglobulin was the tracer used in this work, 3-indoxyl phosphate being a very suitable enzymatic substrate for the electrochemical detection of the corresponding affinity reaction. The hydrolysis of this molecule generates indigo dimer. This product was detected by alternating current voltammetry taking advantage of the adsorptive and inherent electrodic properties that it exhibits. This methodology was applied to the design of two different immunoassays for the determination of human IgG.

Benkert *et al.* (200) developed a creatinine ELISA and an amperometric antibody-based creatinine sensor with a detection limit in the nanomolar range. Creatinine-specific antibodies have been generated and used for highly sensitive and specific immunochemical creatinine determinations. Creatinine was derivatized at N3 and coupled to KLH carrier protein. On the basis of this immunogen, monoclonal antibodies were developed by hybridoma technology. For creatinine determination, the creatinine-containing sample was incubated with B90-AH5 and anti-IgG (mouse)–glucose oxidase conjugate and applied to the measuring cell. After a washing step, glucose was added and the produced hydrogen peroxide was registered at Eappl = +600 mV vs Ag/AgCl. Biosensor-based immunochemical screening assays for the detection of sulphadiazine (SDZ) and sulphamethazine (SMT) in muscle extract from pigs were developed (Bjurling *et al.*, 2000). This simple and straightforward preparation allowed up to 40 samples to be processed and analysed in one day. These limits of detection for the assays were found to be 5.6 ng g^{-1} for SDZ and 7.5 ng g^{-1} for SMT.

Enzymes will clearly be more widely used in the future and this will be reflected in the number of enzymes available on an industrial (and research) scale, the variety of reactions catalysed and the range of environment conditions under which they will operate. Established enzymes will be put to new uses and novel enzymes discovered within their biological niches or produced by design using enzyme technology will be used to catalyse the unexploited reactions. This is just the start of the enzyme technology era.

ENZYME ENGINEERING

A most exciting development over the last few years is the application of genetic engineering techniques to enzyme technology. There are a number of properties which may be improved or altered by genetic engineering including the yield and kinetics of the enzyme, the ease of downstream processing and various safety aspects. Enzymes from dangerous or unapproved microorganisms and from slow-growing or limited plant or animal tissue may be cloned into safe high-production microorganisms. The amount of enzyme produced by a microorganism may be increased by increasing the number of gene copies that code for it. This principle has been used to increase the activity of penicillin G-amidase in *Escherichia coli*. The cellular DNA from a producing strain is selectively cleaved by the restriction endonuclease *Hind*III. This hydrolyses the DNA at relatively rare sites containing the 5′-AAGCTT-3′ base sequence to give identical "staggered" ends. The total DNA is cleaved into about 10,000 fragments, only one of which contains the required genetic information. These fragments are individually cloned into a cosmid vector and thereby returned to *E. coli*. Those colonies containing the active gene are identified by their inhibition of a 6-aminopenicillanic acid-sensitive organism. Such colonies are isolated and the penicillin-G-amidase gene transferred on to pBR 322 plasmids and recloned back into *E. coli*. The engineered cells, aided by the plasmid amplification at around 50 copies per cell, produce penicillin-G-amidase constitutively and in considerably higher quantities than does the fully induced parental strain. Such increased yields are economically relevant not just for the increased volumetric productivity but also because of reduced downstream processing costs, the resulting crude enzyme being that much purer.

Another extremely promising area of genetic engineering is protein engineering. New enzyme structures may be designed and produced in order to improve on existing enzymes or create new activities. Much protein engineering has been directed at subtilisin (from *Bacillus amyloliquefaciens*), the principal enzyme in the detergent enzyme preparation, alcalase. This has been aimed at the improvement of its activity in detergents by stabilizing it at even higher temperatures, pH and oxidant strength.

ARTIFICIAL ENZYMES

A number of possibilities now exist for the construction of artificial enzymes. These are generally synthetic polymers or oligomers with enzyme-like activities, often called synzymes. They must possess two structural entities, a substrate-binding site and a catalytically effective site. It has been found that producing the facility for substrate binding is relatively straightforward but catalytic sites are somewhat more difficult. Both sites may be designed separately but it appears that, if the synzyme has a binding site for the reaction transition state, this often achieves both functions. For a one-substrate reaction the reaction sequence is given by

$$\text{Synzyme} + \text{S} \ \rightleftharpoons \ (\text{Synzyme–S complex}) \ \rightarrow \ \text{Synzyme} + \text{P}$$

Some synzymes are simply derivatized proteins, although covalently immobilized enzymes are not considered here. An example is the derivatization of myoglobin, the oxygen carrier in muscle, by attaching $(\text{Ru(NH}_3)_5)^{3+}$ to three surface histidine residues. This converts it from an oxygen carrier to an oxidase, oxidizing ascorbic acid whilst reducing molecular oxygen. The synzyme is almost as effective as natural ascorbate oxidases.

Antibodies to transition state analogues of the required reaction may act as synzymes. For example, phosphonate esters of the general formula $R-PO_2-OR'-$ are stable analogues of the transition state occurring in carboxylic ester hydrolysis. Monoclonal antibodies raised to immunizing protein conjugates covalently attached to these phosphonate esters act as esterases. The specificities of these catalytic antibodies (also called abzymes) depends on the structure of the side-chains (i.e., R and R' in $(R-PO_2-OR')-$) of the antigens. The K_m values may be quite low, often in the micromolar region, whereas the V_{max} values are low (below 1 s^{-1}), although still 1000-fold higher than hydrolysis by background hydroxyl ions. A similar strategy may be used to produce synzymes by molecular "imprinting" of polymers, using the presence of transition state analogues to shape polymerizing resins or inactive non-enzymic protein during heat denaturation.

REVIEW QUESTIONS

1. What are the biomedical applications of enzymes?

2. Enumerate different methods of enzyme concentration and purification.

3. What are the applications of enzymes in biotechnology?

4. What do you understand by immobilization? What are the different methods of enzyme immobilization?

5. What are the basic principles of biosensors? Enumerate different types of biosensors?

REFERENCES

Albertsen, B.K., Hasle, H., Clausen, N., Schrøder, H. and Jakobsen, P. (2005). "Pain intensity and bioavailability of intramuscular asparaginase and a local anesthetic: A double-blinded study." *Pediatr. Blood Cancer.* 44:255–258.

Ao, Z., Quezada-Calvillo, R., Sim, L., Nichols, B.L., Rose, D.R., Sterchi, E.E., Hamaker, B.R. (2007). "Evidence of native starch degradation with human small intestinal maltase-glucoamylase (recombinant)." *FEBS Lett.* 581:2381–2388.

Balkan, B., Ertan, F. (2005). "Production and properties of alpha-amylase from *Penicillium chrysogenum* and its application in starch hydrolysis." *Prep. Biochem. Biotechnol.* 35:169–178.

Basu, A.K., Chattopadhyay, P., Roychaudhuri, U. and Chakraborty, R. (2006). "Development of biosensor based on immobilized L-glutamate oxidase for determination of monosodium glutamate in food." *Indian J. Exp. Biol.* 44:392–398.

Bechor, O., Smulski, D.R., Van Dyk, T.K., LaRossa, R.A. and Belkin, S. (2002). "Recombinant microorganisms as environment biosensors: pollutants detection by *Escherichia coli* bearing fab A´:: lux fusions." *J. Biotechnol.* 94: 125–132.

Benito, M.J., Connerton, I.F. and Córdoba, J.J. (2006). "Genetic characterization and expression of the novel fungal protease, EPg222 active in dry-cured meat products." *Appl. Microbiol. Biotechnol.* 73:356–365.

Benkert, A., Scheller, F., Schossler, W., Hentschel, C., Micheel, B., Behrising, I., Scharte, G., Stocklein, W. and Warsinke, A. (2000). "Development of a creatinine ELISA and an amperometric and antibody-based creatinine sensor with a detection limit in the nanogram range." *Anal. Chem.* 72: 916–921.

Bjurling, P., Baxter, G.A., Caselunghe, M., Jonson, C., O'Connor, M., Persson, B. and Elliott, C.T. (2000). "Biosensor assay of sulfadiazine and sulfamethazine residues in port." *Analyst.* 125: 1771–1774.

Bookbinder, L.H., Hofer, A., Haller, M.F., Zepeda, M.L., Keller, G.A., Lim, J.E., Edgington, T.S., Shepard, H.M., Patton, J.S. and Frost, G.I. (2006). "A recombinant human enzyme for enhanced interstitial transport of therapeutics." *J. Control Release.* 114:230–241.

Bucke, C. (1998). "The biotechnology of enzyme isolation and purification." In: *Principles of Biotechnology.* (ed.). Wiseman, A. Surrey University Press, UK. p. 143.

Celestino, K.R., Cunha, R.B., Felix, C.R. (2006). "Characterization of a beta-glucanase produced by *Rhizopus microsporus* var. *microsporus*, and its potential for application in the brewing industry." *BMC Biochem.* 2006.7:23.

Cheetham, S.J.P. (1988). "The application of immobilized enzymes and cells and biochemical reactors in biotechnology. Principles of enzyme engineering." In: *Principles of Biotechnology.* (ed.). Wiseman, A. Survey University Press, UK. p. 164.

Fernandez-Sanchez, C., Gonzalez-Garcia, M.B. and Costa-Garcia, A. (2000). "AC voltammetric carbon paste-based enzyme immunosensors." *Biosens. Bioelectron.* 14: 917–924.

Gupta, R. and Ramnani, P. (2006). "Microbial keratinases and their prospective applications: an overview." *Appl. Microbiol. Biotechnol.* 70:21–33.

Kamal Kumar, B., Balakrishnan, H. and Rele, M.V. (2004). "Compatibility of alkaline xylanases from an alkaliphilic *Bacillus* NCL (87-6-10) with commercial detergents and proteases." *J. Ind. Microbiol. Biotechnol.* 31:83–57.

Kozak, I., Kayikcioglu, O.R., Cheng, L., Falkenstein, I., Silva, G.A., Yu, D.X. and Freeman, W.R. (2006). "The effect of recombinant human hyaluronidase on dexamethasone penetration into the posterior segment of the eye after sub-Tenon's injection." *J. Ocul. Pharmacol. Ther.* 22:362–369.

Landbo, A.K., Pinelo, M., Vikbjerg, A.F., Let, M.B. and Meyer, A.S. (2006). "Protease-assisted clarification of black currant juice: synergy with other clarifying agents and effects on the phenol content." *J. Agric. Food Chem.* 54:6554–6563.

Lee, S., Oneda, H., Minoda, M., Tanaka, A. and Inouye, K. (2006). "Comparison of starch hydrolysis activity and thermal stability of two *Bacillus licheniformis* alpha-amylases and insights into engineering alpha-amylase variants active under acidic conditions." *J. Biochem. (Tokyo).* 139:997–1005.

Marazzi, M., Stefani, A., Chiaratti, A., Ordanini, M.N., Falcone, L. and Rapisarda, V. (2006). "Effect of enzymatic debridement with collagenase on acute and chronic hard-to-heal wounds." *J. Wound Care.* 15:222-227

Oh, K.W., Kim, M.J., Kim, H.Y., Kim, B.Y. Baik, M.Y. Auh, A.H. and Park, C.S. (2005). "Enzymatic characterization of a maltogenic amylase from *Lactobacillus gasseri* ATCC 33323 expressed in *Escherichia coli*." *FEMS Microbiol. Lett.* 252:175–181.

Reiff, A., Zastrow, M., Sun, B.C., Takei, S., Mitsuhada, H., Bernstein, B. and Durden, D.L. (2001). "Treatment of collagen induced arthritis in DBA/1 mice with L-asparaginase." *Clin. Exp. Rheumatol.* 19:639–646.

Schult, K., Katerkamp, A., Trau, D., Grawe, F., Cammann, K. and Meusel, M. (1999). "Disposable optical sensor chip for medical diagnostics: new ways in bioanalysis." *Anal. Chem.* 71: 5430–5435.

Semenova, M.V., Sinitsyna, O.A., Morozova, V.V., Fedorova, E.A., Gusakov, A.V., Okunev, O.N., Sokolova, L.M., Koshelev, A.V. Bubnova, I.P. and Sinitsyn, A.P. (2006). "Use of a preparation from fungal pectin lyase in the food industry." *Prikl. Biokhim. Mikrobiol.* 42:681–685.

Shkotova, L.V., Horiushkina, T.B., Slast'ia, E.A., Soldatkin, O.P., Tranh-Minh, S., Chovelon, J.M. and Dziadevych, S.V. (2005). "Amperometric biosensor for lactate analysis in wines and grape must during fermentation." *Ukr. Biokhim. Zh.* 77:123–130.

Thomas, D. and Laval, J.M. "Enzyme technology." In: *Concepts in Biotechnology.* (eds.). Balasubramanian, D., Bryce, C.F.A., Dharmalingam, K., Green, J. and Jayaraman, K. Costed-IBN, India.

Tsiafoulis, C.G., Prodromidis, M.I. and Karayannis, M.I. (2002). "Development of amperometric biosensors for the determination of glycolic acid in real samples." *Anal. Chem.* 74: 132–139.

Van Regenmortel, M.H., Altschuh, D., Chatellier, J., Christensen, L., Rauffer-Bruyere, N., Richalet-Secordel, P., Witz, J. and Zeder-Lutz, G. (1998). "Measurement of antigen-antibody interactions with biosensors." *J. Mol. Recognit.* 11: 163–167.

Vasconcelos, A., Silva, C.J., Schroeder, M., Guebitz, G.M. and Cavaco-Paulo, A. (2006). "Detergent formulations for wool domestic washings containing immobilized enzymes." *Biotechnol. Lett.* 28:725–731.

Watanabe, H., Nishimoto, T., Kubota, M., Chaen, H., Fukuda, S. (2006). "Cloning, sequencing and expression of the genes encoding an isocyclomaltooligosaccharide glucanotransferase and an alpha-amylase from a *Bacillus circulans* strain." *Biosci. Biotechnol. Biochem.* 70:2690–2702.

Wiseman, A. (1988). "Application of the principles of enzymology to biotechnology." In: *Principles of Biotechnology*. (ed.). Wiseman, A. Surrey University Press, UK. p. 136.

Wong, D.W., Robertson, G.H., Lee, C.C. and Wagschal, K. (2007). "Synergistic action of recombinant alpha-amylase and glucoamylase on the hydrolysis of starch granules." *Protein J.* 26:159–164.

Zajoncová, L., Jílek, M., Beranová, V. and Pec, P. (2004). "A biosensor for the determination of amylase activity." *Biosens. Bioelectron.* 20:240–245.

15

GENE THERAPY

INTRODUCTION

Gene therapy, the treatment or prevention of disease by gene transfer, is regarded by many as a potential revolution in medicine. This is because gene therapies are aimed at treating or eliminating the causes of disease, whereas most current drugs treat the symptoms. This radical improvement is possible because the gene-based approach can provide superior targeting and prolonged duration of action. Hence, in comparison with other forms of therapy, it will permit biological effects that are more subtle and better localized to the most appropriate cells. Most of the gene therapy techniques developed so far are of the gene-addition variety; that is, they attempt to provide a good copy of a gene to a cell that harbours a bad one. The hope is that the good, corrective gene will compensate for the bad one and restore the cell to its proper function.

The term **gene therapy** describes any procedure intended to treat or alleviate disease by genetically modifying the cells of a patient. It encompasses many different strategies and the material transferred into patient cells may be genes, gene segments or oligonucleotides. The genetic material may be transferred directly into cells within a patient (*in vivo* gene therapy), or cells may be removed from the patient and the genetic material inserted into them *in vitro*, prior to transplanting the modified cells back into the patient (*ex vivo* gene therapy).

Depending on the basis of pathogenesis, different gene therapy strategies can be considered. One, rather arbitrary, subdivision of gene therapy approaches is as follows:

Classical gene therapy   The rationale of this type of approach is to deliver genes to appropriate target cells with the aim of obtaining optimal expression of the introduced genes. Once inside the desired cells in the patient, the expressed genes are intended to do one of the following:

- produce a product that the patient lacks;

- kill diseased cells directly, e.g. by producing a toxin which kills the cells;

- activate cells of the immune system so as to aid killing of diseased cells.

Non-classical gene therapy   The idea here is to inhibit the expression of genes associated with the pathogenesis, or to correct a genetic defect and so restore normal gene expression.

Current gene therapy is exclusively somatic gene therapy, the introduction of genes into somatic cells of an affected individual. The prospect of human germ-line gene therapy raises a number of ethical concerns, and is currently not sanctioned.

Ex vivo Gene Transfer

This initially involves transfer of cloned genes into cells grown in culture. Those cells which have been transformed successfully are selected, expanded by cell culture *in vitro*, and then introduced into the patient. To avoid immune system rejection of the introduced cells, autologous cells are normally used: the cells are collected initially from the patient to be treated and grown in culture before being reintroduced into the same individual. Clearly, this approach is only applicable to tissues that can be removed from the body, altered genetically and returned to the patient where they will engraft and survive for a long period of time (e.g. cells of the haematopoietic system and skin cells).

In vivo Gene Transfer

Here the cloned genes are transferred directly into the tissues of the patient. This may be the only possible option in tissues where individual cells cannot be cultured *in vitro* in sufficient numbers (e.g. brain cells) and/or where cultured cells cannot be re-implanted efficiently in patients. Liposomes and certain viral vectors are increasingly being employed for this purpose. In the latter case, it is often convenient to implant vector-producing cells (VPCs), cultured cells which have been infected by the recombinant retrovirus *in vitro*: in this case the VPCs transfer the gene to the surrounding disease cells.

The first clinical studies involving gene transfer began in 1990 (Blease *et al.*, 1995) and since then gene therapy has become the focus of a whole new industry. On September 14, 1990 researches at the US National Institute of Health performed the first gene therapy procedure on four-year old Ashanti DeSilva. Born with a rare genetic disease called severe combined immune deficiency (SCID), she lacked a healthy immune system, and was vulnerable to every passing germ. Children with this illness usually develop overwhelming infections and rarely survive to adulthood; a common childhood illness like chickenpox is life-threatening. In Ashanti's gene therapy procedure, doctors removed white blood cells from the child's body, let the cells grow in the lab, inserted the missing gene into the cells, and then infused the genetically modified blood cells back into the patient's bloodstream. Laboratory tests showed that the therapy strengthened Ashanti's immune system.

All clinical studies have involved gene addition rather than the correction or replacement of defective genes, which is technically more difficult. All clinical protocols approved to date involve gene transfer only to somatic cells rather than germ line cells, the latter being the subject of considerable ethical debate at present. Gene transfer to these somatic cells can take place either *ex vivo* or *in vivo*. In the *ex vivo* approach, cells are removed from the patient for transfection, and the therapeutic entity comprises engineered cells. This offers the advantage for more efficient gene transfer and the possibility of cell propagation to generate high cell doses. However, it has the notable disadvantages of being largely patient-specific as a result of cell immunogenicity and more costly because cell manipulation adds manufacturing and quality-control difficulties. The *in vivo* approach involves direct administration of the gene-transfer vector to patients. It is therefore not patient-specific, thus conferring advantages of reduced cost, logistic and infrastructure requirements.

PRINCIPLES OF GENE TRANSFER

Classical gene therapies normally require efficient transfer of cloned genes into disease cells so that the introduced genes are expressed at suitably high levels. In principle, there are numerous different physico-chemical and biological methods that can be used to transfer exogenous genes into human cells. The size of DNA fragments that can be transferred is in most cases comparatively very limited, and so often the transferred gene is not a conventional gene. Instead, an artificial mini gene may be used: a cDNA sequence containing the complete coding DNA sequence is engineered to be flanked by appropriate regulatory sequences for ensuring high-level expression, such as a powerful viral promoter. Following gene transfer, the inserted genes may integrate into the chromosomes of the cell, or remain as extrachromosomal genetic elements (episomes).

Genes Integrated into Chromosomes

The advantage of integrating into a chromosome is that the gene can be perpetuated by chromosomal replication following cell division. As progeny cells also contain the introduced genes, long-term stable expression may be obtained. As a result, gene therapy using this approach may provide the possibility of a cure for some disorders. For example, in tissues composed of actively dividing cells, the key cells to target are stem cells (a minority population of undifferentiated precursor cells which gives rise to the mature differentiated cells of the tissue). This is so because stem cells not only give rise to the mature tissue cells, but during this procedure they also renew themselves. As a result, they are an immortal population of cells from which all other cells of the tissue are derived. High-efficiency gene transfer into stem cells, and subsequent stable high-level expression of a suitable introduced gene, can therefore provide the possibility of curing a genetic disorder.

Non-integrated Genes

Some gene transfer systems are designed to insert genes into cells where they remain as extrachromosomal elements and may be expressed at high levels. If the cells are actively dividing, the introduced gene may not segregate equally to daughter cells and so long-term expression may be a problem. As a result, the possibility of a cure for a genetic disorder may be remote: instead, repeated treatment involving gene transfer will be necessary. In some cases, however, there may be no need for stable long-term expression. For example, cancer gene therapies often involve transfer and expression of genes into cancer cells with a view to killing the cells. Once the malignancy has been eliminated, the therapeutic gene may no longer be needed.

CLINICAL STUDIES

As on year 2000, approximately 300 clinical protocols involving gene transfer have been approved, mostly in the USA. Two-thirds of trials are directed at cancer and most of the reminder at inherited monogenic disorders (especially cystic fibrosis) or infectious diseases (particularly HIV). More than 3500 patients have been administered with experimental gene therapies. Most clinical trials have been small phase I/II studies with the main objectives of demonstrating safety and gene transfer, and obtaining information to guide dose selection for phase II and III efficacy studies. Recently, clinical benefit from gene therapy has been clearly demonstrated for the first time. In these studies, naked DNA encoding the angiogenic protein vascular endothelial growth factor (VEGF) was injected into the skeletal muscles of patients with critical limb ischaemia resulting from an inadequate blood supply. The results are remarkable, with dramatic and long-lasting benefit observed in a large proportion of patients, including those who would otherwise have to face amputation (Baumgartner *et al.*, 1998).

In studying the ethics of gene therapy, one should make a distinction between therapy on the somatic (non-productive) cells and the germ (reproductive) cells of an individual. Only the germ cells carry the genes that will be passed onto the next generation. Many people think that the use of germ-line gene therapy could have an unforeseeable effect on future generations.

TECHNIQUES IN GENE THERAPY

The **somatic cell gene therapy** procedure inserts a normal gene into the DNA of cells in order to compensate for the non-functioning defective gene. This technique involves obtaining blood cells from a person afflicted with a genetic disease and then introducing a normal gene into the defective cell. The normal gene is delivered using a domesticated retrovirus that infects the cell, introducing the properly functioning gene. Retroviruses can infect many types of cell, so it is important to develop gene transfer techniques that allow only retroviruses to deliver genes to a cell and then remain there. For cystic fibrosis, another kind of virus called adenovirus has been used as the vector for the new gene.

Germ-line gene therapy is technically more difficult, and as noted, raises more ethical challenges. The two main methods of performing germ-line gene therapy would be: (1) to treat a pre-embryo that carries a serious genetic defect before implantation in the mother (this necessitates the use of *in vitro* fertilization technique); or (2) to treat the germ cells (sperm or egg cells) of afflicted adult so that their genetic defects would not be passed on to their offspring.

Much attention has been focused on the so-called genetic metabolic diseases in which a defective gene causes an enzyme to be either absent or ineffective in catalysing a particular metabolic reaction effectively. The most likely candidates for future gene therapy trials will be rare diseases such as Lesch–Nyhan syndrome, a distressing disease in which the patients are unable to manufacture a particular enzyme. This leads to a bizarre impulse for self-mutilation, including very severe biting of the lips and fingers. If gene therapy does become practicable, the biggest impact would be on the treatment of disease where the normal gene needs to be introduced into only one organ. One such disease is phenylketonuria (PKU). PKU affects about one in 12,000 white children and if not treated early can result in severe mental retardation. The disease is caused by a defect in a gene producing a liver enzyme.

The types of gene therapy described thus far all have one factor in common: that is, that the tissue being treated are somatic (somatic cells include all the cells of the body, excluding sperm cells and egg cells). In contrast to this is the replacement of defective genes in the germ line cells (which contribute to the genetic heritage of the offspring). Gene therapy in germ line cells has the potential to affect not only the individual being treated, but also his or her children as well. Germ line therapy would change the genetic pool of the entire human species, and future generations would have to live with that change.

CANDIDATE DISEASES FOR GENE THERAPY

Gene therapy is likely to have the greatest success with diseases that are caused by single gene defects. By the end of 1993, gene therapy had been approved for use on such diseases as severe combined immune deficiency, familial hypercholesterolaemia, cystic fibrosis and Gaucher's disease. Most protocols to date are aimed towards the treatment of cancer; a few are also targeted towards AIDS. Numerous disorders are discussed as candidates for gene therapy; Parkinson's and Alzheimer's diseases, arthritis and heart diseases.

Few criteria have been laid out for selection of disease candidates for human gene therapy; (a) the disease is an incurable, life-threatening disease; (b) organ, tissue and cell types affected by the disease have been identified; (c) the normal counterpart of the defective gene has been isolated and cloned; (d) the normal gene can be introduced into a substantial subfraction of the cells from the affected tissue; (e) the gene can be expressed adequately and (f) techniques are available to verify the safety of the procedure.

Prud'homme *et al.* (2007) showed that gene transfer of immunoregulatory molecules can prevent Type I diabetes mellitus (T1D) and other autoimmune diseases. In their studies, non-viral gene transfer was enhanced by *in vivo* electroporation (EP). This technique could be used to perform DNA vaccination against islet cell antigens and when combined with appropriate immune ligands resulted in the generation of regulatory T cells (Tr) and protection against T1D. *In vivo* EP could also be applied for non-immune therapy of diabetes. It could be used to deliver protein drugs such as glucagon-like peptide 1 (GLP-1), leptin or transforming growth factor beta (TGF-beta). These acted in T1D or type II diabetes (T2D) by restoring glucose homeostasis, promoting islet cell survival and growth or improving wound healing and other complications. It was concluded that the non-viral gene therapy and EP represented a safe and efficacious approach with clinical potential.

Intravesical adenovirus-mediated interferon-alpha gene transfer has a potent therapeutic effect against superficial human bladder carcinoma xenografts growing in the bladder of athymic nude mice. Adams *et al.* (2007) treated several human urothelial carcinoma cells with adenovirus-mediated interferon-alpha 2b and monitored its effects on the production of angiogenic factors using real-time reverse-transcription polymerase chain reaction, western blotting, and

immunohistochemical analysis and a gel shift-based transcription factor array. Treatment with adenovirus-mediated interferon 2b increased the angiostatic activity of the bladder cancer microenvironment. This inhibition might prove beneficial for treating superficial bladder cancer with adenovirus-mediated interferon-alpha and contributed to a decreased recurrence rate of this neoplasm.

Sebestyen *et al.* (2007) developed a naked DNA-based gene therapy protocol for the treatment of anaemia. Hydrodynamic limb vein technology has been shown to be an effective and safe procedure for delivering naked plasmid DNA (pDNA) into the skeletal muscles of limbs. The results using a rat anaemia model, provided proof of principle that repeated delivery of small pDNA doses had an additive effect and could gradually lead to the correction of anaemia without triggering excessive haematopoiesis. This simple method provided an alternative approach for regulating EPO expression. EPO expression was also proportional to the injected pDNA dose in non-human primates. In addition, long-term (more than 450 days) expression was obtained after delivering rhesus EPO cDNA under the transcriptional control of the muscle-specific creatine kinase (MCK) promoter. In conclusion, these data suggested that the repeated delivery of small doses of EPO expressing pDNA into skeletal muscle was a promising, clinically viable approach to alleviate the symptoms of anaemia.

GENE TRANSFER SYSTEMS

There are three main types of gene-delivery vectors: viral, non-viral and physical. Many different viruses are being adapted as vectors, but the most advanced are retrovirus (Rv), adenovirus (Ad) and adeno-associated virus (AAV). Substantial effort has also gone into developing poxviruses (especially vaccinia) for genetic vaccines and herpes simplex virus. Non-viral approaches fall into three main categories involving (a) naked DNA, (b) DNA complexed with cationic lipids and (c) particles comprising DNA condensed with cationic polymers. The most important physical methods involve needle-free injectors and electroporation. At the present stage of development, the leading viral vectors generally give the most efficient transfection. Their main disadvantages concern insert-size limitation, immunogenicity and manufacture. Non-viral vectors give less efficient transfection (especially *in vivo*) and more-transient expression, but have no insert-size limitation, are less immunogenic and easier to manufacture. Physical methods give inefficient transfection and have a limited range of applications, but some of these applications are promising and important. The advantages and disadvantages of gene transfer vectors are listed in Table 15.1.

Retrovirus

Retroviruses are RNA viruses which possess a reverse transcriptase function, enabling them to synthesize a complementary DNA form. Following infection (transduction), retroviruses deliver a nucleoprotein complex (pre-integration complex) into the cytoplasm of infected cells. This complex reverse-transcribes the viral RNA genome and then integrates the resulting DNA copy into a single site in the host cell chromosomes. Retroviruses are very efficient at transferring DNA into cells, and the integrated DNA can be stably propagated, offering the possibility of a permanent cure for a disease. Because of these properties, retroviruses were considered the most promising vehicles for gene delivery and currently about 60% of all approved clinical protocols utilize retroviral vectors. The retrovirus vectors that have traditionally been used in gene therapy are derived from simple retroviruses (oncoretroviruses), notably murine leukaemia virus. Unlike adenoviruses they can only be produced at relatively low titres and so it is not possible to get a large number of vector particles to the desired cell type *in vivo*. Since all the viral genes are removed from the vector, the viruses cannot replicate by themselves. They can accept inserts of up to 8 kb of exogenous DNA and require a variety of packaging systems to enclose the viral genome within viral particles.

Retrovirus (Rv) vectors, unlike Ad, transfect by integrating the transgene into the target-cell chromosome. This usually leads to more prolonged transgene expression but at reduced levels. Transgene expression typically ceases within days to weeks and tends to be shorter *in vivo*. This is likely to result from methylation in the vicinity of the promoter and incorporation of the insertion site into condensed chromatin, in which the transgene is inaccessible to the transcription machinery. The Rv provirus penetrates the nucleus only at mitosis, so transfection is restricted to proliferating cells; this has been a major limitation of Rv vectors. Rv vectors have an insert-size limit of 8 kb. Rv gives inefficient gene transfer for most cell types *in vivo*, partly because it is rapidly inactivated by the human complement system. Because of this, it is most extensively used with *ex vivo* applications, for which cells can be cultured to allow efficient transfection and for which sustained expression is generally required. *Ex vivo* applications also require smaller quantities of virus; this is an important consideration because Rv has presented serious manufacturing challenges, with difficulties in achieving high titres and retaining infectivity through concentration and storage. Good progress has been made recently in developing vectors from lentiviruses (Lv) such as HIV and SIV, which are capable of integrative transfection of several quiescent and post-mitotic cell types as well as proliferating cells.

Adenovirus

Adenovirus vectors have been the second most popular delivery system in gene therapy (with extensive applications in gene therapy for cystic fibrosis and certain types of cancer) and have several advantages as gene delivery vectors. They are human viruses which can be produced at very high titres in culture, and they

Table 15.1 Advantages and disadvantages of gene transfer vectors

Vector	Advantages	Disadvantages
Adenovirus	Very high transfection efficiency *ex vivo* and *in vivo*. Transfects proliferating and non-proliferating cells. Substantial clinical experience acquired. Efficient retargeted transfection demonstrated.	Repeat dosing ineffective owing to strong immune response. Insert size limit of 7.5 kb. Manufacture, storage, QC are moderately difficult. Short duration of expression.
Retrovirus	Fairly prolonged expression. High transfection efficiency *ex vivo*. Substantial clinical experience *ex vivo*. Low immunogenicity.	Low transfection efficiency *in vivo*. Insert size limit of 8 kb. Transfects only proliferating cells. Safety concern of insertional mutagenesis. Manufacture, storage and quality control are extremely difficult.
Lentivirus	Transfects proliferating and non-proliferating cells. Transfects haematopoietic stem cells.	Safety concerns from immunodeficiency virus origins. Manufacturing, storage, QC are extremely difficult. Insert size limit of 8 kb. No clinical experience.
Adeno-associated virus (AAV)	Efficiently transfects wide variety of cells *in vivo*. Very prolonged expression *in vivo*. Low immunogenicity.	Insert size limit of 4.5 kb. Manufacture, QC is very difficult. Little clinical experience. Safety concern of insertional mutagenesis. Repeat dosing affected by neutralizing antibody.
Naked DNA	Manufacturing, storage and QC are simple and cheap. Very low immunogenicity. Clinical efficacy demonstrated in clinical limb ischaemia. Very good safety profile.	Very short duration of expression in most tissues. Very efficient transfection *ex vivo* and *in vivo*. Retargeting transfection very difficult.

are able to infect a large number of different human cell types including non-dividing cells. Entry into cells occurs by receptor-mediated endocytosis and transduction efficiency is very high (often approaching 100% *in vitro*). They are large viruses and so have the potential for accepting large inserts. The risk of given the need to administer treatment frequently (because of the inability of immune response to these vectors is negligible. This is an important consideration adenovirus to integrate into chromosomal DNA). They also have the advantage that they can accept much larger inserts (up to 35 kb).

It is clear that for a wide variety of cell types, adenovirus (Ad) gives more efficient gene transfer compared with other systems, especially *in vivo*. Ad vectors can transfer genes to both proliferating and quiescent cells. Following delivery, transgene expression is at a high level, but is transient, being low or undetectable in most tissues after two weeks. This is because Ad vectors do not integrate and for safety reasons are disabled for replication. The expression profile with first generation Ad vectors is not suitable for the long-term correction of chronic diseases but is adequate for direct cell killing, for most immunotherapy strategies and for some acute diseases. The first generation vectors have an insert-size limit of ~7.5 kb.

The promoter used most frequently with Ad (and indeed all other vectors) is derived from cytomegalovirus (CMV), which gives strong expression in many cell types. Ad gives particularly, efficient gene transfer to the liver, such that dissemination from the site of local injection (such as tumours) and consequent liver transfection is the most serious safety concern. The most serious limitation of Ad vectors stems from their tendency to elicit strong immune and (at high doses) inflammatory responses. Single, large doses of Ad provoke neutralizing antibody response directed to proteins of the viral particle, which prevent binding to target cells and abrogate gene transfer upon repeat dosing by systemic administration routes in animals.

Overall, Ad is the easiest viral vector from the manufacturing viewpoint, allowing the production of large quantities with high titre and in relatively robust formulations. Nevertheless, Ad shares with other viral vectors the problem that first-generation vector preparations are contaminated with replication-competent virus (RCV), which arises through recombination between viral sequences in the vector and in the chromosome of the producer cells. Many of the clinical studies giving encouraging signs of efficacy use Ad vectors. The most advanced of these delivers the wild-type gene for the tumour suppressor P53 for induction of tumour-cell killing. A series of phase II studies is underway, testing this recombinant virus alone and in combination with chemotherapies for the local management of various cancers. The steps involved in gene therapy are illustrated in Figure 15.1.

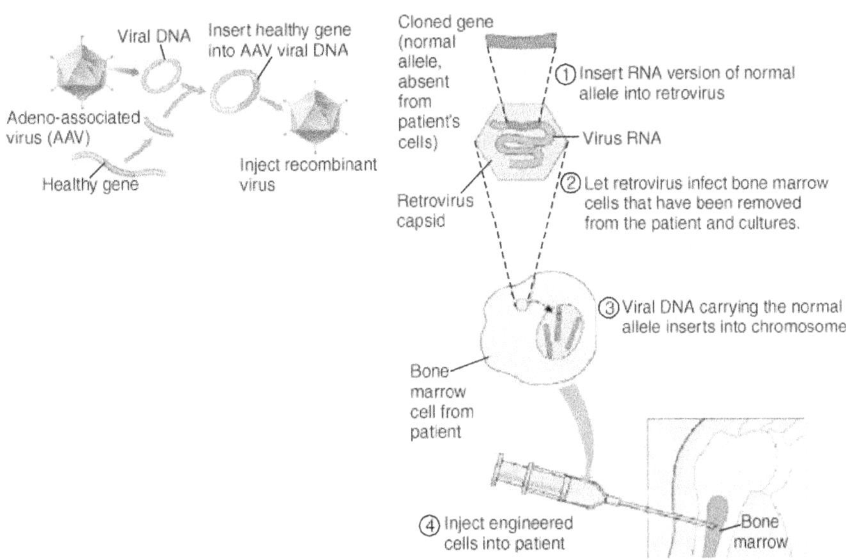

Figure 15.1 Gene therapy method

Adeno-associated Virus

AAV is also capable of integrating into the target-cell chromosome. The integration process is not well characterized by wild-type AAV which integrates exclusively into a single site on human chromosome 19. It appears, however, that AAV recombinants integrate much less efficiently and more randomly, and many show long-term persistence in un-integrated forms (Balague *et al.*, 1997). AAV appears to give sustained transgene expression upon *in vivo* administration, compared with other vectors; expression of homologous genes has been detected two years after injection in mice (Donahue *et al.*, 1999) and several months after injection in dogs (Linden and Woo, 1999) and man (Wagner, 1998). Long-term correction of haemophilia has been achieved in a dog model of the disease with a single injection of an AAV vector expressing a gene for clotting factor IX.

The resilience of AAV particles is a major advantage for *in vivo* applications. Indeed, substantial transfection and long-lasting transgene expression in the epithelial and sub-epithelial cells of the upper intestinal tract have recently been reported following oral administration of an AAV vector. The main disadvanatages of AAV concern insert size and manufacturing; the virus can only accommodate inserts up to 4.5 kb, and manufacturing process have required the use of helper viruses (usually Ad), which present problems of low titre, contamination and costly purification procedures.

Naked DNA

Although naked DNA gives virtually no transfection for cells *ex vivo*, it gives surprisingly efficient gene transfer in several tissues following local injection *in vivo*, notably in muscle and skin. There are many reports of transgene expression lasting several months, following injection of plasmid DNA into skeletal muscles in mice. The observation that injection of plasmids encoding protein antigens into muscle or skin in mice leads to humoral and cellular immune responses directed to the antigen has led to enormous interest in naked DNA vaccines. In a recent phase I clinical study involving intramuscular injection of plasmid DNA encoding a gene of *Plasmodium falciparum*, cytotoxic T-lymphocyte (CTL) responses directed to epitopes of the parasite were detected in most of the vaccines (Wang *et al.*, 1998). The other important application is the injection of naked DNA encoding angiogenic factors into muscles of patients with ischaemic vascular diseases.

Naked DNA can be manufactured simply and cheaply in bacteria, an advantage that is magnified in strategies that require co-delivery of several genes. Its disadvantages include: (a) a gene-delivery efficiency that is much lower than Ad or AAV; (b) very brief expression in most tissues; and (c) unsuitability for targeting.

Herpes Simplex Virus Vectors

HSV vectors are tropic for the central nervous system (CNS) and can establish lifelong latent infections in neurons. They have a comparatively large insert size capacity (>20 kb) but are non-integrating and so long-term expression of transferred genes is not possible. Their major applications are expected to be in delivering genes into neurons for the treatment of neurological diseases such as Parkinson's disease, and for treating CNS tumours.

Lentiviruses

The lentivirus family, which includes HIV (human immunodeficiency virus), are complex retroviruses that infect macrophages and lymphocytes. Unlike oncoretroviruses, lentiviruses are able to transduce non-dividing cells. In the case of HIV, for example, the preintegration complex contains nuclear localization signals that permit its active transport through nuclear pores into the nucleus during interphase. Because of their ability to infect non-dividing cells and to integrate into host cell chromosomes, considerable efforts are now being devoted to making lentivirus vectors for gene therapy.

NON-VIRAL VECTOR SYSTEMS FOR GENE THERAPY

Liposomes

Liposomes are spherical vesicles composed of synthetic lipid bilayers which mimic the structure of biological membranes. The DNA to be transferred is packaged *in vitro* with the liposomes and used directly for transferring the DNA to a suitable target tissue *in vivo*. The lipid coating allows the DNA to survive *in vivo*, bind to cells and be endocytosed into the cells. Cationic liposomes (where the positive charge on liposomes stabilize binding of negatively charged DNA), have become popular vehicles for gene transfer in *in vivo* gene therapy. Unlike viral vectors, the DNA/lipid complexes are easy to prepare and there is no limit to the size of DNA that is transferred. However, the efficiency of gene transfer is low, and the introduced DNA is not designed to integrate into chromosomal DNA. As a result, expression of the inserted genes is transient.

Direct Injection/Particle Bombardment

In some cases, DNA can be injected directly with a syringe and needle into a specific tissue, such as muscle. This approach has been considered, for example, in the case of DMD, where early studies investigated intramuscular injection of a dystrophin mini gene into a mouse model, *mdx* (Acsadi *et al.*, 1991). An alternative direct injection approach uses particle bombardment ("gene gun") techniques: DNA is coated on to metal pellets and fired from a special gun into cells. Successful gene transfer into a number of different tissues has been obtained using this approach. Such direct injection techniques are simple and comparatively safe. However, there is poor efficiency of gene transfer, and a low level of stable integration of the injected DNA. The latter property is particularly disadvantageous in the case of proliferating cells, and would necessitate repeated injections. It may be less of a problem in tissues such as muscle which do not regularly proliferate, and in which the injected DNA may continue to be expressed for several months.

Receptor-mediated Endocytosis

The DNA is coupled to a targeting molecule that can bind to a specific cell-surface receptor, inducing endocytosis and transfer of the DNA into cells. Coupling is normally achieved by covalently linking polylysine to the receptor molecule and then arranging for (reversible) binding of the negatively charged DNA to the positively charged polylysine component. For example, hepatocytes are distinguished by the presence on the cell surface of asialoglycoprotein receptors which clear asialoglycoproteins from the serum. Coupling of DNA to an asialoglycoprotein via a polycation such as polylysine can target the transfer of exogenous DNA into liver cells. The complexes can be infused into the liver either via the biliary tract or vascular bed, whereupon they are taken up by hepatocytes.

GENE THERAPY FOR INHERITED DISORDERS

Over the last two decades molecular genetic technologies have been spectacularly successful in identifying and characterizing novel disease genes, and in devising novel diagnostic tests for inherited disorders. In contrast, the dream of successfully applying molecular genetic technologies on a large scale to curing, or even treating disease has remained unfulfilled.

Recessively Inherited Disorders

Those disorders where the disease results from a simple deficiency of a specific gene product are generally the most amenable to treatment: high level expression of an introduced normal allele should be sufficient to overcome the genetic deficiency. Recessively inherited disorders have been of particular interest as candidates for gene therapy because the mutations are almost always simple loss-of-function mutations. Affected individuals have deficient expression from both alleles and so the disease phenotype is due to complete or almost complete absence of normal gene expression. Heterozygotes, however, have about 50% of the normal gene product and are normally asymptomatic.

Although recessively inherited disorders are, in principle, amenable to gene augmentation therapy, certain disorders are less amenable than others. In addition to the question of accessibility of the disease tissue, some disorders may be difficult to treat for other reasons. A good example is provided by β-thalassemia which results from mutations in the β-globin gene, *HBB*. This is a severe disorder affecting hundreds of thousands of people worldwide, and superficially would appear to be an excellent candidate for gene therapy. The gene is very small and has been characterized extensively, the disorder is recessively inherited and affects blood cells.

The First Gene Therapy Trial in 1990

The first gene therapy trial for an inherited disorder was initiated on 14th September 1990. The patient, Ashanthi DeSilva, was just 4 years old and was suffering from a very rare recessively inherited disorder, adenosine deaminase (ADA) deficiency. ADA is involved in the purine salvage pathway of nucleic acid degradation, and is a housekeeping enzyme which is synthesized in many different types of cells. An inherited deficiency of this enzyme has, however, particularly severe consequences in the case of T lymphocytes, one of the major classes of immune system cells. As a result, ADA⁻ patients suffer from severe combined immunodeficiency. This severe disorder was particularly amenable to gene therapy for a variety of reasons: the *ADA* gene is small, and had previously been cloned and extensively studied; the target cells are T cells which are easily accessible and easy to culture, enabling *ex vivo* gene therapy; the disorder is recessively inherited and, importantly, gene expression is not tightly controlled (enzyme levels in the normal population show huge differences between healthy individuals). The observation that allogeneic bone marrow transplantation can cure the disorder suggested that engraftment of T cells alone may be sufficient, and transfer of normal *ADA* genes into ADA⁻ T cells was noted to result in restoration of the normal phenotype.

Alternative treatments for ADA deficiency do exist. Indeed, the treatment of choice is bone marrow transplantation from a perfectly HLA-matched sibling donor, which provides a cure in about 80% of cases. For children where this is not an option, an alternative is enzyme replacement therapy, consisting of weekly intramuscular injections of ADA conjugated to polyethylene glycol (PEG).

The novel *ADA* gene therapy approach involved essentially four steps:

1. cloning a normal *ADA* gene into a retroviral vector;
2. transfecting the ADA recombinant into cultured ADA⁻ T lymphocytes from the patient;
3. identifying the resulting ADA⁺ T cells and expanding them in culture; and
4. re-implanting these cells in the patient.

This approach was never going to be a cure; instead it was designed to be a form of treatment which would need to be repeated on many occasions. Successful treatment would require high-efficiency gene transfer into bone marrow stem cells and high levels of expression. All patients in the *ADA* gene therapy trials were treated in parallel by conventional enzyme replacement therapy using PEG-ADA. The combined gene therapy plus enzyme replacement therapy appeared to give initially promising results (as assayed by various measures of antibody and T-cell function, and a dramatic decrease in infections compared with the incidence before treatment).

Gene Therapy Trials for a Few Inherited Disorders

Gene therapy has been initiated for only a comparatively few inherited disorders in addition to ADA deficiency (Table 15.2). Different recessively inherited disorders have been targets for *in vivo* or *ex vivo* gene augmentation therapy and, in the one case where a dominantly inherited disorder has been treated, familial hypercholesterolaemia, the patients had the homozygous form of the disease. The following examples are simply illustrative of current progress and difficulties.

Table 15.2 Examples of gene therapy trials for inherited disorders

Disorder	Cells altered	Gene therapy strategy
ADA deficiency	T cells and haemopoietic stem cells	*Ex vivo* GAT* using recombinant retroviruses containing an *ADA* gene
Cystic fibrosis	Respiratory epithelium	*In vivo* GAT using recombinant adenoviruses or liposomes to deliver the *CFTR* gene
Familial hypercholesterolaemia	Liver cells	*Ex vivo* GAT using retrovirus to deliver the LDL receptor gene (*LDLR*)
Gaucher's disease glucocerebrosidase	Haemopoietic stem cells	*Ex vivo* GAT using retroviruses to deliver the gene (*GBA*)

*GAT—gene augmentation therapy.

Familial Hypercholesterolaemia (FH)

This disorder is caused by a dominantly inherited deficiency of low density-lipoprotein (LDL) receptors, which are normally synthesized in the liver, and is characterized by premature coronary artery disease. About 50% of heterozygous affected males die by 60 years of age, unless treated. Because FH is such a common single-gene disorder, homozygotes are occasionally seen. They suffer precocious onset of disease and increased severity, with death from myocardial infarction commonly occurring in late childhood.

Ex vivo gene therapy became a possibility when animal experiments showed that cultured hepatocytes could be injected via the portal venous system— the veins which drain from the intestine directly into the liver—after which they appear to seed in the liver. Gene therapy involved surgical removal of a sizeable portion of the left lobe of the patient's liver, disaggregation of the liver cells and plating in cell culture prior to infection with retroviruses containing a normal human *LDLR* gene (Grossman *et al.*, 1994). The genetically modified cells were infused back into the patient through a catheter implanted into a branch of the portal venous system. The patient's LDL/high density lipoprotein (HDL) ratio subsequently declined from 10–13 before gene therapy to 5–8, and such improvement was maintained over a long period.

Cystic Fibrosis

Cystic fibrosis is an autosomal recessive disorder that results in defective transport of chloride ions through epithelial cells, and results from mutations in a gene, *CFTR*, which encodes a cAMP-regulated chloride channel. The primary expression of the defect is in the lungs: a sticky mucus secretion accumulates, which is prone to chronic infections. Gene therapy trials used adenovirus vectors or liposomes to transfer a suitably sized *CFTR* mini gene, either through a bronchoscope or through the nasal cavity.

The first adenovirus-based protocol began in 1993 and, although preliminary data have confirmed gene transfer into respiratory epithelium *in vivo*, there have been major concerns regarding the safety of the procedure. The first patient to be treated with a high dose of recombinant adenovirus experienced transient pulmonary infiltrates and alterations in vital signs, before recovering uneventfully. This experience prompted recognition of the need to confirm the maximum tolerated adenovirus dose. The liposome-based gene therapy trials are regarded as safer procedures, but the efficiency of gene transfer is much lower. Despite an impressive amount of research, *CF* gene therapy remains ineffective (Boucher, 1999)

Duchenne Muscular Dystrophy

DMD is a severe X-linked recessive disorder: affected males suffer progressive muscle deterioration, are confined to a wheelchair in their teens and die usually by the third decade. The target tissue is skeletal muscle, and initial interest in treatment for this disorder focused on cell therapy because of the unique cell biology of muscle (Miller and Boyce, 1995). Suitable gene therapy approaches have also been difficult to conceive, largely because of the lack of a suitable gene transfer system. Oncoretroviral vectors cannot be used because adult skeletal muscle fibres are post-mitotic and hence not susceptible to oncoretroviral infection. Adenovirus vectors have been used to deliver genes to muscle fibres *in vivo* and, although the post-mitotic state of muscle nuclei allows the expression to persist, the need for expression to continue over the course of a lifetime (which would be required for successful therapy) remains doubtful. Dystrophin has a close relative, utrophin, which is highly expressed during the foetal period and so there is the possibility that up-regulation of utrophin may confer a protective effect. Encouraging results have been obtained in mice where expression of utrophin transgenes in mice with dsytrophin deficiency leads to major improvements in muscle function (Deconinck *et al.*, 1997).

RECENT DEVELOPMENTS AND FUTURE PROSPECTS

The most promising gene-therapy concepts, at the present time, concern the following:

- Direct killing of tumour cells with genes delivered by Ad vectors for local management of cancer;

- Delivery of naked DNA by injection or by the gene gun for preventive vaccination against infectious diseases;

- Naked DNA delivery of genes promoting angiogenesis for cardiovascular disorder, and

- AAV delivery for chronic disorders, such as haemophilia and anaemia.

There is no single vector, at present, with genetic utility for all types of diseases and it is unlikely that such a universal vector will emerge in the next few years. Over this period Ad is likely to remain the most suitable vector for *in vivo* therapeutic concepts and requires only short-term expression, especially where a single dose will suffice for clinical benefit. Most of the gene therapies that have progressed into phase II clinical studies involved gene transfer *in vivo*. The trend observed over the past three years towards an increasing proportion of *in vivo* strategies will continue now that vectors giving improved gene transfer *in vivo* (Ad, AAV) are available. It is likely that there will be renewal of interest in *ex vivo* therapy with stem cells. Preparation of haematopoietic progenitor cells is becoming increasingly routine in medical centres. Considerable advances are being made in identifying and propagating new types of human stem cells, including totipotent stem cells derived from embryonic or foetal tissue (Keller and Snodgrass, 1999).

The approach of developing a non-viral vector with large insert capacity, which exploits the site-specific integration machinery of AAV for safe and sustained transgene expression is very attractive, but is at an early stage of development. In conclusion, the most important recent development in gene therapy is the clear demonstration of efficacy in clinical studies, which might lead to a restoration of public and investor confidence in gene therapy.

ARGUMENTS IN FAVOUR OF GENE THERAPY

The central argument in favour of gene therapy is that it can be used to treat desperately ill patients or to prevent the onset of horrible illnesses. Conventional treatment has failed for the candidate diseases for gene therapy, and for these patients, gene therapy is the only hope for a future. The arguments in favour for gene therapy are as follows:

1. germ-line gene therapy offers a true cure; and not simply palliative or symptomatic treatment;
2. germ-line therapy may be the only effective way of addressing some genetic diseases;
3. by preventing the transmission of disease genes, the expense and risk of somatic cell therapy for multiple generations is avoided;
4. medicine should respond to the reproductive health needs of prospective parents at risk for transmitting serious genetic diseases and
5. the scientific community has a right to free inquiry, within the bounds of acceptable human research.

ARGUMENTS AGAINST GENE THERAPY

Gene therapy patients would need to be under surveillance for decades to monitor long-term effects of the therapy on future generations. Arguments specifically against the development of germ-line therapy techniques include the following:

1. germ-line therapy and experiments would involve too much scientific uncertainty and clinical risks and the long-term effects of such therapy are unknown;
2. such gene therapy would open the door to attempts at altering human traits not associated with disease, which could exacerbate problems of social discrimination;
3. as germ-line gene therapy involves research on early embryos and effects their offspring, such research essentially creates generations of unconsenting research subjects;
4. gene therapy is very expensive, and will never be cost-effective enough to merit high social priority and
5. germ-line gene therapy would violate the rights of subsequent generations to inherit a genetic endowment that has not been intentionally modified.

REVIEW QUESTIONS

1. What do you understand by gene therapy? Discuss the principles involved in gene therapy.
2. What are the two major techniques in gene therapy? Discuss their merits and demerits.

3. What are the different transfer vectors that are used in gene therapy? List their advantages and disadvantages.

4. How has gene therapy been used in human medicine? Explain with suitable examples.

REFERENCES

Acsadi, G., Dickson, G. and Love, D.R. (1991). "Human dystrophin expression in MDX mice after intramuscular injection of DNA constructs." *Nature.* 352: 815–818.

Adam, L., Black, P.C., Kassouf, W., Eve, B., McConkey, D., Munsell, M.F., Benedict, W.F. and Dinney, C.P. (2007). "Adenoviral mediated interferon-alpha 2b gene therapy suppresses the pro-angiogenic effect of vascular endothelial growth factor in superficial bladder cancer." *J. Urol.* 177:1900–1906.

Balague, C., Kalla, M. and Zhang, W.W. (1997). "Adeno-associated virus rep 78 protein and terminal repeats enhance integration of DNA sequences into the cellular genome." *J. Virol.* 71: 3299–3306.

Baumgartner, I., Pieczek, A., Manor, O., Blair, R., Kearney, M., Walsh, K. and Isner, J.M. (1998). "Constitutive expression of ph VEGE$_{165}$ following intramuscular gene transfer promotes collateral vessel development in patients with critical limb ischemia." *Circulation.* 97: 1114–1123.

Blaese, M., Blankenstein, T., Brenner, M., Cohen-Haguenauer, O., Gansbacher, B., Russell, S., Sorrentino, B. and Velu. T. (1995). "Vectors in cancer therapy: how will they deliver?" *Cancer Gene Ther.* 2: 291–297.

Boucher, R.C. (1999). "Status of gene therapy for cystic fibrosis lung disease." *J. Clin. Invest.* 103: 441–445.

Deconinck, N., Tinsley, J.M., De-Backer, F., Fisher, R., Kahn, D., Phelps, S. and Davies, K.E. and Gillis, J.M. (1997). "Expression of truncated utrophin leads to major functional improvements in dystrophin-deficient muscles of mice." *Nature. Med.* 3: 1216–1221.

Donahue, B.A., McArthur, J.G., Spratt, S.K., Bohl, D., Lagarde, C., Sanchez, L., Kaspar, B.A., Sloan, B.A. Lee, Y.L., Danos, O. and Snyder, R.O. (1999). "Selective uptake and sustained expression of AAV vectors following subcutaneous delivery." *J. Gen. Med.* 1: 31–42.

Grossman, M., Raper, S.E., Kozarsky, K., Stein, E.A., Engelhardt, J.F., Muller, D., Lupien, P.J. and Wilson, J.M. (1994). "Successful *ex vivo* gene therapy directed to liver in a patient with familial hypercholesterolaemia." *Nature. Genet.* 6: 335–341.

Keller, G. and Grosveld, F. (1999). "Human embryonic stem cells: the future is now." *Nat. Med.* 5: 151–152.

Linder, R.M. and Woo, S.L. (1999). "AAVant-garde gene therapy." *Nat. Med.* 5: 21–22.

Miller, J.M. and Boyce, F.M. (1995). "Gene therapy by and for muscle cells." *Trends Genet.* 11: 163–165.

Prud'homme, G.J., Draghia-Akli, R. and Wang, Q. (2007). "Plasmid-based gene therapy of diabetes mellitus." *Gene Ther.* 14:553–564.

Sebestyen, M.G., Hegge, J.O., Noble, M.A., Lewis, D.L., Herweijer, H. and Wolff, J.A. (2007). "Progress toward a nonviral gene therapy protocol for the treatment of anemia." *Hum. Gene Ther.* 18:269–285.

Wagner, J.A. (1998). "Efficient and persistent gene transfer of AAV-CFTR in maxillary sinus." *Lancet.* 351: 1702–1703.

Wang, R., Doolan, D.L., Le, T.P., Hedstrom, R.C., Coonan, K.M., Charoenvit, Y., Jones, T.R., Hobart, P., Margalith, M., Ng, J., Weiss, W.R., Sedegah, M., de Taisne, C., Norman, J.A. and Hoffman, S.L. (1998). "Induction of antigen-specific cytotoxic T lymphocytes in humans by a malaria DNA vaccine." *Science.* 282: 476–480.

16

BIOTECHNOLOGY AND MEDICINE

INTRODUCTION

With the advent of improved sanitation and living conditions, together with vaccination and antibodies, life expectancy over the last 150 years has rapidly increased from 35 to almost 80 years, with further improvement expected. Nowadays in advanced societies, infectious diseases are no longer the main threat to life but rather it is the chronic diseases (cancer, cardiovascular diseases, Alzheimer's disease, etc.) that plague our increasingly ageing populations. It is now believed that the solution to these chronic diseases could come through **genetic medicine** and that modern biotechnology will play a major role. While this will undoubtedly be true in part, especially where single gene changes are responsible, it must be recognized that many chronic diseases will most probably not have a single, identifiable genetic cause but rather arise from a complex, cascading series of biological events interacting with environmental factors.

The impact of pharmaceuticals on human health care is an area where biotechnological innovations are likely to have the earlier commercial realization. New medical treatments based on biotechnology are appearing almost daily in the market place. These include:

- therapeutic products (hormones, regulatory proteins and antibiotics)
- prenatal diagnosis of genetic diseases
- vaccines
- immunodiagnostics and DNA probes for disease identification and
- genetic therapy (This is the largest commercially developed area of new biotechnology, with massive present and future markets and can only be selectively examined here.)

PHARMACEUTICALS

The vast bulk of pharmaceutical drugs presently on sale are synthetic chemicals derived either directly by chemical synthesis or by chemically modifying molecules derived from biological sources. There can be little doubt that the techniques of molecular biology/genetic engineering will become a dominating factor of drug discovery, design and development. Biotechnology will also accelerate screening, speed bioassays and the production of new drugs, and also explain more accurately how drugs act in the human system. Biotechnology will almost certainly reduce the huge costs presently incurred in product development of new drugs (e.g. costs of discovery, development, scale-up, clinical trials and regulatory paper work).

ANTIBIOTICS

The discovery in 1928 by Alexander Fleming that a fungus called *Penicillium notatum* could produce a compound which was selectively able to inactivate a wide range of bacteria, without unduly influencing the host, set in motion scientific studies that profoundly altered the relationship of humans to the controlling influence of bacterial diseases. From these studies emerged the fungal antibiotics penicillin and cephalosporin, and the actinomycete antibiotics streptomycin, aureomycin, chloramphenicol, tetracyclines and many others. Many bacterial diseases have largely been brought under control by the use of antibiotics. Pneumonia, tuberculosis, cholera and leprosy, to mention only a few, no longer dominate society and, at least in the developed parts of the world, have been relegated to minor diseases. Griseofulvin, an antibiotic active against fungi, has brought great relief to those infected with debilitating fungal skin diseases such as ringworm.

Antibiotics are antimicrobial compounds produced by living microorganisms and are used therapeutically and sometimes prophylactically in the control of infectious diseases. Over 400 antibiotics have been isolated but only about 50 have achieved wide usage. Antibiotics that affect a wide range of microorganisms are termed broad-spectrum antibiotics, for example, chloramphenicol and the tetracyclines, which can control such unrelated organisms as *Rickettsia*, *Chlamydia* and *Mycoplasma* species. In contrast, streptomycin and penicillin are examples of narrow spectrum antibiotics, being effective against only a few bacterial species. Most antibiotics have been derived from the actinomycetes and the mould fungi.

A disquieting observation has been the gradual evolution of drug resistance in many bacteria. The possibility of this acquired resistance being transmitted to another species of bacterium is now real. For example, gonorrhoea (a venereal disease) resistant to treatment with penicillin is now present in 19 countries. It is recognized that the resistance factors are located on plasmids within the bacterium and because of this can be more easily transmitted between organisms. The very core of gene transfer technology derives from this phenomenon. Antibiotic resistance in many well-known diseases is steadily increasing and must cause serious concern in our society.

The addition of relatively small amounts of certain antibiotics (e.g. bacitracin, chlortetracycline, procaine penicillin) in the feed of livestock and poultry led to the production of animals that were healthier, grew more rapidly and achieved marketable weight faster. However, there is now little doubt that the incorporation of medically important antibiotics into feed has led to increased spread of drug-resistant microorganisms, increased shedding of dangerous *Salmonella* bacteria in animal dung, and the transfer of antibiotic residues into human food. As a consequence of the dangers of using antibiotics of human relevance in animal feed, there has been a massive effort to produce antibiotics specifically for animal feed incorporation, so replacing the medically used antibiotics.

VACCINES AND MONOCLONAL ANTIBODIES

During the last 15 years, we have witnessed the unravelling of the bewildering processes of immune response in human and animal systems. When a foreign substance (e.g. a microorganism) enters an animal system, a remarkable chain of reactions is set in motion, which, if successful, will result in the inactivation and exclusion of the invading organism. The foreign molecule is the antigen, which can elicit a counteracting response, and the antibody is formed in the host system. The antibody-producing cells recognize the shape of particular determinant groups of the antigen and produce specific antibodies in order to neutralize and eliminate the foreign substance.

The ability to stimulate the natural antibodies by vaccines has long been known. Vaccines are preparations of dead microorganisms (or fractions of them), or living attenuated or weakened microorganisms that can be given to humans and animals to stimulate their immunity to infection. In this way they mimic infectious agents without the pathogenic consequences and elicit in the body, protective immune responses. When used on a large scale, vaccines have been a major force in the control of microbial diseases within communities.

Vaccines have been developed against many microbial diseases. However, the success and persistence of the antimicrobial effect varies widely between types of vaccines. Thus, a vaccine for poliomyelitis has almost eliminated this disease on a world scale, while vaccines against typhoid and cholera are still unsatisfactory. A massive global effort is now in progress to develop a vaccine against human immunodeficiency virus (HIV), which is responsible for acquired immunodeficiency syndrome (AIDS).

Vaccines are being developed by recombinant DNA technology against the influenza virus, poliovirus, hepatitis B virus, herpesvirus and more recently HIV. The biggest opportunities exist for diseases such as AIDS or herpes where neither a vaccine nor a cure is yet available.

A significant new development in medically related biotechnology has been the ability to produce monoclonal antibodies. A major advancement of this technique is that when antibody-producing cells are immortalized and stabilized, the secreted antibodies will always be the same from that particular cell line and can be fully characterized to assess their suitability for different applications. In this way, suitable antibodies can be produced, scaled-up in large quantities, allowing much greater standardization for diagnostic applications. Monoclonal antibodies are now finding wide applications in diagnostic techniques requiring highly specific reagents for the detection and measurement of soluble proteins and cell-surface markers in blood transfusions, haematology, histology, microbiology and clinical chemistry.

Monoclonal antibodies may also be used in the treatment of tumours and possibly to carry cytotoxic drugs directly to the tumour site. In particular, monoclonal antibodies have found ready application in *in vitro* diagnostic products that do not need such rigorous safety testing (*See also* Chapter 4).

BIOPHARMACEUTICALS

The vast majority of pharmaceutical products are compounds derived from synthetic chemical processes, from naturally occurring sources (plants, microorganisms), or are combinations of both. Biopharmaceuticals are considered to be recombinant protein drugs, recombinant vaccines and monoclonal antibodies (for therapeutic roles). Biopharmaceuticals are becoming increasingly relevant in biological applications but are still only a small part of the pharmaceutical industry. Genetic engineering is now increasingly being recognized as a practical means of providing some of these scarce molecules in unrestricted quantities. In practice, this involves inserting the necessary human-derived gene construct into suitable host microorganism that will produce the therapeutic protein (biopharmaceutical) in quantities related to the scale of operation (*See also* Chapter 6 Section 7.1).

GENE THERAPY

Undoubtedly, the most far-reaching and controversial area of genetic engineering of humans is gene therapy. This is the treatment of diseases by the transfer and

expression of genetic material in a patient's cells in order to restore normal cellular functions. It is however, essential to distinguish between germ cell gene therapy and somatic cell gene therapy. In germ cell gene therapy, changes are directed at the individual's genetic make-up and can be passed on to the offspring. Ethics and practical wisdom ensures that this type of therapy will not be permitted in any country, in the foreseeable future. In contrast, in somatic cell gene therapy, functioning genes are introduced into body cells that lack them. The effects of this therapy are confined to the person undergoing the treatment and are not passed on to the offspring.

The main thrust of gene therapy has been directed at correcting single-gene defects (mutations), such as cystic fibrosis and haemophilia, that have been observed in families by their Mendelian pattern of inheritance. At present most genetic diseases have no effective treatment and so gene therapy could offer hope for so many people. Somatic cell gene therapy for complex multifactorial diseases, e.g. Parkinson's disease and cancer, must be a long way off. In many of these diseases there can be many genes involved as well as in interaction with environmental factors.

Gene therapy is a complex series of events relying heavily on new biotechnological techniques. Therapy will require a full understanding of the mechanism by which the defective or unusual gene exerts its effect on the individual, an ability to switch off the defective gene and to substitute a healthy gene copy. It is truly a multidisciplinary activity involving skills in molecular biology, cell biology, virology, pharmacology, clinical application and patient interaction (*See also* Chapter 15).

GENETIC MANIPULATION

The vast majority of genetic engineering techniques use bacteria. It is now a totally routine matter to insert novel gene into bacteria, with a view to studying the structure and function of the gene, to monitor the expression or presence of the gene in various cells, to generate large amounts of recombinant protein or to manipulate or alter the gene. This technology has revolutionized the practice of biological and medical science, and has provided the clinician with a new class of therapeutic agent.

Transgenic Animals

In this approach a particular gene is injected into the fertilized egg of the animal, which is then implanted into a female animal (Kappel *et al.*, 1994). In a proportion of eggs the gene integrates into the DNA, and the resulting animals have altered genome that contains the gene. Careful selection and appropriate breeding, of these animals lead to the generation of a transgenic line of animals carrying the gene (*See also* Chapter 8).

Transgenic livestock and transgenic aquatic species have been generated with increased growth rates, enhanced lean muscle mass, enhanced resistance to disease or improved use of dietary phosphorus to lessen the environmental impacts of animal manure. Transgenic poultry, swine, goats, and cattle also have been produced that generate large quantities of human proteins in eggs, milk, blood, or urine, with the goal of using these products as human pharmaceuticals. Examples of human pharmaceutical proteins include enzymes, clotting factors, albumin, and antibodies. The major factor limiting widespread use of transgenic animals in agricultural production systems is the relatively inefficient rate (success rate less than 10 per cent) of production of transgenic animals.

Knock-out Animals

A **gene knockout** is a genetically engineered organism that carries one or more genes in its chromosomes that have been made inoperative (have been "knocked out" of the organism). This is done for research purposes. Also known as **knockout organisms** or simply **knockouts**, they are used in learning about a gene that has been sequenced, but which has an unknown or incompletely known function. Researchers draw inferences from the difference between the knockout organism and normal individuals. A **knockout mouse** is a genetically engineered mouse that has had one or more of its genes made inoperable through a gene knockout. Knockout is a route to learning about a gene that has been but has an unknown or incompletely known function. Mice are the laboratory animal species most closely related to humans in which the knockout technique can be easily performed, so they are a favourite subject for knockout.

In knockout technology, a genetic construct that causes the removal of a particular gene is injected into an embryonic cell line (Galli-Taliadoros *et al.*, 1995). Cells are selected that have had the gene removed (or knocked out), and are then implanted into blastocysts (early embryos). The embryonic cells mature in the blastocysts, where they form part of the new animal. This animal is a chimera, as it is derived from a mixture of the embryonic cell line (containing the knockout) and the cells comprising the blastocysts. Appropriate breeding can lead to the generation of a "knockout" line of animals in which the desired gene is removed.

The utilities of the production of gene knockout mice are many-fold. First, gene knockout mice can be produced as an animal model for human diseases. These gene knockout mice can then be used to explore potential intervention

strategy, including pharmacological and genetic therapy approaches, for treatment of specific metabolic diseases. A classical example is the production of LDL receptor gene knockout mice as an animal model for type II familial hypercholesterolaemia (FH) (Ishibashi *et al.*, 1994). Patients with this genetic disorder suffer coronary heart disease at an early age. The disease is characterized by severely elevated plasma LDL and cholesterol levels due to defective LDL receptor gene expression. The availability of the LDL receptor knockout mice also provided investigators with the opportunity to assess the role of the LDL receptor in chylomicron-remnant metabolism.

A second utility for the production of knockout mice is to explore physiological function and significance of specific genes. A classical example is the apoE (apolipoprotein) gene knockout mice (Plump *et al.*, 1992; Zhang *et al.*, 1992). Previous studies showed an elevation of plasma apoE level in animals when fed an atherogenic high-fat/high- cholesterol diet. The increase in apoE level appears to correlate with development of atherosclerotic lesions in several cholesterol-fed animal models. While these studies suggest a potential contributory role of apoE in atherogenesis, *in vitro* cell culture experiments showed that this apolipoprotein facilitated cholesterol efflux from lipid-laden macrophages/foam cells and increases catabolism of chylomicron remnants and cholesterol-rich VLDL (Mahley 1988). Thus, the *in vitro* studies suggest that apoE may serve a protective role against atherosclerosis, and its increased level after cholesterol feeding may represent a systematic response to the diet. The exact role of apoE in atherosclerosis was clarified by production and characterization of apoE knockout mice in two separate laboratories. Results of both studies indicated that apoE gene deletion resulted in an animal with severe atherosclerosis, thus confirming the protective role of this apolipoprotein against atherosclerosis (Plump *et al.* 1992., Zhang *et al.*, 1992).

Animal biotechnology also can knockout or inactivate a specific gene. Knockout technology creates a possible source of replacement organs for humans. The process of transplanting cells, tissues, or organs from one species to another is referred to as **"xenotransplantation."** Currently, the pig is the major animal being considered as a xenotransplant donor to humans. Unfortunately, pig cells and human cells are not immunologically compatible. Pig cells express a carbohydrate epitope (α-1,3 galactose) on their surface that is not normally found on human cells. Humans will generate antibodies to this epitope, which will result in acute rejection of the xenograft. Genetic engineering is used to knock out or inactivate the pig gene (α-1, 3 galactosyl transferase) that attaches this carbohydrate epitope on pig cells. Other examples of knockout technology in animals include inactivation of the prion-related peptide (PRP) gene that may generate animals resistant to diseases associated with prions (bovine spongiform encephalopathy [BSE], Creutzfeldt–Jakob Disease [CJD], scrapie, etc.).

Xenotransplantation (*xeno-* from the Greek meaning "foreign") is the transplantation of living cells, tissues or organs from one species to another such as from pigs to humans. This includes body fluids, cells, tissues or organs that have had *ex vivo* contact with the live cells, tissues or organs of a different species. Such cells, tissues or organs are called **xenografts** or **xenotransplants**. In contrast, allotransplantation is the transplantation of cells, tissues, and organs between members of the same species.

Human xenotransplantation offers a potential treatment for end-stage organ failure, a significant health problem in parts of the industrialized world. It also raises many novel medical, legal and ethical issues. A continuing concern is that cows and pigs have different lifespans than humans and their tissues age at a different rate. Disease transmission (xenozoonosis) and permanent alteration to the genetic code of animals are a cause for concern. Because there is a worldwide shortage of organs for clinical transplantation, about 60% of patients awaiting replacement organs die on the waiting list. There are only a few published successful xenotransplant procedures. Some patients who were in need of liver transplants were able to use pig livers that were on a trolley by their bedside successfully until a proper donor liver was available. Some recipients of pig neural cells with paralysis due to stroke (CVA) and Parkinson's disease have experienced dramatic improvements.

Potential Animal Organ Donors

Since they are the closest relatives to humans, non-human primates were first considered as a potential organ source for xenotransplantation to humans. Chimpanzees were originally considered to be the best option since their organs are of similar size, and they have good blood type compatibility with humans. However, since chimpanzees are listed as an endangered species, other potential donors were sought out. Baboons are more readily available, however they are also not practical as potential donors. Problems include their smaller body size, the infrequency of blood group O (the universal donor), their long gestation period, and they typically produce few offspring. In addition, a major problem with the use of non-human primates is the increased risk of disease transmission, since they are so closely related to humans (Michler, 1996). Xenotransplantation between baboons and humans raises the issue of xenozoonoses. The organisms of greatest concern are the herpesviruses and retroviruses, which can be screened for and eliminated from the donor pool. Others include *Toxoplasma gondii*, *Mycobacterium tuberculosis*, and encephalomyocarditis virus. Pigs are currently thought to be the best candidates for organ donation. The risk of cross-species disease transmission is decreased because of their increased phylogenetic distance from humans (Dooldeniya and Warrens, 2003). They are readily available, their organs are anatomically comparable in size, and new infectious agents are less likely since they have been in close contact with humans through domestication for many years (Taylor, 2007). Current experiments in xenotransplantation most often use pigs as the donor, and baboons as human models.

Somatic Gene Transfer

The transgenic and knockout technology described above involved the transfer of DNA into the germ line cells (egg or sperm) of the animals. The alterations in the DNA are then inherited from one generation to the next. An alternative approach is to insert genes into somatic cells of animals, that is non-germ line cells that comprise the majority of the animal (Kerr and Mule, 1994). In this technique, a vector—normally a virus—is used to transfer the nucleic acid into the appropriate cells. The genes are expressed only in those cells. Thus, if the vector is targeted to liver cells, only the liver will produce the protein. The expression can be transient; that is; after several weeks the new DNA is thrown out by the cells and they revert to their natural state, or with some viruses, the new DNA can integrate into the cellular DNA and the expression is permanent. The important difference between somatic gene transfer and the transgenic or knockout technologies is that the alterations in the DNA are not present in the germ line of the animal. Thus, the animal is unable to pass the new genotype on to its offspring, and when the animal dies so do the alterations in the DNA.

Another application of animal biotechnology is the use of somatic cell nuclear transfer to produce multiple copies of animals that are nearly identical copies of other animals (transgenic animals, genetically superior animals, or animals that produce high quantities of milk or have some other desirable trait, etc.). This process has been referred to as cloning. To date, somatic cell nuclear transfer has been used to clone cattle, sheep, pigs, goats, horses, mules, cats, rats, and mice. The technique involves culturing somatic cells from an appropriate tissue (fibroblasts) from the animal to be cloned. Nuclei from the cultured somatic cells are then microinjected into an enucleated oocyte obtained from another individual of the same or a closely related species. Through a process that is not yet understood, the nucleus from the somatic cell is reprogrammed to a pattern of gene expression suitable for directing normal development of the embryo. After further culture and development *in vitro*, the embryos are transferred to a recipient female and ultimately will result in the birth of live offspring. The success rate for propagating animals by nuclear transfer is often less than 10 per cent and depends on many factors, including the species, source of the recipient ova, cell type of the donor nuclei, treatment of donor cells prior to nuclear transfer, the techniques employed for nuclear transfer, etc.

APPLICATIONS OF ANIMAL BIOTECHNOLOGY IN MEDICINE

There are four major applications of animal biotechnology in medicine. The first, which to date has been by far the most important has been to use transgenic or knockout animals to investigate physiological and pathological processes, and to develop model systems of disease that allow development of therapeutic or diagnostic reagents. The second application is to use transgenic animals to produce a recombinant protein. The application is to produce material that will be useful for therapy; the most notable example of this is to prepare animals for use as donors of organs for transplantation. Finally biotechnology can be used directly as a form of therapy, so-called **gene therapy**. In these cases somatic gene transfer is used to introduce genes into appropriate somatic cells in the animals and the expression of the recombinant protein is designed to have a therapeutic effect, either by immunizing the animal, replacing a missing protein, or augmenting a natural physiological process.

Animal Biotechnology as a Scientific Tool

The most common use of animal biotechnology has been for the study of natural physiological processes, and for the pathology that occurs when these processes break down. One case in point is the increased understanding of the role of oncogenes in the development of cancer. Oncogenes are normally cellular or virally derived genes that, either through mutation or inappropriate expression, lead to neoplastic transformation of cells. Expression of oncogenes in an inappropriate manner in transgenic mice, and the subsequent analysis of resulting tumours, has proved invaluable in our understanding of the process of tumorigenesis. One example is the oncogene *bcl12*, which is involved in the most common chromosomal translocation is B-cell follicular lymphomas (Bakhshi *et al.*, 1985). Expression of *bcl12* in transgenic animals led to follicular proliferation of the B cells, similar to that seen in human patients. Further analysis of the behaviour of B cells from these mice showed that the role of *bcl12* was not to increase the proliferative rate of the B cells, but to prevent their suicide by apoptosis. Further evidence to this effect was derived by knocking out the *bcl12* gene (Veis *et al.*, 1993). In these animals there were several defects, including extensive apoptosis of B cells. These data helped to demonstrate that the role of *bcl12* is to prevent cell death, and that the development of some malignant states can, in part, be ascribed to the failure of a cellular suicide pathway (Cory, 1995). This has very important implications for the development of novel therapeutic agents for cancer, and in addition, such animals can provide useful models for the development and testing of such molecules.

Biotechnology to Produce Recombinant Pharmaceuticals

One promising application of biotechnology to medical science is in the production of therapeutic and diagnostic reagents. At the present time, many of the proteins used in clinical practice are isolated from material derived from animals (e.g. insulin and most antisera) or humans (e.g. blood products and some antisera). These methods of producing pharmaceutical agents are not without drawbacks; materials isolated from humans, runs the risk of containing pathogenic viruses and other organisms. This risk is serious: there are approximately 10,000 cases of HIV infection in the USA caused by blood transfusion and many haemophiliacs were similarly infected through factor VIII (Darby *et al.*, 1989). In addition, the use of human growth hormone or gonadotrophin prepared from cadavers has been associated with the transmission of Creutzfeldt–Jakob disease (Brown *et al.*, 1994). The use of material derived from animals reduces this risk. However, it is often different from the equivalent human protein, which may not function well in humans, or may be recognized as foreign by the patient's immune system. Thus, patients treated with porcine insulin can develop an immune response against the pig protein, necessitating their conversion to recombinant human insulin.

Much human therapeutics, such as monoclonal antibodies used to treat cancer, require correct configuration and animal-specific modifications to be effective. This means that the production of many human protein drugs cannot be carried out in bacteria or plants, but rather is confined to the cells of mammals. Mammalian cell culture, however, is very expensive and low yields limit the amount and number of different proteins that can be developed. Consequently, researchers are attempting to address this problem by producing therapeutic proteins in the milk of domestic farm animals. The major function of the milk-producing mammary gland is to produce proteins. Transgenic animals, that is animals carrying an additional segment of DNA encoding the therapeutic protein, produce this one additional protein in their milk. These animals and their milk are not intended to enter the food supply, but rather to produce these potentially life-saving proteins in their milk. With hundreds of protein therapeutics currently in clinical trials, transgenic animals may become an important source of these protein drugs as they become available to the patient.

One approach is to use transgenic animals, in which the gene encoding the desired protein has been inserted, to make large amounts of protein. A number of animals are being developed as sources of recombinant protein, including pigs, goats, cows and sheep (Gershon, 1991, Valander *et al.*, 1992; Kumar, 1995). In many cases the gene is modified such that expression is restricted to the lactating breast, and so the protein is secreted into the milk. This has the advantage of reducing potential for harm to the animal and of greatly simplifying the harvesting of the molecule, using well-established agricultural processes. Figure 16.1 shows expression of therapeutic portions in mammary gland and excretion of therapeutic protein in milk.

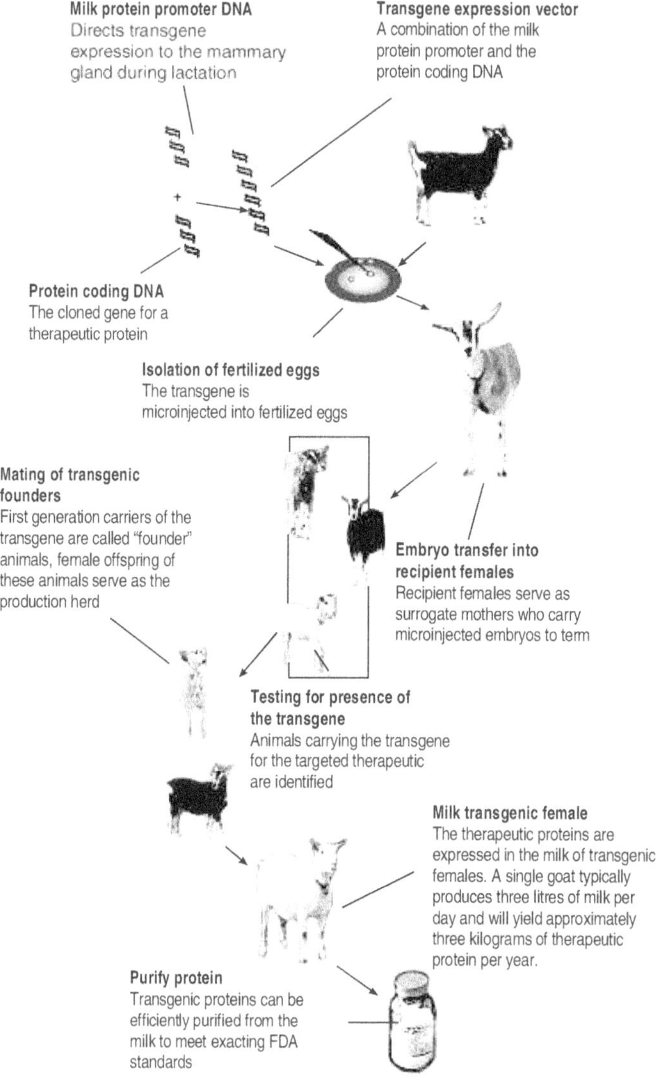

Figure 16.1 Biopharmaceuticals in animals—Expression of therapeutic proteins in mammary cells and excretion of therapeutic protein in milk.

Once the transgenic animal has been produced then the goal is to produce relatively cheap protein in large amounts. Currently a number of drugs are being developed for production in this way, including factor IX for blood clotting, α1 anti-trypsin for emphysema, haemoglobin for the development of blood substitutes and tissue plasminogen activators for treatment of blood clots (Valender *et al.*, 1992; Yom and Bremel 1993; Kumar, 1995). If such pharmaceutical drugs can be produced relatively inexpensively, this will increase their availability and use, as well as reduce health care costs.

Genetically Engineered Animals as Therapeutic Tools

The third area in which animals are just beginning to have an impact in medicine is the direct supply of material for treatment of patients. The main example of this is the use of animals as donors for transplantation (Concar, 1994; Kaufman *et al.*, 1995). The major limitation of transplantation is the supply of organs. The vast majority of organs donated as cadaveric, that is from dead donors frequently those who have died as a result of trauma. However, the supply of organs is insufficient to meet the demand. The other possibility is to use organs from other animals. This is termed xenotransplantation, as the organ transplanted is xenogenic with respect to the host. Following the development of conventional allotransplantation, kidneys, from chimpanzees or baboons were transplanted into patients with renal failure. In most cases the transplanted organ functioned well, in one case for more than 6 months. This gave enough time for a conventional kidney donor to be found.

These, and similar, attempts at heart and liver xenotransplantation all used primates as donors (Bernard *et al.*, 1977; Starzl *et al.*, 1993). This is logical considering the close evolutionary relationship between humans and other primates, which reduces the immunological difference between primate donor organs and the human recipient. The hunt for potential xenogeneic donors has recently concentrated on the pig (Kaufman *et al.*, 1995). This animal is consumed as a food in large amounts, reducing the public concern over such an approach. In addition, pigs are easy to raise, have big litters and grow rapidly. The major trouble with xenotransplantation is the rapid rejection of the organ.

Somatic Gene Transfer

The final application of animal biotechnology is somatic gene transfer (Verma, 1990; Kerr and Mule, 1994). This involves the transfer of genetic material into somatic cells of animals, which are not passed onto the offspring. The main use of this approach will be for gene therapy of disease, though it also has a role as a research tool. Historically, somatic gene transfer has been widely used in humans and other animals to immunize; a virus containing genes from a pathogenic organism is injected into the patient where it infects cells, produces protein derived from the pathogen and elicits a protective immune response against the pathogen.

The area of gene therapy that is exciting more public concern is the transfer of genes that interfere with physiological or pathological processes. The first use of gene therapy to treat disease was for adenosine deaminase (ADA) deficiency. This is a disease in which the patients lack the ability to make the ADA enzymes. As a result, a toxic metabolite builds up, which poisons the cells of the immune system. The affected children are profoundly immunosuppressed and, if not kept in a specialized environment, they die of infection (Cournoyer and Caskey, 1993). The gene therapy trials involve the infection of bone marrow cells from the children with a virus engineered to contain the *ADA* gene, thereby replacing the missing enzyme. This treatment has caused significant improvement in the immune system of patients. Increasingly, such treatments are being devised and tested for a variety of diseases, including genetic disorders such as cystic fibrosis and muscular dystrophy (Kay and Woo, 1994), cancer (Culver and Balaese, 1994), neurological diseases (Friedmann, 1994), infectious diseases (Gilboa and Smith, 1994) and transplantation (Larkin *et al.,* 1996) (*See also* Chapter 15).

ENVIRONMENTAL BIOTECHNOLOGY

Throughout the world there has been increase in urbanization and industrial development, public concern is mounting over the state of the environment and much attention is now being given to improving the environment for future generations. To achieve this, there has been, particularly in developed nations, major environmental legislations directed towards liquid, solid and hazardous wastes. In most developing countries, the situation is less encouraging where financing is limited, or not available, for the construction of water and waste treatment facilities and there is a shortage of trained personnel to operate the systems. Furthermore, in many developing countries, there is a lack of official regulations and control systems, no administrative bodies are responsible for waste control and little obligation for existing and emerging industries to dispose of waste properly.

Environmental biotechnology is when biotechnology is applied to and used to study the natural environment. Environmental biotechnology could also imply that one should try to harness biological process for commercial uses and exploitation. The International Society of Environmental Biotechnology defines environmental biotechnology as "the development, use and regulation of biological systems for remediation of contaminated environments (land, air, water), and for environment-friendly processes (green manufacturing technologies and sustainable development)".

Environmental biotechnology offers ways to make industrial processes work more efficiently and create less pollution.

- Biotechnology may make available biological replacements for synthetic chemicals. For example, it is possible to make a variety of plastics from plant sugar rather than petrochemicals by using specially tailored yeasts and microorganisms. The advantage of these products is that they are biodegradable.

- It is expected that bioleaching will progressively displace chemicals used in recycling or bleaching paper. Many laundry detergents use enzymes (although in some cases these may be produced using gene technology) to replace phosphate detergents.

- Some heavy-duty stain removing bacteria have even been found in heavily-alkaline lakes where they have survived by learning to break down the toxins in their environment.

- Scientists are looking for microorganisms in hostile natural environments (for example very hot, cold or oily places) to use the talents they have developed for industrial uses.

Waste generation is a side-effect of consumption and production activities and tends to rise with the level of economic advancement. Wastes arise from domestic and industrial activity, e.g. sewage, waste waters, agriculture and food wastes from processing, wood wastes and an ever-increasing range of toxic industrial chemical products and by-products.

WASTE-WATER AND SEWAGE TREATMENT

Growth in human populations has generally been matched by a concomitant formation of a wider range of waste products, many of which cause serious environmental pollution if they are allowed to accumulate in the ecosystem. In rural communities recycling of human, animal and vegetable wastes have been practices for centuries, providing in many cases valuable fertilizers or fuel. However, it was also a source of disease to humans and animals by residual pathogenicity of enteric (intestinal) bacteria.

Better Treatments for Solid Waste and Wastewater

- Most sewage treatment plants traditionally use a combination of chemical, physical and microbial treatment to break down waste. It is now possible to use modern biotechnology techniques to analyse the conditions needed to optimise the performance of the bugs and the systems, and to tailor the technology to different uses, from backyard septic tanks to large-scale intensive animal farm such as piggeries.

- The CSIRO has used enzymes to treat run-off from farms that contains pesticides, to avoid contaminating rivers and downstream farms.

- Sometimes the 'treatment' of waste produces valuable products, and modern high-tech composting factories can turn tonnes of organic garbage into precious soil in just days.

Bioremediation—Cleaning up Contamination

- Naturally occurring microorganisms, mainly bacteria and fungi, are being used to help clean up some of Australia's 60,000 sites contaminated by heavy metals, acids, petroleum derivatives, chlorinated solvents and explosives.

- Certain plants have also been found to absorb toxic metals such as mercury, lead and arsenic from polluted soils and water, and scientists are hopeful that they can be used to treat industrial waste.

- Oil spills, from small industrial puddles to massive ocean oil spills, are often treated with oil-eating bacteria.

Tracking the Health of the Environment Through Biomonitoring

Animals react to their environments, and can be used to gauge problems.

- The livers of sand flathead have proven to be useful indicators of municipal and industrial pollution from dioxins and other hazardous chemicals.

- Two frequently harvested species of marine molluscs, the Sydney rock oyster and the mussel, may be useful "bioindicators" for the heavy metal toxins, zinc and cadmium.

Biomass Energy

'Biomass' is plant and animal material that can be used as an energy source, from the traditional wood to waste material such as bagasse from sugar cane, to specially grown energy crops that can be converted to ethanol using modern biotechnology techniques and used with petrol in vehicles.

- Several Australian companies are planning to build biodiesel plants that convert vegetable oil or abattoir by-products into diesel fuel. Another company has a technology that converts green waste from rubbish tips into electricity.

- Even human waste can be converted into a combustible fuel.

Clever Plants

Genetic engineering is not the only tool for creating novel plants with special traits. Conventional breeding is still used for developing useful plants, and which can be done with fewer risks.

- By cross breeding, scientists are developing wheat plants to increase their salt tolerance so they can grow in areas of greater salinity.

- Eucalyptus is also being cross-bred to introduce natural genes from other species to create plants that tolerate salinity and dryland conditions.

Genetic Engineering for Environmental Solutions

Many of the biotechnology applications described above can be varied or refined using genetic engineering.

- CSIRO and Orica Australia Ltd. are using gene technology to develop enzyme products that detoxify pesticide residues, which in Australia would be of particular value to the cotton, horticultural and rice industries.

- In a variation on bioindicators, bacteria have been genetically modified as "bioluminescors" that give off light in response to several chemical pollutants. These are currently being used to measure the presence of some hazardous chemicals in the environment.

- Other genetic sensors that can be used to detect various chemical contaminants are also being trialled. Some can be used to track how pollutants are naturally degrading in ground water.

- Many genetically modified plants have environmental benefits such as reduced pesticide use.

A variety of biological treatment systems have been developed, ranging from cesspits, septic tanks and sewage farms to gravel beds, percolating filters and activated sludge processes coupled with anaerobic digestion. The primary aims of all of these systems or bioreactors is to alleviate health hazards and to reduce the amount of biologically oxidizable organic compound, producing a final effluent or outflow that can be discharged into the natural environment without any adverse effects. Another important means of degrading dilute organic liquid waste is the percolating or trickling filter bioreactor. In this system the liquid flows over a series of surfaces, which may be stones, gravel, plastic sheets, etc., on which attached microbes that remove organic matter for essential growth.

BIOTECHNOLOGY IN THE AGRICULTURE AND FORESTRY INDUSTRIES

Agriculture is the world's largest, single industry and in advanced societies such as the USA, agriculture contributes over 20% of gross natural product. In developed economies, agriculture relies heavily on technology to achieve productivity and profitability. Many aspects of modern biotechnology are now being applied increasingly to agriculture. Genetic engineering is creating a revolution in agriculture allowing an ever-increasing range of plants and animals. Agricultural biotechnology will allow higher quality standards with lower costs of production. The human race is totally dependent on agriculture and as world populations continue to expand there must be continuous reassessment of agricultural practices to optimize their efficiency.

PLANT BIOTECHNOLOGY

Plants are the primary source of food for the human race and only by correct management of plant agriculture can the present human populations continue to be fed. The flow of energy from sunlight through plant photosynthesis is at the heart of the importance of plants in the world economy.

The first practical system for genetically manipulating plants occurred in 1983 with the discovery of the ability of the bacterium *Agrobacterium tumifaciens* to transfer part of its Ti plasmid into the host plant genome. In this way foreign genes can be inserted into the plasmid DNA and then integrated into the plant genome. Transformants can be identified by selection methods, adult plants reconstituted from the transfected cells and the new genetic material transmitted as a Mendelian trait.

Today, biotechnology is being used as a tool to give plants new traits that benefit agricultural production, the environment, and human nutrition and health. The goal of plant breeding is to combine desirable traits from different varieties of plants to produce plants of superior quality. This approach to improving crop production has been very successful over the years. For example, it would be beneficial to cross a tomato plant that bears sweeter fruit with one that exhibits increased disease resistance. To do this, it takes many years of crossing and backcrossing generations of plants to obtain the desired trait. To genetically modify a plant, the thousands of bases of DNA comprising an individual gene are transferred into an individual plant cell where the new gene becomes a permanent part of the cell's genome. This process makes the resulting plant "transgenic." Transfer of DNA into plant cells is done using various "transformation" techniques that are the result of discoveries in basic science.

Input Traits

An "input" trait helps producers by lowering the cost of production, improving crop yields, and reducing the level of chemicals required for the control of insects, diseases, and weeds.

Input traits that are commercially available or being tested in plants are:

- Resistance to destruction by insects
- Tolerance to broad-spectrum herbicides
- Resistance to diseases caused by viruses, bacteria, fungi, and worms
- Protection from environmental stresses such as heat, cold, drought, and high salt concentration

Output Traits

An "output" trait helps consumers by enhancing the quality of the food and fibre products they use.

Output traits that consumers may one day be able to take advantage of are the following:

- Nutritionally enhanced foods that contain more starch or protein, more vitamins, more antioxidants (to reduce the risk of certain cancers), and fewer *trans*-fatty acids (to lower the risk of heart disease)
- Foods with improved taste, increased shelf-life, and better ripening characteristics
- Trees that make it possible to produce paper with less environmental damage
- Nicotine-free tobacco
- Ornamental flowers with new colours, fragrances, and increased longevity

"Value-added" Traits

Genes are being placed into plants that completely change the way they are used.

Plants may be used as "manufacturing facilities" to inexpensively produce large quantities of materials including:

- Therapeutic proteins for disease treatment and vaccination
- Textile fibres
- Biodegradable plastics
- Oils for use in paints, detergents, and lubricants

Plants are being produced with entirely new functions that enable them to do things such as detect and/or dispose of environmental contaminants like mercury, lead, and petroleum products.

Gene transfer into crop plants to impart insect or microbial resistance is a major new area of research into plant protection. Figure 16.2 shows the production of biopharmaceuticals in plants by transgenic methodology.

Figure 16.2 Biopharmaceuticals in plants—Introduction of human and animal genes into plants like sugar cane for their expression in transgenic plants

ANIMAL BIOTECHNOLOGY

Animal agriculture in the form of cattle, pigs, sheep, poultry and fish represents major aspects of food production worldwide. While many of these animals are produced for their meat alone, others contribute to human nutrition by way of milk and egg production. In the developed world, animal production is highly intensified and technologically driven.

Genetic Engineering for Transgenic Animals

Selective breeding is a painfully slow process and especially with larger animals with long gestation periods, can take many years to establish desired phenotypic changes. However, the advent of recombinant DNA technology and its application to animal breeding programmes could greatly increase the speed and range of selective breeding. The first recorded examples of the transfer of a foreign gene into an animal by recombinant DNA technology was the insertion and expression, into the mouse genome of a rat gene for growth hormone. The subsequent progeny were all much larger than the parents. This 'super mouse' gained much public attention as it was the first example of a transgenic animal.

Some of the main opportunities where this new technology can be envisaged with animal breeding programmes are listed below.

- Efficiency of meat production
- Improved quality of meat
- Milk quality and quantity
- Egg production
- Wool quality and quantity
- Disease resistance in animals
- Production of low cost pharmaceuticals and biologicals

Of particular relevance will be improvements in meat production from a wide range of farmed animals including fish, improved milk yields and quality and disease-free animals. Undoubtedly the most unusual but commercially feasible project is the use of certain lactating animals such as sheep, pigs, rabbits and cows to produce novel secretions of human proteins in their milk, which can then be extracted and used pharmaceutically.

How can a novel DNA be incorporated into animal genomes and then stably inherited into the offspring? At the present time the most successful method for gene transfer into livestock is by microinjection into the pronucleus of fertilized eggs. Microinjection techniques make use of finely constructed glass needles that allow the injection of purified DNA into the fertilized eggs of the chosen species. The eggs are then surgically transferred into hormonally synchronized surrogate mothers. Unlike mice and pigs, in sheep and cattle litter size is limited to one or two and, therefore, large numbers of the latter animals have to be employed as recipients for the microinjected eggs.

The sequences necessary to establish transgenic animals include the following:

1. Identification and construction of foreign gene (genetic engineering)
2. Microinjection of DNA directly into pronucleus of a single fertilized egg
3. Implantation of these engineered cells into surrogate mothers
4. Bringing the developing embryo to term
5. Providing that the foreign DNA has been stably and heritably incorporated into the DNA of at least some of the newborn offspring
6. Demonstrating that the gene is regulated well enough to function in its new environment

Transgenic pigs, sheep and cattle have been obtained, although the frequency of success is only about 1% compared with 2–5% with mice. This low efficiency of the technology will continue to exert some limitation to wider acceptance. However, with fish, the eggs are fertilized externally, thus eliminating many of the complicated techniques required in mammals to harvest ova, fertilize them and then introduce the embryos into foster mothers. Successful fish transgenics can be as high as 70%.

A novel and commercially realistic use of transgenic animals is the production of human proteins/pharmaceuticals in transgenic lactating animals. Transgenic constructs that allow the mammary glands of lactating animals to secrete high value human proteins are now possible and will undoubtedly be the first truly commercial use of transgenic animals for product formation. The animals will in fact become bioreactors producing pharmaceutical products previously only produced in culture by transgenic microorganisms. Gene constructs for human coagulation factor IX (some haemophiliacs lack a blood-clotting agent called factor IX) have been successfully inserted into the sheep genome and while expression levels are still low, factor IX is present and the trait is heritable. The potential of transgenic animals to secrete a wide range of commercially valuable healthcare products is almost unlimited and should be realized in the future.

Genetically Engineered Hormones and Vaccines

The pituitary gland of animals secretes growth hormones that can have major influence on how the animal grows and, in lactating animals, on milk production. In the 1980s the gene responsible for bovine growth hormone (somatotropin or BST) production was successfully isolated and transferred into bacterial cells to produce large quantities of BST. When cows were injected with about 30 mg of BST, there was significant increase in milk production (10–30%) but continued increased yields depended on regular injections.

BST is the first genetically engineered product in agriculture that has been intensively examined for its economic impact. This is due mainly to the vast importance of milk to most Western economies and the positive product image of milk of human health. Animal welfare is also a major current concern. From a scientific point of view, BST has been clearly demonstrated to be a safe product and is now permitted in many countries, in particular the USA, where it is marketed by Monsanto under the trade name Posilac. But many consumer organizations continue to oppose the use of this method.

Production of genetically engineered animal vaccines has been a major, if somewhat unheralded, success story in biotechnology. Numerous vaccines have been developed for specific cattle, pig poultry, sheep and fish diseases. The commercial success is considerable, though not dramatic, and has led to much reduction in animal hardship.

DIAGNOSTICS IN AGRICULTURE

Immunoassays in general, but specifically those using monoclonal antibodies, are widely recognized for their commercial success in clinical and veterinary diagnostics. Indeed, these diagnostic tests make up a major part of new biotechnological products presently on the market. Nucleic acid probe technology is based on the principle of hybridization of complementary sequences of DNA or of DNA and RNA. The respective nucleotide strands must have exact corresponding sequences of nucleotides for exact hybridization or alignment to occur; thus a given strand can hybridize only with its complementary strand. This high level of specificity has now been directed to identify microorganisms in complex mixtures—the DNA probe or hybridization assay.

Using these diagnostic methods it is now possible to detect microbial diseases in humans at very low levels of infection in body fluids or tissues and to be able to isolate individuals before it becomes infectious. Early diagnosis can be an essential prerequisite for containment and elimination of infectious diseases. One of the largest areas of application for diagnostic kits is in measuring fertility hormones in animal blood or milk, e.g. progesterone, oestrogen sulphate and equine gonadotrophin. Illegal use of growth hormones and antibodies can also be monitored.

Immunochemical technology using monoclonal antibodies is now widely used for the analysis of pesticide residues in foods, together with toxic microbial products such as mycotoxins.

REVIEW QUESTIONS

1. In which areas could biotechnology be useful in human medicine?
2. Write in detail the applications of animal biotechnology in medicine.
3. How could biotechnological methods be used in the control of environmental pollution?

REFERENCES

Bakhshi, A., Jensen, J.P., Goldman, P., Wright, J.J., McBride, O.W., Epstein, E.L. and Korsmeyer, S.J. (1985). "Cloning the chromosomal breakpoint of t(14:18) human lymphomas: clustering around J_H on chromosome 14 and near a transcriptional unit on 18." *Cell.* 41: 899–906.

Bernard, C.N., Wolpowitz, A. and Losman, J.G. (1977). "Heterotopic cardiac transplantation with a xenograft for assistance of the left heart in cardiogenic shock after cardiopulmonary bypass." *South African Med. J.* 52: 1035–1038.

Brown, P., Cervenakova, L., Goldfarb, L.G., McCombie, W.R., Rubenstein, R., Wills, R.G., Pocchiari, M., Martinez-Large, J.F., Scalici, C., Masullo, C., Graupera, G., Ligan, J. and Gajdusek, D.C. (1994). "Latrogenic Creutzfeldt-Jakob disease: An example of the interplay between ancient genes and modern medicine." *Neurology.* 44: 291–293.

Concar, D. (1994). "The organ factory of the future." *New Scientist.* 82: 7439–7443.

Cournoyer, C. and Caskey, C.T. (1993). "Gene therapy of the immune system." *Ann. Rev. Immunol.* 11: 297–329.

Cory, S. (1995). "Regulation of lymphocyte survival by the *bcl12* gene family." *Ann. Rev. Immuol.* 13: 513–543.

Culver, K.W. and Balaese, R.M. (1994). "Gene therapy for cancer." *Trends in Genetics.* 10: 174–178.

Darby, S.C., Rizza, C.R., Doll, R., Spooner, R.J.D., Stratton, I.M. and Thakrar, B. (1989). "Incidence of AIDS and excess of mortality associated with HIV in haemophiliacs in the UK: Report on behalf of the directors of haemophilia centers in the UK." *Brit. Med. J.* 298: 1064–1068.

Dooldeniya, M. and Warrens, A. (2003). "Xenotransplantation: Where are we today?" *J. R. Soc. Med.* 96: 11–117.

Friedmann, T. (1994). "Gene therapy for neurological disorders." *Trends in Genetics.* 10: 210–214.

Galli-Taliadoros, L.A., Sedgwick, J.D., Wood, S.A. and Korner, H. (1995). "Gene knock-out technology: A methodological overview for the interested novice." *J. Immunol. Method.* 181: 1–15.

Gershon, D. (1991). "Biotechnology. Will milk shake up industry?" *Nature.* 353: 7.

Gilboa, E. and Smith, C. (1994). "Gene therapy for infectious diseases: The AIDS model." *Trends in Genetics.* 10: 139–144.

Ishibashi, S., Goldstein, J.L., Brown, M.S., Herz, J. and Burns, D.K. (1994). "Massive xanthomatosis and atherosclerosis in cholesterol-fed low density lipoprotein receptor-negative mice." *J. Clin. Invest.* 93: 1885–1893.

Kappel, C.A., Bieberich, C.J. and Jay, G. (1994). "Evolving concepts in molecular pathology." *FASEB J.* 8: 583–592.

Kaufman, C.L., Gaines, B.A. and Ildstad, S.T. (1995). "Xenotransplantation." *Ann. Rev. Immunol.* 13: 339–367.

Kay, M. and Woo, S.L.C. (1994). "Gene therapy for metabolic disorders". *Trends in Genetics.* 13: 339–367.

Kerr, W.G. and Mule, J.J. (1994). "Gene therapy: Current status and future prospects." *J. Leuk. Biol:* 56: 210–214.

Kumar, R. (1995). "Recombinant haemoglobins as blood substitutes: A biotechnology perspective." *Proc. Soc. Exp. Biol. Med.* 208: 150–158.

Larkin, D.F.P., Orla, H.B., Ring, C.J.A., Lemoine, N.R. and George, A.J.T. (1996). "Adenovirus-mediated gene delivery to the corneal endothelium." *Transplantation.* 61: 363–370.

Mahley, R.W. (1988). "Apolipoprotein E: Cholesterol transport protein with expanding role in cell biology." *Science.* 240: 622–630.

Michler, R. (1996). "Xenotransplantation: Risks, Clinical Potential, and Future Prospects." EID 2(1). http://www.cdc.gov/ncidod/eid/vol2no1/michler.htm

Plump, A.S., Smith, J.D., Hayek, T. Aalto-Setala, K., Walsh, A., Verstuyft, J.G., Rubin E.M. and Breslow, J.L. (1992). "Severe hypercholesterolemia and atherosclerosis in apolipoprotein E-deficient mice created by homologous recombination in ES cells." *Cell.* 71: 343–353.

Starzl, T.E., Fung, J., Tzakis, A., Todo, S., Demetris, A.J., Marino, I.R., Doyle, H., Zeevi, A., Warty, V., Michaels, M., Kusne, S., Rudert, W.A. and Trucoo, M. (1993). "Baboon-to-human liver transplantation." *Lancet.* 341: 65–71.

Taylor, L. (2007). "Xenotransplantation. Emedicine online journal." http://www.emedicine.com/med/topic3715.htm

Valander, W.H., Johnson, J.L., Page, R.L., Russell, C.G., Subramaniam, A., Wilkins, T.D., Gwazdauskas, F.C., Pittus, C. and Drohan, W.N. (1992). "High-level expression of a heterologous protein in the milk of transgenic swine using the cDNA encoding human protein C." *Proc. Natl. Acad. Sci. USA.* 89: 12003–12007.

Veis, D.J., Sorenson, C.M., Shutter, J.R. and Korsmeyer, S.J. (1993). "*Bcl12* deficient mice demonstrate fulminant lymphoid apoptosis, polycystic kidneys and hypopigmented hair." *Cell.* 75: 229–240.

Verma, I.M. (1990). "Gene therapy." *Scientific American.* 163: 34–41.

Yom, H.C. and Bremel, R.D. (1993). "Genetic engineering of milk composition: modification of milk components in lactating transgenic animals." *Amer. J. Clin. Nutr.* 58: 299S–306S.

Zhang, S.H., Reddick, R.L., Piedrahita, J.A. and Maeda, N. (1992). "Spontaneous hypercholesterolemia and arterial lesions in mice lacking apolipoprotein." *E. Science.* 258: 468–471.

17

INTELLECTUAL PROPERTY RIGHTS AND BIOSAFETY IN BIOTECHNOLOGY

INTELLECTUAL PROPERTY RIGHTS

INTRODUCTION

The legal characterization and treatment of trade-related biotechnological processes and products are described as Intellectual Property. Its protection (Intellectual Property Protection, IPP) and the Rights (Intellectual Property Rights, IPR) available to protect this property has been the subject of discussion in recent years. The term property is often found associated with physical objects only, such as household goods or land, for which ownership and associated rights are guaranteed and protected by law prevalent in a country. This property is described as tangible. Intellectual property, on the other hand, is intangible and includes 'patents', 'trade secrets', 'copyrights' and 'trademarks'. The right to protect this property prohibits others form making, copying, using or selling the proprietary subject matter. Under biotechnology, one of the most important examples of intellectual property is the processes and products, which result from the development of genetic engineering techniques through the use of restriction enzymes to create recombinant DNA.

Another example of intellectual property is the development of crop varieties which are protected through 'plant breeder's rights' (PBR). Through PBR, the plant breeder who developed a variety enjoys the exclusive right for marketing the variety, although use of the variety for further breeding or for replantation of seed saved by a farmer is permissible. More recently, however, utility patents for genetic materials, both plants and animals, have been allowed in some countries, so that the patented material can neither be used for further breeding, nor will the farmers be allowed to save and use the seed for cultivation, without paying a fee to the patent holder. Similarly, if patents on superior animal breeds are allowed, a dairy farmer will find that a calf born to his hybrid cow will belong to the company, which sold him the animal. There are also arguments against patenting life forms like transgenic animals and plants, because these patents will work as impediments in free exchange of genetic materials for improvement of crops and livestock. IPRs may also affect (i) food security, (ii) use of evolved agricultural practices, (iii) biological diversity and ecological balance and (iv) the livelihood of the poor in developing countries.

PATENTS

Patents are granted by the Government for the commercial exploitation of an invention for a specific period of time in consideration of the disclosure of the invention so that on expiry of the terms of the patent, the information can benefit the public at large.

The patent system provides a social benefit as it bestows monetary reward for revealing technological innovation along with accolades for the inventor. Patent is an award for the inventor and a reward for the investor. The grant of patent for an invention attracts investment because the commercial exploitation of the invention is possible to its fullest extent during the term of patent. Another major advantage of the patent system is that it promotes 'invent around' concept. Patent is granted only when the invention and its operation or use and the method by which it is to be performed are fully disclosed. When the patentee launches the product (in which the invention is incorporated) in the market, his competitors may lose the market if the product is technically advanced and cheap as compared to the existing one. The patentee can prevent others from manufacturing the same product without his authorization and can resort to legal means to enforce his right. But the competitors have an option to 'invent around' the patented product by conducting further research around to bring out a better invention, which may result in cheaper and better product. It paves the way for healthy competition among manufacturers that results in day-to-day improvement of technology. Ultimately, it contributes to the economic growth of the country, thereby enhancing living standard of the people.

By virtue of the grant, patentee gets the exclusive right to prevent the third parties (not having his consent) from the act of making, using, offering for sale, selling or importing the patented product or process within the territory of grant.

The granting of special exclusive rights (for trading new articles) has been a practice to encourage innovations. As an example, monopoly rights (only to inventors) were granted in some countries like Europe, as an incentive to develop new articles that would be of benefit to the society. Under the USA law, a patent means grant of "right to exclude others from making, using or selling" an invention for a 17-year period. Patents are usually allowed for a specified period. Before 1970, when the discovery of an oil-eating bacterium (*Pseudomonas*) by a non-resident Indian (Dr. Chakrabarty), was patented in the USA by a multinational corporation, the life forms could not be patented. A later patent issued for 'oncomouse' was another milestone in patenting of life forms. In India, The Indian Patents Act of 1970 allows process patents, but no product patents for food, chemicals, drugs and pharmaceuticals. The duration of the patent in India is five years from the date of grant of patent or seven years from the date of filing the application, whichever is less.

How can biotechnology products and processes be protected and the due financial profits returned to the rightful inventors and industrial developers? Inventors in the area of biotechnology can be protected by way of different titles of protection including patents for inventions, plant breeder's rights and trade secrets. In the context of biotechnology, inventions can be in the form of products or processes.

Products These can be considered either as: (a) living entities of natural or artificial origin such as animals, plants and microorganisms, cell lines, organelles, plasmids and DNA sequences; or (b) naturally occurring substances, primary or secondary, derived from living system.

Processes These can include those of isolation, cultivation, multiplication, purification and bioconversion. Such processes can be involved in the isolation or the creation of the above products, e.g. antibiotic production; the production of substances through bioconversion of products, e.g. enzymatic conversion of sugar to alcohol; the use of the products for many purposes, e.g. monoclonal antibodies for analysis or diagnosis; and the use of microbes for biocontrol of pathogens.

PATENT PROTECTION

What is a patent? A patent is a legal right that owes its existence to a granting act by a government authority, i.e., a patent office. With the granting of the patent, the holder or patentee is given the right to exclude, for a limited time period, all others within the territory of application of the patent for commercial utilization of the patent invention. In return for this monopoly situation, the patentee discloses the details of the invention to the public so that, at the end of the monopoly period, the invention may be worked freely by the public (i.e., other competitors). To obtain patent rights in Europe, USA or Japan, the patent application must be made in each country. Each country has set its own standards for granting patents to biotechnological inventions.

After the patent application has been scrutinized and granted, the patent is in the form of a letter patent, which *inter alia* contains the name of the inventor, the name of the patentee (if different), a description of the patent and the relevant claims. Generally, an invention is patentable, while a discovery is not. Patents can be granted for inventions that

- are seen to be novel;

- involve an inventive step;

- can lead to industrial application and

- are properly disclosed in the patent specification.

Unlike other fields of technology, inventions in biotechnology most often relate to living material, which can raise some unique difficulties in their legal protection. Problems can arise in how to describe an invention relating to living material for the purpose of obtaining patent protection and most basically whether living matter should be protectable under traditional schemes of industrial property protection. For example, the National Institute of Health in the USA was refused patent rights on segments of DNA isolated from the human genome.

In this context, the difference between discovery (not patentable) and invention for a biotechnological substance has been highlighted by the European Patent Office.

To find a substance freely occurring in nature is a mere discovery and therefore unpatentable. However, if a substance found in nature has first to be isolated, from its surroundings and a process for obtaining it is developed, that process is patentable. Moreover, if the substance can be properly characterized, either by its chemical structure, by the process by which it is obtained or by other parameters and if it is 'new' in the absolute sense of having no previous recognized existence, then the substance per se may be patentable.

A human-made, genetically engineered bacterium capable of breaking down multiple components of crude oil was "not hitherto unknown natural phenomena, but to a non-naturally occurring manufacture or composition of matter, a product of human ingenuity having distinctional name, character and use". The patent claim was for a new bacterium that had potential for significant industrial use and had different characteristics from any found in nature, thereby qualifying this bacterium as a patentable subject—it was not nature's handiwork, but that of the patentee.

Another interesting example for patent consideration was the **Oncomouse**, a mouse genetically manipulated so that it was more predisposed to succumb to carcinogens than an ordinary mouse. The European Patent Convention excludes inventions 'in respect of plant or animal varieties or essentially biological processes for the production of plants or animals; this provision does not apply to microbiological and, therefore, *prima facie* patentable. Microbiological processes involving plants and animals are excluded. A patent was awarded for the Oncomouse on the narrowest of grounds, in part because the processes of production could not occur in nature and therefore, the process was not essentially biological.

Many biotechnology companies prefer to use trade secrets to protect their products or processes rather than to apply for patents. The forms of information that can be protected include, for example, hybridoma cell lines for monoclonal antibody production, ideas, formulae and production details, experimental procedures, etc.

PATENTABLE INVENTIONS

A patent can be granted for an invention which may be related to any process or product. The word "Invention" has been defined under the Patents Act 1970 as amended from time to time.

"An **invention** means a new product or process involving an inventive step and capable of industrial application."

"**New invention**" is defined as any invention or technology which has not been anticipated by publication in any document or used in the country or elsewhere in the world before the date of filing of patent application with complete specification, i.e., the subject matter has not fallen in public domain or that it does not form part of the state of the art *capable of industrial application, in relation to an invention, means that the invention is capable of being made or used in an industry.*

Therefore, the criteria for an invention to be patentable are as follows:

1. An invention must be novel.
2. It must have an inventive step.
3. It should be capable of industrial application.

Novelty of the invention   A novel invention is one which has not been disclosed, in the prior art where prior art means everything that has been published, presented or otherwise disclosed to the public on the date of patent (The prior art includes documents in foreign languages disclosed in any format in any country of the world.) For an invention to be judged as novel, the disclosed information should not be available in the 'prior art'. This means that there should not be any prior disclosure of any information contained in the application for patent (anywhere in the public domain, either written or in any other form, or in any language) before the date on which the application is first filed, i.e., the 'priority date'.

Industrial applicability   An invention is capable of industrial application if it satisfies three conditions, cumulatively.

- It can be made.
- It can be used in at least one field of activity.
- It can be reproduced with the same characteristics as many times as necessary.

Different categories of independent claims stating unity of invention

- Product, process for its manufacture and use of the product
- Process and apparatus for carrying out the process
- Product, process for its manufacture and apparatus for carrying out the process

Example 1   If one has invented a new kind of spray bottle, patents may be granted for

- the bottle itself (a product)
- a chemical in the plastic (chemical composition)
- the spraying mechanism (an apparatus)
- how you extruded the plastic (a process).

Example 2   Similarly if a new gene sequence has been invented, patents may be granted for

- new gene sequence A
- a method of expressing sequence A
- an antibody made to the protein of sequence A
- a kit made from the antibody to sequence A.

All of these claims are linked by the inventive concept that sequence A is new and inventive. Therefore anything based on sequence A must share this property too.

Sufficiency of disclosure   "Every complete specification shall

   (a) fully and particularly describe the invention and its operation or use and method by which it is to be performed;

   (b) disclose the best method of performing the invention which is known to the applicant and for which he is entitled to claim protection; and
........................."

Clarity of claims   "The claim or claims of a complete specification shall be clear and succinct and shall be fairly based on the matter disclosed in the specification."

BIOTECHNOLOGY PATENTS IN AUSTRALIA, USA AND EUROPE

As mentioned earlier, countries differ even in the subject matter they view as "inventable" and "discoverable". Australia's approach to biotech patents is one of the most liberal. In 1976, the Australian Patent Office (APO) held that living organisms are patentable, implying that they are inventable. The APO considers all living organisms excluding human beings as potentially patentable subject matter.

The US Patent Law has followed Australia's liberal approach towards biotech patents. The United States Patent Office (USPTO) has since issued patents for over 6000 genes, and about 1000 of these relate to human genes. Currently, there are more than 20,000 patent applications related to genes pending in the US. In contrast, Europe adopts a more cautious approach towards granting biotech patents.

INDIAN PERSPECTIVE

India is a storehouse of biological resources and one of the world's richest biodiversity countries. In the past two years, there has been a rise in the investment in the biotech-oriented industries. It is estimated that over the next five years, biotechnology can offer opportunities for fresh investment of Rs. 7–8 billion in India. Biotechnology is poised to take India to a different playing field where it can dominate the world market.

The first provision in the nature of patent rights in India, which was at that time under the British rule, was enacted in 1856. Under this Act, the monopolies were styled "exclusive privileges". This Act was repealed *in toto* in 1857 as it was introduced without the prior sanction of the Queen. Soon after that in 1859, another Act free from the defects of 1857 was passed. In 1872 the provision of the Act of 1859 were further added by the provision of "The Patents and

Designs Protection Act", which solely related to designs. The Act of 1872 was further supplemented in 1883. In 1888, all the Acts of 1859, 1872 and 1883 were superseded by Act V of 1888. This was further revised and replaced by the Indian Patents and Designs Act, 1911. This was amended from time to time in 1920, 1930, and 1945.

In 1957, Govt. of India further appointed Justice N. Rajagopala Ayyangar to examine and review the Patent law in India who submitted his report in September 1959 recommending the retention of Patent System despite its shortcomings. The Patent Bill, 1965, based mainly on his recommendations and incorporating a few changes, in particular relating to Patents for food, drug, medicines, was introduced in the lower house of Parliament on 21st September, 1965. The bill was passed by the Parliament and the Patents Act 1970 came into force on 20th April 1972 along with Patent Rules 1972. The Patents Act, 1970, is a landmark in the industrial development of India. The basic philosophy of the Act is that patents are granted to encourage inventions and to secure that these inventions are worked on a commercial scale without undue delay; and patents are granted not merely to enable patentee to enjoy a monopoly for the importation of the patented article into the country.

Uruguay round of GATT negotiations paved the way for WTO. Therefore India was put under the contractual obligation to amend its Patents Act in compliance with the provisions of TRIPS. India had to meet the first set of requirements on 1-1-1995. This was to give a pipeline protection till the country starts giving product patent. It came to force on 26th March 1999 retrospective from 1-1-1995. India amended its Patents Act again in 2002 to meet with the second set of obligations (Term of Patent, etc.), which had to be effected from 1-1-2000. This amendment, which provides for 20 years term for the patent, reversal of burden of proof, etc. came into force on 20th May, 2003. The Third Amendment of the Patents Act 1970, by way of the Patents (Amendment) Ordinance 2004 came into force on 1st January, 2005 incorporating the provisions for granting product patent in all fields of Technology including chemicals, food, drugs and agrochemicals and this Ordinance is replaced by the Patents (Amendment) Act 2005 which is in force now having effect from 1-1-2005.

Under the Patents Act, the Government of India is empowered to make rules for implementing the Act and regulating the Patent Administration. Accordingly, the Government brought into force Patents Rules, 1972 w.e.f. 20.4.1972. These Rules were amended on 2.6.99 and replaced by the Patents Rules 2003 w.e.f. 20.5.2003 and further it was amended by the Patents (Amendment) Rules 2005, which is in force now; this includes provisions relating to time-lines with a view to introducing flexibility and reducing processing time gradually for patent applications, and simplifying and rationalizing procedure for grant of the patent. There are four Schedules to the Patents (Amendment) Rules 2005; the First Schedule prescribes the fees to be paid; the Second Schedule specifies the list of forms, and the texts of various forms required in connection with various activities under the Patents Act are set out in this schedule. These forms are to be used wherever required and if needed, they can be modified with the consent of the Controller. The Third Schedule prescribes form of Patent to be issued on Grant of the Patent. The Fourth Schedule prescribes costs to be awarded in various proceedings before the Controller under the Act.

Establishment of Patent Administration in India

Patent system in India is administered under the superintendence of the Controller General of Patents, Designs, Trademarks and Geographical Indications. The Office of the Controller General functions under the Department of Industrial Policy and Promotion, Ministry of Commerce and Industry. Controller General's office is in Mumbai. There are four patent offices in India. The Head Office is located at Kolkata and other Patent Offices are located at Delhi, Mumbai and Chennai. The Controller General delegates his powers to Sr. Joint Controller, Joint Controllers, Deputy Controllers and Assistant Controllers. Examiners of patents in each office discharge their duties according to the direction of the Controllers.

What is Patentable in Biotechnology?

This is the question that India must decide with regard to new biotechnology (genomics, proteomics, bioinformatics, and genetic engineering). As per the Indian Patent Act, the following are patentable:

Microorganisms   Under the Indian Patent Act, microbiological processes can be patented. Also patentable are processes for producing new microorganisms through genetic engineering and the products that result out of this process, such as microorganisms including plasmids and viruses if they are non-living.

Cell lines   A cell line is patentable if artificially produced.

rDNA, mRNA, amino acid   If the end result is non-living, it is patentable.

Hybridoma technology   Patents are also allowed on hybridoma technology, but not on protoplast fusion.

Expressed sequence tags or ESTs   They are small fragments of genetic material obtained by reverse transcriptions of messenger RNA (mRNA) from expressed genes. The gene sequence, or expressed sequence tags (ESTs), can be patented if it has a use, such as if it works as a probe.

What is not Patentable in Biotechnology?

Additionally, the Indian Patent Act defines what is not patentable in biotechnology: inventability does not apply to plant or animals. Accordingly, a method of producing a new form of a known plant or tissue culture method for production of plant variety is not patentable, nor is a method of treatment of a human body by surgery or operation for diagnosis. Nor is a method of improving or changing the appearance of the human body or parts of it patentable.

To compete worldwide, India must decide whether to remain with the more conservative European approach, or if the Australia/US or some other approach better suits the needs of its emerging economy. In either case, Indian companies, inventors, and investors venturing into the biotech sector must be well-informed and well-aware of Indian laws, as well as the laws of other countries as they seek to join the biotechnology headlines.

There are some products and processes, which are not patentable in India. They are classified into two categories in the Patent Act

(a) those which are not inventions

(b) invention relating to atomic energy

Various types of **non-patentable inventions** under Section 3 are as follows:

3(a) An invention which is frivolous or which claims anything obvious contrary to well established natural laws.

3(b) An invention the primary or intended use or commercial exploitation of which could be contrary to public order or morality or which causes serious prejudice to human, animal or plant life or health or to the environment.

3(c) The mere discovery of a scientific principle or the formulation of an abstract theory or discovery of any living thing or non-living substances occurring in nature.

3(d) The mere discovery of a new form of a known substance which does not result in the enhancement of the known efficacy of that substance or the mere discovery of any new property or new use for a known substance or of the mere use of a known process, machine or apparatus unless such known process results in a new product or employs at least one new reactant.

3(e)   A substance obtained by a mere admixture resulting only in the aggregation of the properties of the components thereof or a process for producing such substance.

3(f)   The mere arrangement or re-arrangement or duplication of known devices each functioning independently of one another in a known way.

3(h)   A method of agriculture or horticulture.

3(i)   Any process for the medicinal, surgical, curative, prophylactic diagnostic, therapeutic or other treatment of human being or any process for a similar treatment of animals to render them free of disease or to increase their economic value or that of their products.

3(j)   Plants and animals in whole or any part thereof other than microorganisms but including seeds, varieties and species and essentially biological processes for production or propagation of plants and animals.

Example   Clones and new variety of plants are not patentable. But process/method of preparing genetically modified organisms are patentable subject matter.

3(k)   A mathematical or business method or a computer program per se or algorithms.

3(l)   A literary, dramatic, musical or artistic work or any other aesthetic creation whatsoever including cinematographic works and television productions.

3(m) A mere scheme or rule or method of performing mental act or method of playing game.

3(n)   A presentation of information.

3(o)   Topography of integrated circuits.

3(p)   An invention which in effect, is traditional knowledge or which is an aggregation or duplication of known properties of traditionally known component or components.

Obligations with Patent Applications

Before a patent can be issued, the following specific conditions must be met: (a) the invention must be new (novelty) and should have utility; (b) it must be inventive (which means it should not be obvious but should represent a real advance made through the insight of the inventor); (c) it must be disclosed in a way, which enables a person of normal skill to reproduce it; (d) the scope of protection to be granted must be in proportion to the invention; (e) it must relate to a technology where patents are permitted (patentable). The application of these conditions can vary to some extent in different countries. A fundamental requirement of patent law, thus is that in the application, the inventor should describe fully the invention.

Implications of Patenting

The patent application, with its accompanying description and the deposited material, is held secret in the early stages only for a short time, after which the application must be published. The publication gives advantage to applicant's competitors, who can use this information for developing an improved process or product. Publication of application also makes the deposited material available to the public.

Persons Entitled to Apply for a Patent in India

An application for a patent for an invention may be made by any of the following **persons** either **alone** or **jointly with another**

    (a) True and first inventor

    (b) His/her assignee

    (c) Legal representative of deceased inventor or assignee.

Where to Apply?

Application for the patent has to be filed in the respective patent office as mentioned below where the territorial jurisdiction is decided based on whether any of the following occurrences falls within the territory.

    (a) Place of residence, domicile or business of the applicant (first- mentioned applicant in the case of joint applicants)

    (b) Place from where the invention actually originated.

    (c) Address for service in India given by the applicant when he has no place of business or domicile in India.

Examples of Patents

Amgen Inc. holds a US patent for preparation of erythropoietin (which stimulates production of RBCs) by recombinant method. In this case, DNA sequences, vector and transformed cells were protected, but not the product, since US patent office usually does not allow patents on the naturally occurring products.

Among higher plants, an important example is the patent for 'tryptophan-overproducing maize' obtained through tissue culture. The patent known as

"Hibberd Patent" was issued to Molecular Genetics Research and Development. A similar patent was allowed in US for a transgenic animal, popularly known as 'oncomouse patent'. Patent was also allowed for a polyploid oyster produced by the application of hydrostatic pressure to zygotes.

BIOSAFETY IN BIOTECHNOLOGY

INTRODUCTION

All the microorganisms used in biotechnology are non-pathogenic to humans and other animals. Biotechnology has made major advances in public health by the control of communicable diseases with vaccines and the improvement in the quality of the environment by the continued improvements in the biological waste treatment processes.

Many microorganisms can infect humans, animals and plants and cause disease. Most microorganisms used by industry are harmless and many are indeed used directly for the production of human or animal foods. Only a small number of potentially dangerous microorganisms have been used by industry in the manufacture of vaccines or diagnostic reagents, e.g. *Bordetella pertussis* (whooping cough), *Mycobacterium tuberculosis* (TB) and the virus that causes foot-and-mouth disease. Stringent containment practices have been the norm.

In recent years, there have been many scientific advances permitting alterations to the genetic make-up of microorganisms. Recombinant DNA techniques have been the most successful but have also been the cause of much concern to the public.

The term "containment" is used in describing safe methods for managing infectious materials in the laboratory environment where they are being handled or maintained. The purpose of containment is to reduce or eliminate exposure of laboratory workers, other persons, and the outside environment to potentially hazardous agents.

Primary containment, the protection of personnel and the immediate laboratory environment from exposure to infectious agents, is provided by both good microbiological technique and the use of appropriate safety equipment. The use of vaccines may provide an increased level of personal protection. Secondary containment, the protection of the environment external to the laboratory from exposure to infectious materials, is provided by a combination of facility design and operational practices. Therefore, the three elements of containment include laboratory practice and technique, safety equipment, and facility design. The risk assessment of the work to be done with a specific agent will determine the appropriate combination of these elements.

LABORATORY PRACTICE AND TECHNIQUE

The most important element of containment is strict adherence to standard microbiological practices and techniques. Persons working with infectious agents or potentially infected materials must be aware of potential hazards, and must be trained and proficient in the practices and techniques required for handling such material safely. Each laboratory should develop or adopt a biosafety or operations manual that identifies the hazards that will or may be encountered, and that specifies practices and procedures designed to minimize or eliminate exposures to these hazards. Personnel should be advised of special hazards and should be required to read and follow the required practices and procedures. Laboratory personnel, safety practices, and techniques must be supplemented by appropriate facility design and engineering features, safety equipment, and management practices.

SAFETY EQUIPMENT (PRIMARY BARRIERS)

Safety equipment includes biological safety cabinets (BSCs), enclosed containers, and other engineering controls designed to remove or minimize exposures to hazardous biological materials. The biological safety cabinet (BSC) is the principal device used to provide containment of infectious splashes or aerosols generated by many microbiological procedures. Three types of biological safety cabinets (Class I, II, III) are used in microbiological laboratories. Open-fronted Class I and Class II biological safety cabinets are primary barriers which offer significant levels of protection to laboratory personnel and to the environment when used with good microbiological techniques. The Class II biological safety cabinet also provides protection from external contamination of the materials (e.g. cell cultures, microbiological stocks) being manipulated inside the cabinet. The gas-tight Class III biological safety cabinet provides the highest attainable level of protection to personnel and the environment. Safety equipment also may include items for personal protection, such as gloves, coats, gowns, shoe covers, boots, respirators, face shields, safety glasses, or goggles. Personal protective equipment is often used in combination with biological safety cabinets and other devices that contain the agents, animals, or materials being handled.

FACILITY DESIGN AND CONSTRUCTION (SECONDARY BARRIERS)

The design and construction of the facility contributes to the laboratory workers' protection, provides a barrier to protect persons outside the laboratory, and protects persons or animals in the community from infectious agents which may be accidentally released from the laboratory. Laboratory management is responsible for providing facilities commensurate with the laboratory's function and the recommended biosafety level for the agents being manipulated. The recommended secondary barrier(s) will depend on the risk of transmission of specific agents. For example, the exposure risks for most laboratory work in Biosafety Levels 1 and 2 facilities will be direct contact with the agents, or inadvertent contact exposures through contaminated work environments. Secondary barriers in these laboratories may include separation of the laboratory work area from public access, availability of a decontamination facility (e.g. autoclave), and hand washing facilities. When the risk of infection by exposure to an infectious aerosol is present, higher levels of primary containment and multiple secondary barriers may become necessary to prevent infectious agents from escaping into the environment. Such design features include specialized ventilation systems to ensure directional air flow, air treatment systems to decontaminate or remove agents from exhaust air, controlled access zones, airlocks as laboratory entrances, or separate buildings or modules to isolate the laboratory.

Table 17.1 Classification of microorganisms according to pathogenicity

Class 1   Microorganisms that have never been identified as causative agents of disease in human beings and that offer no threat to the environment.

Class 2   Microorganisms that may cause human disease and might therefore offer a hazard to laboratory workers. They are unlikely to spread in the environment. Prophylactics are available and treatment is effective.

Class 3   Microorganisms that offer a server threat to the health of laboratory workers but a comparatively small risk to the population at large. Prophylactics are available and treatment is effective.

Class 4   Microorganisms that cause severe illness in human beings and offer a serious hazard to laboratory workers and to people at large. In general, effective prophylactics are not available and no effective treatment is known.

Class 5   This group contains microorganisms that offer a more severe threat to the environment than to people. They may be responsible for heavy economic losses. National and international lists or regulations concerning these microorganisms are already in existence in contexts other than biotechnology (e.g. for phytosanitary purposes).

BIOSAFETY LEVELS

Four biosafety levels (BSLs) are described, which consist of combinations of laboratory practices and techniques, safety equipment, and laboratory facilities. Each combination is specifically appropriate for the operations performed, the documented or suspected routes of transmission of the infectious agents, and the laboratory function or activity. The recommended biosafety level(s) for the organisms represent those conditions under which the agent ordinarily can be safely handled. Generally, work with known agents should be conducted at the biosafety level recommended. When specific information is available to suggest that virulence, pathogenicity, antibiotic resistance patterns, vaccine and treatment availability, or other factors are significantly altered, more (or less) stringent practices may be specified.

Biosafety Level 1 (BSL-1)

Biosafety Level 1 practices, safety equipment, and facility design and construction are appropriate for undergraduate and secondary educational training and teaching laboratories, and for other laboratories in which work is done with defined and characterized strains of viable microorganisms not known to consistently cause disease in healthy adult humans. *Bacillus subtilis, Naegleria gruberi*, infectious canine hepatitis virus, and exempt organisms under the NIH Recombinant DNA Guidelines are representative of microorganisms meeting these criteria. Many agents not ordinarily associated with disease processes in humans are, however, opportunistic pathogens and may cause infection in the young, the aged, and immunodeficient or immunosuppressed individuals. Vaccine strains that have undergone multiple *in vivo* passages should not be considered avirulent simply because they are vaccine strains. Biosafety Level 1 represents a basic level of containment that relies on standard microbiological practices with no special primary or secondary barriers recommended, other than a sink for hand washing.

The following standard and special practices, safety equipment and facilities apply to agents assigned to Biosafety Level 1.

A. *Standard microbiological practices*

1. Access to the laboratory is limited or restricted at the discretion of the laboratory director when experiments or work with cultures and specimens are in progress.

2. Persons wash their hands after they handle viable materials, after removing gloves, and before leaving the laboratory.

3. Eating, drinking, smoking, handling contact lenses, applying cosmetics, and storing food for human use are not permitted in the work areas. Persons who wear contact lenses in laboratories should also wear goggles or a face shield. Food is stored outside the work area in cabinets or refrigerators designated and used for this purpose only.

4. Mouth pipetting is prohibited; mechanical pipetting devices are used.

5. Policies for the safe handling of sharps are instituted.

6. All procedures are performed carefully to minimize the creation of splashes or aerosols.

7. Work surfaces are decontaminated at least once a day and after any spill of viable material.

8. All cultures, stocks, and other regulated wastes are decontaminated before disposal by an approved decontamination method such as autoclaving. Materials to be decontaminated outside of the immediate laboratory are to be placed in a durable, leakproof container and closed for transport from the laboratory. Materials to be decontaminated outside of the immediate laboratory are packaged in accordance with applicable local, state, and federal regulations before removal from the facility.

9. A biohazard sign can be posted at the entrance to the laboratory whenever infectious agents are present. The sign may include the name of the agent(s) in use and the name and phone number of the investigator.

10. An insect and rodent control program is in effect.

B. *Special practices none*

C. *Safety equipment (primary barriers)*

1. Special containment devices or equipment such as a biological safety cabinet is generally not required for manipulation of agents assigned to Biosafety Level 1.

2. It is recommended that laboratory coats, gowns, or uniforms be worn to prevent contamination or soiling of street clothes.

3. Gloves should be worn if the skin on the hands is broken or if a rash is present. Alternatives to powdered latex gloves should be available.

4. Protective eyewear should be worn for conduct of procedures in which splashes of microorganisms or other hazardous materials is anticipated.

D. *Laboratory facilities (secondary barriers)*

1. Laboratories should have doors for access control.

2. Each laboratory contains a sink for hand washing.

3. The laboratory is designed so that it can be easily cleaned. Carpets and rugs in laboratories are not appropriate.

4. Bench tops are impervious to water and are resistant to moderate heat and the organic solvents, acids, alkalis, and chemicals used to decontaminate the work surface and equipment.

5. Laboratory furniture is capable of supporting anticipated loading and uses. Spaces between benches, cabinets, and equipment are accessible for cleaning.

6. If the laboratory has windows that open to the exterior, they are fitted with fly screens.

Biosafety Level 2 (BSL-2)

Biosafety Level 2 practices, equipment, and facility design and construction are applicable to clinical, diagnostic, teaching, and other laboratories in which work is done with a broad spectrum of indigenous moderate-risk agents that are present in the community and associated with human disease of varying severity. With good microbiological techniques, these agents can be used safely in activities conducted on the open bench, provided the potential for producing splashes or aerosols is low. Hepatitis B virus, HIV, the salmonellae, and *Toxoplasma* spp. are representative of microorganisms assigned to this containment level. Biosafety Level 2 is appropriate when work is done with any human-derived blood, body fluids, tissues, or primary human cell lines where the presence of an infectious agent may be unknown.

Primary hazards to personnel working with these agents relate to accidental percutaneous or mucous membrane exposures, or ingestion of infectious materials. Extreme caution should be taken with contaminated needles or sharp instruments. Even though organisms routinely manipulated at Biosafety Level 2 are not known to be transmissible by the aerosol route, procedures with aerosol or high splash potential that may increase the risk of such personnel exposure must be conducted in primary containment equipment, or in devices such as a BSC or safety centrifuge cups. Other primary barriers should be used as appropriate, such as splash shields, face protection, gowns, and gloves. Secondary barriers such as hand washing sinks and waste decontamination facilities must be available to reduce potential environmental contamination.

The following standard and special practices, safety equipment, and facilities apply to agents assigned to Biosafety Level 2:

A. *Standard microbiological practices—as given for biosafety level 1*

B. *Special practices*

1. Access to the laboratory is limited or restricted by the laboratory director when work with infectious agents is in progress. In general, persons

who are at increased risk of acquiring infection, or for whom infection may have serious consequences, are not allowed in the laboratory or animal rooms. For example, persons who are immunocompromised or immunosuppressed may be at increased risk of acquiring infections. The laboratory director has the final responsibility for assessing each circumstance and determining who may enter or work in the laboratory or animal room.

2. The laboratory director establishes policies and procedures whereby only persons who have been advised of the potential hazards and meet specific entry requirements (e.g. immunization) may enter the laboratory.

3. A biohazard sign must be posted on the entrance to the laboratory when aetiologic agents are in use. Appropriate information to be posted includes the agent(s) in use, the biosafety level, the required immunizations, the investigator's name and telephone number, any personal protective equipment that must be worn in the laboratory, and any procedures required for exiting the laboratory.

4. Laboratory personnel receive appropriate immunization or tests for the agents handled or potentially present in the laboratory (e.g. hepatitis B vaccine or TB skin testing).

5. When appropriate, considering the agent(s) handled, baseline serum samples for laboratory and other at-risk personnel are collected and stored. Additional serum specimens may be collected periodically, depending on the agents handled or the function of the facility.

6. Biosafety procedures are incorporated into standard operating procedures or in a biosafety manual adopted or prepared specifically for the laboratory by the laboratory director. Personnel are advised of special hazards and are required to read and follow instructions on practices and procedures.

7. The laboratory director ensures that laboratory and support personnel receive appropriate training on the potential hazards associated with the work involved, the necessary precautions to prevent exposures, and the exposure evaluation procedures. Personnel receive annual updates or additional training as necessary for procedural or policy changes.

8. A high degree of precaution must always be taken with any contaminated sharp items, including needles and syringes, slides, pipettes, capillary tubes, and scalpels.

    a. Needles and syringes or other sharp instruments should be restricted in the laboratory for use only when there is no alternative, such as parenteral injection, phlebotomy, or aspiration of fluids from laboratory animals and diaphragm bottles. Plasticware should be substituted for glassware whenever possible.

b. Only needle-locking syringes or disposable syringe-needle units (i.e. needle is integral to the syringe) are used for injection or aspiration of infectious materials. Used disposable needles must not be bent, sheared, broken, recapped, removed from disposable syringes, or otherwise manipulated by hand before disposal; rather, they must be carefully placed in conveniently located puncture-resistant containers used for sharps disposal. Non-disposable sharps must be placed in a hard-walled container for transport to a processing area for decontamination, preferably by autoclaving.

c. Syringes which re-sheathe the needle, needleless systems, and other safety devices are used when appropriate.

d. Broken glassware must not be handled directly by hand, but must be removed by mechanical means such as a brush and dustpan, tongs, or forceps. Containers of contaminated needles, sharp equipment, and broken glass are decontaminated before disposal, according to any local, state, or federal regulations.

9. Cultures, tissues, specimens of body fluids, or potentially infectious wastes are placed in a container with a cover that prevents leakage during collection, handling, processing, storage, transport, or shipping.

10. Laboratory equipment and work surfaces should be decontaminated with an effective disinfectant on a routine basis, after work with infectious materials is finished, and especially after overt spills, splashes, or other contamination by infectious materials. Contaminated equipment must be decontaminated according to any local, state, or federal regulations before it is sent for repair or maintenance or packaged for transport in accordance with applicable local, state, or federal regulations, before removal from the facility.

11. Spills and accidents that result in overt exposures to infectious materials are immediately reported to the laboratory director. Medical evaluation, surveillance, and treatment are provided as appropriate and written records are maintained.

12. Animals not involved in the work being performed are not permitted in the lab.

C. *Safety equipment (primary barriers)*—As given in earlier section

D. *Laboratory facilities (secondary barriers)*—As given in earlier section

Biosafety Level 3 (BSL-3)

Biosafety Level 3 is applicable to clinical, diagnostic, teaching, research, or production facilities in which work is done with indigenous or exotic agents which may cause serious or potentially lethal disease as a result of exposure by the inhalation route. Laboratory personnel have specific training in handling pathogenic and potentially lethal agents, and are supervised by competent scientists who are experienced in working with these agents. All procedures involving the manipulation of infectious materials are conducted within biological safety cabinets or other physical containment devices, or by personnel wearing appropriate personal protective clothing and equipment. The laboratory has special engineering and design features.

It is recognized, however, that some existing facilities may not have all the facility features recommended for Biosafety Level 3 (i.e., double-door access zone and sealed penetrations). In this circumstance, an acceptable level of safety for the conduct of routine procedures, (e.g. diagnostic procedures involving the propagation of an agent for identification, typing, susceptibility testing, etc.), may be achieved in a Biosafety Level 2 facility, providing the following: 1) the exhaust air from the laboratory room is discharged to the outdoors, 2) the ventilation to the laboratory is balanced to provide directional airflow into the room, 3) access to the laboratory is restricted when work is in progress, and 4) the recommended Standard Microbiological Practices, Special Practices, and Safety Equipment for Biosafety Level 3 are rigorously followed.

The following standard and special safety practices, equipment and facilities apply to agents assigned to Biosafety Level 3:

A. *Standard microbiological practices*—As given under Biosafety Level 1

B. *Special practices*—As given under Biosafety Level 1

C. *Safety equipment (primary barriers)*

    1. Protective laboratory clothing such as solid-front or wrap-around gowns, scrub suits, or coveralls are worn by workers when in the laboratory. Protective clothing is not worn outside the laboratory. Reusable clothing is decontaminated before being laundered. Clothing is changed when overtly contaminated.

    2. Gloves must be worn when handling infectious materials, infected animals, and when handling contaminated equipment.

    3. Frequent changing of gloves accompanied by hand washing is recommended. Disposable gloves are not reused.

    4. All manipulation of infectious materials, necropsy of infected animals, harvesting of tissues or fluids from infected animals or embryonated

eggs, etc., are conducted in a Class II or Class III biological safety cabinet (*see* Appendix A).

5. When a procedure or process cannot be conducted within a biological safety cabinet, then appropriate combinations of personal protective equipment (e.g. respirators, face shields) and physical containment devices (e.g. centrifuge safety cups or sealed rotors) are used.

6. Respiratory and face protection are used when in rooms containing infected animals.

D. *Laboratory facilities (secondary barriers)*

1. The laboratory is separated from areas that are open to unrestricted traffic flow within the building, and access to the laboratory is restricted. Passage through a series of two self-closing doors is the basic requirement for entry into the laboratory from access corridors. Doors are lockable (*see* Appendix F). A clothes change room may be included in the passageway.

2. Each laboratory room contains a sink for hand washing. The sink is hands-free or automatically operated and is located near the room exit door.

3. The interior surfaces of walls, floors, and ceilings of areas where BSL-3 agents are handled are constructed for easy cleaning and decontamination. Seams, if present, must be sealed. Walls, ceilings, and floors should be smooth, impermeable to liquids and resistant to the chemicals and disinfectants normally used in the laboratory. Floors should be monolithic and slip-resistant. Consideration should be given to the use of coved floor coverings. Penetrations in floors, walls, and ceiling surfaces are sealed or are capable of being sealed to facilitate decontamination. Openings such as those around ducts and the spaces between doors and frames are capable of being sealed to facilitate decontamination.

4. Bench tops are impervious to water and are resistant to moderate heat and the organic solvents, acids, alkalis, and those chemicals used to decontaminate the work surfaces and equipment.

5. Laboratory furniture is capable of supporting anticipated loading and uses. Spaces between benches, cabinets, and equipment are accessible for cleaning. Chairs and other furniture used in laboratory work should be covered with a non-fabric material that can be easily decontaminated.

6. All windows in the laboratory are closed and sealed.

7. A method for decontaminating all laboratory wastes is available in the facility and utilized, preferably within the laboratory (i.e., autoclave,

chemical disinfection, incineration, or other approved decontamination method). Consideration should be given to means of decontaminating equipment. If waste is transported out of the laboratory, it should be properly sealed and not transported in public corridors.

8. Biological safety cabinets are required and are located away from doors, from room supply louvres, and from heavily-travelled laboratory areas.

9. A ducted exhaust air ventilation system is provided. This system creates directional airflow which draws air into the laboratory from "clean" areas and toward "contaminated" areas. The exhaust air is not recirculated to any other area of the building. Filtration and other treatments of the exhaust air are not required, but may be considered based on site requirements, and specific agent manipulations and use conditions. The outside exhaust must be dispersed away from occupied areas and air intakes, or the exhaust must be HEPA-filtered. Laboratory personnel must verify that the direction of the airflow (into the laboratory) is proper. It is recommended that a visual monitoring device that indicates and confirms directional inward airflow be provided at the laboratory entry. Consideration should be given to installing an HVAC control system to prevent sustained positive pressurization of the laboratory. Audible alarms should be considered to notify personnel of HVAC system failure.

10. HEPA-filtered exhaust air from a Class II biological safety cabinet can be recirculated into the laboratory if the cabinet is tested and certified at least annually. When exhaust air from Class II safety cabinets is to be discharged to the outside through the building exhaust air system, the cabinets must be connected in a manner that avoids any interference with the air balance of the cabinets or the building exhaust system (e.g. an air gap between the cabinet exhaust and the exhaust duct). When Class III biological safety cabinets are used, they should be directly connected to the exhaust system. If the Class III cabinets are connected to the supply system, it is done in a manner that prevents positive pressurization of the cabinets (*see* Appendix A).

11. Continuous flow centrifuges or other equipment that may produce aerosols are contained in devices that exhaust air through HEPA filters before discharge into the laboratory. These HEPA systems are tested at least annually. Alternatively, the exhaust from such equipment may be vented to the outside if it is dispersed away from occupied areas and air intakes.

12. Vacuum lines are protected with liquid disinfectant traps and HEPA filters, or their equivalent. Filters must be replaced as needed. An

alternative is to use portable vacuum pumps (also properly protected with traps and filters).

13. An eyewash station is readily available inside the laboratory.

14. Illumination is adequate for all activities, avoiding reflections and glare that could impede vision.

15. The Biosafety Level 3 facility design and operational procedures must be documented. The facility must be tested for verification that the design and operational parameters have been met prior to operation. Facilities should be re-verified, at least annually, against these procedures as modified by operational experience.

16. Additional environmental protection (e.g. personnel showers, HEPA filtration of exhaust air, containment of other piped services and the provision of effluent decontamination) should be considered if recommended by the agent summary statement, as determined by risk assessment, the site conditions, or other applicable federal, state, or local regulations.

Biosafety Level 4 (BSL-4)

Biosafety level 4 practices, safety equipment, and facility design and construction are applicable for work with dangerous and exotic agents that pose a high individual risk of life-threatening disease, which may be transmitted via the aerosol route and for which there is no available vaccine or therapy. Agents with a close or identical antigenic relationship to Biosafety Level 4 agents also should be handled at this level. Viruses such as Marburg or Congo-Crimean haemorrhagic fever are manipulated at Biosafety Level 4. The primary hazards to personnel working with Biosafety Level 4 agents are respiratory exposure to infectious aerosols, mucous membrane or broken skin exposure to infectious droplets, and autoinoculation. All manipulations of potentially infectious diagnostic materials, isolates, and naturally or experimentally infected animals, pose a high risk of exposure and infection to laboratory personnel, the community, and the environment. The laboratory worker's complete isolation from aerosolized infectious materials is accomplished primarily by working in a Class III BSC or in a full-body, air-supplied positive-pressure personnel suit. The Biosafety Level 4 facility itself is generally a separate building or completely isolated zone with complex, specialized ventilation requirements and waste management systems to prevent release of viable agents to the environment.

The following standard and special safety practices equipment, and facilities apply to agents assigned to Biosafety Level 4:

A. *Standard microbiological practices*—As given under section Biosafety Level 3.

B. *Special practices*

1. Only persons whose presence in the facility or individual laboratory rooms is required for program or support purposes are authorized to enter. Persons who are immunocompromised or immunosuppressed may be at risk of acquiring infections. Therefore, persons who may be at increased risk of acquiring infection or for whom infection may be unusually hazardous, such as children or pregnant women, are not allowed in the laboratory or animal rooms.

   The *supervisor* has the final responsibility for assessing each circumstance and determining who may enter or work in the laboratory. Access to the facility is limited by means of secure, locked doors; accessibility is managed by the laboratory director, biohazard control officer, or other person responsible for the physical security of the facility. Before entering, persons are advised of the potential biohazards and instructed as to appropriate safeguards for ensuring their safety. Authorized persons comply with the instructions and all other applicable entry and exit procedures. A logbook, signed by all personnel, indicates the date and time of each entry and exit. Practical and effective protocols for emergency situations are established.

2. When infectious materials or infected animals are present in the laboratory or animal rooms, hazard warning signs, incorporating the universal biohazard symbol, are posted on all access doors. The sign identifies the agent, lists the name of the laboratory director or other responsible person(s), and indicates any special requirement for entering the area (e.g. the need for immunization or respirators).

3. The *laboratory director* is responsible for ensuring that, before working with organisms at Biosafety Level 4, all personnel demonstrate a high proficiency in standard microbiological practices and techniques, and in the special practices and operations specific to the laboratory facility. This might include prior experience in handling human pathogens or cell cultures, or a specific training program provided by the laboratory director or other competent scientist proficient in these unique safe microbiological practices and techniques.

4. Laboratory personnel receive available immunization for the agents handled or potentially present in the laboratory.

5. Baseline serum samples for all laboratory and other at-risk personnel are collected and stored. Additional serum specimens may be periodically collected, depending on the agents handled or the function of the laboratory. The decision to establish a serologic surveillance program takes into account the availability of methods for the assessment

of antibody to the agent(s) of concern. The program provides for the testing of serum samples at each collection interval and the communication of results to the participants.

6. A biosafety manual is prepared or adopted. Personnel are advised of special hazards and are required to read and follow instructions on practices and procedures.

7. Laboratory and support personnel receive appropriate training on the potential hazards associated with the work involved, the necessary precautions to prevent exposures, and the exposure evaluation procedures. Personnel receive annual updates or additional training as necessary for procedural changes.

8. Personnel enter and leave the laboratory only through the clothing change and shower rooms. They take a decontaminating shower each time they leave the laboratory. Personnel use the airlocks to enter or leave the laboratory only in an emergency.

9. Personal clothing is removed in the outer clothing change room and kept there. Complete laboratory clothing, including undergarments, pants and shirts or jumpsuits, shoes, and gloves, is provided and used by all personnel entering the laboratory. When leaving the laboratory and before proceeding into the shower area, personnel remove their laboratory clothing in the inner change room. Soiled clothing is autoclaved before laundering.

10. Supplies and materials needed in the facility are brought in by way of the double-doored autoclave, fumigation chamber, or airlock, which is appropriately decontaminated between each use. After securing the outer doors, personnel within the facility retrieve the materials by opening the interior doors of the autoclave, fumigation chamber, or airlock. These doors are secured after materials are brought into the facility.

11. A high degree of precaution must always be taken with any contaminated sharp item, including needles and syringes, slides, pipettes, capillary tubes, and scalpels.

    a. Needles and syringes or other sharp instruments are restricted in the laboratory for use only when there is no alternative, such as for parenteral injection, phlebotomy, or aspiration of fluids from laboratory animals and diaphragm bottles. Plastic ware should be substituted for glassware whenever possible.

    b. Only needle-locking syringes or disposable syringe-needle units (i.e., needle is integral to the syringe) are used for injection or aspiration of infectious materials. Used disposable needles must not be bent, sheared, broken, recapped, removed from disposable syringes, or otherwise manipulated by hand before disposal; rather, they must be carefully placed in conveniently located puncture-resistant containers used for sharps disposal. Non-disposable sharps must be placed in a hard-walled container for transport to a processing area for decontamination, preferably by autoclaving.

c. Syringes that re-sheath the needle, needleless systems, and other safety devices are used when appropriate.

d. Broken glassware must not be handled directly by hand, but must be removed by mechanical means such as a brush and dustpan, tongs, or forceps. Containers of contaminated needles, sharp equipment, and broken glass must be decontaminated before disposal, according to any local, state, or federal regulations.

12. Biological materials to be removed from the Class III cabinet or from the Biosafety Level 4 laboratory in a viable or intact state are transferred to an unbreakable, sealed primary container and then enclosed in an unbreakable, sealed secondary container. This is removed from the facility through a disinfectant dunk tank, fumigation chamber, or an airlock designed for this purpose.

13. No materials, except biological materials that are to remain in a viable or intact state, are removed from the Biosafety Level 4 laboratory unless they have been autoclaved or decontaminated before they leave the laboratory. Equipment or material that might be damaged by high temperatures or steam may be decontaminated by gaseous or vapour methods in an airlock or a chamber designed for this purpose.

14. Laboratory equipment is decontaminated routinely after work with infectious materials is finished, and especially after overt spills, splashes, or other contamination with infectious materials. Equipment is decontaminated before it is sent for repair or maintenance.

15. Spills of infectious materials are contained and cleaned up by appropriate professional staff or others properly trained and equipped to work with concentrated infectious material. A spill procedure is developed and posted within the laboratory.

16. A system is established for reporting laboratory accidents and exposures and employee absenteeism, and for the medical surveillance of potential laboratory-associated illnesses. Written records are prepared and maintained. An essential adjunct to such a reporting-surveillance system is the availability of a facility for the quarantine, isolation, and medical care of personnel with potential or known laboratory-associated illnesses.

17. Materials not related to the experiment being conducted (e.g., plants, animals, and clothing) are not permitted in the facility.

C. *Safety equipment (primary barriers)*

All procedures within the facility are conducted in the Class III biological safety cabinet or in Class II biological safety cabinets used in conjunction with one-piece positive pressure personnel suits ventilated by a life support system.

D. *Laboratory facility (secondary barriers)*

There are two models for Biosafety Level 4 laboratories: (a) the Cabinet Laboratory where all handling of the agent is performed in a Class III Biological Safety Cabinet, and (b) the Suit Laboratory where personnel wear a protective suit. Biosafety Level-4 laboratories may be based on either model or a combination of both models in the same facility. If a combination is used, each type must meet all the requirements identified for that type.

(A) Cabinet Laboratory

1. The Biosafety Level 4 facility consists of either a separate building or a clearly demarcated and isolated zone within a building. The rooms in the facility are arranged to ensure passage through a minimum of two doors prior to entering the room(s) containing the Class III biological safety cabinet (cabinet room). Outer and inner change rooms separated by a shower are provided for personnel entering and leaving the cabinet room. A double-door autoclave, dunk tank, fumigation chamber, or ventilated anteroom for decontamination is provided at the containment barrier for passage of those materials, supplies, or equipment that are not brought into the cabinet room through the change room.

2. Daily inspections of all containment parameters (e.g. directional airflow) and life support systems are completed before laboratory work is initiated to ensure that the laboratory is operating according to its operating parameters.

3. Walls, floors, and ceilings of the cabinet room and inner change room are constructed to form a sealed internal shell which facilitates fumigation and is resistant to entry and exit of animals and insects. Floors are integrally sealed and coved. The internal surfaces of this shell are resistant to liquids and chemicals to facilitate cleaning and decontamination of the area. All penetrations in these structures and surfaces are sealed. Openings around doors into the cabinet room and inner change room are minimized and are capable of being sealed to facilitate decontamination. Any drain in the cabinet room floor is

connected directly to the liquid waste decontamination system. Sewer vents and other service lines contain HEPA filters and protection against vermin.

4. Bench tops have seamless or sealed surfaces which are impervious to water and are resistant to moderate heat and the organic solvents, acids, alkalis, and chemicals used to decontaminate the work surfaces and equipment.

5. Laboratory furniture is of simple open construction, capable of supporting anticipated loading and uses. Spaces between benches, cabinets, and equipment are accessible for cleaning and decontamination. Chairs and other furniture used in laboratory work should be covered with a non-fabric material that can be easily decontaminated.

6. A hands-free or automatically operated hand washing sink is provided near the door of the cabinet room(s) and the outer and inner change rooms.

7. If there is a central vacuum system, it does not serve areas outside the cabinet room. In-line HEPA filters are placed as near as practicable to each use point or service cock. Filters are installed to permit in-place decontamination and replacement. Other liquid and gas services to the cabinet room are protected by devices that prevent backflow.

8. If water fountains are provided, they are automatically or foot-operated and are located in the facility corridors outside the laboratory. The water service to the fountain is isolated from the distribution system supplying water to the laboratory areas and is equipped with a backflow preventer.

9. Access doors to the laboratory are self-closing and lockable.

10. All windows are breakage-resistant and sealed.

11. Double-door autoclaves are provided for decontaminating materials passing out of both the Class III biological safety cabinet(s) and the cabinet room(s). Autoclaves that open outside of the containment barrier must be sealed to the wall of the containment barrier. The autoclave doors are automatically controlled so that the outside door can only be opened after the autoclave "sterilization" cycle has been completed.

12. Pass-through dunk tanks, fumigation chambers, or equivalent decontamination methods are provided so that materials and equipment that cannot be decontaminated in the autoclave can be safely removed from both the Class III biological safety cabinet(s) and the cabinet room(s).

13. Liquid effluents from the dirty-side inner change room (including toilets) and cabinet room sinks, floor drains (if used), autoclave chambers, and other sources within the cabinet room are decontaminated by a proven method, preferably heat treatment, before being discharged to the sanitary sewer. Effluents from showers and clean-side toilets may be discharged to the sanitary sewer without treatment. The process used for decontamination of liquid wastes must be validated physically and biologically.

14. A dedicated non-recirculating ventilation system is provided. The supply and exhaust components of the system are balanced to ensure directional airflow from the area of least hazard to the area(s) of greatest potential hazard. The differential pressure/directional airflow between adjacent areas is monitored and alarmed to indicate any system malfunction. An appropriate visual pressure monitoring device that indicates and confirms the pressure differential of the cabinet room is provided and located at the entry to the clean change room. The airflow in the supply and exhaust components is monitored and the HVAC control system is designed to prevent sustained positive pressurization of the laboratory. The Class III cabinet should be directly connected to the exhaust system. If the Class III cabinet is connected to the supply system, it is done in a manner that prevents positive pressurization of the cabinet.

15. The supply air to and exhaust air from the cabinet room, inner change room, and anteroom pass through HEPA filter(s). The air is discharged away from occupied spaces and air intakes. The HEPA filter(s) are located as near as practicable to the source in order to minimize the length of potentially contaminated ductwork. All HEPA filters need to be tested and certified annually. The HEPA filter housings are designed to allow for *in situ* decontamination of the filter prior to removal, or removal of the filter in a sealed, gas-tight primary container for subsequent decontamination and/or destruction by incineration. The design of the HEPA filter housing should facilitate validation of the filter installation. The use of pre-certified HEPA filters can be an advantage. The service life of the exhaust HEPA filters can be extended through adequate prefiltration of the supply air.

16. The Biosafety Level 4 facility design and operational procedures must be documented. The facility must be tested for verification that the design and operational parameters have been met prior to operation. Facilities should be re-verified annually against these procedures as modified by operational experience.

17. Appropriate communication systems are provided between the laboratory and the outside (e.g. voice, fax, computer).

(B) Suit Laboratory

1. The Biosafety Level 4 facility consists of either a separate building or a clearly demarcated and isolated zone within a building. The rooms in the facility are arranged to ensure passage through the changing and decontamination areas prior to entering the room(s) where work is done with BSL-4 agents (suit area). Outer and inner change rooms separated by a shower are provided for personnel entering and leaving the suit area. A specially designed suit area is maintained in the facility to provide personnel protection equivalent to that provided by Class III biological safety cabinets. Personnel who enter this area wear a one-piece positive pressure suit that is ventilated by a life-support system protected by HEPA filtration. The life support system includes redundant breathing air compressors, alarms and emergency backup breathing air tanks. Entry to this area is through an airlock fitted with airtight doors. A chemical shower is provided to decontaminate the surface of the suit before the worker leaves the area. An automatically starting emergency power source is provided at a minimum for the exhaust system, life support systems, alarms, lighting, entry and exit controls, and BSCs. The air pressure within the suit is positive to the surrounding laboratory. The air pressure within the suit area is lower than that of any adjacent area. Emergency lighting and communication systems are provided. All penetrations into the internal shell of the suit area, chemical shower, and airlocks, are sealed.

2. A daily inspection of all containment parameters (e.g. directional airflow, chemical showers) and life support systems is completed before laboratory work is initiated to ensure that the laboratory is operating according to its operating parameters.

3. A double-doored autoclave is provided at the containment barrier for decontaminating waste materials to be removed from the suit area. The autoclave door, which opens to the area external to the suit area, is sealed to the outer wall of the suit area and is automatically controlled so that the outside door can be opened only after the autoclave "sterilization" cycle. A dunk tank, fumigation chamber, or ventilated airlock for decontamination is provided for passage of materials, supplies, or equipment that are not brought into the suit area through the change room. These devices can be also used for the safe removal of materials, supplies, or equipment from the laboratory that cannot be decontaminated in the autoclave.

4. Walls, floors, and ceilings of the suit area are constructed to form a sealed internal shell, which facilitates fumigation and is animal- and insect-prohibitive (*see* Appendix G). The internal surfaces of this shell are resistant to liquids and chemicals, facilitating cleaning and decontamination of the area. All penetrations in these structures and surfaces are sealed. Any drain in the floor of the suit area contain traps filled with a chemical disinfectant of demonstrated efficacy against the target agent, and they are connected directly to the liquid waste decontamination system. Sewer vents and other service lines contain HEPA filters.

5. Internal facility appurtenances in the suit area, such as light fixtures, air ducts, and utility pipes, are arranged to minimize the horizontal surface area.

6. Bench tops have seamless surfaces which are impervious to water and are resistant to moderate heat and the organic solvents, acids, alkalis, and chemicals used to decontaminate the work surfaces and equipment.

7. Laboratory furniture is of simple open construction capable of supporting anticipated loading and uses. Non-porous materials are preferable. Spaces between benches, cabinets, and equipment are accessible for cleaning and decontamination. Chairs and other furniture used in laboratory work should be covered with a non-fabric material that can be easily decontaminated.

8. A hands-free or automatically operated hand washing sink is provided in the suit area(s); hand washing sinks in the outer and inner change rooms should be considered based on the risk assessment.

9. If there is a central vacuum system, it does not serve areas outside the suit area. In-line HEPA filters are placed as near as practicable to each use point or service cock. Filters are installed to permit in-place decontamination and replacement. Other liquid and gas services to the suit area are protected by devices that prevent backflow.

10. Access doors to the laboratory are self-closing and lockable. Inner and outer doors to the chemical shower and inner and outer doors to airlocks are interlocked to prevent both doors from being opened simultaneously.

11. All windows are breakage-resistant and are sealed.

12. Liquid effluents from sinks, floor drains (if used), autoclave chambers and other sources within the containment barrier are decontaminated by a proven method, preferably heat treatment, before being discharged to the sanitary sewer. Effluents from showers and toilets may be discharged to the sanitary sewer without treatment. The process used

for decontamination of liquid wastes must be validated physically and biologically.

13. A dedicated non-recirculating ventilation system is provided. The supply and exhaust components of the system are balanced to ensure directional airflow from the area of least hazard to the area(s) of greatest potential hazard. Redundant supply fans are recommended. Redundant exhaust fans are required. The differential pressure/directional airflow between adjacent areas is monitored and alarmed to indicate malfunction of the system. An appropriate visual pressure monitoring device that indicates and confirms the pressure differential of the suit area must be provided and located at the entry to the clean change room. The airflow in the supply and exhaust components is monitored and an HVAC control system is installed to prevent positive pressurization of the laboratory.

14. The supply air to the suit area, decontamination shower, and decontamination airlock is protected by passage through a HEPA filter. The general room exhausts air from the suit area, decontamination shower and decontamination airlock is treated by a passage through two HEPA filters in series prior to discharge to the outside. The air is discharged away from occupied spaces and air intakes. The HEPA filters are located as near as practicable to the source in order to minimize the length of potentially contaminated ductwork. All HEPA filters need to be tested and certified annually. The HEPA filter housings are designed to allow for *in situ* decontamination of the filter prior to removal. Alternatively, the filter can be removed in a sealed, gas-tight primary container for subsequent decontamination and/or destruction by incineration. The design of the HEPA filter housing should facilitate validation of the filter installation. The use of pre-certified HEPA filters can be an advantage. The service life of the exhaust HEPA filters can be extended through adequate prefiltration of the supply air.

15. The positioning of the supply and exhaust points should be such that dead air space in the suit room is minimized.

16. The treated exhaust air from Class II biological safety cabinets, located in a facility where workers wear a positive pressure suit, may be discharged into the room environment or to the outside through the facility air exhaust system. If the treated exhaust is discharged to the outside through the facility exhaust system, it is connected to this system in a manner that avoids any interference with the air balance of the cabinets or the facility exhaust system.

17. The Biosafety Level 4 facility design and operational procedures must be documented. The facility must be tested for verification that the design and operational parameters have been met prior to operation. Facilities should be re-verified annually against these procedures as modified by operational experience.

18. Appropriate communication systems should be provided between the laboratory and the outside.

BIOSAFETY CABINETS (BSC)

Class I BSC These are negative pressure, ventilated cabinets designated for general microbiological research work with low and moderate risk agents. Figure 17.1 shows the class I biosafety cabinet.

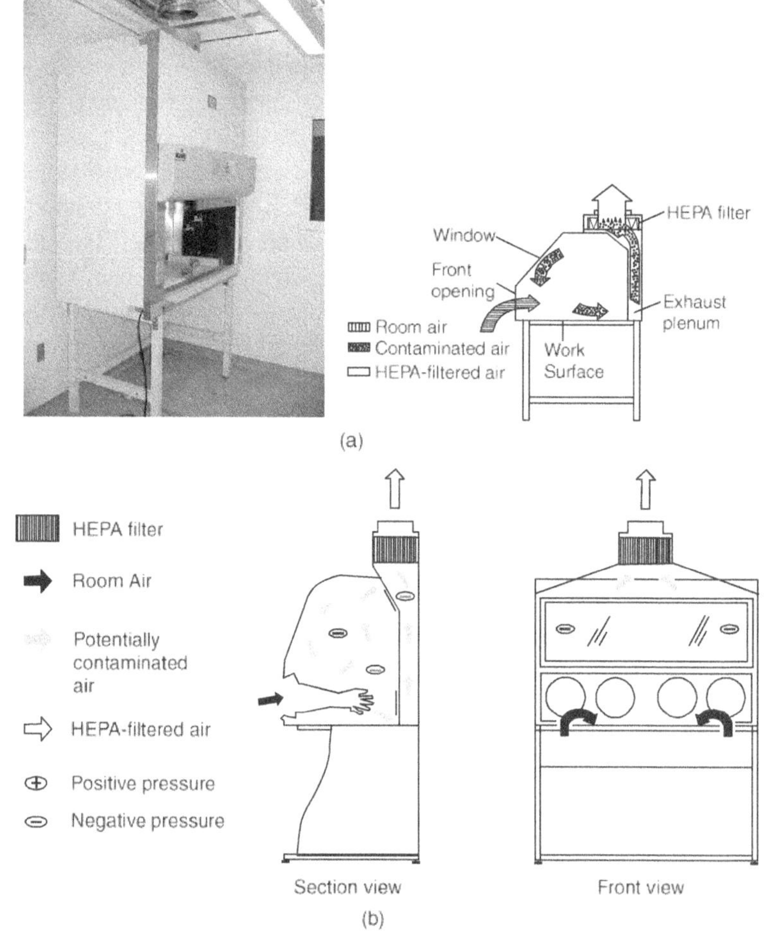

Figure 17.1 Class I biosafety cabinet

Class II BSC   These are designed with inward air flow to protect personnel. HEPA-filtered downward vertical laminar air flow of product protection. Class II BSC are classified into two types, A and B, based on construction, air flow velocities and patterns and exhaust systems. Figure 17.2 shows class II biosafety cabinet.

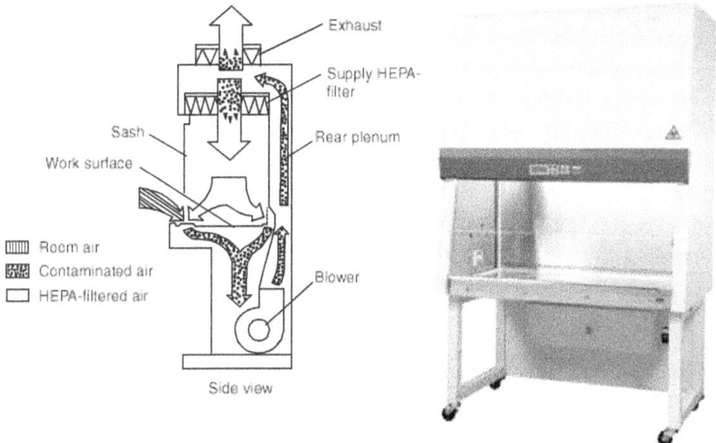

Figure 17.2   Class II biosafety cabinets

Type A cabinets   are suitable for microbiological research in the absence of volatile or toxic chemicals and radionucleotides, since air is recirculated within the cabinet (Figure 17.3). **Type B cabinets** are subtyped into B1, B2 and B3. The B cabinets are connected to the building exhaust system and contain negative pressure. This allows working with toxic chemicals and radioisotopes (Figure 17.4).

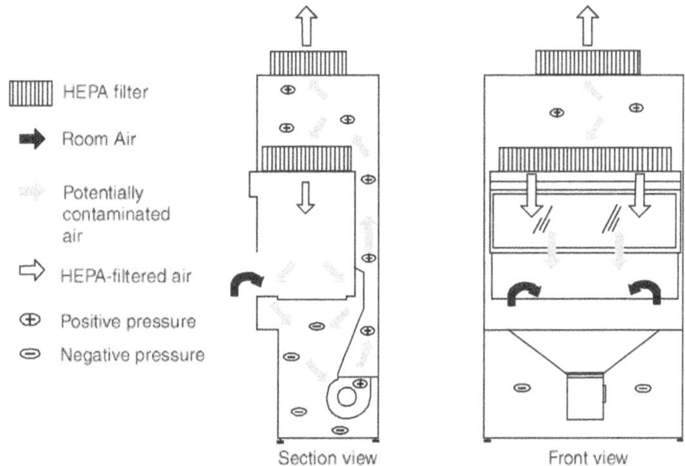

Figure 17.3   Class II type A1 biosafety cabinet

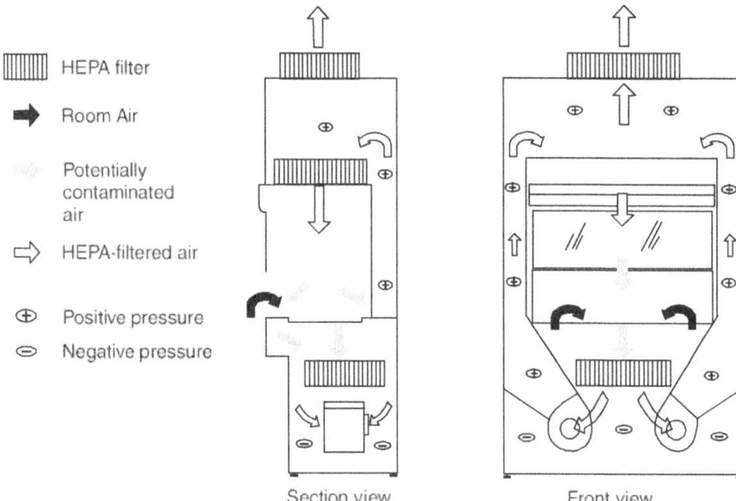

Figure 17.4 Class II type B1 biosafety cabinet

Class III BSC   These are totally enclosed, ventilated cabinets of gas-tight construction and offer highest degree of personal and environmental protection. This cabinet is operated through negative pressure. Supply air is HEPA-filtered and the cabinet exhaust air is filtered through two HEPA filters in series. Figure 17.5 shows class III biosafety cabinet.

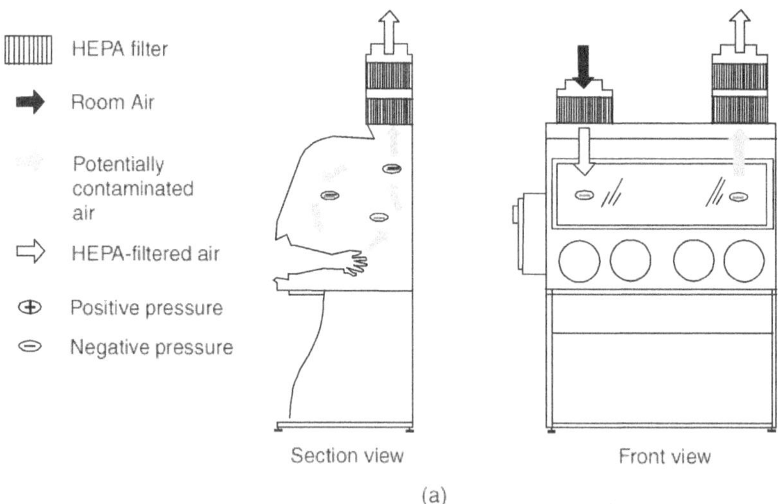

Figure 17.5 Class III biosafety cabinet (*Continues*)

(b)

Figure 17.5 Class III biosafety cabinet

GENETIC ENGINEERING—SAFETY, SOCIAL, MORAL AND ETHICAL CONSIDERATIONS

Genetic modification involves the transfer of genetic material between different organisms by artificial methods and is now increasingly being used beneficially in the fields of health care, agriculture, food production, industrial enzyme production and environmental management. While there are undoubted benefits from the use of transgenic organisms, it is important to ensure that such systems do not cause problems of safety to people and environment or create unacceptable social, moral and ethical issues.

RELEASE OF GENETICALLY MANIPULATED ORGANISMS TO THE ENVIRONMENT

Many people fear that genetically engineered microorganisms could escape from the laboratory into the environment, with unpredictable and perhaps catastrophic consequences. It was believed that such released microorganism could 'upset the balance of nature' or that foreign DNA in the new microorganisms could alter its metabolic activity in unpredictable and undesirable ways.

In response to these concerns, guidelines were established to ensure safe working practices and levels of containment based on potential hazards. Many important medical products such as insulin and human growth hormone and some industrial enzymes are manufactured in large-scale containment fermentation processes, involving specific GMOs. The final products from these processes are free from the genetically manipulated host organism and, therefore, do not constitute a release problem. Quite recently, rennet (chymosin) for cheese manufacture has been produced from genetically manipulated microorganisms. Recombinant microorganisms are now being considered for deliberate release into the environment, where they cannot be contained, e.g. biological control, inoculants in agriculture, live vaccines, bioremediation, baker's and brewer's yeasts. Increased pathogenicity of microorganisms or microbial ability to destroy essential raw materials are often cited as potential problems of GMOs. Organisms with any possibility of unusual pathogenicity will never be permitted to be used. Where microorganisms are released to be used for biocontrol of, for example, insects, care must be taken that they will not influence other life forms. The use of a recombinant rabies vaccine in baits in Belgium has significantly reduced the level of rabies in wild animals. The public were informed and in general approved this worthwhile use of the technology.

The overall aim of the food industry with respect to genetic engineering will be to improve the quantity and to increase the quality and properties of existing food productions, to produce new products and, of course, to improve financial returns. New biotechnology now offers a major opportunity to tailor food products to public demand.

While the public have readily accepted medical products produced from GMOs they are much less willing to accept such procedures with food. Genetic engineering is seen as 'unnatural' and unnecessary in food production.

Transgenesis is seen by some as a fundamental breach in natural breeding barriers that nature set up through the process of evolution to prevent genetic interplay between unlike species. In this way the species is seen as 'sacred'. In the viewpoint of many molecular biologists the gene has become the ultimate unit of life—the gene is merely a unique aggregation of organic molecules (common to all types of cell) available for manipulation. Consequently, they see no ethical problem in transferring genes between species and genera.

Risks associated with GM crops and foods have been identified (Stewart *et al.*, 2000). Ecological biosafety research has identified potential risks associated with certain crop-transgene combinations, such as intra- and interspecific transgene flow, persistence and the consequences of transgenes in unintended hosts. Resistance management strategies for insect resistance transgenes and non-target effects of these genes have also been studied. Food biosafety research has focused on transgenic product toxicity and allergenicity.

Currently, the public interest in biosafety issues has focused on the discussions surrounding the use of genetically modified organisms, very specifically on the use of transgenic plants in agriculture. Attention is now focused on the problem of control of new and re-emerging infectious diseases, the need for new vaccines, control of transport and routes of dissemination, biosafety information exchange and networking, where research results are dearly needed. In the area of modern biotechnology, new applications such as gene therapy and transgenic animals will be on the list of future priorities for biosafety-related activities and research (Doblhoff-Dier and Collins, 2001).

REVIEW QUESTIONS

1. What do you understand by the term patent? How can patent be protected?
2. What are patentable and non-patentable items or inventions?
3. What are the different biosafety levels and their requirements?

REFERENCES

Doblhoff-Dier, P. and Collins, C.H. (2001). "Biosafety: future priorities for research in health care." *J. Biotechnol.* 85: 227–239.

Stewart, C.N. Jr., Richards, H.A. and Halfhill, M.D. (2000). "Transgenic plants and biosafety: science, misconceptions and public perceptions." *Biotechniques.* 29: 832–836, 838–843.

GLOSSARY

Adenine A purine base, 6-aminopurine, occurring in ribonucleic acid (RNA) as well as in deoxyribonucleic acid (DNA) and a component of adenosine diphosphate (ADP) and adenosine triphosphate (ATP). Adenine pairs with thymine in DNA and uracil in RNA.

Adeno-associated virus (AAV) The smallest of known human viruses. There is no disease which has been to date associated with AAV. The virus causes very mild immune response and can infect non-dividing cells. It incorporates into the host cell's genome, but there is no evidence that it can cause malignant transformation. Because of these features it presents a very attractive subject for creating vectors for gene therapy.

Adenosine Refers to the **nucleoside** (i.e., hybrid-with-ribose or -deoxyribose) **form** of adenine.

Adenosine triphosphate (ATP) The major carrier of chemical energy in the cells of all living things on this planet. A ribonucleoside 5´- triphosphate functioning as a phosphate-group donor in the energy cycle of the cell. ATP contains three phosphate/oxygen molecules linked together. When a phosphate–phosphate bond in ATP is broken (hydrolysed), energy that the cell can use to carry out its functions is produced. Thus, ATP serves as the universal medium of biological energy storage and exchange in living cells.

Adenovirus A type of virus that can infect humans. Like all viruses, it can reproduce only inside living cells (of other host organisms). Adenovirus causes a protein (metabolite) to be made that disables the *p53* gene. Because the *p53* gene then cannot perform its usual function (i.e., prevention of uncontrolled cell growth caused by virus/DNA damage), the adenovirus thus "takes over" and causes the cell to make numerous copies of the virus until the cell dies (thus, releasing the virus copies into the body of the host organism to cause further infection).

Adult stem cell This term referred historically to as stem cell that is (extracted or) derived from the bone marrow tissue of adults; but scientists are now discovering additional types of "adult" stem cells within other tissues of older-than-infant humans. In addition to bone marrow, these adult stem cells have been found in liver, skin, adipose (fat), testicle, intestine, menstrual blood, and brain tissue. Similar to human embryonic stem cells, adult stem cells can—under certain conditions—differentiate/proliferate into cell types specific to many of the human body's 210 different types of tissue.

Affinity chromatography A method of separating a mixture of proteins or nucleic acids (molecules) by specific interactions of those molecules with a component known as a ligand, which is immobilized on a support. If a solution of, say, a mixture of proteins is passed over (through) the column, one of the proteins binds to the ligand on the basis of specificity and high affinity (they fit together like a lock and key). If there is no naturally occurring "lock" inherent on the desired protein molecule (to go with the ligand's "key"), then the scientist can add an affinity tag during the synthesis of the protein (e.g. in a cell-free gene expression system) to act as that protein molecule's "key".

AFLP Acronym for amplified fragment length polymorphism. Also known by its acronym AFLP, it is "DNA marker" utilized in a "genetic mapping" technique which utilizes the specific sequence of bases (nucleotides) in a piece of DNA (from an organism). Since the specific sequence of bases in their DNA molecules is different for each species, strain, variety, individual (due to DNA polymorphism), AFLP can be utilized to "map" those DNA molecules (e.g. to assist and speed up plant breeding programs).

Agarose A highly purified form of agar. Used as a stationary phase (substrate) in some chromatographic and electrophoretic methods.

Alpha-amylase Although found in many tissues, amylase is most prominent in pancreatic juice and urine which each have their own isoform of human α-amylase. They behave differently on isoelectric focusing, and can also be separated in testing by using specific monoclonal antibodies. In humans, all amylase isoforms link to chromosome 1p21.

Amperometric biosensors Biosensors that function by the production of a current when a potential is applied between two electrodes. They generally have response times, dynamic ranges and sensitivities similar to the potentiometric biosensors. The simplest amperometric biosensors in common usage involve the Clark oxygen electrode.

Amplicon A specific sequence of DNA that is produced by a DNA amplification technology such as the polymerase chain reaction (PCR) technique.

Amplification The production of additional copies of a chromosomal sequence, found in the form of either intrachromosomal or extrachromosomal DNA.

Amplimer *See Amplicon.*

Amylase A term that is used to refer to a category of enzymes that catalyses the chemical reaction in which amylose (starch) molecules are hydrolytically cleaved ('broken') to molecular pieces (e.g. the polysaccharides maltose, maltotriose, α-dextrin, etc.).

Anneal The process by which the complementary base pairs in the strands of DNA combine.

Antibiotic Coined by Selman Waksman during the 1940s, this term refers to organic compounds that are naturally formed and secreted by various species of microorganisms and/or plants. It has a defensive function and is often toxic to other species (e.g. penicillin, originally produced by bread mould, is toxic to numerous human pathogens). Antibiotics generally act by inhibiting protein synthesis, DNA replication, synthesis of cell wall (cytoskeleton) constituents, inhibition of required cell (e.g. bacteria) metabolic processes, and nucleic acid (DNA and RNA) biosynthesis; hence killing the (targeted bacteria) cells involved.

Antibiotic resistance A property of a cell (e.g. pathogenic bacteria) that enables it to avoid the effect of an antibiotic that had formerly killed or inhibited that cell.

Antibody Also called immunoglobulin, Ig. A large defense protein that consists of two classes of polypeptide chains, light (L) chains and heavy (H) chains. A single antibody molecule consists of two identical copies of the L chain and two of the H chain. They are synthesized (i.e., made) by the immune system (B lymphocytes) of the organism. The antibody is composed of four proteins linked together to form a Y-shaped bundle of proteins (looks somewhat like a slingshot, or two hockey sticks taped together at the handles).

Anticodon A specific sequence of three nucleotides in a transfer RNA (tRNA), complementary to a codon (also three nucleotides) for an amino acid in a messenger RNA.

Antigen Also called an immunogen. Any large molecule or small organism whose entry into the body provokes synthesis of an antibody or immunoglobin (i.e., an immune system response).

Antigenic determinant The antigenic determinant is the epitope of an antigen. These are the immunologically active regions of an immunogen that bind to antigen-specific membrane receptors on lymphocytes or to secreted antibodies.

Anti-idiotype antibodies These are antibodies which recognize the binding sites of other antibodies. Their binding sites are complementary to the binding site of another immunoglobulin.

Anti-idiotypes Antibodies to antibodies. In other words, if a human antibody is injected into rabbits, the rabbit immune systems will recognize the human antibodies as foreign (regardless of the fact that they are antibodies) and produce antibodies against them. To the rabbit the foreign antibodies represent just another invader or non-self to be targeted and destroyed. Anti-idiotypes mimic antigens in that they are shaped to fit into the antibody's binding site (in lock-and-key fashion). As such, anti-idiotypes can be used to create vaccines that stimulate production of antibodies to the antigen (that the anti-idiotype mimics). This confers disease resistance (to the pathogen associated with that antigen) without the risk that a vaccine using attenuated pathogens entails (i.e., that the pathogen "revives" to cause the disease).

Antioxidants Compounds (e.g. phytochemicals) that act to prevent lipids from oxidizing (e.g. to plaque), breaking down (e.g. to carcinogenic compounds), or that act to capture and halt singlet oxygen (O^-) free radicals; which can damage DNA in cells (i.e., causing mutations). Since oxidation of lipids in the blood is the initial step in atherosclerosis, consumption of large amounts of certain antioxidants (e.g. flavonoids, melanoidins, etc.) may help prevent atherosclerosis.

Antisense (DNA sequence) A strand of DNA that produces a messenger RNA (mRNA) molecule which (when reversed end-for-end) has the same sequence as (i.e., is complementary to) the unwanted ("bad") messenger RNA. The Sense (i.e., forward) and Antisense (i.e., backward) mRNA strands hybridize (i.e., tightly bond to each other), which prevents the bonded-pair from leaving the cell's nucleus, so that bonded-pair is rapidly degraded (destroyed) by nucleases within the cell nucleus.

Apoptosis Also called "programmed cell death," it is a series of programmed steps that cause a cell to die via "self digestion" without rupturing and releasing intracellular contents (e.g. nucleus, chromosomes, refractile bodies, etc.) into the local (i.e., surrounding tissue) environment. Manifestations of cell apoptosis include shrinking of the cell's cytoplasm and chromatin condensation, and the presence of phosphatidyl serine on the exterior surface of cell's plasma membrane.

Aptamers   Single-stranded RNA molecules that form extended three-dimensional structures which bind (i.e., "stick to") other, specific molecules (e.g. proteins) and sometimes inactivate the molecules they "stick" to. Aptamer is from the Latin *aptus* ("to fit").

Artificial enzymes   Synthetic, organic molecules prepared to re-create the active site of an enzyme.

Ascites   Liquid accumulations in the peritoneal cavity. Used as an input in one of the methods for producing monoclonal antibodies.

Autoradiography   A technique to detect radioactively labelled molecules by creating an image on photographic film. The slab of gel or other material in which the molecules are held (suspended) is placed on top of a piece of photographic film. The two are then securely fastened together such that movement is eliminated and the film is exposed for a period of time. The exposed (to the radiation) film is subsequently developed and the radioactive area is seen as a dark (black) area. Among other uses, autoradiography has been used to track the spread of (radioactively labelled) viruses in a living plant. After treatment (i.e., the radioactive labelling process), the whole plant (in a slab) is placed on top of a piece of photographic film. When the film is subsequently developed, the "picture" seen is of a plant, with darker areas indicating regions of greater virus concentration.

Avidin   A protein that is naturally present in egg white, oilseed protein (e.g. soybean meal), and grain (e.g. corn/maize). The protein is 70 kilodaltons in mass (weight) and has a high affinity for biotin (i.e., it "sticks" tightly to the biotin molecule).

Avidity (of an antibody)   The "tightness of fit" between a given antibody's combining site and the antigenic determinant that it combines with. It is the firmness of the combination of antigen with antibody.

Bacterial artificial chromosomes (BAC)   Pieces of DNA (e.g. plant DNA) that have been cloned (made) inside living bacteria (e.g. by plant researchers who need to "manufacture" some pieces of plant DNA). They can be utilized as vectors (for genetic engineering), to carry (inserted) genes into certain organisms.

Bacterial expressed sequence tags   These are ESTs (expressed sequence tags) that are based on sequenced/mapped bacterial genes instead of the genes of ("traditional" EST) *C. elegans* nematode. They are utilized to "label" a given gene (i.e., in terms of that gene's function/protein).

Bacteriophage   Discovered in 1917 by Felix d'Herelle (fr. "bacteria eaters"), a bacteriophage is a virus that attaches to, injects its DNA into, and multiplies inside bacteria, which eventually causes the bacteria to die. Upon the bacteria's death, 10–200 more new bacteriophages are released to infect other bacteria. Often abbreviated as simply **phage**.

Baculovirus   A class of virus that infects lepidopteran insects (e.g. cotton bollworm or gypsy moth larva). Baculoviruses can be modified via genetic engineering to insert new genes into the larva, causing those larva to then produce proteins desired by man (e.g. pharmaceuticals).

Baculovirus expression vector (BEV)   Refers to vectors (used by researchers to carry new genes into insect cells) in which the agent is a baculovirus (i.e., a virus that infects certain types of insect cells only). A genetically engineered BEV is commonly utilized to carry a new gene into the insect cells within a baculovirus expression vector system (BEVS), to induce cell culture production of a protein desired by man.

Base pair (bp)   Two nucleotides that are in different nucleic acid chains and whose bases pair (interact) by hydrogen bonding. In DNA, the nucleotide bases are adenine (which pairs with thymine) and guanine (which pairs with cytosine).

Base pairing   Refers to the hydrogen-bonding driven matching of complementary base pairs (e.g. components of DNA molecules) during replication of DNA in cells, or during hybridization (e.g. of a DNA sample fragment to a DNA probe).

BHK cells   Abbreviation for baby hamster kidney cells. This refers to cell lines propagated/grown in cell culture (e.g. in petri dishes) that were originally removed from the kidney of baby hamster(s).

Biofilm   A layer of microorganisms growing on a surface, in a bed of polymer material which they themselves have made. Biofilms tend to form wherever a surface on which bacteria can grow is exposed to some suitable growth medium and to a supply of bacteria.

Biofilter   A large-scale filtering system where the surface area of the filter provides a support for organisms. Chemicals in the fluid flowing through the filter are broken down by the organisms. This is mainly used for waste disposal.

Biofuels   Fuels made from bulk biological materials such as cane sugar or wood pulp. There are a range of ways of converting these rather bulky, inconvenient fuel materials into fuels which are useful for industrial or transport use or as starting materials for the chemical industry.

Biogas   Any fuel gas generated by biological means. It usually refers to methane (natural gas) produced by fermenting waste, and particularly sewage. Waste is incubated with suitable bacteria in a digester in the complete absence of air (an anaerobic fermenter). The organic matter in the waste is converted mainly into methane and carbon dioxide, and the methane can be burned to provide power or heat.

Bioinformatics   This is the collection, storage and organization of information of biological interest. In particular, it is concerned with organizing biomolecular databases, in getting useful information out of such databases, and in integrating information from disparate sources.

Bioluminescence   The production of light from biological systems, either enzymes or organisms. It is of substantial interest from a technical point of view, and also very pretty.

Biomass   Any bulk biological material, and by extension any large mass of biological matter. Fermentations generate biomass as well as the fermentation product, and it is important to control the amount of biomass to maintain the rate of fermentation.

Biopharmaceuticals   This has come to mean two things—a type of product and a type of company. In its original meaning, biopharmaceuticals meant pharmaceuticals that were inherent biological, and hence pharmaceutical proteins. Also these are the companies discovering and developing pharmaceuticals.

Bioreactor   A vessel in which a biological reaction or change takes place, usually a fermentation or biotransformation. Bioreactors, and indeed fermentation and biotransformation, are central to much of biotechnology.

Bioremediation   The use of organisms (e.g. plants, bacteria, fungi, etc.) to consume or otherwise help remove (e.g. biorecovery) materials (e.g. toxic chemical wastes, metals, etc.) from a contaminated site (e.g. remove toluene from the land and ponds on site of an old refinery, etc.). It is the use of biological systems—usually microorganisms—to clean up a contaminated site (environment).

Biosafety   The establishment and maintenance of safe conditions in a biological research laboratory to ensure that pathogenic microbes are contained (and not released to workers or the environment).

Biosensors (chemical)   Chemically-based devices that are able to detect and/or measure the presence of certain molecules (e.g. DNA, antigens, glucose, active ingredients of pesticides, etc.). The (chemical) biosensor consists of an immobilized enzyme (to bind the trace chemical) combined with a colour reagent (to indicate visually the presence of the trace chemical).

Biotechnology   The application of knowledge of living systems to use those systems or their components for industrial purposes. In other words, it is the means or way of manipulating life forms (organisms) to provide desirable products for man's use. For example, beekeeping and cattle breeding could be considered to be biotechnology-related endeavours. The word biotechnology was coined in 1919 by Karl Ereky, to apply to the interaction of biology with human technology.

Biotin   A B-complex vitamin, also known as vitamin H, which is essential (i.e., required) for life of many grain-eating insects; and is also essential for many of the metabolic pathways (i.e., series of chemical reactions) involved in milk production by cattle.

Blots   A range of molecular biological techniques are called "blots". They are all based on transferring biological material from a gel onto a porous membrane in a way that preserves how they were separated in the gel. There are a variety of ways of doing the transfer—passive diffusion, wicking, suction (vacuum blotting), or electric field (electroblotting).

Blunt-end DNA   A segment of DNA that has both strands terminating at the same base pair location, that is, fully base-paired DNA. No sticky ends.

Blunt-end ligation   A method of joining blunt-ended DNA fragments using the enzyme T4 ligase which can join fully base-paired, double-stranded DNA.

B lymphocytes   A class of white blood cells originating in the bone marrow and found in blood, spleen, and lymph nodes. They are the precursors of (blood) plasma cells (B cells) that secrete antibodies (IgG) directed against invading antigens (e.g. of pathogenic bacteria).

Calorimetric biosensor   These rely on biological molecules for transduction or detection; enzymes (or even cells) immobilized on a surface; enzymes catalyse a reaction which generates heat and heat is proportional to amount of substrate present.

Catalase Catalase (human erythrocyte catalase: PDB 1DGF, EC 1.11.1.6) is a common enzyme found in living organisms. Its functions include catalysing the decomposition of hydrogen peroxide to water and oxygen. Catalase has one of the highest turnover rates of all enzymes; one molecule of catalase can convert millions of molecules of hydrogen peroxide to water and oxygen per second. Catalase is a tetramer of four polypeptide chains, each over 500 amino acids long. It contains four porphyrin haeme (iron) groups which allow the enzyme to react with the hydrogen peroxide. The optimum pH for catalase is approximately neutral (pH 7.0), while the optimum temperature varies by species.

cDNA In genetics, complementary DNA (cDNA) is a single-stranded DNA synthesized from a mature mRNA template in a reaction catalysed by the enzyme reverse transcriptase. cDNA is often used to clone eukaryotic genes in prokaryotes.

C-DNA Also known as copy DNA, it is a helical form of DNA. It occurs when DNA fibres are maintained in 66 per cent relative humidity in the presence of lithium ions. It has fewer base pairs per turn than B-DNA.

cDNA clone A DNA molecule synthesized ("made") from an mRNA sequence via sequential use of reverse transcriptase (acting on mRNA) and DNA polymerase. cDNA is copy DNA (or complementary DNA). It is a DNA copy of an RNA. A collection of such cloned molecules, which represents all the genetic information expressed by a given cell or by a given tissue type, is referred to as a cDNA library.

Cell culture The *in vitro* (i.e., outside of body, in a test tube or vat) propagation of cells isolated from living organisms.

Cell differentiation The process whereby descendants of a common parental cell achieve and maintain specialization of structure and function. In humans, for instance, all the different types of cells (e.g. muscle cells, bone cells, etc.) differentiate from the zygote (itself formed by union of the simple sperm and egg). In humans, the various blood cell types (e.g. red blood cells, white blood cells, etc.) differentiate from stem cells in the bone marrow. Cell differentiation is caused/triggered/assisted by micro-RNAs, colony stimulating factors (CSFs), growth factors (GFs), and certain other proteins (e.g. hedgehog proteins).

Cell fusion The combining of cell contents of two or more cells to become a single cell. Fertilization is such a process (fusing of gametes' cells).

Cellular immune response Also called cell-mediated immunity. The immune response that is carried out by specialized cells, in contrast to the response carried out by soluble antibodies. The specialized cells that make up this group include cytotoxic T lymphocytes (CTL), helper T lymphocytes, macrophages, and monocytes. This system works in concert with the humoral immune response.

Central Dogma (old) The historical organizing principle of molecular genetics; it states that genetic information flows from DNA to RNA to protein. Stated in another way, DNA makes RNA which makes protein. This principle was first stated by Watson and Crick.

Centromere A constricted region of a chromosome that includes the site of attachment to the mitotic or meiotic spindle.

Chaperones Protein molecules inside living cells that assist with correct protein folding as the protein molecule emerges from the cell's ribosomes. Also, they help to convey those protein(s) to their ultimate destination(s) in the organism.

Chelating agent A molecule capable of "binding" metal atoms. The chelating agent/metal complex is held together by coordination bonds which have a strong polar character. One example of a common chelating agent is ethylenediamine tetraacetate (EDTA) which tightly and reversibly binds Mg^{2+} and other divalent cations (positively charged ions). If a chelate is allowed to bind to metal ions required for enzyme activity, the enzyme will be inactivated (inhibited). Cobalamin (vitamin B_{12}), EDTA and the iron–porphyrin complex of haeme (which provides the red colour of blood) are other examples of chelates.

Chemiluminescence Chemiluminescence (sometimes "chemoluminescence") is the emission of light (luminescence) without emission of heat as a result of a chemical reaction.

Chemiluminescent immunoassay (CLIA) An immunoassay (i.e., an antibody-based bioassay) which utilizes a signal that is generated by light-releasing chemical reactions (e.g. triggered by the binding of antibody to analyte).

Chemotaxis Sensing of, and movement toward or away from a specific chemical agent by living, freely moving cells (e.g. bacteria, macrophages, neutrophils, etc.).

Chemotherapy When this term was first coined by Paul Ehrlich in 1905, it was defined as any therapy (to cure diseases) via chemically synthesized drugs.

Chimera   An organism consisting of tissues or parts of diverse genetic constitution. An example of a chimera would be a centaur the half-man, half-goat figure of Greek mythology. The word "chimera" is from the mythological creature by that name which possessed the head of a lion, the body of a goat, and the tail of a serpent. The word chimera is very general and may be applied to any number of entities. For example, chimeric antibodies may be produced by cell cultures in which the variable, antigen-binding regions are of murine (mouse) origin while the rest of the molecule is of human origin.

Chimeric antibody   A (genetically engineered) antibody which combines characteristics of antibodies from two different sources. For example, the complementarity-determining (i.e., antigen-binding) portion of an animal antibody (e.g. raised against a specific antigen) with human monoclonal antibody.

Chimeric DNA   (Recombinant) DNA containing spliced genes from two different species.

Chimeric proteins   Fused proteins from different species, produced from the chimeric DNA template.

Chinese hamster ovary cells (CHO cells)   This refers to cell line(s) propagated/grown in cell culture (e.g. in petri dishes) which were originally removed from a chinese hamster. Such cell culturing of CHO cells has been done by scientists since the 1960s, to study genetics, gene expression, nutrition, etc.

Chromatids   Copies of a chromosome produced by replication within a living eukaryotic cell during the prophase (i.e., the first stage of mitosis). They are compact cylinders consisting of DNA coiled around flexible rods of histone protein.

Chromatin   From the Greek word *chroma* = colour. Named by Walter Flemming in 1882, due to the fact that chromatin's band-like structures stained darkly, chromatin is the complex of DNA and (histone) protein of which the chromosomes are composed. Consisting of fibrous swirls of unravelled DNA molecules in the nucleus of the interphase (i.e., the prolonged period of cell growth between cell division phases) eukaryotic cell. Chromatin DNA gradually coils itself around flexible rods of histone protein during the prophase (i.e., the first stage of mitosis), forming two parallel compact cylinders (called chromatids) connected by a knot-like structure (called a centromere) at their middle.

Chromatography   Coined by Mikhail S. Tswett in 1906, this word refers to a process by which complex mixtures of different molecules may be separated from each other. This is accomplished by subjecting the mixture to many repeated partitionings between a flowing phase and a stationary phase. Chromatography constitutes one of the if not the most fundamental, separation techniques used in the biochemistry/biotechnology arena to date.

Chromatography was originally developed as a way of separating pigments from plants by dissolving them and then "wicking" the solution through paper. The same basic ideas apply to all chromatographic separations: dissolve the mixture in some suitable solvent and then pass it slowly over some solid material. Depending on whether the molecules in the sample stick to the solid material (stationary or solid phase) or dissolve in the solvent (mobile phase), they either move over the solid or stay put.

Chromosome map   The chart of the linear array of genes on a chromosome. A chromosome map can also refer to the visual appearance of a chromosome when stained and examined under a microscope. Particularly important are visually distinct regions called light and dark bands, which give each of the chromosomes a unique appearance. This feature allows a person's chromosomes to be studied in a clinical test known as a karyotype, which allows scientists to look for chromosomal alterations.

Chromosome painting   The use of fluorescent-tagged chromosome-specific dispersed repeat DNA sequences to visualize specific chromosomes or chromosome segments by *in situ* DNA hybridization and fluorescence microscopy. *See also* FISH.

Chromosomes   Genes are arranged in cells on long DNA molecules called chromosomes that may contain a few dozen genes in a few tens of kilobases in the chromosome of a virus to tens of thousands of genes in hundreds of megabases of DNA in the chromosomes of higher plants. Discrete units of the genome carrying many genes, consisting of (histone) proteins and a very long molecule of DNA. Found in the nucleus of every plant and animal cell.

Chromosome walking   A methodology for determining the location and sequencing a given gene (within an organism's DNA) by sequencing [specific DNA sequences which overlap and span collectively] that gene's location within the organism's DNA.

Cistron   Synonymous with gene, it refers to a specific DNA sequence which codes for the synthesis (by ribosome) of a single protein (polypeptide molecular chain).

Clades   The taxonomic subgroups within cladistics.

Cladistics   Initially popularized by Willi Hennig's 1950 book entitled *Phylogenetic Systematics*, cladistics is a system of taxonomic classification of organisms (and/or their specimens) that is based upon (determined) similar lines of selected shared traits.

Clone (a molecule) To create copies of a given molecule via various methods. Gene clone means collection of organisms (usually bacteria) which all contain the same piece of recombinant DNA.

Clone (an organism) A group of individual organisms (or cells) produced from one individual cell through asexual processes that do not involve the interchange or combination of genetic material. As a result, members of a clone have identical genetic compositions.

Coding sequence The region within a DNA molecule (i.e., between the start and stop codons) that encodes the amino acid sequence of a protein.

Codon A triplet of nucleotides [three nucleic acid units (residues) in a row] within messenger RNA (mRNA) that code for an amino acid (triplet code) or a termination signal.

Co-enzyme The term cofactor is used almost interchangeably with co-enzyme in most contexts. A co-enzyme is a non-peptide molecule which is needed by an enzyme to work, and is changed in the chemical reaction catalysed by the enzyme. However, it is recycled after reaction so that it is ready to be used again.

Cohesive ends A single-stranded end to a linear duplex DNA molecule which can hydrogen-bond with a complementary single-strand base sequence from the end of the same or another DNA molecule.

Colony hybridization A technique using *in situ* hybridization to identify bacterial colonies carrying inserted DNA that is homologous with some particular sequence (probe).

Comparative genomics This is the comparison of some genes or whole genomes of organisms that are related by descent from a common ancestor, to gain understanding of what the genes in one organism do by comparison with those of another. It is related to phylogeny (the study of how organisms are related) and specifically to phylogenetics (how genes are related to descent).

Conjugation A process akin to sexual reproduction occurring in bacteria; mating in bacteria. A process that involves cell-to-cell contact and the one-way transfer of DNA from the donor to the recipient. In contrast to some other DNA-transfer processes of bacteria, conjugation may involve the transfer of large portions of the genome. The discovery caused considerable controversy at the time.

Constitutive genes Expressed as a function of the interaction of RNA polymerase with the promoter, without additional regulation. They are sometimes also called "household genes" in the context of describing functions expressed in all cells at a low level.

Copy number The number of molecules (copies) of an individual plasmid or plastid that is typically present in a single (e.g. bacterial for *plasmid*, plant for *plastid*) cell. Each plasmid has a characteristic copy number value ranging from 1 to 50 or more. Higher copy numbers result in a higher yield of the protein encoded for by the plasmid gene in each cell.

Crossing over The reciprocal exchange of material between chromosomes that occurs during meiosis. The event is responsible for genetic recombination. The process involves the natural breaking of chromosomes, the exchange of chromosome pieces, and the reuniting of DNA molecules.

Cryopreservation Preservation of things by keeping them cold. There are several variations of relevance to biotechnology. This is usually carried out by freezing. Just putting something in the fridge or freezer is fine for many biological materials. However, physically delicate materials, such as animal cells, are destroyed by the ice crystals that form inside them on freezing. To prevent damage to cells on freezing, cells are often frozen in a mixture of a watery material (their usual growth medium) and another material called a cryoprotectant which mixes with water and stops it forming ice crystals. Glycerol is a favourite for bacteria and dimethyl sulphoxide (DMSO) for animal cells.

Culture Any population of cells (e.g. bacteria, algae, protozoa, virus, yeasts, plant cells, mammalian cells, etc.) growing on, or in a medium that supports their growth. Typically used to refer to a population of the cells of a single species or a single strain. A medium which contains only one specific organism (e.g. *E. coli* bacteria) is known as a pure culture. A culture may be preserved (i.e., stored alive) via freezing, drying (in which the cells go dormant), subculturing on an agar medium, or other preservation methods.

Culture medium Any nutrient system for the artificial cultivation of bacteria or other cells. It usually consists of a complex mixture of organic and inorganic materials. For example, the classic culture (growth) medium used for bacteria consists of nutrients (required by that bacteria) plus agar to solidify or semi-solidify the nutrient-containing mass.

Cystic fibrosis (CF) Also called mucoviscidosis, this is a hereditary disease that affects the entire body, causing progressive disability and early death. Formerly known as cystic fibrosis of the pancreas, this entity has increasingly been labelled simply "cystic fibrosis". Life expectancy is on average 37.5 years old.

Cytokines　A large class of glycoproteins similar to lymphokines but produced by non-lymphocytic cells such as normal macrophages, fibroblasts, keratinocytes and a variety of transformed cell lines. They participate in regulating immunological and inflammatory processes, and can contribute to repair processes and to the regulation of normal cell growth and differentiation.

Cytomegalovirus (CMV)　A virus that infects different groups of people in varying amount, depending on their behaviour. For example, 40–90 per cent of American heterosexuals, and about 95 per cent of homosexuals are infected with CMV. CMV normally produces a latent (non-clinical, non-obvious) infection, but when AIDS or other events cause immune system suppression, CMV produces a febrile (fever-causing) illness that is usually mild in nature but can become retinitis (eye infection).

Cytoplasmic DNA　The DNA within an organism (e.g. plant) that is not inside the cell's nucleus. Cytoplasmic DNA (i.e., located in the cells' mitochondria and the chloroplasts) is not transferred from plant to plant via pollen, as nuclear DNA is.

Cytoplasmic membrane　The membrane that surrounds a cell's cytoplasm, separating it from the environment. It consists of a double layer of phospholipids and has proteins embedded in it.

Cytotoxic T cells　Also called killer T cells. These are T cells that have been created by stimulated helper T cells. The T refers to cells of the cellular system rather than to cells of the humoral system (B cells). Cytotoxic T cells detect and destroy infected body cells by the use of a special type of protein. The protein attaches to the infected cell's membrane and forms holes in it. This allows the uncontrolled leakage of ions out of and water into the cell, causing cell death. In general, the loss of the integrity of the cell membrane leads to death. The cytotoxic T cells also transmit a signal to the (leaking) infected cells that causes the cell to "chew up" its DNA. This includes its own DNA as well as that of the virus.

Deamination　The removal of amino groups from molecules (e.g. in an animal's food) via the energy-consuming metabolism of "excess" amino acids eaten by that animal. For example, when livestock are fed more lysine (amino acid) than their body needs in a given day (i.e., animals' bodies can only utilize the essential amino acids in precise amounts/ratios of

their daily diet), that exess lysine is metabolized to urea, then excreted in the animal's urine.

Degenerate codons　Two or more codons that code for the same amino acid. For example, isoleucine is specified by the AUU, AUC, and AUA triplets. Since in this case more than one triplet codes for isoleucine, the codons are called degenerate.

Deletions　Loss of a section of the genetic material from a chromosome. The size of a deleted material can vary from a single nucleotide to sections containing a number of genes.

Demethylation　The chemical process resulting in the removal of a methyl group (CH_3) from a molecule. In biochemical systems, this process is often catalysed by an enzyme such as one of the cytochrome P450 (CYP) family of liver enzymes.

Denaturation　The loss of the native conformation of a macromolecule resulting, for instance, from heat, extreme pH (i.e., by acidity or basicity) changes, chemical treatment, etc. It is accompanied by loss of biological activity.

Denatured DNA　DNA that has been converted from double-stranded to single-stranded form by a denaturation process such as heating the DNA solution. In the case of heat denaturation, the solution becomes very gelatinous and viscous.

Denaturing gradient gel electrophoresis　One particular method of gel electrophoresis, which can be utilized to separate different segments of double-stranded DNA or RNA from each other.

Denaturing polyacrylamide gel electrophoresis　The use of PAGE (polyacrylamide gel electrophoresis) in order to separate and analyse DNA fragments (sequences) after that DNA is first denatured. This methodology can be utilized to scan DNA in order to detect point mutations.

Dendrimers　Polymers (i.e., molecules composed of repeating atomic units within the molecule) that repeatedly branch (while "growing" due to addition of more atoms in a repeating pattern) until that branching is stopped by the physical constraint of contacting itself (i.e., having formed a complete, hollow sphere). Name based on the Greek word for *tree*.

Dendrites　Highly-branched structures that extend from the (nucleus of) neurons to (synapse junctions with) other neurons (e.g. in human brain tissue). The primary purpose of dendrites is to "process" signals that are generated/received at the synapses (e.g. from the dendrites of adjoining neurons).

Dendritic cells　Rare white blood cells, which act to stimulate the human immune system lymphocytes (i.e., 'naive' T cells, or 'naive' B cells) to become effector T cells or effector B cells and combat certain pathogens by "presenting" the antigens of those pathogens to those naive T cells or naive B cells.

Deoxyribonucleic acid (DNA) Discovered by Frederick Miescher in 1869, it is the chemical basis for genes, i.e., the chemical building blocks (molecules) of which genes (i.e., paired nucleotide units that code for a protein to be produced by a cell's machinery, such as its ribosomes) are constructed. Every inherited characteristic has its origin somewhere in the code of the organism's complement of DNA. The code is made up of subunits called nucleic acids. The sequence of the four nucleic acids is interpreted by certain molecular machines (systems) to produce the proteins required by an organism. The structure of the DNA molecule was elucidated in 1953 by James Watson, Francis Crick, and Maurice Wilkins. The DNA molecule is a linear polymer made up of deoxyribonucleotide repeating units (composed of the sugar 2-deoxyribose, phosphate, and a purine or pyrimidine base). The bases are linked by a phosphate group, joining the 3´ position of one sugar to the 5´ position of the next sugar. Most molecules are double-stranded and anti-parallel, resulting in a right-handed helical structure that is held together by hydrogen bonds between a purine on one chain and pyrimidine on the other chain. DNA is the carrier of genetic information, which is encoded in the sequence of bases; it is present in chromosomes and chromosomal material of cell organelles such as mitochondria and chloroplasts, and also present in some viruses.

Diagnostics This usually means chemically based diagnostic tests for disease. Biotechnology helps develop diagnostics in two ways. It finds out more about living systems in health and disease, so we can discover the "markers" which a test subsequently detects. It can also develop reagents to do the tests, notably monoclonal antibodies. DNA probe-based tests are another rapidly growing area of diagnostics which is based on biotechnology.

DNA amplification This is the use of enzymes to take a piece of DNA and multiply it in a test tube into many thousands or millions of copies. The best known and by far the most commonly used system is polymerase chain reaction.

DNA-dependent RNA polymerase *See* RNA polymerase.

DNA fingerprinting Also known as DNA profiling it is a way of making a unique pattern from the DNA of an individual, which can then be used to distinguish that individual from another. The technologies used can be based on DNA probes, PCR or any other DNA detection technology.

DNA fragmentation The cleavage (i.e., "chewing up") of DNA (within a cell) at inter-nucleosomal sites on that DNA molecule.

DNA helicase *See* Helicase.

DNA ligase Discovered during the 1960s by Baldomero Olivera, and is also called T4 DNA ligase. It is an enzyme that creates a phosphodiester bond between the 3´ end of one DNA segment and the 5´ end of another, while they are base-paired to a template strand, i.e., it joins two double-helical DNA molecules together to make a single, longer one. The enzyme seals (joins) the ends of single-stranded DNA in a duplex DNA chain. DNA ligase constitutes a part of the DNA repair mechanism available to the cell.

DNA marker *See* Marker (DNA marker).

DNA melting temperature *See* Melting Temperature.

DNA methylation Refers to a process resulting in a DNA molecule that is saturated with methyl groups (i.e., methyl sub-molecule groups —CH_3 have attached themselves to the DNA molecule's "backbone" at all possible locations on that DNA molecule).

DNA microarray Initially developed by Patrick Brown during the 1980s, these microarrays enable analysis of the levels of expression of genes in an organism, or comparison of gene expression levels (e.g. between diseased and non-diseased tissues) via hybridization of messenger RNA (mRNA) to its counterpart DNA sequence when biological samples containing DNA (e.g. in liquid) are passed over the array surface.

DNA polymerase Discovered in 1956 by Arthur Kornberg, it is an enzyme that catalyses the synthesis of DNA. It must have a DNA molecule to copy (the template) and a short DNA molecule to start with (the primer). It then adds bases onto the primer, copying the template until it gets to the end. It does this by catalysing the addition of the deoxyribonucleotide residues to the free 3´-hydroxyl end of a DNA molecular chain, starting from a mixture of the appropriate triphosphorylated bases, which are dATP, dGTP, dCTP and dTTP. This chemical reaction is reversible and, hence, DNA polymerase also functions as an exonuclease.

DNA probe Also called gene probe or genetic probe, these are short, specific (complementary to desired gene) artificially produced segments of DNA used to combine with and detect the presence of specific genes (or shorter DNA segments) within a chromosome. If a DNA probe of known composition and length is mingled with pieces of DNA (genes) from a chromosome, the probe will cling to its exact counterpart in the "chromosomal DNA pieces" (genes), forming a stable double-stranded hybrid. The presence of this (now) "labelled" probe is detected visually or with the aid of another detection instrument.

DNA profiling   Invented in 1985 by Alec Jeffreys, it is a technique used by forensic (i.e., crime-solving) chemists to match biological evidence (e.g. a blood stain) from a crime scene to the person (e.g. the assailant) involved in that particular crime. DNA profiling involves the use of RFLP (restriction fragment length polymorphism) analysis or ASO/PCR (allele-specific oligonucleotide/polymerase chain reaction) analysis to analyse the specific sequence of bases (i.e., nucleotides) in a piece of DNA taken from the biological evidence. Since the specific sequence of bases in DNA molecules is different for each individual (due to DNA polymorphism), a criminal's DNA can be matched to that of the evidence to prove guilt or innocence. Biological evidence may include among other things blood, hair, nail fragments, skin, and sperm.

DNA–RNA hybrid   A double helix that consists of one chain of DNA hydrogen-bonded to a chain of RNA by means of complementary base pairs.

DNase (Deoxyribonuclease)   An endonuclease enzyme "family" that degrades (cuts up) DNA molecules. DNase I is produced and secreted by the salivary glands, intestines, liver, and pancreas of animals. It has optimal activity (i.e., greatest ability to cut up DNA molecules) at neutral pH (i.e., neither acidic nor basic).

DNA sequencing   Determining the sequence of the bases in DNA is one of the mainstays of gene cloning technology. At root, this is a chemical procedure. By far the most common chemistry used the Sanger technique (di-deoxy method, chain termination method). This uses the enzymes to make a new DNA chain on the target you want to sequence, using the "dideoxy" reagents to stop the chain randomly as it grows.

DNA vaccines   Products in which "naked" genes (i.e., pieces of bare DNA) are used to stimulate an immune response (e.g. either a cellular immune response, humoral immune response, or otherwise raising antibodies against the pathogen from which the naked genes have arisen/been derived).

Double helix   The natural coiled conformation of two complementary, anti-parallel DNA chains. This structure was first put forward by Watson and Crick in 1953.

Down regulating   Phrase utilized to refer to regulatory sequences, chemical compounds (e.g. transcription factors), mutations (e.g. down promoter mutations), etc. that cause a given gene to express less of the protein that it normally codes for.

Downstream processing   The general term for all the things which happen in a biotechnological process after the biology, be it an enzyme reaction, fermentation of a microorganism, or growth of a plant. It is particularly relevant to fermentation processes, which produce a large amount of a dilute mixture of substrates, products and microorganisms. These must be separated, the product concentrated and purified and converted into a product which is useful.

dsDNA   Acronym for the double-stranded structure of DNA molecule.

dsRNA   Acronym for the double-stranded structure of RNA molecule. Among its other functions, dsRNA can induce degradation of its counterpart (i.e., 'matching') mRNA; thereby causing RNA interference.

Duplex   The double-helical structure of DNA (deoxyribonucleic acid).

E. coli   *See Escherichia coli.*

Edible vaccines   Edible substances, bearing antigens, that cause activation of an animal's immune system via that animal's GALT (gut-associated lymphoid tissues). These "edible vaccines" are derived from transgenic plants (e.g. grains, tubers, fruits, etc.) or eggs (i.e., via the activation of the hen's immune system to cause that hen to secrete desired molecule(s) into the eggs it lays).

Editing   A term with several different meanings:

- In transcription, it is the process that removes the intron sequences during synthesis of mRNA from DNA, and joins together the exon sequences.

- During DNA recombination, it is the process of ligating two segments of DNA together.

- During some types of gene therapy, as in the process of utilizing zinc finger proteins (coupled with relevant nucleases), editing means to "correct" a "wrong" DNA sequence (e.g. a disease-causing SNP) within cells of a living organism.

EDTA (Ethylenediamine tetraacetate) An organic molecule which, due to the chemical groups it contains and their juxtaposition within that molecule, is able to chelate (bind) certain other molecules such as divalent metal cations. EDTA thus inhibits some enzymes requiring such ions for activity.

Electrophoresis A technique for separating molecules based on the differential movement of charged particles through a matrix when subjected to an electric field. The term is usually applied to large ions of colloidal particles dispersed in water. The most important use of electrophoresis (currently) is in the analysis of proteins, and then a technique known as gel electrophoresis is used. Since the proportion of proteins varies widely in different diseases, electrophoresis can be used for diagnostic purposes.

Electroporation Electroporation, also called electroporesis, or electropermeabilization, this technique uses a brief direct-current electrical pulse to cause formation of "micropores" (tiny holes) in the surface of cells [or protoplasts, in the case of plants; e.g. suspended in a solution containing DNA sequences (genes)]. After the gene(s) enter cell via the temporarily created micropores, the electrical pulse ceases, and the micropores close so that the gene(s) cannot depart the cell. The cell then incorporates (some) of the new genetic material (genes) into its genetic complement (genome), and produces whatever product (i.e., a protein) the newly introduced gene codes for.

ELISA Acronym for enzyme-linked immunosorbent assay, this is a test for proteins and can readily measure less than a nanogram (10^{-9}g) of a protein. This assay is more sensitive than simple immunoassay (tests) because one of the two antibodies used to bind and quantitate (measure) the protein's antigen, based on two concurrent epitopes within the protein, is attached to an enzyme. The enzyme can rapidly convert an added colourless substrate into a coloured product, or a nonfluorescent substrate into an intensely fluorescent product (thus enabling finer quantitation).

Embryonic stem cells *See* Human embryonic stem cells.

Embryo technology A genetic name for any manipulation of mammalian embryos. This encompasses animal cloning, *in vitro* fertilization, gamete and embryo storage and stem cell technology.

Endocytosis A process of cellular ingestion by which the plasma membrane folds inward to bring substances into the cell.

Endodermal adult stem cells Certain stem cells present within (adult) bodies of organisms, that can be differentiated (via chamical signals) to give rise to cells of tongue, tonsils, the bladder/urethra, digestive tract, liver, pancreas, lung tissues, etc.

Endonucleases A class of enzymes capable of hydrolysing (breaking) the interior phosphodiester bonds of DNA or RNA chains, as opposed to cleavage (by exonucleases) at the terminal bonds (ends) of the molecular chain.

Entrapment Entrapment within a 3-D gel matrix is a common method of enzyme immobilization. Numerous matrices have been employed but the most favoured have been alginate, cross linked with linear chains of Ca^{2+} ions, gelatin and polyacrylamide. With due attention to the degree of cross linking and the nature of the gelling process, i.e., minimizing the concentration of free radicals, gel entrapment can be applied to any enzyme.

Environmental biotechnology A general term covering any biotechnology product or process which can be considered to be helpful to "the environment". Usually, this means control, reduction or disposal of waste, removal of chemical pollutants, or reduction in power use, especially in industry. The most commonly discussed topics in environmental biotechnology are bioremediation, soil amelioration, waste disposal, and creating alternative energy sources.

Enzyme engineering An exciting development over the last few years is the application of genetic engineering techniques to enzyme technology. There are a number of properties which may be improved or altered by genetic engineering including the yield and kinetics of the enzyme, the ease of downstream processing and various safety aspects. Enzymes from dangerous or unapproved microorganisms and from slow-growing or limited plant or animal tissue may be cloned into safe high-production microorganisms.

Enzyme immobilization This is linking an enzyme to a solid support material, typically polymeric beads in a column or bed. Typical immobilizing materials could be silica or carbohydrate polymers such as sepharose.

Enzyme immunoassay (EIA) *See* ELISA.

Enzyme-linked immunosorbent assay The Enzyme-linked immunosorbent assay, or ELISA, is a biochemical technique used mainly in immunology to detect the presence of an antibody or an antigen in a sample. The ELISA has been used as a diagnostic tool in medicine and plant pathology, as well as a quality control check in various industries.

Performing an ELISA involves at least one antibody with specificity for a particular antigen. The sample with an unknown amount of antigen is immobilized on a solid support (usually a polystyrene microtiter plate) either non-specifically (via adsorption to the surface) or specifically (via capture by another antibody specific to the same antigen, in a "sandwich" ELISA). After the antigen is immobilized the detection antibody is added, forming a complex with the antigen. The detection antibody can be covalently linked to an enzyme, or can itself be detected by a secondary antibody which is linked to an enzyme through bioconjugation. Between each step the plate is typically washed with a mild detergent solution to remove proteins or antibodies that are not specifically bound. After the final wash step the plate is developed by adding an enzymatic substrate to produce a visible signal, which indicates the quantity of antigen in the sample.

Enzymes   The core of traditional biotechnology, and a key feature of the new biotechnology of gene cloning, is the use of enzymes. For practical purposes, these can be considered to be catalytic proteins, although recent work has shown that RNA can act as an enzyme.

Erythropoietin (EPO)   A glycoprotein cytokine produced in the kidneys that stimulates pluripotent stem cells in the bone marrow to differentiate, then increase the number of red blood cells. Erythropoietin can be used to help correct a variety of aneamias.

EST   *See* Expressed Sequence Tags.

Eukaryotes   Animals, plants, fungi, and protists are eukaryotes organisms whose cells are organized into complex structures by internal membranes and a cytoskeleton. The most characteristic membrane-bound structure is the nucleus. This feature gives them their name, (also spelled "eucaryote,"). In the nucleus the genetic material, DNA, is arranged in chromosomes. Many eukaryotic cells also contain membrane-bound organelles such as mitochondria, chloroplasts and Golgi bodies. Eukaryotes often have unique flagella made of microtubules in a $9+2$ arrangement.

Expressed sequence tags (EST)   Molecular tags consisting of a DNA sequence approximately 200 base pairs long, that are utilized to "label" a given gene (i.e., in terms of that gene's function/protein). Physically, the EST was historically composed of mRNA [i.e., the gene's "message" after the "junk DNA" (introns) has been edited-out]—produced by the analogous gene in (simple) model organisms such as (traditionally) *Caenorhabditis elegans* nematode—which has been sequenced/mapped. Functions of the "labelled" genes were (at least initially) inferred from (known function) *C. elegans* genes.

Expression library   A gene library that actually "expresses" the genes that have been cloned. Usually this means making the proteins from them. It can also mean that those proteins are allowed to perform whatever function we want, usually an enzymatic reaction.

Expression profiling   The generation of a profile of which genes are making how much of what RNA and protein in a sample at any one time. Usually it is RNA that is measured, as this is easier to do, and so the term "expression profiling" usually refers to RNA profiling—finding how much of all the different RNAs there are in a cell or tissue.

Expression systems   Combinations of host cell and vector which provide a "genetic context" which makes the gene function in the host cell. Usually, this means making a protein at high levels. They are therefore a way of making a recombinant protein, either a replica or the natural protein made from a recombinant gene, or a protein that is itself altered.

Ex vivo   This is used in reference to medical therapy strategies, and means that you are going to treat something outside the body rather than inside it, where it normally resides. Usually this term is applied to blood or bone marrow cells, which taken out of the body, treated *ex vivo*, and then put back in.

FACS   *See* Fluoresence Activated Cell Sorter.

Factor VIII   Also known as antihaemophilic globulin (AHG) or antihaemophilic factor VIII. A protein factor in the blood serum that is instrumental in the "cascade" of chemical reactions (involving 17 blood components in the intrinsic pathway) that leads to clot formation following a cut or other wound to body tissue. Also, a deficiency of AHG is the cause of the classical type of haemophilia sometimes known as haemophilia AM (approximately 85% of all haemophilia patients).

Factor IX   A protein factor in the blood serum that is instrumental in the cascade of chemical reactions (involving 17 blood components) that leads to clot formation, following a cut or other wound to body tissue.

Fermentation   A term first used with regard to the foaming that occurs during the manufacture of wine and beer. The process dates back to at least 6,000 BC when the Egyptians made wine and beer by fermentation. From the Latin word *fermentare meaning*, "to cause to rise." The term "fermentation" is now used to refer to so many different processes that fermentation is no longer accepted for use in most scientific publications.

Fertilization   The union of the (haploid) male and (haploid) female germ cells (sex cells or gametes) to produce a diploid zygote. Fertilization marks the start of development of a new individual (organism), the beginning of cell differentiation.

Fingerprinting   In biotechnological terms, fingerprinting means making a characteristic chemical profile of something, to identify it. There are many variants like DNA fingerprinting, protein fingerprinting, peptide fingerprinting and chemical fingerprinting.

FISH   Acronym for Fluorescence *in situ* Hybridization.

Flow cytometry   Flow cytometry is a technique for counting, examining, and sorting microscopic particles suspended in a stream of fluid. It allows simultaneous multiparametric analysis of the physical and/or chemical characteristics of single cells flowing through an optical and/or electronic detection apparatus.

Fluorescence   The reaction of certain molecules (known as fluorophores) upon absorption of specific amount/wavelength of light, in which those molecules emit (reradiate) light energy possessing a longer wavelength than the original light absorbed. All cells will naturally fluoresce, at least a bit. Human colon cancer cells, and precursor cells, fluoresce much more (and emit much more red light when they fluoresce) than non-cancerous cells. This may lead to a new and better means of early detection.

Fluorescence activated cell sorter (FACs)   A machine or MEMS/Lab-On-A-Chip that is used to sort specific cells from a mixed group of cells (e.g. to remove only the cells of one type of tissue, or cells into which a new gene has been inserted via genetic engineering techniques, etc.). It was initially invented by Leonard A. Herzenberg during the late 1960s. The desired cells are first labelled with a specific fluorescent dye, or a gene for a fluorophore (e.g. green fluorescent protein) is inserted; then the cells (e.g. as part of a mixture) are passed through a flow chamber that is illuminated by a laser beam, which causes the labelled cells to fluoresce (i.e., glow).

Fluorescence *in situ* hybridization (FISH)   A method for detecting the presence of particular genes (e.g. in a biological sample), which utilizes a number of fluorescein-"tagged" DNA probes. When those DNA probes hybridize to each of their respective particular genes (i.e., that they were selected to be

complementary-to), each DNA probe's "tag" fluoresces at a different wavelength (different 'colour'), thereby indicating positively the presence in the sample of that particular gene.

Fluorescence mapping   Refers to use of a special microscope/light of selected wavelength (i.e., to induce fluorescence of 'targets') in order to scan (e.g. in tissue) two-dimensional planes at varying depths, in order to thoroughly 'map' in three dimensions all of the molecules of interest which fluoresce (e.g. when a pharmaceutical compound binds to each 'target' molecule, such as a cell receptor).

Fluorescence resonance energy transfer (FRET)   Refers to (fluorescence-induced) resonance which occurs when two different molecular (fluorescent) labels are in very close proximity to each other. That resonance causes the fluorescence-excitation energy to transfer from one to the other (causing the second one to fluoresce) or else (the two in combination) to emit a third colour [wavelength]; but the two revert (to original two colours) when some event moves the two labels apart.

Fluorophore   Refers to any substance that is fluorescent.

Footprinting   A technique used by researchers to determine precisely *where* (on DNA molecule) certain DNA-binding proteins make specific contact with that DNA molecule. For example, certain types of drugs act by binding tightly to certain DNA molecules in specific locations (e.g. in order to halt cancerous growth of cells, etc.).

Freeze-drying   A common technique, also called lyophilization, for preserving biomolecules and microorganisms. The sample is frozen, often in a solution containing another material such as lactose or trehalose which acts to stabilize it. It is then put into a chamber attached to a vacuum pump and, while the sample is still frozen, the chamber is evacuated. The ice sublimes under vacuum and the water vapour is removed. After a while all the water in the sample has been removed, and what is left is a dry powder or pellet of material.

FRET   Acronym for Fluorescence Resonance Energy Transfer.

Functional genomics   The functional genomics movement aims to add high-throughput biology to the genomics programme of high-throughput gene identification and sequencing, to identify what all these genes do. The approach is to link gene identification and sequencing with a tool or set of tools like proteome technology, animal genetics, knock-out mutants and cell biology, which give clues about the gene's function.

Fusion protein   A protein consisting of all or part of the amino acid sequences (known as the "domain") of two or more proteins.

Gel filtration   Also known as exclusion chromatography. An effective technique for separating molecules (such as peptide mixtures) on the basis of size. This is accomplished by passing a solution of the molecules to be separated over a column of Sephadex®, for example; which is a polymerized carbohydrate derivative that contains tiny holes. The holes are of such a size that some of the smaller molecules diffuse into them and are in this way retained (held back) while the larger molecules are not able to get into the holes and pass on by the solid phase (Sephadex®, in this example). This, simplistically, is how separation is effected.

Gene   A natural unit of the hereditary material, which is the physical basis for the transmission of the characteristics of living organisms from one generation to another. The basic genetic material is fundamentally the same in all living organisms: it consists of chain-like molecules of nucleic acids—deoxyribonucleic acid (DNA) in most organisms and ribonucleic acid (RNA) in certain viruses—and is usually associated in a linear arrangement that (in part) constitutes a chromosome.

Gene amplification   The copying of segments (e.g. genes) within the DNA or RNA molecule. This can be done by man (e.g. polymerase chain reaction), can be caused by certain chemical carcinogens (e.g. phorbol ester), or occur naturally (e.g. in prokaryotes and certain lower eukaryotes). The five primary techniques that are used by man to perform gene amplification are: 1) polymerase chain reaction (PCR), 2) ligase chain reaction (LCR), 3) Self-sustained Sequence Replication (SSR), 4) Q-beta replicase technique, and 5) strand displacement amplification (SDA).

Gene chips   A type of DNA array, that is built into a very small area in a "biochip" type device. The term actually covers many different types of devices, but they all have many DNA probes packaged into a small device of the same general size as a large semiconductor 'chip'. Gene array and DNA array are effectively synonyms.

Gene expression   Conversion of the genetic information within a gene, into an actual protein (or cell process).

Gene fusion   Refers to the technology/methods utilized to fuse together two or more genes. When such a "fused gene" is then inserted into a genome (e.g. the DNA of a plant), it causes production (in plant's ribosomes) of protein(s) consisting of all or part of the amino acid sequences (known as the "domain") of the two proteins typically coded for by those two genes. This fusion is often done in order to put expression of the "second" (fused) gene under the control of the (strong) promoter of the 'first' gene.

Gene library   A collection of gene clones, which contains the entire DNA that is present in some source, but split up and joined onto suitable vector DNAs. It is also sometimes called a gene bank. If the original source of the DNA was the original DNA from a living organism, then the library seeks to include clones of all that DNA: it is called a genome gene library, because it contains all the DNA from that organism's genome. If the DNA is from some other source such as cDNA, then the maker of the library seems to include representative clones from that entire source, and, in this case, would be called a cDNA library.

Gene silencing   The suppression of gene expression (e.g. of the gene for polygalacturonase which causes fruit to ripen, of the gene for P34 protein in soybeans, etc.) via a variety of methods (e.g. via RNA interference (RNAi), chemical genetics, effect of certain viruses, "zinc finger proteins", sense or antisense genes, etc.). Also occurs with some genes in an organism as the organism matures (e.g. from an embryo to a seedling/juvenile).

Gene splicing   The enzymatic attachment (joining) of one gene (or part of a gene) to another; also removal of introns and splicing of exons during mRNA synthesis.

Gene therapy   Changing the genetic make-up of a human. Gene therapies had been administered to more than 4000 patients by the start of 2001: however, nearly all of these were phase I trials, testing to see if the approach was safe, and whether gene transfer took place. There are two approaches, germ-line gene therapy and somatic cell gene therapy. The former changes the "germ cells", the cells which make sperm or ova. This has a permanent effect on all the individuals which are the descendants of whoever had the therapy. Somatic therapy alters the somatic cells—all the non-germ cells in the body. Changing them does not affect the germ cells, but does affect the engineered person.

Genetic code   The set of triplet code words in DNA coding for all of the amino acids. There are more than 20 different amino acids and only four bases (adenine, thymine, cytosine, and guanine). The mRNA code is a triplet code, that is, each successive "frame" of three nucleotides (sometimes called a codon) of the mRNA corresponds to one amino acid of the protein. This rule of correspondence is the genetic code. The genetic code consists of 64 entries—the 64 triplets possible when there are four possible nucleotides, each of which can be at any of three places ($4 \times 4 \times 4 = 64$). A triplet code was required because a doublet code would have only been able to code for ($4 \times 4 = 16$) sixteen amino acids. A triplet code allows for the coding of 64 theoretical amino acids. Since only a little over 20 exist, there is some redundancy in the system. Hence some certain amino acids are coded for by two or three different triplets.

Genetic engineering The selective, deliberate alteration of genes (genetic material) by man. This term has come to have a very broad meaning including the manipulation and alteration of the genetic material (constitution) of an organism in such a way as to allow it to produce endogenous proteins with properties different from those of the traditional (historic/typical), or to produce entirely different (foreign) proteins altogether. Some other words often applicable to the same process are gene splicing, gene manipulation, or recombinant DNA technology (techniques).

Genetic map A description of how genes lie along a chromosome. It looks rather like a ladder with very erratic rungs. There are two types of maps—the map which shows the distance between any two points in terms of how much DNA lies between them, it is called physical map. A map can also represent the recombination distance between two genes. This is the chance that two genes at those two points will be separated by recombination during meiosis. The resulting map is called a genetic map or a recombination map, with distances measured in centimorgans (cM), which is the distance apart genes have to be to have a 1% chance of being recombined during meiosis.

Genome The entire hereditary material (which was proven by Oswald Avery in 1944 to be DNA) in a cell. In addition to the DNA contained in cell nucleus (known as nuclear DNA), an organism's cells contain DNA in other locations within those cells like in plasmids (bacteria), plastids (plants) and mitochondria (animals).

Genomics Coined in 1986 by Tom Roderick, by combining gene and "-omics" (from the Greek word for "all"), this term refers to the scientific study of all the genes and their roles in an organism's structure, growth, health, disease (and/or resistance to disease, etc.). For example, how the (approximately) 3,000 genes in a given strain of bacteria, or the (approximately) 6,000 genes in a given strain of yeast, contribute to the shape, function, and the development of those whole organisms.

Genotype The total genetic, or hereditary, constitution that an individual receives from its parents. An individual organism's genotype is distinguished from its phenotype, which is its appearance or observable character.

Germ-line gene therapy The introduction of genes into reproductive cells or embryos to correct inherited genetic defects that can cause disease.

Glycosylation Putting sugar molecules on things, almost always other molecules and usually proteins: glycosylated proteins are called glycoproteins. Most of the proteins present on the surface of cells, viruses, and in the blood in animals are glycosylated and this is important for their function, so some biopharmaceuticals also have to be glycosylated to have the same function as their natural counterparts.

Green fluorescent protein Discovered by Osamu Shimomura, it is a protein that is naturally present within the jellyfish *Aequorea victoria*. Green fluorescent protein (GFP) has been utilized since 1994 by scientists to "label" certain protein molecules which are of interest to scientists (e.g. in cell samples); help visualize thin layers of biological tissue in fluorescence microscopy and to "mark" certain endpoints in experiments (at which point the green light signals that endpoint was reached).

Growth hormone (GH) A hormone produced by the anterior pituitary gland. This hormone is a protein (somatotropin) and can be obtained from the bodies of animals, or produced by genetically engineered microorganisms. Its major action in humans (human growth hormone) is a generalized stimulation of skeletal growth. However, human growth hormone (hGH) is also known to affect the growth of other tissues, to be important in fat, protein, and carbohydrate metabolism, and to enhance the effects of various other hormones.

Gyrase *See* Helicase.

Hairpin loop A section of highly curving, single-stranded DNA or RNA formed when a long piece (string) of the DNA or RNA bends back on itself and hydrogen-bonds (is able to base-pair) in some regions to form double-stranded regions. The structure can be visualized by taking a human hair, bending it back on itself and holding it in such a way as to halve its original length. The section where the two ends of hair lie next to each other represents the section of double-stranded DNA or RNA. At one end the hair will have to make a sharp turn and will form a loop. This loop represents the single-stranded hairpin loop.

Haplotype A subgroup (e.g. an ethnic minority, all members of a genetically-related family group, etc.)

of organisms (e.g. humans) whose phenotype results in their body responding in the same way to a physical agent (e.g. a certain pharmaceutical, a toxin, a food, etc.); or are predisposed to particular diseases. For example, more than 70% of black people in North America are lactose-intolerant (e.g. their bodies cannot metabolize the lactose sugar in cow's milk), but fewer than 19% of Caucasian people in North America are lactose intolerant.

H-bonding   *See* Hydrogen Bonding.

Helicase   An enzyme which "unwinds" the DNA molecule's double helix structure during DNA replication.

HER-2 gene   Abbreviation often utilized for Human epidermal growth factor receptor-2 gene/*neu*, which is an oncogene that is responsible for approximately 30% of breast cancers (i.e., in those women whose body over-expresses) that particular oncogene, and it spreads via metastaticism.

Heredity   Transfer of genetic information from parent cells to progeny.

Heritability   The fraction of variation (of an individual's given trait) that is due to genetics. For example, if a pig's trait (e.g. weight at birth) is 30% heritable, that means that 30% of the (birthweight) difference between that individual pig and its (statistically representative) group of contemporaries (pigs) is due to genetics. The other 70% would be due to factors such as nutrition of the mother during pregnancy, etc.

Herpes simplex virus   Herpes simplex virus 1 and 2 (HSV-1 and HSV-2) are two strains of the herpesvirus family, Herpesviridae, which cause infections in humans. HSV-1 and 2 are also referred to as human herpesvirus 1 and 2 (HHV-1 and HHV-2).

Human artificial chromosomes (HAC)   Chromosomes that have been synthesized (made) from chemicals that are identical to chromosomes within human cells.

Human chorionic gonadotropin   A human hormone. In 1986, Mark Bogart discovered that elevated levels of human chorionic gonadotropin in pregnant women are correlated with babies (later) born with Down syndrome.

Human embryonic stem cells   Those cells (in the early embryo's inner cell mass) from which each of the human body's 210 different types of tissues arise via differentiation, proliferation, and growth processes.

Humoral immune response   Refers to the rapid manufacture and secretion by the body of the soluble blood serum components like antibodies, complement proteins, lymphokines and cecrophins, in response to an infection.

Hybridization (molecular genetics)   DNA hybridization is the formation of a double helix of DNA from two single DNA strands. The two separate strands of DNA will come together to make a double helix if their bases are complementary, so that wherever there is an A in one strand there is a T in the other, a G in one strand and a C in the other.

The pairing (tight physical bonding) of two complementary single strands of RNA and/or DNA to give a double-stranded molecule.

Hybridoma   The cell line produced by fusing a myeloma (tumour cell) with a lymphocyte (which makes antibodies); it continues indefinitely to express the immunoglobulins (antibodies) of both parent cells.

Hydrolase   Any of a class of enzymes which catalyse the hydrolysis of proteins, nucleic acids, starch, fats, phosphate esters, and other macromolecular substances.

Idiotype   The region of the antibody molecule (i.e., antigen combining site) that enables each antibody to recognize a specific foreign structure (i.e., epitope or hapten) is said to have an idiotype (for that epitope or hapten). An identifying characteristic (or property) of the epitope or hapten that one is talking about.

Immortalization   Immortalization of a cell type is its genetic change into a cell line which can proliferate indefinitely. Cells taken from a mammal, called primary cells, will divide in culture for 20–60 divisions, but then stop dividing. This is not because they have run out of room or nutrients to grow, but rather because they have become incapable of dividing any more. This is called cell senescence. To avoid this, cells are "immortalized", put through a treatment which allows them to overcome senescence and divide indefinitely, keeping whatever differentiated features they had to start with.

Immunoconjugate   A compound which is a combination of an antibody molecule (or part of one) and another molecule. There are several types. Immunotoxins are conjugates of an antibody and a protein toxin. Antibody–enzyme conjugate is the complex of antibody chemically linked to an enzyme. Common conjugates are horseradish peroxidase and alkaline phosphatase coupled with antibody.

Immunodiagnostics/Immunoassays   The medical diagnostic methods use antibodies. The antibody only binds its target antigen, and does so at very low concentrations, and so can be a very sensitive test.

The antibodies can be labelled in various ways. Immunodiagnostics can use a variety of labels in *in vitro* assays. They usually have different names:

- ELISA (Enzyme-linked immunosorbent assays)—uses an enzyme label on the antibody.

- RIA (Radioimmunoassay)—uses a radioactive label on antibody or antigen.

- FIA (Fluorescent immunoassay)—uses a fluorescent tag on antibody or antigen.

- CLIA (Chemiluminescent immunoassay)—there are variety of luminescent labels that can be used.

- Immunogold —Labelling the antibody with colloidal gold particles.

Immunogen A molecule or an organism (e.g. pathogenic bacteria) which is specifically "recognized" by the immune system (e.g. of humans it has entered) and triggers an immune response.

Immunoglobulin A class of (blood) serum proteins (IgA, IgE, IgG and IgM) representing antibodies. Often used, along with the more specific monoclonal antibodies, in health diagnostic reagents. In certain people who are genetically predisposed to food-borne allergies, immunoglobulin-E (IgE) initiates an immune system response to antigen(s) present on protein molecule(s) in the particular food that the person is allergic to. Severe allergic reactions to foods may lead to death.

Immunosensor A biosensor with a selected antibody attached, which can sense when a given molecule (from sample) binds (i.e., "attaches to") that antibody.

Inclusion bodies Nuclear or cytoplasmic aggregates of stainable substances, usually proteins. They typically represent sites of viral multiplication in a bacterium or a eukaryotic cell and usually consist of viral capsid proteins.

Induction In biotechnological terms this means getting an organism to make a protein by exposing it to some stimulus, usually a chemical trigger (an inducer). In nature, the protein is usually an enzyme, and the inducer is the substrate for that enzyme. A classic example is the induction of beta galactosidase (the *lac* gene) lactose (the sugar that breaks down) in E. *coli*.

Information RNA (iRNA) Refers to an RNA molecule (within cell) which does not code for production of a protein, but only provides some "information" to regulate one or more cell functions (e.g. protein synthesis).

Integrins A class of proteins that is found on the surface (membranes) of cells, and that function as cellular adhesion receptors. For example, integrin $\alpha_v\beta_3$ is a receptor on the surface of endothelial cells in growing blood vessels (e.g. the new blood vessels forming in a body with cancer to supply blood to growing tumours). It binds angiogenic endothelial cells, enabling them to form new blood vessels.

Interfering RNAs Small interfering RNA (siRNA), sometimes known as short interfering RNA or silencing RNA, are a class of 20–25 nucleotide-long double-stranded RNA molecules that play a variety of roles in biology. Most notably, siRNA is involved in the RNA interference (RNAi) pathway where the siRNA interferes with the expression of a specific gene. In addition to their role in the RNAi pathway, siRNAs also act in RNAi-related pathways, e.g. as an antiviral mechanism or in shaping the chromatin structure of a genome; the complexity of these pathways is only now being elucidated.

Interferons Discovered in 1957 by Alick Isaacs and J. Lindenman, they are a family of small (cytokines) proteins (produced by vertebrate cells following a virus infection) that interfere with (i.e., block) translation of viral DNA.

Interleukins A class of 24 different cytokines that "carry a signal" **between different leucocyte populations** within the immune system of an organism.

Interleukin-1 (IL-1) A cytokine (glycoprotein) released by activated macrophages, during the inflammatory stage of immune system response to an infection, which promotes the growth of epithelial (skin) cells and white blood cells. Research has indicated that too much IL-1 is linked to the development of rheumatoid arthritis, diabetes, inflammatory bowel disease, and other autoimmune diseases.

Interleukin-2 (IL-2) Also known as T cell growth factor. A cytokine (glycoprotein) secreted by (immune system response) stimulated helper T cells which promotes the proliferation/differentiation of more helper T cells, and promotes the growth of lymphocytes to combat an infection. Interleukin-2 also stimulates the lymphocytes to produce gamma interferon. It is gamma interferon that prompts the cytotoxic T cells to attack virus-infected cells and kill the virus within them. The structure of the gene that codes for synthesis of IL-2 (by immune system cells) was determined by Tadatsugu Taniguchi in 1983.

Interleukin-3 (IL-3) A haematologic growth factor (glycoprotein) cytokine that stimulates the proliferation of a wide range of white blood cells (to combat an infection).

Interleukin-4 (IL-4)   A cytokine (glycoprotein) that stimulates production of antibody-producing B cells, immunoglobulin-E (I$_g$E), and promotes cytotoxic T cell (i.e., killer T cells) growth.

Interleukin-5 (IL-5)   A cytokine (glycoprotein) that stimulates eosinophil growth.

Interleukin-6 (IL-6)   A cytokine (glycoprotein) that is pleiotropic (i.e., stimulates several different types of immune system cells); and is a haematopoietic growth factor.

Interleukin-7 (IL-7)   A cytokine (glycoprotein) synthesized in the bone marrow that stimulates early (foetal) proliferation and differentiation of B cells and T cells. May be useful in regenerating lymphoid cells in patients whose immune systems have been devastated by cancer chemotherapy.

Interleukin-8 (IL-8)   A basic polypeptide (glycoprotein) with heparin-binding activity. Endogenous endothelial IL-8 appears to regulate transvenular traffic during acute inflammatory responses.

Interleukin-9 (IL-9)   A cytokine (glycoprotein) that is released at sites in the body where inflammation has occurred.

Interleukin-12 (IL-12)   A cytokine (glycoprotein) produced by the body, which serves to activate the immune system against certain tumours and pathogens.

Intron   Discovered in 1977, an intron is a (intervening sequence) segment of deoxyribonucleic acid (DNA) within a gene that is transcribed, but is removed from within the mRNA transcript by splicing together the sequences (exons) on either side of it (in the molecule) by snRNP during final step of the transcription process. In the past, it was generally considered to be a "non-functioning" portion of the DNA molecule.

Invention   A new device, method, or process developed from study and experimentation: the phonograph, an invention attributed to Thomas Edison.

In vitro   In an unnatural position (e.g. outside the body, in the test tube). "*In vitro*" is Latin for "in glass." For example, the testing of a substance, or the experimentation in (using) a "dead" cell-free system.

In vitro vs. in vivo   These Latinisms are widely used when scientists are talking about doing something "simple" in the laboratory and then taking the result and applying it to a more complicated, living system. *In vivo* literally means "in the living", and means in a

living system, such as a complete animal. It is contrasted with *in vitro*, literally meaning "in glass", which is translated to "in the test tube", although no one in biology has used test tubes for 20 years. It means "in the laboratory" and is taken to mean the opposite of "*in vivo*". Doing experiments in cultured cells is usually taken to be *in vitro*.

In vivo   Latin for "*in living*" (e.g. the testing of a new pharmaceutical substance or experimentation in (using) a living, whole organism). An *in vivo* test is one in which an experimental substance is injected into an animal such as a rat in order to ascertain its effect on the organism.

In silico   This means "done in a computer, not in real life". Most biology cannot be done in a computer—it has to be done using the much more expensive, much messier and much slower techniques of laboratory experiment. *In silico* methods include computer analysis of large databases of information to supplement or replace experiment and modelling many aspects of biology that are slow or expensive to measure.

In situ   This means doing something in place, rather than taking out of its usual context. Sterilization *in situ* is sterilizing some piece of equipment in place in the overall machine, rather than taking it out, cleaning it, and then putting it back.

Ion-exchange chromatography   Separation of ionic compounds (which include nucleic acids and proteins) in a chromatographic column containing a polymeric resin (i.e., the stationary phase) having fixed charge groups. The process works in that the charges of the column (stationary phase) interact with the opposite charges of the material dissolved in the solution that is flowing through the column (mobile phase). The charge interaction between the column material and, say, the protein has the effect of slowing down the rate of movement of the protein through the column. The other molecules, meanwhile, which do not interact with the column, flow right on through. This then constitutes the separation process.

iRNA   Acronym for information RNA. *See* Information RNA.

Isoelectric focusing (IEF)   An electrophoresis methodology in which protein molecules are moved (via application of an electrical charge/potential) through a pH gradient (e.g. in a 2-D gel until they reach their individual isoelectric points).

Isoelectric point   Refers to that point (e.g. in a 2-D gel) at which the charge/mass of a given protein is exactly matched by the electrical charge/potential applied to that 2-D gel. Because the isoelectric point is different for virtually every protein (e.g. in a sample applied to the 2-D gel), this enables separation of individual proteins from a (mixed) sample.

Isomerase   One of a group of enzymes that catalyses the conversion of one isomer into another.

Isotope   Refers to one of the several "varieties" of atoms that exist, of the same element, that differ from each other in the number of neutrons in the atom's nucleus. For example, the element chlorine exists primarily in two forms (isotopes) in nature with 18 neutrons (76% of the time) and with 20 neutrons (24% of the time). The chemical properties of isotopes of a given element are virtually identical.

Junk DNA   A term historically utilized by some to refer to portions of an organism's DNA that were not obviously genes (i.e., not transcribed into mRNA; thus not part of the DNA "tagged"/labelled with ESTs, etc.). However, it has recently been discovered that at least some of what was formerly called "junk DNA" (e.g. introns) helps enable more than one specific protein molecule to be expressed from certain genes.

Karyotype   A size-order alignment of an organism's chromosome pairs in the format of a (photo-micrograph) chart. It enables the connecting of chromosomes to symptoms (e.g. of genetic diseases in the organism) and traits.

Kilobase pairs (Kbp)   A unit of DNA equal to 1,000 base pairs.

Kilodalton (Kd)   A unit of mass equal to 1,000 Daltons.

Knockout   Refers to one of the following:

- (a scientist's) alteration of a particular gene within an organism, so that the organism loses a (specific) function (e.g. the ability to produce a given needed clotting factor in its blood, the ability to produce a given allergen in its seeds, etc.).

- the altered organism itself (i.e., in which the particular gene has been inactivated as detailed above).

- (a scientist's) alteration of a particular protein (e.g. within an organism's cell) so that protein loses its biological activity (e.g. the ability to cause blood clotting). This can enable detailed study of which proteins within a cell are responsible for particular diseases, etc.

Kozak sequence   The DNA sequence which "surrounds" (both ends of) the ATG start signal (for translation of mRNA).

Lactase   A member of the β-galactosidase family of enzymes, is a glycoside hydrolase involved in the hydrolysis of the disaccharide lactose into constituent galactose and glucose monomers. In humans, lactase is present predominantly along the brush border membrane of the differentiated enterocytes lining the villi of the small intestine. Lactase is essential for digestive hydrolysis of lactose in milk. Deficiency of the enzyme causes lactose intolerance.

Lambda phage   A bacteriophage that infects *Escherichia coli (E. coli)*. It is commonly used as a vector in recombinant DNA (deoxyribonucleic acid) research.

Leader sequence (mRNA)   The non-translated sequence at the 5´ end of mRNA that precedes the initiation codon.

Lentivirus   Any of a group of animal viruses that cause diseases having an unusually long incubation period, as Creutzfeldt–Jakob disease. Also called slow virus.

Leucocytes   A diverse 'family' of nucleated white blood cells (including mast cells) that has many immunological functions.

Ligand in chromatography   This is a term used to describe a substance (the ligand) that has the capacity for specific and non-covalent (reversible) binding to some protein. A ligand may be a co-enzyme for a specific enzyme. The ligand can be covalently attached (immobilized) by means of the appropriate chemical reaction to the surface of certain porous column material. When a mixture of proteins containing the enzyme to be isolated is passed through the column, the enzyme, which is capable of tightly binding to the ligand, does so, and is in this manner held to the column. The other proteins present, which have no specific affinity for the ligand, pass on through the column. The protein/ligand complex is then dissociated and the enzyme eluted from the column, which may be accomplished by passing more free (unbound) co-enzyme through the column. The ligand may be hormones (i.e., used to isolate receptor molecules) or any other type of molecule that is capable of binding specifically and reversibly to the desired protein or protein complex.

Ligase   An enzyme used to catalyse the joining-together (i.e., "ligating") of two separate molecules, in an energy-requiring process, e.g. the joining-together of two single-stranded DNA segments.

Ligation   The formation of a phosphodiester bond to link two adjacent bases separated by a nick in one strand of a double helix of DNA (deoxyribonucleic acid).

Linkage map   A depiction of gene loci (on chromosomes) based on the frequency of recombination (of linked genes) in the offspring's genome. Close (linked) genes tend to be inherited together. When that does not happen (i.e., fewer linked-together genes inherited across generations than would be expected from mathematical prediction), the phenomenon is known as linkage disequilibrium.

Linker   A short synthetic duplex oligonucleotide containing the target site for some restriction enzyme and used to join disparate DNA molecules together. To do the actual joining, DNA ligase is needed. The linker may be added to the ends of a DNA (deoxyribonucleic acid) fragment prepared by cleavage with some other enzyme reconstructions of recombinant DNA.

Lipases   Enzymes which break down lipids into their component fatty acid and "head group" moieties. The lipases used in biotechnology are almost invariably digestive lipases, meant to break down the fats in food.

Liposome   Originally coined to mean small droplets of fat in a cell, this word has been taken over by biotechnology to mean a closed shell of lipid molecules. In solution, lipids tend to form droplets with the polar "heads" pointing outwards into the watery solution and the apolar "tails" sticking together into the middle. If this is a droplet of lipid, it is called a micelle: many lipids form micelles if their concentration in solution rises above the critical micelle concentration. If the droplet is filled with water, then a two-layer structure called a lipid bilayer is formed. Such a lipid-walled ball is a liposome.

Live vaccines   Vaccines containing living organisms or intact viruses rather than inactivated (killed) organisms or extracts of them. They can cause better immunity in patients, but have the potential drawback that, unless they are thoroughly "crippled" in some way, they may cause disease.

Locus   The position of a gene on a chromosome.

Luminescence   The production of light by chemicals. It is gaining increasing use as a readout system for assays. There are two ways of generating light using chemicals:

1. Chemiluminescence which uses specific chemical groups that, when reacted give out light. They can be attached to many other chemicals (e.g. proteins, DNA).

2. Bioluminescence which is the use of biological systems to generate light.

Lyase   Any of a group of enzymes that catalyse the formation of double bonds by removing chemical groups from a substrate without hydrolysis, or catalyse the addition of chemical groups to double bonds.

Lymphocyte   A type of cell found in the blood, spleen, lymph nodes, etc. of higher animals. They are formed very early in foetal life, arising in the liver by the sixth week of human gestation. There exist two subclasses of lymphocytes: B lymphocytes and T lymphocytes.

Lymphokines   Peptides and proteins secreted by (immune system response) stimulated T cells. These hormone-like (peptide and protein) molecules direct the movements and activities of other cells in the immune system. Some examples of lymphokines are interleukin-1, interleukin-2, tumour necrosis factor (TNF), gamma interferon, colony stimulating factors, macrophage chemotactic factor, and lymphocyte growth factor. The suffix "-kine" comes from the Greek word kinesis, meaning movement.

Lysogeny   The ability of a bacteriophage to be able to itself become a part of the DNA of a bacterium.

Lysosome   A membrane-surrounded organelle within the cytoplasm of eukaryotic cells, that contains many hydrolytic enzymes and was discovered by Christian de Duve. The lysosome internalizes and digests foreign proteins as well as cellular debris. The protein fragments (epitopes) are "presented" to T cells by the major histocompatibility complex (MHC) proteins on the surface of the eukaryotic cell.

Lysozyme   An enzyme, naturally produced by some animals, which possesses antibacterial (i.e., bacteria killing) properties. Discovered in 1922 by Alexander Fleming, in his nasal mucus, Fleming named it (from the Greek) *lyso*—due to its ability to lyse (cut) bacteria and *zyme*—due to its being an enzyme. Lysozyme lyses certain kinds of bacteria by dissolving the polysaccharide components of the bacteria cell wall. When that cell wall is weakened, the bacterial cell then bursts because osmotic pressure (inside that bacterial cell) is greater than the weakened cell wall can contain. Tears and egg whites both contain significant amounts of lysozyme, as agents to prevent bacterial infections (e.g. against bacteria entering body via eye openings, against bacteria entering chicken embryo through the eggshell).

Lytic infection   A viral infection in which the final act of the infection is to lyse (i.e., burst, or destroy) the cell. This then releases the new (progeny) viruses, so they can go on to infect other cells.

MAbs   *See* Monoclonal Antibodies.

Macrophage   A phagocytic cell that is the counterpart of the monocyte. A monocyte which has left the bloodstream and has moved into the tissues. Macrophages have basically the same functions as monocytes, but they carry these out in the tissues. In summary, they engulf and kill microorganisms, present antigen to the lymphocytes, kill certain tumour cells, and their secretions (e.g. leukotrienes) regulate inflammation. Macrophages utilize nitric oxide and hydrogen peroxide (which they synthesize) to kill the microorganisms they engulf (via oxidation), and the nitric oxide also helps to regulate the immune system.

Major histocompatibility antigen-class I A 'family' of glycoproteins which appear on the surfaces of most cells of an organism; which help enable that organism's immune system to distinguish "self" (cells) from "non-self" (e.g. invading pathogens).

Major histocompatibility antigen-class II A "family" of glycoproteins which appear only on the surface of specific lymphocyte cells (dendritic cells) and on the surface of certain macrophages, within an organism.

Major histocompatibility complex (MHC) A genetic loci, or chromosomal region (approximately 3,000 Kb) which encodes for three classes of transmembrane (cell) proteins. MHC I proteins (located on the surface of nearly all cells) present foreign epitopes (i.e., fragments of antigens that have been ingested; peptides) to cytotoxic T cells (killer T cells). MHC II proteins (located on the surface of immune system lymphocyte/dendritic cells and phagocytes) present foreign epitopes to helper T cells. That presenting of epitopes induces the organism's immune response.

MALDI-TOF-MS Acronym for matrix-associated laser desorption ionization time of flight mass spectrometry. A mass spectrometry methodology/ technology that was initially developed by Franz Hillenkamp, for analysis of biological molecules. MALDI-TOF-MS can establish, in seconds, the identity, purity, etc. of a sample of proteins, oligonucleotide, or (poly)peptides. Also the identification of gram-positive microorganisms, or characterization of genetic materials (e.g. DNA, RNA, etc.) on hybridization surfaces.

Mammalian cell culture Technology to artificially cultivate cells, of mammal origin, in a laboratory or production-scale device (i.e., *in vitro*). Can be either a batch or continuous process device. The first mammalian cell culture was performed by a neurobiologist named R. G. Harrison in 1907, when he added chopped-up spinal cord tissue to clotted (blood) plasma in a humidified growth chamber. The nerve cells from this spinal cord tissue successfully grew, divided, and extended long fibres into the clot. Many improvements to cell culture process have been made over the years, including special growth media (fluids that bathe the cultured cells with the right amounts of amino acids, salts, and other minerals).

Marker assisted selection (MAS) The utilization of DNA sequence "markers" by commercial breeders to select the organisms (e.g. crops, livestock, etc.) which possess gene(s) for a particular performance trait (e.g. rapid growth, high yield, etc.) desired; for subsequent breeding/propagation. Marker assisted selection has been utilized in many plant (e.g. crop) breeding programs since the mid-1970s.

Marker (DNA marker) A DNA fragment of known size used to calibrate an electrophoretic gel.

MAS *See* Marker Assisted Selection.

Mast cells Fixed (non-circulating) leucocyte cells that are present in many different kinds of body tissues. When two IgE molecules of the same antibody "dock" at adjacent receptor sites on a mast cell, then (the two IgE molecules) capture an allergen (e.g. a particle of pollen) between them, a chemical-energetic signal is sent to the interior (inside mast cell) portion of receptor molecules, which causes that interior portion of molecule to change (i.e., transduction). That signal transduction causes a protein named "syk" to set off a chemical chain reaction inside the mast cell, thereby causing that mast cell to release leukotrienes, histamine, serotonin, bradykinin, and "slow reacting substance." Release of these chemicals into the body causes the blood vessels to become more permeable (leaky) and causes runny nose, and itchy and watery eyes. These chemicals also cause smooth muscle contraction causing sneezing, breath constriction, coughing, wheezing, etc.

Medium A substance used to provide nutrients for cell growth. It may be liquid (e.g. broth) or solid (e.g. agar).

Megabase A unit of length for DNA, equal to 1,000,000 base pairs.

Mega-yeast artificial chromosomes (mega YAC) A large (i.e., greater than 500 base pairs in length) piece of DNA that has been cloned (made) inside a living yeast cell. While most bacterial vectors cannot carry DNA pieces that are larger than 50 base pairs, and "standard" YACs typically cannot carry DNA pieces that are larger than 500 base pairs, mega YACs can carry DNA pieces (chromosomes) as large as one million base pairs in length.

Melting (of DNA)   Melting DNA means to heat-denature it. When this happens, the hydrogen bonds holding the DNA molecule together in the normal way are disrupted, allowing a more random polymer structure to exist.

Melting temperature (of DNA) (T_m)   The midpoint of the temperature range over which DNA is denatured.

Mesenchymal stem cell (MSC)   A pluripotent stem cell which differentiates (and migrates where needed) to become connective tissue within the body of an organism.

Mesodermal adult stem cells   Certain stem cells present within (adult) bodies of organisms, that can be differentiated (via chemical signals) to give rise to bone, muscle, and/or fat cells.

Messenger RNA (mRNA)   Messenger ribonucleic acid, first identified by Francis Crick, Sydney Brenner, and Matthew Meselson. mRNA is the intermediary molecule between DNA and ribosomes (in a cell) which synthesize (i.e., manufacture) those proteins coded for by the cell's DNA. Upon receiving the "message" encoded in the DNA, the messenger RNA passes through the ribosomes like a reel of punched paper passes through an old player piano (pianola) giving the ribosomes the specifications for making the coded-for proteins.

Microarray (testing)   A piece of glass, plastic, or silicon onto which has been placed a large number of biosensors at known, specific locations. These microarrays (sometimes called "biochips" or "DNA chips") can then be utilized to test a single biological sample for a variety of attributes or effects.

Also microarray is an array of chemicals in a very small area used as a test or analytical tool. Nearly all microarrays are DNA arrays, where hundreds to millions of DNA probes are bound to a surface and used to test a sample—if the sample contains DNA complementary to any of the probes, that DNA binds to the array in the specific place where that probe is located.

Microgram 10^{-6} gram, or 3.527×10^{-8} ounce (avoirdupois).

Micron   Also called micrometre. A unit of length convenient for describing cellular dimensions; the Greek letter μ is used as its symbol. A micron is equal to 10^{-3} mm (millimetre) or 10^{4} (Angstroms) or 0.00003937 inch.

Microorganism   Any organism of microscopic size (i.e., requires a microscope to be seen by man). First viewed by Antoni van Leeuwenhoek in 1676.

Microparticles   The metal particles (<1 micron in diameter) that are coated with gene(s) and are "shot" into cells with the Biolistic® gene gun.

Microsatellite DNA   Pieces of the same small segment (i.e., a DNA sequence) which are "repeated" (appear repeatedly in sequence within the DNA molecule) adjacent to a specific gene within the DNA molecule. Thus, these "microsatellites" are linked to that specific gene.

Mitochondrial DNA   The DNA within an organism's (e.g. human) cells that is located inside the mitchondria (organelles); not inside the cell nucleus. In virtually all eukaryotes, the mitochondrial DNA is only passed down from mother to offspring; not from father to offspring, as nuclear DNA is.

Molecular beacon   Term that is used to refer to specific oligonucleotides possessing a "hairpin loop" and bearing a fluorescent dye. A "quencher dye" located on a nearby portion of the hairpin loop prevents fluorescence until the hairpin loop is opened up.

Molecular biology   A term coined by Vannevar Bush during the 1940s that eventually came to mean the study and manipulation of molecules that constitute, or interact with, cells. Molecular biology as a distinct scientific discipline originated largely as a result of a decision to provide "support for the application of new physical and chemical techniques to biology" during the 1930s by Warren Weaver, Director of the Biology (funding) Program at America's Rockefeller Foundation (a philanthropic organization).

Molecular weight   The sum of the atomic weights of the constituent atoms in a molecule.

Monoclonal antibodies (MAbs)   Discovered and developed in the 1970s by Cesar Milstein and Georges Kohler, monoclonal antibodies is the name for antibodies derived from a single source or clone of cells that recognize only one kind of antigen. Made by fusing myeloma cancer cells (which multiply very fast) with antibody-producing cells, then spreading the resulting conjugate colony so thin that each cell can be grown into a whole, separate colony (i.e., cloning). In this way, one gets whole batches of the same (monoclonal) antibody, which are all specific to the same antigen. Monoclonal antibodies have found markets in diagnostic kits, pharmaceuticals (e.g. trastuzumab, rituximab, bevacizumab, adalimumab, infliximab, etc.), imaging agents, and in purification processes.

Monocytes   Also called monocyte macrophages. The round-nucleated cells that circulate in the blood.

In summary they engulf and kill microorganisms, present antigen to the lymphocytes, kill certain tumour cells, and are involved in the regulation of inflammation. These cells are often the first to encounter a foreign substance or pathogen or normal cell debris in the body. When they do, the material is taken up (engulfed) and degraded by means of oxidative and hydrolytic enzymatic attack. Peptides that result from the degradation of foreign protein are then bound to a monocyte protein called class II MHC (major histocompatibility complex) and this self-foreign complex then migrates to the surface of the cell where it is embedded into the cell membrane in such a way as to present the peptide to the outside of the cell. This positioning allows T lymphocytes to recognize (inspect) the peptide. Whereas self-peptides derived from normal cellular debris are ignored, foreign peptides activate precursors of helper T cells to further mature into active, lymphokine-secreting helper T lymphocytes, also known as T_H cells. When monocytes move out of the bloodstream and into the tissues they are called macrophages.

Motifs   Short strings of bases or amino acids that crop up time and again in different genes and proteins. Usually, they are significant because they denote that some bit of the molecule has a particular function. Thus, "zinc finger motifs" in proteins suggest that the protein has a section that binds to DNA. In DNA, the "TATAA" motif is suggestive of a promoter sequence in eukaryotic cells.

mRNA   *See* Messenger RNA.

mtDNA   Acronym for mitochondrial DNA. *See* Mitochondrial DNA.

Multi-copy plasmids   Plasmids that are present inside bacteria in quantities greater than one plasmid per (host) cell.

Multiplex assay   An assay which generates more than one data point in each (assay) evaluation that is performed. For example, fluorescence mapping (i.e., scanning an x-y plane within tissue at varying depths with a microscope/light of selected wavelength) designed to cause/detect any fluorescence resulting from a biological event of interest (e.g. the binding of a particular protein to a given cell receptor, the expression of particular gene(s), etc.).

Multipotent adult stem cell   Certain stem cells present within (adult) bodies of organisms, that can be differentiated (via chemical signals) to give rise to a variety of different cell/tissue types (e.g. bone, cartilage, fat, muscle, red blood cells, B cells, T cells, etc.). For example, adipose (body fat) cells have been

removed from the bodies of some mammals by researchers and subsequently coaxed into expressing genes; and otherwise exhibiting characteristics of (differentiated) bone cells, cartilage cells, muscle cells, etc.

Mutation   From the Latin term *mutare*, meaning "to change". Any change that alters the sequence of the nucleotide bases in the genetic material (DNA) of an organism or cell, with alteration occurring either by displacement, addition, deletion, cross-linking, or other destruction. The mutation alteration to the DNA sequence would alter its meaning, that is, its ability to produce the normal amount or normal kind of protein, so the organism or cell is itself altered. Such an altered organism is called a mutant.

Naked gene   A bare gene (strand of DNA that codes for a protein) which has been extracted from an organism, or otherwise derived (e.g. synthesized from sequence data).

Nanobodies   The smallest possible portion of an antibody which will bind to an antigen/hapten. Although nanobodies are not made naturally (i.e., the body always makes complete/full-size antibodies), nanobodies [approximately 120 amino acids in length] can be made via genetically engineered cells grown via cell culture. Because nanobodies are able to survive highly acidic conditions (e.g. passage through the stomach) without losing their biological activity, they could potentially be utilized in orally administered pharmaceuticals.

Nanobots   Very small "robots" whose dimensions would be measured in terms of nanometres (nm), and could perform specific tasks. For example, during 2005, Ben L. Feringa created a nanoscale rotor which is powered by light, by attaching a chiral helical alkene molecule onto a piece of gold.

Nanocapsules   Nanometer-scale, hollow, spherically-shaped objects that can be utilized to encapsulate small amounts of pharmaceuticals, enzymes or other catalysts, etc.

Nanocrystals   A term that is utilized to refer to any crystalline structure possessing dimensions (e.g. overall width) measured in terms of nanometres.

Nanofibres   Refers to fibres (created by man) possessing diameter(s) of less than 100 nanometres. For example, during 2003, Sam Stupp created self-assembling nanofibres that promote (re)growth of nerve cells in laboratory rats. These nanofibres consist of peptide amphiphiles with hydrophilic peptide portions covering the exterior of the fibres, while their hydrophobic portions are in the centre of the fibres.

Nanogram (ng) 10^{-9} gram, or 3.527×10^{-11} ounce (avoirdupois).

Nanolithography   Refers to the practice of using an atomic force microscope tip to apply specific (e.g. DNA) molecules to surfaces such as metals, oxides, etc. Nanolithography can be used to direct (via DNA interactions) the assembly of tiny structures such as gene chips, catalysts, nanoscale circuits, etc.

Nanometres (nm) 10^{-9} metre. Often used to express wavelengths of light (e.g. in a spectrophotometer), or to express dimensions of nanocomposites, devices (e.g. of miniature "machines" called nanoelectromechanical systems), etc. in the field of nanotechnology.

Nanoparticles   Term utilized to refer to a variety of nanometer-scale particles (e.g. nanocrystals, nanoshells, quantum dots, etc.). Depending on the materials used to construct nanoparticles, their uses include imaging tissues within the body, detection of pancreatic cancer cells, detection of cancer metastasis and detection of DNA hybridization.

Nanopore   A device that can distinguish between different DNA strands (molecules) that differ from each other by a single nucleotide (in the make-up of those molecular strands). Developed by Hagan Bayley, David Deamer and Mark Akeson in 2001, it consists of an artificial membrane (lipid bilayer) with a "hole" (nanopore) punctured in that membrane by the protein alpha-haemolysin.

Nanoscience   A term utilized to refer to the science underlying nanotechnology, nanocrystals, nanocrystal molecules, nanocomposites, quantum dots, nanoelectromechanical systems (NEMS), nanovalves, nanostructured materials, etc.

Nanoshells   Refers to nanometer-scale crystalline structures which form into the shape of hollow balls. For example, nanoshells can be manufactured by surrounding cobalt nanoparticles with sulphur (in a 9 : 8 ratio). When the proper reaction conditions are subsequently applied to this 9 : 8 cobalt–sulphur mixture, the Kirkendall effect causes the cobalt atoms within the nanoparticle to diffuse out to (and react with) the sulphur atoms faster than the sulphur atoms diffusing into the nanoparticle. The end result is a spherical nanoshell comprised of the compound Co_9S_8.

Nanotechnology   From the Latin *nanus* = "dwarf", so it literally means "dwarf technology." The word was originally coined by Norio Taniguchi in 1974, to refer to high precision machining. However, Richard Feynman and K. Eric Drexler later popularized the concept of nanotechnology as a new and developing technology in which man manipulates objects whose dimensions are approximately 1 to 100 nanometres. Theoretically, it is possible that in the future a variety of man-made "nano-assemblers" [i.e., tiny (molecular) machines smaller than a grain of sand] would manufacture those things that are produced today in factories. For example, enzyme molecules function essentially as jigs and machine tools to shape large molecules as they are formed in biochemical reactions. The technology also encompasses nanostructured materials, biochips and biosensors, and involves manipulating atoms and molecules in order to form (build) bigger, but still vanishingly small functional structures and machines.

Nanotube   Refers to a tiny tube whose diameter is measured in nanometres. These have been constructed from a variety of materials such as carbon, tungsten sulphide, titanium dioxide, amino acids, etc.

Nanovalve   Refers to any nanometer-scale device which allows the release of molecules in a manner that can be controlled by man. For example, during 2005 Fraser Stoddart and colleagues were able to create a nanovalve that is "switchable" via chemical energy (i.e., it can be opened when desired, via nearby addition of appropriate chemicals that react to yield an electron which causes the nanovalve to open, thereby releasing a molecule). That nanovalve consisted of a rotaxane molecule that is attached directly over a pore to the surface of porous glass. When a relevant molecule (e.g. a pharmaceutical) is first inserted into the pore and the rotaxane molecule is subsequently attached above the open end of the pore, this results in a system via which the pharmaceutical molecule can be released at the time desired by man (e.g. when needed to treat a disease, stimulate an organ such as a weakened heart, etc.).

Nanowire   Term utilized to describe (relatively) long and narrow electrical conductors whose dimensions are measured in nanometers (nm). For example, during 2003, Susan L. Lindquist utilized yeast amyloid proteins (which self-assemble into 60–300 nanometer-long fibres) to create nanowires by subsequently coating those fibres with gold and silver.

Natural killer T cells (NKT cells)   A special kind of lymphocytes that bridges the adaptive immune system with the innate immune system and are involved in tumour surveillance. Unlike conventional T cells that recognize peptide antigen presented by major histocompatibility complex (MHC) molecules, NKT cells recognize glycolipid antigen presented by a molecule called CD1d. Once activated, these cells can perform functions ascribed to both T_H and T_C cells (i.e. cytokine production and release of cytolytic/cell-killing molecules). They also kill virus-laden cells.

Nested PCR A specific PCR (polymerase chain reaction) technique of two consecutive-run PCRs, in which the second PCR amplifies (i.e., makes multiple copies of) a DNA-sequence within the product (amplicon) of the first PCR.

Neutrophils Phagocytic (ingesting, scavenging) white blood cells that are produced in the bone marrow. They ingest and destroy invading microorganisms and facilitate post-infection tissue repair.

Nonsense codon A triplet of nucleotides that does not code for an amino acid. Any one of three triplets (UAG, UAA, or UGA) that cause termination of protein synthesis (in ribosome), and thus the release from ribosome of a (completely translated) protein molecule.

Northern blotting A research test/methodology used to transfer RNA fragments from an agarose gel (e.g. following gel electrophoresis) to a filter paper without changing the relative positions of the RNA fragments (e.g. re-electrophoresis separation grid).

nt An abbreviation for nucleotide. *See* Nucleotide.

Nuclear DNA The DNA that is contained within the nucleus of a cell.

Nucleic acid probes *See* DNA Probe.

Nucleic acids A nucleotide polymer. A large, chain-like molecule containing phosphate groups, sugar groups, and purine and pyrimidine bases; two types are ribonucleic acid (RNA) and deoxyribonucleic acid (DNA). The bases involved are adenine, guanine, cytosine, and thymine (uracil in RNA).

Nucleoid The compact body that contains the genome in a bacterium.

Nucleolus A round, granular structure situated in the nucleus of eukaryotic cells. It is involved in rRNA (ribosomal RNA) synthesis and ribosome formation.

Nucleoside A 'hybrid' molecule consisting of a purine (adenine, guanine) or pyrimidine (thymine, uracil, or cytosine) base covalently linked to a five-membered sugar ring (ribose in the case of RNA and deoxyribose in the case of DNA).

Nucleotide An ester of a nucleoside and phosphoric acid. Nucleotides are nucleosides that have a phosphate group attached to one or more of the hydroxyl groups of the sugar (ribose or deoxyribose). In short, a nucleotide is a hybrid

molecule consisting of a purine or pyrimidine base covalently linked to a five-membered sugar ring which is covalently linked to a phosphate group. While (polymerized) nucleotides are the structural units of a nucleic acid, free nucleotides that are not an integral part of nucleic acids are also found in tissues and play important roles in the cell, e.g. ATP and cyclic AMP.

Nucleus Discovered by Robert Brown in the early 1800s, it is the usually spherical body within each living cell that contains its hereditary biological material (e.g. DNA, genes, chromosomes, etc.) and controls the cell's life functions (e.g. metabolism, growth, and reproduction). The nucleus is a highly differentiated, relatively large organelle lying in the cytoplasm of the cell. The nucleus is surrounded by a (nuclear) membrane which is quite similar to the plasma (cell) membrane, except that the nuclear membrane contains holes or pores. It is characterized by its high content of chromatin, which contains most of the cell's DNA. That chromatin is normally (when cell is not in the process of dividing) distributed throughout the nucleus in a diffuse manner.

Nutraceuticals Coined in 1989 by Stephen DeFelice, this term is used to refer to either a food or portion of food (e.g. a vitamin, essential amino acid, etc.) that possesses medical or health benefits (to the organism that consumes that nutraceutical). For example, saponins (present in beans, spinach, tomatoes, potatoes, alfalfa, clover, etc.) possess some cancer-prevention properties. Also sometimes called pharmafoods, functional foods, or designer foods, these are food products that have been designed to contain specific concentrations and/or proportions of certain nutrients (e.g. vitamins, amino acids, etc.) that are critical for good health.

Oligonucleotide Synonymous with oligodeoxyribonucleotide, they are short chains of nucleotides (i.e., single-stranded DNA or RNA) that have been synthesized (i.e., made by man or inside living cells) by chemically linking together a number of specific nucleotides.

Oligonucleotide probes Short-chain fragments of DNA that are used in various gene analysis tests (e.g. the single base change in DNA that causes sickle-cell anaemia).

Oncogene Genes that are believed to be necessary for cancers to develop. There are a large number of them and, as would be expected from the variety of cancer types, they act in many different ways. Most are present in normal cells as proto-oncogenes, that is, versions of the gene which are benign, and indeed are essential to the body's normal development. A mutation turns them into the malignant oncogene. Oncogenes are of interest to biotechnology because of the importance of cancer as a cause of morbidity and mortality in Western societies.

Oocytes   The cells, produced by an organism's ovaries, that eventually become an ovum ("egg cell") via meiosis.

Open reading frame (ORF)   Region of a gene (DNA) that contains a series of triplet (bases) coding for amino acids without any termination codons. The ORF sequence is potentially translatable into a protein, but the presence of an open reading frame (sequence) does not guarantee that a protein molecule will be produced (by cell ribosome).

Operator   Also known as the "o locus". The site on the DNA to which a repressor molecule binds to prevent the initiation of transcription. The operator locus is a distinct entity and exists independently of the structural genes and the regulatory gene. It is the structural/biochemical "switch" with which the operon is turned on or off, and it controls the transcription of an entire group of coordinately induced genes. One type of mutation of the operator locus is called operator constitutive mutants. Constitutive mutants continually churn out the protein characteristic for that operon because the operon unit cannot be turned off by the repressor molecule.

Operon   A gene unit consisting of one or more genes that specify a polypeptide and an operator unit that regulates the structural gene, that is, the production of messenger RNA (mRNA) and hence, ultimately, of a number of proteins. Generally an operon is defined as a group of functionally related structural genes mapping (that is, being) close to each other in the chromosome and being controlled by the same (one) operator. If the operator is "turned on", then the DNA of the genes comprising the operon will be transcribed into mRNA, and down the line specific proteins are produced. If, on the other hand, the operator is "turned off" then transcription of the genes does not occur and the production of the operon-specific proteins does not occur.

Optical biosensor   There are two main areas of development in optical biosensors. These involve determining changes in light absorption between the reactants and products of a reaction, or measuring the light output by a luminescent process. The former usually involves the use of colorimetric test strips. These are disposable single-use cellulose pads impregnated with enzyme and reagents. The most common use of this technology is for whole-blood monitoring in diabetes control.

Organ culture   The growth of whole organs or parts of organs *in vitro*. Organs consist of several different cell types, as opposed to tissues which are made up of uniform cells.

Overlapping gene   A gene whose sequence at least partially overlaps that of another gene (adjacent to the first within the DNA of an organism).

Oxidoreductase   A member of a group of enzymes involved in redox reactions. Oxidoreductases were formerly known as oxidases or dehydrogenases.

PAGE   *See* Polyacrylamide Gel Electrophoresis.

Palindrome   A DNA molecule sequence that is the same when one strand of the molecule is read left to right and the other strand is read right to left.

Patents   Most biotechnology is a "knowledge industry"—being able to do something is what makes a company valuable. It is therefore, imperative for a biotechnology company to be able to protect its knowledge, known as "intellectual property" and the most common way to do this is to patent it. In principle, patents are based on claims of novelty, utility, and enablement: a patent must describe something new, something useful, and must describe it in a way that enables someone else to duplicate it.

pBR 322   An *Escherichia coli (E. coli)* plasmid cloning vector that contains the ampicillin resistance and tetracycline resistance genes. It consists of a circle of double-stranded DNA.

PCR   *See* Polymerase Chain Reaction.

PEG (Polyethylene glycol)   A water-soluble polymer used in many guises in biotechnology. The biotechnological uses include: large-scale purification of proteins and viruses through precipitation, as a fusogen to make mammalian cells fuse and as a molecule to conjugate onto proteins (a process sometimes called Pegylation).

Peptide mapping (fingerprinting)   The characteristic pattern of peptides (i.e., pieces that make up a protein molecule) resulting from partial hydrolysis (cleavage, digestion) of a protein. The pattern (fingerprint) is obtained by separating the peptides via two-dimensional chromatography, in which the peptides are first subjected to chromatography using one solution which separates many, but not all peptides. The chromatogram is then turned 90 degrees, and is again chromatographed using a second solution, which then separates all of the peptides thereby producing the final "fingerprint" of the protein.

Peyer's patches   A specific set of lymphoid organs found in the intestinal wall of many mammals. These patches filter out antigens that enter the intestine in food or come from bacteria growing in the intestine, and "present" those intact antigens to adjacent lymphoid tissues via special M cells of the Peyer's patches. This activates the lymphocytes in the patches, which then migrate out of the node and into the blood where they float in the tissue spaces just inside the intestinal lining. There they secrete antibodies (primarily IgA), which are then transported into the lumen (contents) of the gut and subsequently attack (bind) the antigens.

Phage   Abbreviation for bacteriophage, which is another name for a specific type of virus. It is a virus that attacks bacteria (and "consumes" those bacteria via the making of more copies of that virus) and so known as a bacteriophage. Phage is derived from the Greek word meaning "to eat".

Phage display   A methodology in which capsid proteins (i.e., on surface of bacteriophages) are attached to selected peptides. Because determination of each bacteriophage's DNA (gene) sequence thus determines the peptide (sequence) that is "displayed" on its surface, large 'libraries' of phage-displayed peptides can be created (e.g. to be utilized by scientists in the screening of candidate-compounds, etc. in search for new pharmaceuticals). For example, the pharmaceutical ADALIMUMAB (Humira™) was developed using phage display technology.

Phenotype   Coined in 1909 by Wilhelm Johannsen, this term refers to the outward appearance (structure) or other visible characteristics of an organism (which of course, is determined by the DNA of its genotype). This also includes (and/or determines) how that organism's body responds to a given physical agent (e.g. a pharmaceutical, a toxin, sunlight, etc.). For example, genetically fair-skinned people tend to get sunburned easier/faster than other people do.

Phosphorylation   The introduction of a phosphate group into a molecule. Formation of a phosphate derivative of a biomolecule, usually by enzymatic transfer of a phosphate group from ATP. In some forms of cancer, certain very specific phosphorylations are produced (e.g. in post-translational modification of protein molecules).

Phylogenetic profiling   A research methodology that is utilized to try to predict the function of a protein molecule within a larger, complex organism (e.g. man) from the function of a similar/related protein molecule in a smaller/simple organism (e.g. a "model organism") which is easier to study.

Physical map (of genome)   A diagram showing the linear order of genes or genetic markers on the genome, with units indicating the actual distance between the genes or markers.

Phytoremediation   Bioremediation (clean-up of waste and environmental contamination) using plants. The use of transgenic plants has advanced rapidly here, as the concerns about the accidental escape of a plant are far less than the escape of a genetically engineered bacterium.

Picogram (pg) 10^{-12} gram or 3.527×10^{-14} ounce (avoirdupois).

piRNAs   Abbreviation for piwi-interacting RNAs, because they are RNAs that interact with members of the piwi sub-class of Argonaute proteins. These are RNA segments of approximately 24–31 nucleotides in length.

Plant cell culture   Like any living organism, plants are composed of cells, which are capable of growing and dividing outside the plant given the right conditions. As with animal cell culture, it is essential to keep the cells free from any other contaminating organism like a bacterium or fungus. Plant cell culture has a wide range of applications in biotechnology including plant cloning, plant genetic engineering and protein production in plants.

Plaque   Refers to deposits of (oxidized) cholesterol intermixed with smooth-muscle cells and some macrophages, lining the inside of certain blood vessels. These deposits can result in the disease atherosclerosis, and/or adversely increasing blood platelet aggregation (e.g. clotting/thrombosis).

Plasma cell   Also called plasma B cells or plasmocytes, these are cells of the immune system that secrete large amounts of antibodies. They differentiate from B cells upon stimulation by $CD4^+$ lymphocytes. The B cell acts as an antigen-presenting cell (APC), consuming an offending pathogen. That pathogen gets taken up by the B cell by phagocytosis, and broken down within phagosomes after fusion with lysosomes releasing proteolytic enzymes onto the pathogen.

Plasma membrane   From the Greek word *plasm* = "something formed". Also sometimes known as a plasmalemma, it is a thin structure that completely surrounds the cell as a sort of "skin". This membrane may be seen with the aid of an electron microscope. The entire membrane appears to be about 100 Angstroms (Å; 0.1 mm) thick and is composed of two dark lines each about 30 Å thick which are, however, separated by a lighter area. This trilaminar "sandwich" structure is referred to as the unit membrane.

Plasmid   An independent, stable, self-replicating piece of DNA in bacterial cells that is not a part of the normal cell genome and that never becomes integrated into the host chromosome. This is in contrast to a similar genetic element known as an episome plasmid which may exist independently of the chromosome or may become integrated into the host chromosome. Plasmids are known to confer resistance to antibiotics and may be transferred by cell-to-cell contact (by conjugation via the sex pilus) or by viral-mediated transduction. Plasmids are commonly used in recombinant DNA experiments as acceptors of foreign DNA. Known forms of plasmids include both linear and circular molecules.

Plasmocyte   Another name for a blast cell.

Plastid   From the Greek word *plastis* = "builder"; it is an independent, stable, self-replicating organelle containing a piece of DNA, inside a plant cell's cytoplasm. Plastids, which include both chloroplasts and chromoplasts, are not a part of the reproduction cell genome (i.e., in nucleus). Because there can exist up to 10,000 plastids in a given plant cell, the insertion of a gene (e.g. via genetic engineering) into plastids can result in a higher yield (of the specific protein coded for by that gene) than is achieved via insertion of the gene into the cell's nuclear DNA.

Platelets   Disc-shaped blood cells that stick to the (microscopically 'jagged') edges of wounds. The aggregation of platelets at the wound site leads to blood clotting, forming a temporary wound covering. During this blood clotting process, the platelets release platelet-derived growth factor (PDGF) which attracts fibroblasts to the wound area (for subsequent healing process).

Pluripotent stem cells   Stem cells which can develop into any of the three major tissue types: endoderm (interior gut lining), mesoderm (muscle, bone, blood), and ectoderm (epidermal tissues and nervous system). Pluripotent stem cells can eventually specialize in any bodily tissue, but they cannot themselves develop into a human being.

Point mutation   A mutation consisting of a change of only one nucleotide in a DNA molecule. At "hot spots" (i.e., certain locations on the DNA within some organisms), numerous point mutations can occur. In the case of single-nucleotide polymorphisms (SNPs), the same point mutation occurs at the same location (on the DNA within some organisms) across a population of individuals of that organism.

Polyacrylamide gel electrophoreis (PAGE)   A form of chromatography in which molecules are separated on the basis of size and charge. The stationary phase (the polyacrylamide gel) is a polymerized version of acrylamide monomers. The gel looks and feels like Jello™. On a molecular basis, it consists of an intertwined and cross-linked mesh of polyacrylamide strings. As can be imagined, there are tiny "holes" in the gel (like in a plastic mesh bag) and with enough cross-linking the size of the holes begins to approach the size of the molecules which are to be separated. Since some molecules will be larger and some smaller, some of them will be able to pass through the gel matrix more easily than others. This is part of the basis for separation. It should be noted at this point that if the gel is cross-linked enough and because of this the holes in that gel are smaller than the molecules to be separated, then the molecules will not be able to penetrate into the gel and no separation can occur. The charge on the molecule also plays a role in the separation. Functionally, the gel serves to hold and separate the molecules. Although details are not presented here, after the gel has been prepared (poured and cross-linked) a small amount of the solution containing the molecules to be separated is placed into wells (grooves to hold the liquid) on the gel and the system is subjected to an electric current. Over the course of minutes to hours, molecules bearing different charge/mass separate.

Polyadenylation   The addition of a sequence of polyadenylic acid to the 3′ end of a eukaryotic mRNA after its transcription (post-transcriptional).

Polyclonal antibodies (used in humans)   A mixture of antibody molecules that are specific for given antigen(s), which have been purified from an immunized (to that given antigen) animal's blood. Such antibodies are polyclonal in that they are the products of many different populations of antibody-producing cells (within the animal's body). Hence they differ somewhat in their precise specificity and affinity for the antigen.

Polymerase   An enzyme that catalyses the assembly of nucleotides into RNA (RNA polymerase) and of deoxynucleotides into DNA (DNA polymerase).

Polymerase chain reaction (PCR)   A reaction that uses the enzyme DNA polymerase to catalyse the formation of more DNA strands from an original one by the execution of repeated cycles of DNA synthesis. Functionally, this is accomplished by heating and melting double-stranded (hydrogen-bonded) DNA into single-stranded (non-hydrogen-bonded) DNA and producing an oligonucleotide primer complementary to each DNA strand. The primers bind to the DNA and mark it in such a way that the addition of DNA polymerase and deoxynucleoside triphosphates cause a new strand of DNA to form which is complementary to the target section of DNA. The process described previously is repeated (trait, product, etc.) again and again to produce millions of copies (amplicons) of the desired strand of DNA. PCR and its registered trademarks are the property of F. Hoffmann-La Roche & Co. AG, Basel, Switzerland.

Polymorphism (genetic)   A name applied to a condition in which a species of plant or animal is represented by several distinct, non-integrating forms or types unrelated to age or sex. The differences are often in colouration, though any characteristic of the organism may be involved (e.g. nuclei shape for polymorphonuclear leucocytes).

Post-transcriptional processing (Modification) of RNAs   The enzyme-catalysed processing or structural modifications that RNAs such as mRNAs, rRNAs, and tRNAs must undergo before they are functionally finished products. For example, in eucaryotes a block of poly(A) containing at least 200 AMP residues is enzymatically attached to the 3′ end of mRNA in the nucleus of the cell. The mRNAs with the "tail" are then transferred to the cytoplasm and the tail enzymatically removed to form the functional mRNAs. It is believed that the poly(A) tail aids in the transfer of the complex and/or targets the complex to the cytoplasm.

Post-translational modification of protein   Enzymatic processing of a polypeptide chain (i.e., protein) after its translation from its mRNA transcript, which includes:

- Glycosylation—addition of carbohydrate moieties to the protein molecule;

- Phosphorylation—addition of a phosphate molecular group to the protein molecule;

- Sulphation—addition of a sulphate molecular group to the protein;

- Acetylation—addition of an acetyl molecular group to the protein;

- Ribosylation—addition of a ribose molecular group to the protein;

- Deamidation—the loss of their side-chain molecular groups by some of the glutamine and asparagine portions of a given protein molecule and

- Cleavage or the removal of a portion of the polypeptide chain in order to produce a functional protein in the correct environment.

Potentiometric biosensors   Biosensors that make use of ion-selective electrodes in order to transduce the biological reaction into an electrical signal. In the simplest terms this consists of an immobilized enzyme membrane surrounding the probe from a pH meter, where the catalysed reaction generates or absorbs hydrogen ions. The reaction occurring next to the thin sensing glass membrane causes a change in pH which may be read directly from the pH meter's display.

Pribnow box   The consensus sequence TATAATG centred about 10 base pairs before the starting point of bacterial genes. It is a part of the promoter and is especially important in binding RNA polymerase.

Primary structure   The sequence of amino acids in a protein "molecular" chain, or the linear sequence of nucleotides in a polynucleotide (RNA or DNA) molecular chain.

Primer (DNA)   A short sequence deoxyribonucleic acid (DNA) that is paired with one strand of the template DNA, in the polymerase chain reaction (PCR) technique. In PCR testing (e.g. a paternity test), the primer is selected to be complementary to the analytically relevant sequence of DNA. It is the growing end of the DNA chain and it simply provides a free 3′- OH end at which the enzyme DNA polymerase adds on deoxyribonucleotide units (monomers). Which deoxyribonucleotide is added is dictated by base-pairing to the template DNA chain. Without a DNA primer sequence, a new DNA chain cannot form since DNA polymerase is not able to initiate DNA chains.

Prion   Proteinaceous structures (molecules) found in the plasma membrane (surface) of cells, in the brains and various other tissues of all vertebrate animals. In addition to a role in signalling, one of the functions of (normal) prions [PrPc] is to help "capture" and deactivate oxygen-free radicals (i.e., oxygen atoms bearing an extra electron, thus high in energy (e.g. those which are sometimes generated in a biological system such as within the body of an organism).

Probe   A relatively small molecule that can be used to sense the presence and condition of a specific protein, DNA fragment, RNA fragment, or nucleic acid by a unique interaction with that macromolecule.

Probiotics   Compounds that (generally) act to stimulate growth of beneficial types of bacteria within the digestive system of animals (e.g. livestock). For example, organic acids (e.g. propionic acid, acetic acid, lactic acid, citric acid, etc.) act to inhibit the growth/multiplication of pathogens (i.e., disease-causing microorganisms) in the digestive system of monogastric (i.e., single-stomach) animals such as poultry and swine. Those acids are able to pass through the outer cell membrane (i.e., plasma membrane) of pathogenic bacteria and fungi. Once inside those pathogens' cells, the acids dissociate, and acidify the cell interior (which disrupts the cell's protein synthesis, growth, and replication of the pathogen).

Prokaryotes Simple organisms that lack a distinct nuclear membrane and other membrane-bound organelles. Most are unicellular, but some prokaryotes are multicellular). The prokaryotes are divided into two domains: the bacteria and the archaea. Archaea or Archaebacteria are a newly appointed kingdom of life. These organisms were originally thought to live only in inhospitable conditions such as temperature, pH-extremes, and radiation, but have since been found in all types of habitat. Many structural systems are different between prokaryotes and eukaryotes including the DNA arrangement, composition of membranes, the respiratory chain, the photosynthetic apparatus, ribosome size, the presence or lack of cytoplasmic streaming, the cell wall, flagella, the mode of sexual reproduction, and the presence or lack of vacuoles. Some representative prokaryotes are the bacteria and blue-green algae.

Promoter The region on DNA to which RNA polymerase binds and initiates transcription (of RNA). The promoter "promotes" the transcription (expression) of that gene, but the promoter's impact on the timing/degree of gene expression is itself regulated by the molecules that bind to the promoter. For example, the "binding" of RNA polymerase causes transcription of RNA to begin, and the "binding" to promoter of other STATs (i.e., signal transducers and activators of transcription) can regulate the degree to which a given gene is expressed.

Proof-reading Any mechanism for correcting errors in nucleic acid synthesis that involves scrutiny of individual (chemical) units after they have been added to the (molecular) chain. This function is carried out by a $3'$ to $5'$ exonuclease, among others. Proof-reading dramatically increases the fidelity of the base-pairing mechanism.

Proteases Enzymes which break up proteins. There are four distinct uses for proteases in biotechnology. Their use depends partly on how cheap they are to make, and partly on how specific they are, that is, whether they chop up all proteins indiscriminately or only a few proteins at specific points.

Protein engineering This is the design, production, analysis and use of altered non-natural proteins. Almost all protein engineering is actually the engineering of genes using recombinant DNA techniques, and then expressing the proteins from the genes. Protein engineering has a number of aims like improving protein stability, altering antibody properties, altering the substrate specificity of an enzyme and altering pharmacological action.

Protein folding The complex interactions of a polypeptide molecular chain with its environment and itself and other protein entities, which cause the polypeptide molecule to fold up into a highly organized, tightly packed, three-dimensional structure. It was proven to occur spontaneously, by Christian B. Anfinsen during the 1960s: for protein molecules outside of living cells. This ability of polypeptide chains to fold into a great variety of topologies, combined with the large number of sequences (in the molecular chain) that can be derived from the 20 common amino acids in proteins, confers on protein molecules their great powers of recognition and selectivity. How a protein folds up determines its chemical function. During the 1990s, it was discovered that inside living cells, "chaperone" molecules are needed for proper protein folding to occur. These chaperones are protein molecules (e.g. certain heat-shock proteins) that form a loosely bound complex to suppress incorrect protein folding as the protein molecule is emerging from the cell's ribosome, so protein folding is both complete and correct as soon as the newly formed protein molecule is released from the cell's ribosome.

Protein microarrays A piece of glass, plastic, or silicon, onto which has been attached a number of capture agents (e.g. antibodies, aptamers, enzymes, antigens, receptors, ligands, or other molecules of other chemical compounds that bind/interact with proteins in a specific manner) at specific/known locations on the microarray. These microarrays (sometimes called "biochips", protein biochips, protein arrays, etc.) can then be utilized to test (e.g. a single sample) for a wide variety of attributes or effects (on, or by the protein molecules in the sample that is exposed to that microarray).

Proteomics The scientific study of an organism's proteins and their role in an organism's structure, growth, health, disease (and/or the organism's resistance to disease, etc.). Those roles are predominantly due to each protein molecule's tertiary structure/conformation, but some are also due to some proteins' interaction with the organism's DNA (e.g. transcription factors).

Protoplasts Plant, fungal and most bacterial cells are surrounded by a tough, thick cell wall. A protoplast is such a cell from which we have removed the cell wall, leaving the cell naked and surrounded only by its plasma membrane.

Pyrimidine   A heterocyclic organic compound containing nitrogen atoms at (molecular ring) positions 1 and 3. Naturally occurring derivatives are components of nucleic acids and coenzymes, uracil, thymine, and cytosine.

QPCR   Acronym for quantitative polymerase chain reaction.

qRT-PCR   Acronym for quantitative real-time reverse transcription polymerase chain reaction.

Quantitative trait loci (QTL)   Individual specific DNA sequences that are related to known traits (e.g. litter size in animals, egg production in birds, yield in crop plants.)

Radioactive isotope   An isotope with an unstable (atomic) nucleus that spontaneously emits radiation. The radiation emitted includes alpha particles, nucleons, electrons, and gamma rays.

Radioimmunoassay   Invented by Rosalyn Yarlow and Solomon Berson in 1959, it is a very sensitive method of quantitating a specific antigen using a specific radiolabelled antibody. Functionally, the antibody is made radioactive by the covalent incorporation of radioactive iodine. The radioimmunoprobe thus prepared is exposed to its antigen (which may be a protein, or a receptor, etc.) in excess (the exact amount will have to be determined). The radiolabelled probe then binds to the antigen and the unbound, free probe is washed away. The radioactivity is then determined (counted) and by comparison to a standard plot which has been constructed previously, the amount of antigen (binding) is determined.

Random amplified polymorphic DNA (RAPD) technique   A genetic mapping methodology that utilizes as its basis the fact that specific DNA sequences (polymorphic DNA) are "repeated" (i.e., appear in sequence) with the gene of interest. Thus, the polymorphic DNA sequences are linked to that specific gene. Their linked presence serves to facilitate genetic mapping (i.e., "location" of specific gene(s) on an organism's genome).

Ranibizumab   A monoclonal antibody (fragment) approved by the U.S. Food and Drug Administration (FDA) in 2006 for use as a pharmaceutical treatment to inhibit the growth of abnormal blood vessels in front of the eye's retina/macula. Such abnormal blood vessels are a cause of approximately ten per cent of the cases of the disease known as age-related macular degeneration (AMD).

rasiRNA   Abbreviation for repeat-associated small interfering RNA. *See* Small interfering RNA.

rDNA (Recombinant DNA)   A form of artificial DNA which is engineered through the combination or insertion of one or more DNA strands, thereby combining DNA sequences which would not normally occur together. In terms of genetic modification, recombinant DNA is produced through the addition of relevant DNA into an existing organismal genome, such as the plasmid of bacteria, to code for or alter different traits for a specific purpose, such as immunity. It differs from genetic recombination, in that it does not occur through processes within the cell or ribosome, but is exclusively engineered.

Reading frame   The particular nucleotide sequence that starts at a specific point and is then partitioned into codons. The reading frame may be shifted by removing or adding a nucleotide(s). This would cause a new sequence of codons to be read. For example, the sequence CATGGT is normally read as the two codons: CAT and GGT. If another adenosine nucleotide (A) were inserted between the initial C and A, producing the sequence CAATGGT, then the reading frame would have been shifted in such a way that the two new (different) codons would be CAA and TGG, which would code for something completely different.

Real-time PCR   Refers to the use of PCR to attempt quantitative (determination of a given DNA sequence within a sample), via coupling of a "molecular beacon" (with a "quencher molecule" attached to it) to the PCR probe. Thus, at the same time the PCR reaction (cycling) is producing copies of the relevant DNA sequence, the molecular beacon (i.e., fluorescent marker) is "un-quenched" so it fluoresces in direct proportion to the amount of DNA present (which can theoretically be back-calculated to infer the original amount of that particular DNA present in sample prior to initiation of PCR cycling).

Reassociation (of DNA)   The pairing of complementary single strands (of the molecule) to form a double helix (structure).

Recessive allele   Discovered by Gregor Mendel in the 1860s, this refers to an allelic gene whose existence is obscured in the phenotype of a heterozygote by the dominant allele. In a heterozygote the recessive allele does not produce a polypeptide; it is "switched off". In this case, the dominant allele is the one producing the polypeptide chain (via cell's ribosome).

Recombinant DNA (rDNA)   DNA formed by the joining of genes (genetic material) into a new combination.

Recombinant DNA technology This is the blanket term for the technologies of artificially breaking, rejoining, and amplifying DNA. Recombinant DNA techniques allow the biotechnologist to isolate and amplify a single gene out of all the genes in an organism, so that it can be studied, altered and put into another organism. The technique is also known as gene cloning (because you produce a lot of genetically identical genes), and the result is sometimes called a gene clone or simply a clone. An organism manipulated using recombinant DNA techniques is called a genetically manipulated organism (GMO).

Recombination The joining of genes, sets of genes, or parts of genes, into new combinations, either biologically or through laboratory manipulation (e.g. genetic engineering).

Regulatory genes Genes whose primary function is to control the state of synthesis of the products of other genes.

Regulatory sequence A DNA sequence involved in regulating the expression of a gene, e.g. a promoter or operator region (in the DNA molecule).

Regulatory T cells One class of T cells (thymus-derived lymphocytes) which was formerly called suppressor T cells, and which consitutes less than 10% of the total T cells in the body of the average human. Discovered by Tomio Tada in 1971, these T cells suppress B cell activity (i.e., after the body's immune system has fought off an infection). Research has shown that regulatory T cells can reduce inflammation, and regulate the body's other immune system cells involved in some autoimmune diseases (e.g. resulting when those other immune system cells are overly active, are active against the body's own cells, etc.). Absence of regulatory T cells has been shown to lead to some autoimmune diseases (e.g. immune dysregulation, polyendocrinopathy, enteropathy, X-linked syndrome—also known as IPEX).

Remediation The clean-up or containment (if chemicals are moving) of a hazardous waste disposal site to the satisfaction of the applicable regulatory agency [e.g. the environmental protection agency (EPA)]. Such clean-up can sometimes be accomplished via use of microorganisms that have been adapted (naturally or via genetic engineering) to consume those chemical wastes that are present in the disposal site.

Renaturation The return to the natural structure of a protein or nucleic acid from a denatured (more random coil) state. For example, a protein may be denatured [lose its native (natural) structure] by exposure to surfactants such as SDS or to changes in the pH of the medium, etc. If the surfactant is slowly removed or the pH is slowly readjusted to the optimum for the protein, it will refold (snap) back into its original (native) form.

Replication fork The point at which strands of parental duplex DNA are separated in a Y shape. This region represents a growing point in DNA replication.

Replication (of DNA) Reproduction of a DNA molecule (inside a cell). This process can be viewed as occurring in stages, in which the first stage consists of a helicase enzyme "unwinding" the double helix of the DNA molecule at a replication origin, forming a replication fork. At the replication fork, the two separated (DNA) strands serve as templates for new DNA synthesis. That new DNA synthesis is accomplished on each strand via enzymes known as DNA polymerase, which travel along each (single) strand making a second complementary strand by catalysing the addition of DNA bases (to the new, growing strands). The result is two new double helices (DNA molecules), each of which has one chain from the original DNA molecule and one chain that was newly synthesized by the DNA polymerase enzymes.

Replicon Refers to the non-replicating viral RNA particles (e.g. derived by scientists from the poliovirus) utilized to induce apoptosis (i.e., "programmed cell death") in brain tumours. Replicons are able to cross the blood-brain barrier (BBB), they preferentially infect cancer/tumour cells, and they then produce proteins which cause tumour cells to die via apoptosis.

Reporter gene In molecular biology, a reporter gene (often simply reporter) is a gene that researchers attach to another gene of interest in cell culture, animals or plants. Certain genes are chosen as reporters because the characteristics they confer on organisms expressing them are easily identified and measured, or because they are selectable markers. Reporter genes are generally used to determine whether the gene of interest has been taken up by or expressed in the cell or organism population.

Restriction endonucleases A class of enzymes that cleave (i.e., cut) DNA at a specific and unique internal location along its length. These enzymes are naturally produced by bacteria that use them as a defence mechanism against viral infection. The enzymes chop up the viral nucleic acids and hence their function is destroyed. Discovered in 1970 by Werner Arber, Hamilton Smith, and Daniel Nathans, restriction endonucleases are an important tool in genetic engineering, enabling the biotechnologist to splice new genes into the location(s) of a molecule of DNA where a restriction endonuclease has created a gap (via cleavage of the DNA).

Restriction enzyme Enzymes which cut double-stranded DNA at very specific base sequences, and nowhere else. Thus, they cut cloned DNA into a few pieces only. The place at which they cut is called a restriction site, and the map of all such sites on a clone is called a "restriction map".

Restriction fragment length polymorphism (RFLP) technique A "genetic mapping" technique which analyses the specific sequence of bases (i.e., nucleotides) in a piece of DNA (from an organism). Since the specific sequence of bases in their DNA molecules is different for each species, strain, variety, and individual (due to DNA polymorphism), RFLP can be utilized to "map" those DNA molecules (e.g. for plant breeding purposes, for criminal investigation purposes, etc.).

Restriction map A pictorial representation of the specific restriction sites (i.e., nucleotide sequences that are cleaved by given restriction endonucleases) in a DNA molecule (e.g. plasmid or chromosome).

Restriction site A nucleotide sequence (of base pairs) in a DNA molecule that is "recognized," and cleaved by a given restriction endonuclease.

Retroviral vectors Certain retroviruses that are used by genetic engineers to carry new genes into cells. These molecules become part of that cell's protoplasm.

Retroviruses (From the Latin word *retrovir*, which means "backward man") Oncogenic (i.e., cancer-producing), single-stranded, diploid RNA (ribonucleic acid) viruses that contain (+) RNA in their virions and propagate through a double-helical DNA intermediate. They are known as retroviruses because their genetic information flows from RNA to DNA (reverse of normal). That is, the viruses contain an enzyme that allows the production of DNA using RNA as a template. Retroviruses can only infect cells in which DNA is replicating, such as tumour cells (since they are constantly replicating) or cells comprising the lining of the stomach (since that lining must replace itself every few days).

Reverse osmosis Process by which a solvent such as water is purified of solutes by being forced through a semi-permeable membrane through which the solvent, but not the solutes, may pass.

Reverse transcriptases Also known as RNA-directed DNA polymerases, reverse transcriptases were discovered by Howard Martin Temin and David Baltimore in 1970. They are a class of enzymes first discovered to be present in RNA tumour virus, which allows the synthesis of DNA (complementary to the RNA) using the RNA present in the virus as a template. This is the reverse of what normally happens and hence the name.

RFLP *See* Restriction Fragment Length Polymorphism.

RIA *See* Radioimmunoassay.

Ribonucleic acid (RNA) A long-chain, usually single-stranded nucleic acid consisting of repeating nucleotide units containing four kinds of heterocyclic, organic bases: adenine, cytosine, guanine, and uracil. These bases are conjugated to the pentose sugar ribose and held in sequence by phosphodiester (chemical) bonds. The primary function of RNA is protein synthesis within a cell. However, RNA is involved in various ways in the processes of expression and repression of hereditary information. The three main functionally distinct varieties of RNA molecules are: (1) messenger RNA (mRNA) which is involved in the transmission of DNA information, (2) ribosomal RNA (rRNA) which makes up the physical machinery of the synthetic process, and (3) transfer RNA (tRNA) which also constitutes another functional part of the machinery of protein synthesis.

Ribosomal RNA (rRNA) The primary constituent of ribosomes making up approximately two-thirds of the mass of the bacteria *Escherichia coli* ribosome, and approximately one-half of the mass of mammalian ribosomes. Ribosomes are the protein-manufacturing organelles of cells and exist in the cytoplasm. rRNA is transcribed from DNA, like all RNA. Ribosomal proteins are transported into the nucleus and assembled together with rRNA before being transported through the nuclear membrane. This type of RNA makes up the vast majority of RNA found in a typical cell. While proteins are also present in the ribosomes, solely rRNA is able to form peptides. Ribosomal RNA accounts for nearly 80 per cent of the RNA content of the bacterial cell.

Ribosomes The molecular "machines" within cells that coordinate the interplay of tRNAs, mRNA, and proteins in the complex process of protein synthesis (manufacture). RNA constitutes nearly two-thirds of the mass of these large (mega-Dalton) molecular assemblies, which are technically ribozymes (i.e., an enzyme in which the catalysis is performed by RNA).

Ribozyme Also called catalytic RNA, these are RNA molecules which catalyse chemical reactions, often the breakdown of other RNAs. Ribozymes have potential in two areas. They are widely used as potential pharmaceutical agents, as their action against other RNAs can be extremely specific. They could attach a viral RNA without affecting the normal RNAs in a cell, thus acting as anti-viral agents. The other area is to use ribozymes as industrial catalysts, selecting suitable catalytic activities through Darwinian cloning.

RNA *See* Ribonucleic acid.

RNA interference (RNAi) Coined by Andrew Fire and Craig Mello in 1998, this term refers to what happens when short strands of (complementary) double-stranded RNA (dsRNA) are introduced into living cells. That interaction can be done either by physical insertion of the dsRNA or by genetic engineering of the organism so the organism's cell(s) themselves produce that (new) dsRNA. For example, genetic engineers can utilize T7 RNA polymerase to cause the production of such dsRNA within living cells. Viral infection (i.e., insertion of viral dsRNA) and also micro-RNAs can also cause RNA interference.

RNA polymerase Discovered by Severo Ochoa in 1955, it is an enzyme that catalyses the synthesis of a complementary mRNA (messenger RNA) molecule from a DNA (deoxyribonucleic acid) template in the presence of a mixture of the four ribonucleotides (ATP, UTP, GTP, and CTP). Also called transcriptase.

RNase A category of enzymes, which catalyse the destruction of nucleic acids within cells.

RNA transcriptase *See* RNA polymerase.

RT-PCR Acronym for reverse transcriptase polymerase chain reaction, a PCR technique which starts with cellular RNA (transcript), then utilizes reverse transcriptase to create its counterpart DNA; this is then amplified/copied by PCR (polymerase chain reaction) technique.

Salting out A technique used for forcing (dissolved) proteins out of a solution by increasing the concentration of salt in the solution. The Na^+ and Cl^- ions derived from the salt compete for and "tie up" water molecules that are solubilizing the protein molecules thereby rendering them insoluble or more insoluble.

Salt tolerance Refers to the trait (of a plant) which enables a plant to grow/survive in soil that contains a high level of salt. For example, during 2001, Eduardo Blumwald and Hong-Xia Zhang inserted an *AtNHX1* gene from *Arabidopsis thaliana* into a tomato plant (*Lycopersicon esculentum*) and thereby made that tomato plant resistant to salt concentrations up to 200 mM (i.e., far higher than it could previously survive).

Satellite DNA Many tandem repeats (identical or related) of a short basic repeating unit (in the DNA molecule).

Sequence map A pictorial representation of the sequence of amino acids in a protein molecule, the sequence of nucleic acids in a DNA molecule, or the sequence of oligosaccharide components in a glycoprotein/carbohydrate molecule.

Sequence (of a DNA molecule) The specific nucleic acids (and the order in which they occur) that comprise a given segment of a DNA molecule.

Sequence (of a protein molecule) The specific amino acids (and the order in which they are coupled together) that comprise a given segment of a protein molecule.

Sequencing (of DNA molecules) The process used to obtain the sequential arrangement of nucleotides in the DNA backbone. The cleavage into fragments (followed by separation of those fragments, which can then be achieved of DNA molecules by one of several methods: (1) a chemical cleavage method followed by polyacrylamide gel electrophoresis (PAGE) or capillary electrophoresis, (2) a method consisting of controlled interruption of enzymatic replication methods followed by PAGE, (3) a dideoxy method utilizing fluorescent "tag" atoms attached to the DNA fragments, followed by use of spectrophotometry to identify the respective DNA fragments by their differing "tags" (which fluoresce at different wavelengths). This (fluorescent tag) variant of the dideoxy method can be automated to "decipher" large DNA molecules (i.e., genomes). Such automated machines are sometimes called "gene machines".

Short hairpin RNA (shRNA) Specific segments of dsRNA (i.e., that bend into a 'hairpin-like' molecular shape after they self-assemble) that can either be chemically synthesized by man to cause RNA interference, or are formed inside cells when the applicable DNA-directed RNA interference methodology is utilized.

Short interfering RNA (siRNA) Specific short sequences of double-stranded RNA (dsRNA) 19–21 base pairs (bp) in length, which trigger degradation of messenger RNA (mRNA) possessing the same sequence (as those siRNAs) within a cell, as part of the cellular process known as RNA interference (RNAi). Because that degradation of mRNA thereby shuts down (quells) production of the corresponding protein, siRNA (via RNAi) constitutes a pathway that cells utilize to regulate/silence gene expression. Relevant promoters within the DNA are silenced via DNA methylation and/or chromatin remodelling. The siRNA can be utilized by man to cause gene silencing/knockout.

Shotgun cloning method   A technique for obtaining the desired gene that involves "chopping up" the entire genetic complement of a cell using restriction enzymes, then attaching each (resultant) DNA fragment to a vector and transferring it into a bacterium, and finally screening those (engineered) bacteria to locate the bacteria that are producing the desired product (e.g. a protein).

Shotgun sequencing   Sometimes called whole-genome shotgun sequencing, it was invented by J. Craig Venter and Hamilton O. Smith during the mid-1990s. Shotgun sequencing is a technology for rapid sequencing of (eukaryotic and prokaryotic) DNA, in which an organism's genome (DNA) is first fragmented ("broken up"), and then randomly selected pieces of the DNA are individually sequenced.

Shuttle vector   A vector capable of replicating in two unrelated species. A vector (e.g. a plasmid) constructed in such a way that it can replicate in at least two different host species (e.g. a prokaryote and a eukaryote). A DNA recombined into such a vector can be tested or manipulated in several cell types. It can pass like a shuttle from one place to another.

Single-cell protein (SCP)   Protein that is derived from single-celled organisms with a high protein content. Yeast is an example. Generally used in regard to those organisms that are edible by domesticated animals, or humans.

Single-domain antibodies (dAbs)   "Very heavy chains" (VH) (portion of antibody molecules) produced by genetically engineered *Escherichia coli* cells that act to bind antigens in a manner similar to antibodies or monoclonal antibodies (MAbs). Similar to MAbs, dAbs can be produced in large quantities, to be used as human or animal therapeutics (e.g. to combat diseases).

Single-nucleotide polymorphisms (SNPs)   Variations (in individual nucleotides) that occur within DNA at the rate of approximately one in every 1,300 base pairs in most organisms (approximately one in every 1,200 base pairs in human DNA). SNPs usually occur in the same genomic location (e.g. on the organism's DNA) in different individuals. These variations account for diversity within a given species (e.g. black cattle and white cattle, different human eye colours, different strains/serotypes within a given bacterial species, etc.) and some genetic diseases [e.g. the disease cystic fibrosis is due to one SNP, the disease sickle cell anemia is due to one SNP, the disease known as familial dysautonomia is due to one SNP, the disease known as (Duchenne) muscular dystrophy is due to one SNP, Huntington's disease is due to one SNP, the disease known as neuro-fibromatosis is due to one SNP, Tay–Sachs disease is due to one SNP, etc.].

Single-stranded DNA   Molecules of "unwound" DNA (i.e., half of the double helix DNA).

siRNA   Acronym for short interfering RNA.

Small interfering RNA   *See* Short Interfering RNA.

Small RNA   *See* Short Interfering RNA.

Somatic gene therapy   The introduction of genes into tissue or cells to treat a genetic related disease in an individual. Somatic gene therapy is provided by introducing a new therapeutic gene (transgene) into the diseased cells of a patient. The modified cells express the introduced gene and their new phenotype provides some advantage to the patient.

Southern blot analysis   Invented in 1975 by Edwin Mellor Southern, it is a test that is performed on biological samples such as restriction endonuclease-digested (i.e., fragmented) plant DNA (e.g. to ascertain if genetically engineered "inserted" DNA is present in particular plant cells). Gel electrophoresis is used to separate (the DNA fragments) according to the size of those fragments, and then those are transferred to a filter (blot).

ssDNA   Acronym for single-stranded DNA.

ssRNA   Acronym for single-stranded RNA.

Stacked genes   The insertion of two or more (synthetic) genes into the genome of an organism. One example of that would be a plant into which has been inserted a gene from *Bacillus thuringiensis* (B.t.) and a gene for resistance to a specific herbicide.

Starch hydrolysis   Excretion of the enzyme amylase out of the cells (an exoenzyme) into the surrounding media, catalysing the breakdown of starch into sugars.

Stem cells   Certain cells—present in the bodies of mammals even prior to birth, although also present in adult mammals—that can grow/differentiate into different cells/tissues of the (adult organism) body. e.g. (bone marrow (stem) cells) some of which eventually mature into red blood cells or white blood cells. The stem cells that remain in the bone marrow maintain their own numbers by self-renewal divisions, yielding more (adult stem cell) cells to start the maturation process. This maturation process is stimulated and controlled by stem cell growth factor (SCF), granulocyte colony stimulating factor

(G-CSF), and by granulocyte-macrophage colony stimulating factor (GM-CSF).

Streptavidin   A bacterial origin protein which possesses natural anti-cancer properties (e.g. it causes human promyelocytic leukaemia cells to die when pure streptavidin enters those cells). Streptavidin has a specific and high affinity for biotin (i.e., it "sticks" tightly to the biotin molecule).

Structural gene   A gene that codes for any RNA (ribonucleic acid) or protein product other than a regulator molecule. It determines the primary sequences (i.e., the amino acid sequences) of a polypeptide (protein).

Structural genomics   Study of, or discovery of, where (gene) sequences are located within the genome, and what (DNA) subunits comprise those sequences.

Suppressor gene   A gene that can reverse the effect of a specific type of mutation in other genes, such as a premature termination sequence.

***Taq* DNA polymerase**   A 94-kilodalton DNA polymerase, which was originally isolated from the thermophilic archaean *Thermus aquaticus* and is commonly utilized to catalyse PCR reactions due to its heat resistance (needed for thermal cycles utilized in the PCR technique).

TATA homology   An adenine-thymidine-rich (gene) sequence present 20 to 30 nucleotides "upstream" of the transcription start site on most eukaryotic protein-coding genes; it is required for correct expression. Recent research indicates that blocking this portion of the (gene) sequence may inhibit ability of the AIDS virus to reproduce.

T cells   A class of (thymus-derived) lymphocytes which include helper T cells (also known as T helper cells or T_H cells), suppressor T cells, and cytotoxic T cells (also known as killer cells or CTL for cytotoxic T lymphocyte). These cells mediate (i.e., control/direct) the cellular response of the human immune system in very complex ways (e.g. synthesis of leukotrienes). T cells are involved in the activation of B cells.

Telomerase   An enzyme that enables the "repair" of telomeres (thereby stabilizing their length, and preventing "shortening" of the telomeres). The telomerase enzyme is only present in cancerous cells (thereby enabling the "immortality" of cancerous cells). Human telomerase contains an RNA component and a catalytic protein component (i.e., a member of the reverse transcriptase "family" of enzymes).

Telomeres   Assemblies consisting of protein and DNA sequences (that do not code for proteins), which are located at the (end) tips of chromosomes. Telomeres protect the ends of chromosomes, prevent chromosomes from fusing to each other, and serve to limit the maximum number of times that a given cell divides (in the cell's "lifetime"). Telomeres' DNA consist of the sequence GGGGTT repeated many times.

Telophase   The fourth of the four phases of eukaryotic mitosis (i.e., cell replication via division) during which a new membrane is made to envelop each set of the newly-divided chromosomes (each of which now constitutes a new cell nucleus).

Termination codon   Also known as terminator sequence. One of three triplet sequences (UAG, UAA, or UGA) found in DNA molecules (genes) that cause termination of protein synthesis; they are also called nonsense codons. The sequences cause the termination of the peptide chain and its release in free form.

Tertiary structure   The three-dimensional folding of the polypeptide (i.e., protein) molecular chains that characterizes a protein molecule in its native state.

Therapeutic cloning   The name given to the still unachieved goal of cloning tissues or organs from an individual for therapy. The object here is specifically not to "clone" an individual, but to generate a replacement organ that is genetically identical to them, so that it is not rejected when it is transplanted into them.

Ti plasmid   Abbreviation for tumour-inducing plasmid or tumour induction plasmid. It is the plasmid of *Agrobacterium tumefaciens* bacteria that naturally has a part of its DNA (T DNA) transferred to a plant when *Agrobacterium tumefaciens* infects that plant (e.g. via a wound in the plant). After it has been transferred into the plant, that Ti plasmid DNA segment (now known as T-DNA or transferred DNA) inserts itself into the plant's DNA where it causes cells to grow into tumour-like structures known as galls.

Tissue culture   The growth and maintenance (by researchers) of cells from higher organisms *in vitro*, that is, in a sterile test tube or petri dish environment which contains the nutrients necessary for cell growth. One use of tissue culture is to produce disease-free offspring from certain (valuable, high quality) crop plants.

Tissue engineering   Refers to the technologies utilized to induce (injected) liver, cartilage, etc. cells to grow (within recipient organism's body) and form entire (integral) tissues and (extant) cells within the body to grow and form desired (integral) tissues, via precise injection of relevant compounds (e.g. certain growth factors, growth hormones, etc.).

Tissue plasminogen activator (tPA) A glycoprotein that possesses thrombolytic (i.e., blood clot-dissolving) activity. It is used as a drug to dissolve clots and acts by first binding to fibrin (clots). It then activates (i.e., proteolytically cleaves) plasminogen (molecules) to yield plasmin, a blood-borne enzyme that itself cleaves molecular bonds in the fibrin clot. The plasmin molecules diffuse through the fibrin clot and cause the clot to dissolve rapidly. With the dissolution of the clot, blood flow to the formerly blocked blood vessel (e.g. the heart) is restored.

Tobacco mosaic virus (TMV) One of the smallest viruses, consisting of some 2,200 chains of identical polypeptides and a molecule of RNA. All of the genetic/heredity information of the tobacco mosaic virus is contained in its RNA. The first discovery of a self-assembling, active biological structure occurred in 1955, when Heinz Frankel-Conrat and Robley Williams showed that TMV, will reassemble into functioning, infectious virus particles (after the TMV has been dissociated into its components via immersion in concentrated acetic acid. The TMV virus infects the leaves of tomato and tobacco plants, causing disease.

Totipotency The ability to grow/differentiate into all of the types of cells/tissues constituting an (adult) organism's body.

Totipotent stem cells Bone marrow cells that (when signalled) mature into both red blood cells and white blood cells. Receptors on the surface of totipotent stem cells "grasp" passing blood cell growth factors (e.g. interleukin-7, stem cell growth factor, etc.), bringing them inside these stem cells and thus causing the maturation and differentiation into red and white blood cells. These receptors are called FLK-Z receptors.

Transcript Term used to refer to the various segment(s) of messenger RNA (mRNA) that result from transcription of a gene.

Transcription The enzyme-catalysed process whereby the genetic information contained in one strand of DNA (deoxyribonucleic acid) is used as a template to specify and produce a complementary mRNA strand. Transcription may be thought of as a rewriting of the information contained in DNA into RNA. The language, however, is the same—both are nucleic acid-based. This is in contrast to translation, in which the information is translated from one language (RNA, nucleic acid-based) into another language (protein, amino acid-based).

Transcription activators Refers to transcription factors (proteins and/or other molecules) which interact with regulatory sequences within DNA (in cell). By binding directly to those regulatory sequences (usually at multiple sites on the sequence), and "recruiting" modifying molecules (chromatin remodelling elements) to also come to the site(s) on the DNA, transcription activators cause transcription (of a given gene) to begin or to increase.

Transcriptional repressor A regulatory sequence [segment of DNA] which 'binds' (i.e., adheres to) a DNA transcription control sequence, and thereby represses [decreases/halts] the transcription of a gene.

Transcriptome The entire (complete, possible) set of all gene transcripts (i.e., mRNA segments resulting from gene transcription process) in a given organism. Also the knowledge of their roles in that organism's structure, growth, health, disease (and/or that organism's resistance to disease), etc. Those roles are predominantly due to the impact of each protein molecule (i.e., resulting from the mRNA segments being translated in cells' ribosomes); which is itself due to the protein molecule's composition and its tertiary conformation (which determines the protein's impact in the organism's tissues, metabolism, etc.).

Transduction (gene) The transfer of bacterial genes (DNA) from one bacterium to another by means of a (temperate or defective) bacterial virus (bacteriophage). There exist two kinds of transduction: specialized and general.

Transfection This term has several different meanings, depending on the context in which it is used:

- A word utilized most generally to refer to insertion of DNA segments (genes) into cells (e.g. via electroporation, endocytosis, etc.). For example, insertion of a gene that codes-for green fluorescent protein (GFP) into a cell in such a manner that it causes the cell (or class of cells) to fluoresce under certain conditions/illumination.

- A word utilized since 1998 to refer to insertion of certain double-stranded RNA (dsRNA) segments into cells (via electroporation, certain RNA polymerases, certain viral infections, etc.); to cause RNA interference (RNAi)/knockout/silencing.

- 🐦 A word utilized to refer to insertion of (complementary-to-constitutive mRNA) antisense oligonucleotides, to cause co-suppression/knockdown.

- 🐦 A word utilized to refer to insertion of a DNA-biologically-active protein (e.g. transcription factors, STATs, etc.) or other DNA/RNA-biologically-active small molecule.

- 🐦 A special case of transformation in which an appropriate recipient strain of bacteria is exposed to (free) DNA isolated from a transducing phage with the "take-up" of that DNA by some of the bacteria and consequent production and release of complete virus particles. The process involves the direct transfer of genetic material from donor to recipient.

Transferase Any of various enzymes that catalyse the transfer of a chemical group, such as a phosphate or an amine, from one molecule to another.

Transfer RNA (tRNA) First hypothesized by Francis Crick and discovered in 1957 by Mahlon Bush Hoagland, they are a class of relatively small RNA (73–93 nucleotides) molecules of molecular weight 23,000 to about 30,000. tRNA molecules act as carriers of specific amino acids during the process of protein synthesis. Each of the 20 amino acids found in proteins has at least one specific corresponding tRNA. It has a 3′ terminal site for amino acid attachment. This covalent linkage is catalysed by an aminoacyl tRNA synthetase. It also contains a three-base region called the anticodon that can base-pair to the corresponding three base codon region on mRNA. Each type of tRNA molecule can be attached to only one type of amino acid, but because the genetic code contains multiple codons that specify the same amino acid, tRNA molecules bearing different anticodons may also carry the same amino acid. Attachment of an amino acid to a tRNA molecule forms an active (amino-acyl) tRNA, which then functions as a ribosomal adaptor (tRNA adaptor).

Transformation For bacteria, this means getting the bacterium to take up DNA which the experimenter has added to its medium. Bacteria that has been treated to make them able to do this are called "competent". Demonstrating transformation was one of the key proofs that DNA was the genetic material.

Transgenic A transgenic organism is one which has been altered to contain a gene from another organism, usually from another species. While this would suggest that any genetically engineered organism could be called "transgenic", the term is usually only applied to animals or plants. Bacteria and yeasts are always called "recombinant" and with plants you can use either term, or call them "genetically engineered" or "genetically manipulated".

Translation The process via which protein molecules are synthesized (made) whereby the genetic information present in an mRNA molecule directs the order of incorporation of specific amino acids, and hence the growth of the polypeptide chain during protein synthesis. One can think of translation as the process of translating one language into another. In this particular case the nucleic acid-based language represented by mRNA is translated into the amino acid-based language of proteins.

Translocation Genetic mutation in which a section of a chromosome "breaks off" and moves to a new (abnormal) position in that (or a different) chromosome.

Transposase An enzyme that is required for transposition to occur (i.e., this enzyme assists movement of a transposon from one location to another within a cell's DNA).

Transposon A DNA (deoxyribonucleic acid) sequence (segment of molecule) able to replicate and insert one copy (of itself) at a new location in the genome (i.e., a transposition of location). It was discovered in 1950 by geneticist Barbara McClintock in corn (maize) plants (*Zea mays* L.), and in bacteria a decade later by Joshua Lederberg. Transposons can either carry genes along one organism's genome, or even into another organism's genome (e.g. via sexual conjugation, in bacteria). By such sexual conjugation, transposons can carry genes that confer new phenotypic properties (e.g. resistance to certain antibiotics, for a given bacterial cell).

Trastuzumab A ("humanized") monoclonal antibody against *HER-2* gene that was approved by the U.S. Food and Drug Administration (FDA) during 2002 for use as a pharmaceutical in conjunction with chemotherapy against metastatic breast cancer.

tRNA (Transfer RNA) *See* Transfer RNA.

Type I diabetes The form of diabetes disease that usually strikes young people (thus, it was formerly known as juvenile or insulin-dependent diabetes). This disease is characterized by the body's immune system (antibodies) destroying the insulin-producing cells (beta cells) of the pancreas.

Type II diabetes The form of diabetes disease that usually strikes people who are older than 40 years old. Also known as adult-onset diabetes or non-insulin-dependent diabetes, this disease is characterized by the body's tissues becoming insensitive to insulin. Effects on the body include increased likelihood of blindness, atherosclerosis, coronary heart disease, heart attack, stroke, and kidney disease.

Ultrafiltration A filtration process in which particles of colloidal size are retained by a filter medium while solvent and accompanying low-molecular-weight solutes are allowed to pass through. Ultrafilters are used (1) to separate colloid from suspending medium, (2) to separate particles of one size from particles of another size, and (3) to determine the distribution of particle sizes in colloidal systems by the use of filters of graded pore size.

Uracil A pyrimidine base, important as a component of ribonucleic acid (RNA). Its hydrogen-bonding counterpart in DNA is thymine.

Uridine A nucleoside form of uracil.

Urokinase A thrombolytic (i.e., clot-dissolving) enzyme used as a bio-pharmaceutical.

Vaccine Preparations which, when given to a patient, elicit an immune response which subsequently protects the patient from a disease. Usually, the vaccine consists of the organisms that cause the disease (suitably attenuated or killed). Any substance, bearing antigens on its surface, that cause activation of an animal's immune system without causing actual disease is a vaccine. The animals' immune system components (e.g. antibodies) are then prepared to quickly vanquish those particular pathogens when they later enter the body.

Vaccinia A non-pathogenic virus that is believed to be a (modified) form of the virus that causes cowpox. Vaccinia readily accepts genes (inserted into its genome via genetic engineering) from pathogenic viruses so it can be used to make vaccines that do not possess the risk inherent in attenuated virus vaccines (i.e., that the attenuated virus "revives" and causes disease). Such genetically engineered vaccinia codes for (presents) the proteins of the pathogenic virus on its surface, which activates the immune system (e.g. of vaccinated animal) to produce antibodies against that pathogenic virus.

van der Waals forces The relatively weak forces of attraction between molecules that contribute to intermolecular bonding (i.e., binding-together two or more adjacent molecules). Historically, it was thought that van der Waals forces were always weaker than the hydrogen bond forces responsible for intramolecular bonding. However, in 1995,

Dr. Alfred French discovered that van der Waals forces are primarily responsible for holding together a mass of cellulose molecules, with hydrogen bonding playing a lesser role.

Vector The agent used (by researchers) to carry new genes into cells. Plasmids currently are the biological vectors of choice, though viruses and other biological vectors such as *Agrobacterium tumefaciens* bacteria or BACs are increasingly being used for this purpose.

Western blot test A test that is performed on biological samples such as blood (after centrifugation to remove red blood cells from the blood) to detect proteins (e.g. AIDS-related antibodies) individually. Gel electrophoresis is used to separate the AIDS antigen proteins of killed (known) AIDS viruses. Next the protein bands (resulting from the gel electrophoresis) are exposed to the blood being tested and (AIDS) antibodies stick to specific individual antigens (bands), which are then identified (as being present in the tested blood) via dyes.

Whole genome amplification (WGA) A methodology that is required for sequencing an entire genome (i.e., the entire DNA of an organism); that of first amplifying (i.e., making copies of) every sequence within the selected organism's DNA. Typically, the genome (i.e., DNA molecule) is first fragmented into numerous small, overlapping segments. Next, relevant chemicals are added which convert those overlapping segments into a "library" of DNA segments. Then, conventional PCR (polymerase chain reaction) technique is utilized to amplify (i.e., make copies of) the entire library.

X chromosome A sex chromosome that usually occurs paired in each female cell, and single (i.e., unpaired) in each male cell in those species in which the male typically has two unlike sex chromosomes (e.g. humans).

Xenobiotic A chemical that would not normally be found in a given environment, and usually means a toxic chemical which is entirely artificial such as a chlorinated aromatic compound or an organo-mercury compound. All chemically synthesized drugs are xenobiotics.

Xenograft A graft of a tissue which comes from another species. A related idea is a xenotransplant, where a whole organ from a non-human species is transplanted to a human: the two terms are often used interchangeably.

Xenotransplant From the Greek word xenos, meaning "stranger." Xenotransplant is the implantation of an organ or limb from one species to another organism in a different species. When performed in animals, "rejection" of the transplant by the recipient's immune system is a common response.

Y chromosome   A sex chromosome that is characteristic of male zygotes (and cells) in species in which the male typically has two unlike sex chromosomes.

Yeast   A fungus of the family Saccharomycetaceae that is used by man especially in the making of alcoholic liquors and as a leavening agent in bread making. Some strains of yeast cells are also commonly used in bioprocesses, because they are relatively simple to genetically engineer (via recombinant DNA) and relatively easy to propagate (via fermentation) to yield desired products (e.g. proteins).

Yeast artificial chromosomes (YAC)   Pieces of DNA (usually human DNA) that have been cloned (made) inside living yeast cells. While most bacterial vectors cannot carry DNA pieces that are larger than 50 base pairs, YACs can typically carry DNA pieces that are as large as several hundred base pairs. They consist of those pieces of DNA which define the ends (telomeres) and the middle (centromere) of a chromosome in yeast. Both these elements are needed to allow a chromosome to be replicated in yeast cells.

Yeast episomal plasmid (YEP)   A cloning vehicle used for introduction of constructions (i.e., genes and pieces of genetic material) into certain yeast strains at high copy number. YEP can replicate in both *Escherichia coli* and certain yeast strains.

Z-DNA   A left-handed helix (molecular structure) of DNA, in contrast to A-DNA and B-DNA which are right-handed helix structures. The difference is in the direction of the double-helix twist. Z-DNA has the most base pairs per turn (in the helix), and so has the least twisted structure; it is very "skinny" and its name is taken from the zig-zag path that the sugar–phosphate "backbone" follows along the helix. This is quite different from the smoothly curving path of the backbone of B-DNA. The Z-form of DNA has been found in polymers that have an alternating purine–pyrimidine sequence.

INDEX

A

Adeno-associated virus 467
Adenovirus 465
Adenovirus vectors 193
Adsorption 441
Affinity purification 422
Alkali fixation 144
Allele-specific PCR 355
Amperometric biosensors 446
Anaerovibrio lipolytica 323
Anchored PCR 345
Androgens 305
Antibiotic resistance 188
Anti-idiotype antibody 103
Anti-idiotype antibody vaccines 73
Anti-platelet therapy 111
Arbitrarily primed PCR 352
Artificial enzymes 451
Ascitic fluid preparation 94
Asymmetric PCR 349
Automated nucleic acid sequencing 400

B

Bacteriophage P1 174
Bacteriophages 170
Bacteroides ruminicola 322
batch culture 44
Biocatalysis 19
Biochemical pathways 255
Biomass energy 494
Biopharmaceuticals 482
Bioreactor systems 33
Bioremediation 494
Biosafety 517
Biosafety levels 519

Biosensors 443
Biotechnology in
    bioprocess technology 11
    environmental technology 11
    enzyme technology 11
    health care 11
    plant and animal agriculture 11
    renewable resources technology 11
    waste technology 11
Blunt end ligation 183
Bovine papilloma virus vectors 193
Breast cancer gene *BRCA1* 198

C

Calorimetric biosensors 445
Carcass composition 252
Catalase
    in the food industry 434
Cell lines 27, 29
Chain termination sequencing 390
Chemical degradation sequencing 397
Cloning cDNA 183
Cloning vectors 165
Co-enzymes 414
Cofactors 414
Continuous cell lines 28
Continuous culture 44
Covalent binding 441
Coverslip culture 35
Cryopreservation 36, 214
Cryopreservatives 36
Cystic fibrosis 473

D

Decapacitation factor 225
Defaunation 326

Defined media 24
Degenerate PCR 354
Deletion mutant vaccines 72
Denaturation 343
Deoxynucleoside triphosphates 337
Dimethyl sulphoxide (DMSO) 36
Direct cycle sequencing 396
Disease models 248
Disease resistance 263
DNA arrays and chips 152
DNA fingerprinting 159, 377
DNA polymorphism 373
DNA probes 129
DNA/RNA probes 5
Donor 211
Dot/slot blots 151
Downstream processing 41
Duchenne muscular dystrophy 473

E

EMBL 3 173
EMBL 4 173
Embryo collection 212
Embryonic stem cells 240
Embryo sexing 215
Embryo splitting 214
Embryo transfer 226, 239
Embryo transfer (ET) technology 209
End labelling 131
Endocytosis 469
Enhanced
    chemiluminescence (ECL) 147
Entrapment 442
Environmental biotechnology 492
Enzyme concentration 420
Enzyme engineering 450
Enzyme immobilization 439
Enzyme production 438
Enzyme purification 421
Exonucleases 176
Expression of target antigens 65
    in prokaryotic hosts 65

in eukaryotic hosts 66
    Avian fowlpox virus 68
    Avian leucosis virus 69
    Herpesvirus of turkeys 68
    Insect cells 67
    Mammalian cells 66
    Yeasts 66
Ex vivo gene transfer 458

F

Familial hypercholesterolaemia 472
Fed-batch culture 44
Fermentation 3, 20, 39
Fibrobacter succinogenes 323
Filter hybridization 148
Flask cultures 35
Fluorescence *in situ* hybridization 218
Fred Sanger 389
Fuels 18
Fusion protein vaccines 70

G

Galactopoiesis 316
Gel chromatography 421
Gene bank 181
Gene library 181
Gene pharming 261
Gene tagging 376
Gene therapy 204, 482
Genetically engineered reassortants 72
Genetic engineering 165
Genetic medicine 479
Germ-line gene therapy 461
Gestagens 305
Glucose oxidase 434
Glycerol 36
Gordon and Ruddle 234
Growth hormone 253, 303
Growth hormone releasing
    hormone 303

H

H-Y antigen 217
HAT 92

Herpes simplex virus vectors 468
HGPRT 92
Hybridization 137
Hybridomas 88
Hybridoma technology 13
Hydrolases 414
Hydroxyapatite chromatography 148

I

Idiophase 44
Immobilization 441
Immunoscintigraphy 100
Immunosensors 448
Insertion vector 172
Insertional inactivation 188
In situ hybridization 151
In situ PCR 356
In vitro fertilization 221
In vitro maturation 224
In vivo gene transfer 458
Insulin 303
Intellectual Property Protection 505
Intellectual Property Rights 505
Interferons 197
Inverse PCR 344
Ion-exchange chromatography 422
Isomerases 415

J

Jeffreys probes 281

K

Kary mullis 334
Knockout animals 484
Kohler 87

L

lac promoter 191
Lactases 431
    in the dairy industry 431
Lactogenesis 316
Lambda phage 170
Laparoscopic ovum pick-up 225

Lentiviruses 468
lgt 10 172
Ligases 415
Ligation of DNA 185
Liposomes 469
Lyases 414

M

M13 phage 173
Major histocompatibility complex 297
Mammalian cell culture 14
Mammogenesis 315
Manipulation of growth 303
Manipulation of wool growth 316
Medical applications 434
Membrane confinement 442
Microbial biomass 45
Microcarrier cultures 33
Microinjection 238
Microsatellite probes 284
Milstein 87
Minisatellite probes 283
Minisatellites 281
Molecular mapping 376
Monoclonals
    as *in vivo* reagents 107
Multiplex PCR 349
Mutation 49

N

Naked DNA 468
Natural media 24
Nested PCR 348
Nick translation 130
Nitrocellulose (NC) filters 142
Non-patentable inventions 514
Non-radioactive probes 131
Northern blots 150
Nuclear transfer 237, 244
Nuclease S1 digestion 148
Nucleic acid probes 13, 126
Nucleic acid vaccines 75

O

Oestrogens 305
Oligonucleotide fingerprinting 285
Oligonucleotide probes 129
Oncogenes 250
Oocyst 223
Optical biosensors 447
Optimization 337
Organ culture 35
Organ donors 486
Oxidoreductases 414

P

Pasteurization 3
Patent protection 507
Patentable inventions 509
Patents 506
pBR 322 166
PCR
    buffers and $MgCl_2$ 337
    enzymes 338
    inhibitors and enhancers of 341
PEG 92
Peptide vaccines 69
Pfu DNA polymerase 340
Phagemids 174
Plant biotechnology 496
Plant cell culture 16
Plasmids 166
Polymerase chain reaction 6, 219
Potentiometric biosensors 446
Precipitation 420
Prehybridization 144
Prenatal diagnosis 377
Prevotella ruminicola 322
Primary barriers 518
Primary culture 26
Primary explantation technique 34
Primary metabolites 46
Primer annealing 343
Primer design 337
Primer-dimer 336

Primer extension 343
Primers 336
Probiotics 308
Process engineering 20
Proteases 428, 431
    in the food industry 428
    in the leather industry 431
    in the wool industry 431
pUC 167

Q

Quantitative PCR 353

R

Radioactive probes 130
Radioisotope 128
Random priming 130
RAPD markers 374
Real-time PCR 356
Recombinant 59
Recombinant DNA 165
Recombinant DNA technology 4
Recombinant insulin 195
Recombinant vaccines 14, 201
Recombination 49
Rep-PCR 355
Replacement vectors 172
Replicating plasmids 169
Reproductive techniques 14
Restriction endonucleases 176
Restriction fragment
    length polymorphisms 286
        in cattle 287
        in dogs and cats 296
        in horse 294
        in pigs 293
        in poultry 294
        in sheep and goats 291
Restriction site-specific PCR 354
Retriever vectors 172
Retroviral vectors 194, 241
Retrovirus 464
Reverse osmosis 420

Reverse transcriptase PCR 346
RNA probes 129
RPMI 1640 25
Rumen microbial digestive system 317
Ruminobacter amylophillus 323
Ruminococcus albus 324

S

Safety equipment 518
Screening assays for MAb 94
Secondary barriers 518
Secondary metabolites 46
Semi-nested PCR 348
Sex chromosomal analysis 216
Sexing of embryo 377
Sickle cell anaemia 159
Simple tandem repeats 284
Single cell protein (SCP) 4, 7
Solid-state fermentation 50
Soluble enzymes 419
        clarification of 419
Solution hybridization 147
Somatic cell gene therapy 461
Somatic cell nuclear transfer 16
Somatic gene transfer 487
Somatomedins 303
Somatostatin 305
Somatotropin 303
Southern blots 150
SRIF 305
SSCP-PCR 353
Starch hydrolysis 423
Superovulation 211
Suspension cultures 33
SV-40 vectors 192

T

TA cloning® 175
tac promoter 192
taq DNA polymerase 338
Test tube cultures 35

Therapeutic monoclonal antibodies 108
Thyroid hormones 305
Tissue culture 35
TOPO® cloning 175
Transduction 187
Transfection 187
Transferases 414
Transferred gene 376
Transformation 186
Transgenic 234
Transgenic animals 15, 483
Transgenic chicken 269
Trophophase 44
trp promoter 191
Tth DNA polymerase 340
Tumour markers 99

U

UITma™ DNA polymerase 340
Ultrafiltration 420

V

Vaccines 59
Vaccinia virus vector 194
Vectorate PCR 353
Vent™ DNA polymerase 340

W

Walter Gilbert 389
Waste treatment 19

X

X-gal system 189
X-linked enzymes 218
Xenotransplant 267
Xenotransplantation 486

Y

Yeast artificial chromosome vector 169
Yeast centromere plasmids 169
Yeast episomal plasmids 168
Yeast integrative plasmid 169